Lecture Notes in Physics

Springer-Verlag Berlin Heidelberg GmbH

Physics and Astronomy ONLINE LIBRARY

http://www.springer.de/phys/

The Editorial Policy for Proceedings

The series Lecture Notes in Physics reports new developments in physical research and teaching – quickly, informally, and at a high level. The proceedings to be considered for publication in this series should be limited to only a few areas of research, and these should be closely related to each other. The contributions should be of a high standard and should avoid lengthy redraftings of papers already published or about to be published elsewhere. As a whole, the proceedings should aim for a balanced presentation of the theme of the conference including a description of the techniques used and enough motivation for a broad readership. It should not be assumed that the published proceedings must reflect the conference in its entirety. (A listing or abstracts of papers presented at the meeting but not included in the proceedings could be added as an appendix.)
When applying for publication in the series Lecture Notes in Physics the volume's editor(s) should submit sufficient material to enable the series editors and their referees to make a fairly accurate evaluation (e.g. a complete list of speakers and titles of papers to be presented and abstracts). If, based on this information, the proceedings are (tentatively) accepted, the volume's editor(s), whose name(s) will appear on the title pages, should select the papers suitable for publication and have them refereed (as for a journal) when appropriate. As a rule discussions will not be accepted. The series editors and Springer-Verlag will normally not interfere with the detailed editing except in fairly obvious cases or on technical matters.
Final acceptance is expressed by the series editor in charge, in consultation with Springer-Verlag only after receiving the complete manuscript. It might help to send a copy of the authors' manuscripts in advance to the editor in charge to discuss possible revisions with him. As a general rule, the series editor will confirm his tentative acceptance if the final manuscript corresponds to the original concept discussed, if the quality of the contribution meets the requirements of the series, and if the final size of the manuscript does not greatly exceed the number of pages originally agreed upon. The manuscript should be forwarded to Springer-Verlag shortly after the meeting. In cases of extreme delay (more than six months after the conference) the series editors will check once more the timeliness of the papers. Therefore, the volume's editor(s) should establish strict deadlines, or collect the articles during the conference and have them revised on the spot. If a delay is unavoidable, one should encourage the authors to update their contributions if appropriate. The editors of proceedings are strongly advised to inform contributors about these points at an early stage.
The final manuscript should contain a table of contents and an informative introduction accessible also to readers not particularly familiar with the topic of the conference. The contributions should be in English. The volume's editor(s) should check the contributions for the correct use of language. At Springer-Verlag only the prefaces will be checked by a copy-editor for language and style. Grave linguistic or technical shortcomings may lead to the rejection of contributions by the series editors. A conference report should not exceed a total of 500 pages. Keeping the size within this bound should be achieved by a stricter selection of articles and not by imposing an upper limit to the length of the individual papers. Editors receive jointly 30 complimentary copies of their book. They are entitled to purchase further copies of their book at a reduced rate. As a rule no reprints of individual contributions can be supplied. No royalty is paid on Lecture Notes in Physics volumes. Commitment to publish is made by letter of interest rather than by signing a formal contract. Springer-Verlag secures the copyright for each volume.

The Production Process

The books are hardbound, and the publisher will select quality paper appropriate to the needs of the author(s). Publication time is about ten weeks. More than twenty years of experience guarantee authors the best possible service. To reach the goal of rapid publication at a low price the technique of photographic reproduction from a camera-ready manuscript was chosen. This process shifts the main responsibility for the technical quality considerably from the publisher to the authors. We therefore urge all authors and editors of proceedings to observe very carefully the essentials for the preparation of camera-ready manuscripts, which we will supply on request. This applies especially to the quality of figures and halftones submitted for publication. In addition, it might be useful to look at some of the volumes already published. As a special service, we offer free of charge LaTeX and TeX macro packages to format the text according to Springer-Verlag's quality requirements. We strongly recommend that you make use of this offer, since the result will be a book of considerably improved technical quality. To avoid mistakes and time-consuming correspondence during the production period the conference editors should request special instructions from the publisher well before the beginning of the conference. Manuscripts not meeting the technical standard of the series will have to be returned for improvement.
For further information please contact Springer-Verlag, Physics Editorial Department II, Tiergartenstrasse 17, D-69121 Heidelberg, Germany

Series homepage – http://www.springer.de/phys/books/lnpp

H. Gausterer H. Grosse L. Pittner (Eds.)

Geometry and Quantum Physics

Proceedings of the
38. Internationale Universitätswochen
für Kern- und Teilchenphysik,
Schladming, Austria, January 9-16, 1999

Springer

Editors

H. Gausterer
L. Pittner
Institut für Theoretische Physik, Karl-Franzens-Universität
Universitätsplatz 5, 8010 Graz, Austria

Harald Grosse
Institut für Theoretische Physik, Universität Wien
Boltzmanngasse 5, 1090 Wien, Austria

Supported by the Österreichische Bundesministerium für Wirtschaft, Verkehr und Kunst, Vienna, Austria

Library of Congress Cataloging-in-Publication Data applied for.
Die Deutsche Bibliothek - CIP-Einheitsaufnahme

Geometry and quantum physics : proceedings of the 38. Internationale Universitätswochen für Kern- und Teilchenphysik, Schladming, Austria, January 9 - 16, 1999 / H. Gausterer ... (ed.).
(Lecture notes in physics ; 543)
ISBN 978-3-662-14287-5 ISBN 978-3-540-46552-2 (eBook)
DOI 10.1007/978-3-540-46552-2
ISSN 0075-8450
ISBN 978-3-662-14287-5

Typesetting: Camera-ready by the authors/editors
Cover design: *design & production*, Heidelberg

Printed on acid-free paper
SPIN: 10720725 55/3144/du - 5 4 3 2 1 0

Preface

In many areas of theoretical physics, geometrical and functional analytic methods are used simultaneously. During the course of our studies and education, these methods appeared to be extremely different. For classical field theories, a differential complex over smooth manifolds, smooth sections of vector bundles over manifolds, principal fibre bundles, connections, curvatures, etc. are introduced. For quantum field theory, singular operator-valued distributions act on Fock space and lead to all the well-known difficulties. After the work of Alain Connes and the many applications of non-commutative geometry to various subjects, we became familiar with, for example, the fact that on any associative superalgebra at least one differential calculus exists. But clearly there are many more aspects. We selected seven subjects and lecturers to cover symplectic geometry, quantum gravity, string theory, an introduction to non-commutative geometry, and the application to fundamental interactions, to the quantum Hall effect as well as to physics on q-deformed spaces.

In the first contribution, Anton Alexseev uses techniques of symplectic geometry to evaluate certain integrals over manifolds that have special symmetries. For them the stationary phase approximation becomes exact. Starting from simple examples, he introduces equivariant cohomology and sketches the proof of localization formulas of Duistermaat - Heckman type. The Weil model, well-known from the BRST formalism, is introduced, and a non-commutative generalization is used to prove group-valued localization.

John Baez elucidates the present status of quantum geometry of space-time. Since the implementation of constraints within the Hamiltonian approach to Einstein gravity is complicated, he explains in detail the simplifying BF system connected to the Chern - Simons model. We learn about spin net works and spin foams, become familiar with triangulations of four-manifolds, and calculate spectra of quantum tetrahedra. His lecture notes not only survey the subject, but also give an extensive list of references arranged according to nine different subjects.

Cesar Gomez then gives an introduction to string theory. Starting from the mode expansion he explains T-duality, a symmetry relating closed strings of radius R and $1/R$. Next D-brane ideas are represented.

Both quantum gravity and string theory have their own way of treating space-time. But there is a further approach that goes by the name of non-commutative geometry. This is the subject of the chapter by Daniel Kastler.

John Madore introduces a differential complex over an arbitrary associative algebra. We learn that, depending on the procedure, a regularization effect may result. Attempts to add a gravitational field are also reviewed.

This introduction is useful to follow the lectures of Daniel Kastler. He spoke about Connes' approach to the standard model as well as recent ideas for including gravity as an external field. He presents the interesting idea that additional dimensions finally lead via the Higgs effect to masses of particles, as well as the recent attempts to use $SU_q(2)$ for q being a third root of unity to "deduce" the gauge group of the standard model.

Julius Wess applies ideas of non-commutative geometry to the q-deformed Heisenberg algebra. The spectra of position and momentum are discrete. Phase space gets a lattice-like structure. Higher-dimensional analogues are investigated too.

Finally, Ruedi Seiler explains geometrical properties of transport in quantum Hall systems. The integer effect is treated in detail.

Although we learn here about many different applications of non-commutative geometry, the hope is that a unique picture of a quantum space-time may finally result. Quantum theory changes our ideas on geometry, and further surprises may come along soon.

We have tried to cover many subjects and were lucky to be supported by so many excellent lecturers. We hope that the school helped young people seeking an entry to these subjects. They will hopefully remember the excellent atmosphere of the Schladming Winter School 1999.

At this point we also want to express our thanks to the main sponsor of the School, the Austrian Ministry for Science and Transportation. In addition, we grateful acknowledge the generous support by the Government of Styria and the Town of Schladming. Valuable help towards the organization was received from the Wirtschaftskammer Steiermark (Sektion Industrie), Steyr-Daimler-Puch AG, Ricoh-Austria, and Styria Online.

Organizing the 1999 Schladming Winter School, and making it the successful event that it was, would not have been possible without the help of a number of colleagues and graduate students from our institute. Without naming them all we want to acknowledge the traditionally good cooperation within the organizing committee and beyond, which once again guaranteed the smooth running of all organizational, technical, and social matters. Thanks are due to Wolfgang Schweiger for his valuable technical assistance. Finally we should like to express our sincere thanks to Miss Sabine Fuchs for carrying out the secretarial work and for finalizing the text and layout of these proceedings.

Graz, November 1999 *H. Gausterer, H. Grosse, L. Pittner*

Contents

Notes on Equivariant Localization

Anton Alekseev

Institutionen för Teoretisk Fysik, Uppsala Universitet, Box 803, S-751 08, Uppsala, Sweden

Abstract. We review the localization formula due to Berline-Vergne and Atiyah-Bott, with applications to the exact stationary phase phenomenon discovered by Duistermaat-Heckman. We explain the Weil model of equivariant cohomology and recall its relation to BRST. We show how to quantize the Weil model, and obtain new localization formulas which, in particular, apply to Hamiltonian spaces with group valued moment maps.

1 Introduction

The purpose of these lecture notes is to present the *localization formulas* for equivariant cocycles. The localization phenomenon was first discovered by Duistermaat and Heckman in [DH], and then explained in the works of Berline-Vergne [BV] and Atiyah-Bott [AB]. The main idea of the localization formulas is similar to the residue formula: a multi-dimensional integral is evaluated exactly by summing up a number of the *fixed point contributions.*

In Section 2 we review the localization formula of [BV] and [AB]. We use an elementary example of the sphere S^2 as an illustration. Then, we outline the relation between the localization formulas and Hamiltonian Mechanics, and recover the Duistermaat-Heckman formula [DH].

In Section 3 we discuss the relations between the localization formulas and the group actions. In the case of the Duistermaat-Heckman formula, localization is intimately related to the symmetry group of the underlying Hamiltonian system. In particular, we compare the equivariant differential to the BRST differential.

In Section 4 we explain how to *quantize* the equivariant cohomology. This Section is based on the papers [AMM], [AM], [AMW1] and [AMW2]. We end up by presenting the new localization formula which is derived in [AMW2]. Some simple applications of this new formula can be found in [P]. Section 4 is based on the joint works with A.Malkin, E.Meinrenken and C.Woodward.

These notes do not touch upon various applications of localization formulas in Physics. Usually, one proceeds by extrapolating the localization phenomenon to path integrals. Some of the most exciting examples of this approach can be found in [MNP], [W2], [G], [BT], [MNS]. In fact, [W2] was the original motivation for the formulas of Section 3.

I am grateful to the organizers and participants of the 38th Schladming Winter School for the inspiring atmosphere!

2 Localization Formulas

In this Section we review the *localization formula* due to Berline-Vergne [BV] and Atiyah-Bott [AB]. It is then used to derive the *exact stationary phase* formula due to Duistermaat and Heckman [DH]. The presentation is illustrated at the elementary example of sphere S^2.

2.1 Stationary phase method

In this section we recall the stationary phase method. It applies when one is interested in the asymptotic behavior at large s of the integral

$$I(s) = \int_{-\infty}^{\infty} \mathrm{d}x e^{isf(x)} g(x). \tag{1}$$

Here we assume that functions $f(x)$ and $g(x)$ are real, and sufficiently smooth.

At large $s > 0$ the leading contribution into the integral (1) is given by the neighborhood of the critical points of $f(x)$, where its derivative in x vanishes. Let x_0 be such a critical point. Then, one can approximate $f(x)$ near x_0 by the first two terms of the Taylor series,

$$f(x) = f(x_0) + \frac{1}{2} f''(x_0)(x - x_0)^2 + \ldots,$$

where ... stand for the higher order terms.

The leading contribution of the critical point x_0 into the integral $I(s)$ is given by a simpler integral

$$I_0(s) = g(x_0) e^{isf(x_0)} \int_{-\infty}^{\infty} \mathrm{d}x e^{\frac{i}{2} s f''(x_0)(x-x_0)^2}.$$

This integral is Gaussian, and can be computed explicitly,

$$I_0(s) = g(x_0) e^{i(sf(x_0) + \varepsilon \frac{\pi}{4})} \left(\frac{2\pi}{s|f''(x_0)|} \right)^{\frac{1}{2}}$$

Here ε is the sign on the second derivative $f''(x_0)$.

A similar formula holds for multi-dimensional integrals,

$$I(s) = \int \mathrm{d}^n x g(x) e^{isf(x)}. \tag{2}$$

Again, the leading contribution into the asymptotics at large s is given by the critical points of $f(x)$, where its gradient vanishes $\nabla f = 0$. At the critical point x_0 one can expand $f(x)$ into the Taylor series,

$$f(x) = f(x_0) + \frac{1}{2} \sum_{i,j=1}^{n} f''_{ij}(x_0)(x - x_0)_i (x - x_0)_j + \ldots, \tag{3}$$

where

$$f''_{ij} = \frac{\partial^2 f}{\partial x_i \partial x_j}.$$

We assume that the critical point is non-degenerate, that is, the matrix f''_{ij} is invertible. Then, the leading contribution of x_0 into the integral $I(s)$ is of the form,

$$I_0(s) = g(x_0)e^{isf(x_0)} \left(\frac{2\pi}{s}\right)^{\frac{n}{2}} \frac{e^{i\sigma\frac{\pi}{4}}}{|\det(f''(x_0))|^{\frac{1}{2}}}. \tag{4}$$

Here $\sigma = \sigma_+ - \sigma_-$ is the signature of the matrix f''_{ij}, σ_+ and σ_- are numbers of positive and negative eigenvalues, respectively.

In general, one can have several critical points. Then, on can add the leading contributions (4) to obtain the approximate answer for $I(s)$,

$$I(s) \approx \left(\frac{2\pi}{s}\right)^{\frac{n}{2}} \sum_i g(x_i)e^{isf(x_i)} \frac{e^{i\sigma_i\frac{\pi}{4}}}{|\det(f''(x_i))|^{\frac{1}{2}}} \tag{5}$$

Of course, there is no reason for the right hand side to be the exact answer for $I(s)$. But sometimes this is the case! Such a situation is called *exact stationary phase*, and will be studied in these notes.

Example: sphere S^2 The simplest example of the exact stationary phase phenomenon is the computation of the following integral. Consider the unit sphere S^2 defined by equation $x^2 + y^2 + z^2 = 1$. We choose $g(x, y, z) = 1$ and $f(x, y, z) = z$. Then, the integral $I(s)$ is of the form,

$$I(s) = \int_{S^2} \mathrm{d}A \; e^{isz}, \tag{6}$$

where $\mathrm{d}A$ is the area element normalized in the standard way, $\int_{S^2} \mathrm{d}A = 4\pi$.

The critical points of the function $f(x, y, z) = z$ are the North and the South poles of the sphere. At both points one can use x and y as local coordinates to obtain,

$$z \approx 1 - \frac{1}{2}(x^2 + y^2)$$

near the North pole, and

$$z \approx -1 + \frac{1}{2}(x^2 + y^2)$$

near the South pole. Thus, for the stationary phase approximation one obtains,

$$I(s) \approx \frac{2\pi}{s} \left(-ie^{is} + ie^{-is}\right) = 4\pi \, \frac{\sin(s)}{s}. \tag{7}$$

Here we have used that at both North and South poles $\det(f_{ij}) = 1$, and that $\sigma_N = -2$ and $\sigma_S = 2$.

Now we can compare the 'approximate' result (7) with the exact calculation. It is convenient to use polar angles $0 < \theta < \pi, 0 < \phi < 2\pi$. Then, the coordinate function $z = \cos(\theta)$, and the area element is $\mathrm{d}A = \mathrm{d}\cos(\theta)\mathrm{d}\phi$. The simple calculation gives,

$$I(s) = \int \mathrm{d}\cos(\theta)\mathrm{d}\phi e^{is\cos(\theta)} = 2\pi\,\frac{e^{is} - e^{-is}}{is} = 4\pi\,\frac{\sin(s)}{s}. \tag{8}$$

This expression coincides with the stationary phase result (7).

In the following sections we shall see that the equality of the exact and approximate results (7) and (8) is not a coincidence. In fact, this is the simplest example of *equivariant localization.*

2.2 Equivariant cohomology

Stokes's theorem and residue formula The main tool in proving the localization formula will be the generalization of the Stokes's integration formula. The latter states that given an exact differential form α, $\alpha = \mathrm{d}\beta$, its integral over the domain D can be expressed as an integral of β over the boundary of D,

$$\int_D \mathrm{d}\beta = \int_{\partial D} \beta. \tag{9}$$

As a warm up exercise we prove the standard residue formula using the Stokes's formula (9). Given a function $f(z)$ analytic on the complex plane with the exception of some finite number of poles, its integral over a closed contour C is given by the sum of residues at the poles inside C,

$$\frac{1}{2\pi i}\int_C f(z)\mathrm{d}z = \sum_i \mathrm{res}_{\mathrm{z}_i}\mathrm{f}. \tag{10}$$

We naturally choose β in the form,

$$\beta = \frac{1}{2\pi i}\, f(z)\mathrm{d}z,$$

the domain D is the interior of C, and its boundary is C. The form α is given by equation,

$$\alpha = \mathrm{d}\beta = \frac{1}{2\pi i}\mathrm{d}(f(z)\mathrm{d}z) = \frac{1}{2\pi i}(\bar{\partial} f)\mathrm{d}\bar{z}\wedge\mathrm{d}z,$$

where $\bar{\partial} f$ is the partial derivative in $\bar{z}$. The function $f(z)$ is analytic. Hence, α vanishes everywhere except for the poles. We conclude, that α is a distribution supported at some number of points. Such a distribution is a sum of δ-functions and its derivatives. The only terms which contribute into the integral of α over D are δ-functions at the poles. They give rise to the residues,

$$\bar{\partial}\,\frac{\mathrm{res}_{\mathrm{z}_i}\mathrm{f}}{2\pi i(z - z_i)} = (\mathrm{res}_{\mathrm{z}_i}\mathrm{f})\delta(\mathrm{z} - \mathrm{z}_i).$$

Now we use the Stokes's formula,

$$\frac{1}{2\pi i}\int_C f(z)\mathrm{d}z = \int_D \sum_i (\mathrm{res}_{z_i} f)\delta(z - z_i) = \sum_i \mathrm{res}_{z_i} f,$$

and recover the residue formula.

S^1-action and fixed points The derivation of the localization formula requires more structure on the integration domain. In particular, the notion of symmetry plays an important role. We assume that our symmetry is continuous. In particular, this may be the action of the circle group S^1, which is our main example in this Section. For instance, in the case of S^2 there is an action of S^1 by rotations around the z-axis.

We always choose the integration domain to be a compact manifold M without boundary (as in the case of S^2). The S^1-action defines a vector field $v = \partial/\partial\phi$ on M. Zeroes of v correspond to fixed points of the circle action. For simplicity, we assume that all the fixed points are isolated. This is only possible if the dimension of the manifold is even, $n = 2m$. Given such a fixed point x_0 one can write the action near this point as

$$x^i(\phi) = x_0^i + R^i_j(\phi)(x - x_0)^j + \ldots,$$

where ... stand for higher order terms in $x - x_0$. It is easy to see that one can linearize the action (drop higher order terms). The matrix R gives a linear representation of S^1,

$$R(\phi_1)R(\phi_2) = R(\phi_1 + \phi_2),$$

and satisfies condition $R(2\pi) = \mathrm{id}$. By the appropriate choice of the basis such a matrix can always be represented as a direct sum of 2×2 blocks, each block of the form

$$\begin{pmatrix} \cos(\nu\phi) & \sin(\nu\phi) \\ -\sin(\nu\phi) & \cos(\nu\phi) \end{pmatrix},$$

where ν is an integer. We can denote the corresponding local coordinates by x_i, y_i where $i = 1 \ldots m$. In these local coordinates the vector field v has the form,

$$v = \sum_{i=1}^m \nu_i \left(x_i \frac{\partial}{\partial y_i} - y_i \frac{\partial}{\partial x_i} \right).$$

The integers λ_i are called indices of the vector field v at the point x_0.

In fact the indices ν_i are defined up to a sign: the flip of coordinates x_i and y_i changes the sign of the corresponding index. In what follows we shall need a product of all indices corresponding to the given fixed point, $\nu_1 \ldots \nu_m$. It is well defined is the tangent space at the fixed point is oriented: one should choose the coordinate system $(x_1, y_1, \ldots, x_m, y_m)$ with positive orientation. This condition determines the product of indices in a unique way. In particular, if the manifold is oriented, the products of indices at fixed points are well defined.

In the following we shall use that one can always choose an S^1-invariant metric on M. Indeed, given any metric, one can always average it over the S^1-action. In particular, this metric at the fixed point x_0 can always be chosen in the form,

$$g = \sum_{i=1}^{m} (\mathrm{d}x_i^2 + \mathrm{d}y_i^2). \tag{11}$$

Equivariant differential Now we are ready to define the equivariant differential, and equivariant cohomology. We define the space of *equivariant forms* on M as S^1 invariant differential forms with values in functions of one variable, which we denote by ξ. Typically, such a differential form is a polynomial,

$$\alpha(\xi) = \sum_{s=0}^{N} \alpha^s \xi^s,$$

where α^s are S^1-invariant differential forms. We shall also need equivariant forms with more complicated ξ-dependence.

Sometimes it is convenient to decompose equivariant forms according to the degree,

$$\alpha(\xi) = \sum_{j=0}^{n} \alpha_j(\xi),$$

where $\alpha_j(\xi)$ is a form of degree j which takes values in functions of ξ.

The differential on the space of equivariant forms is defined by formula,

$$\mathrm{d}_{S^1} = \mathrm{d} + i\xi\iota_v, \tag{12}$$

where ι_v is the contraction with respect to the vector field v. One can assign to parameter ξ degree 2 in order to make the equivariant differential homogeneous. Unfortunately, this arrangement is only meaningful for equivariant differential forms polynomial in ξ.

It is the basic property of the differential (12) that it squares to zero on the space of equivariant forms. Indeed,

$$\mathrm{d}_{S^1}^2 = (\mathrm{d} + i\xi\iota_v)^2 = i\xi(\mathrm{d}\iota_v + \iota_v\mathrm{d}) = i\xi L_v,$$

where we have used Cartan's formula for L_v. The Lie derivative L_v vanishes on equivariant forms, and so does $\mathrm{d}_{S^1}^2$.

Using the differential (12) one can define equivariantly closed forms, $\mathrm{d}_{S^1}\alpha = 0$, and equivariantly exact forms $\alpha = \mathrm{d}_{S^1}\beta$. Because $\mathrm{d}_{S^1}^2 = 0$, equivariantly exact forms are automatically equivariantly closed, and one can define equivariant cohomology $H_{S^1}(M)$ as the quotient of the space of (equivariantly) closed forms by the space of (equivariantly) exact forms. If $\alpha(\xi)$ is an equivariant cocycle, it satisfies the closedness condition,

$$(\mathrm{d} + i\xi\iota_v)\alpha(\xi) = 0,$$

which implies a number of equations for the forms $\alpha_k(\xi)$,

$$\mathrm{d}\alpha_{k-2}(\xi) + i\xi\iota_v\alpha_k(\xi) = 0. \tag{13}$$

Note that this recurrence relation has step 2. Hence, odd and even degree parts of an equivariant cocycle are also equivariant cocycles. If the manifold M is even dimensional, the closedness condition relates the top degree component $\alpha_n(\xi)$ and the function $\alpha_0(\xi)$. We shall see that exactly this relation is used in the localization formula.

The Stokes' integration formula generalizes to equivariantly exact forms. Indeed,

$$\int_D (\mathrm{d} + i\xi\iota_v)\beta = \int_{\partial D} \beta,$$

where $\int_D \iota_v\beta = 0$ because the integrand has a vanishing top degree component (at least one degree is eaten up by ι_v). In particular, if the integration domain has no boundary, the integral of any equivariantly exact form vanishes, and the integration map descends to equivariant cohomology. That is, given a class $[\alpha] \in H_{S^1}(M)$ one can choose any representative α in integrate it over M. The result is a function of ξ which is independent of the representative: the representatives differ by an exact form, and the integral of an exact form vanishes.

The localization formula is a tool of computing integral of equivariant cocycles in terms of fixed points. This formula was discovered by Berline-Vergne [BV] and by Atiyah-Bott [AB]. For an equivariant cocycle $\alpha(\xi)$, its integral over M is given by,

$$\int_M \alpha(\xi) = \left(\frac{2\pi}{i\xi}\right)^{\frac{n}{2}} \sum_p \frac{(\alpha_0(\xi))(x_p)}{\nu_1^p \dots \nu_m^p}, \tag{14}$$

where the index p labels fixed points of the circle action (we assume that all of them are isolated), and $\nu_1^p, \dots, \nu_m^p$ are indices of the p's fixed point.

Note that the integral on the left hand side of (14) depends only on the top degree component $\alpha_n(\xi)$ of the cocycle $\alpha(\xi)$. At the same time the right hand side is expressed in terms of the zero degree component $\alpha_0(\xi)$. This is possible because $\alpha_n(\xi)$ and $\alpha_0(\xi)$ are related by the recurrence relations (13) expressing closedness of $\alpha(\xi)$.

Also note that even if the equivariant form $\alpha(\xi)$ is smooth at $\xi = 0$, the right hand side of (14) contains the divergent factor $\xi^{-n/2}$. The singularity at $\xi = 0$ is canceled by the sum of contributions of fixed points, which has a zero of degree $n/2$ at $\xi = 0$. This idea leads to *reside formulas* [JK].

2.3 Proof of localization formula

In this section we give a proof of the localization formula (14). This proof was suggested by Witten in [W1]. The idea is as follows. Choose an S^1-invariant metric g on M, with behavior at fixed points given by (11), and define a 1-form

$$\psi = g(v, \cdot),$$

such that $\iota_u\psi = g(v,u)$ for any vector field u on M. The form ψ is S^1-invariant (because the metric is S^1-invariant). Define an equivariantly exact form

$$\beta(\xi) = \mathrm{d}_{S^1}\psi = \mathrm{d}\psi + i\xi\iota_v\psi = \mathrm{d}\psi + i\xi g(v,v).$$

Note that near a fixed point ψ is of the form

$$\psi \approx -\frac{1}{2}\sum_{k=1}^{m} \nu_k(x_k \mathrm{d}y_k - y_k \mathrm{d}x_k),$$

and the form $\beta(\xi)$ is given by

$$\beta(\xi) \approx -\sum_{k=1}^{m} \nu_k \mathrm{d}x_k \wedge \mathrm{d}y_k + \frac{i\xi}{2}\sum_{k=1}^{m} \nu_k^2(x_k^2 + y_k^2).$$

Let us consider the equivariant form

$$e^{is\beta(\xi)} - 1 = \sum_{k=1}^{\infty} \frac{(is)^k}{k!} (\mathrm{d}_{S^1}\psi)^k = \mathrm{d}_{S^1}\left(\sum_{k=1}^{\infty} \frac{(is)^k}{k!} \psi(\mathrm{d}_{S^1}\psi)^{k-1}\right).$$

It is equivariantly exact, and hence,

$$\int_M \alpha(\xi) = \int_M \alpha(\xi) e^{is\beta(\xi)} \tag{15}$$

for any value of the parameter s. In particular, if $\xi > 0$, one van consider the limit $s \to +\infty$. On one hand, the asymptotics of the integral (15) can be computed using the stationary phase method. But on the other hand, the answer does not depend on s. Hence, it is sufficient to extract the term in the asymptotics which does not depend on s, and this will be the exact answer for the integral!

At large s the form

$$e^{is\beta(\xi)} = e^{is\mathrm{d}\psi - s\xi g(v,v)}$$

is exponentially small everywhere except for the small neighborhoods of the fixed points where $v = 0$ and $g(v,v) = 0$. So, the fixed points are at the same time the critical points which give contributions into the stationary phase asymptotics. The leading contribution in s of the critical point x_p is given by

$$(\alpha_0(\xi))(x_p) \prod_{k=1}^{m} \left(is\nu_k^p \int \mathrm{d}x_k \mathrm{d}y_k e^{-\frac{s\xi}{2}\sum(\nu_k^i)^2(x_k^2+y_k^2)} \right) \tag{16}$$

Here we have used the fact that the 2-form $is\mathrm{d}\psi$ which enters the integrand is proportional to s, and, hence, it gives the leading contribution into the integration measure. The integral in (16) is Gaussian. It yields

$$\left(\frac{2\pi}{i\xi}\right)^m \frac{(\alpha_0(\xi))(x_p)}{\nu_1^p \ldots \nu_m^p}$$

which is indeed independent of s. Summing up these contributions for all fixed points x_p we obtain the localization formula (14).

2.4 Duistermaat-Heckman formula

Perhaps, the most well-known application of the localization formulas is the Duistermaat-Heckman formula in symplectic geometry [DH]. In fact, it was discovered before the general localization principle was formulated.

The framework is as follows: we have a closed 2-form ω,

$$\mathrm{d}\omega = 0,$$

which satisfies the nondegeneracy condition. That is,

$$\iota_u \omega = 0$$

for some vector field u on M implies $u = 0$. The pair (M, ω) is called a symplectic manifold. A standard example is R^{2m} with coordinates p_i and q_i and the 2-form

$$\omega = \sum_{i=1}^{m} \mathrm{d}p_i \wedge \mathrm{d}q_i.$$

A vector field v is called Hamiltonian if there exists a function H such that

$$\iota_v \omega + \mathrm{d}H = 0.$$

The function H is called the Hamiltonian of v. A symplectic manifold is always even dimensional (otherwise ω is necessarily degenerate), and has the volume form

$$\mathcal{L} = \frac{\omega^m}{m!} = [e^{\omega}]_{top}$$

called the Liouville form. The volume form $\mathcal{L}$ is invariant with respect to all Hamiltonian vector fields.

Now assume that M is compact, and carries a circle action. In addition, let the corresponding vector field v be Hamiltonian, with Hamiltonian H. Then, one can define the following integral,

$$I(\xi) = \int_M \mathcal{L}\, e^{i\xi H}, \tag{17}$$

which is called the Duistermaat-Heckman integral, and can be evaluated using localization theorem.

First, we define the equivariant extension of the symplectic form ω,

$$\omega(\xi) = \omega + i\xi H.$$

It is an equivariantly closed form,

$$\mathrm{d}_{S^1}\omega(\xi) = (\mathrm{d} + i\xi\iota_v)(\omega + i\xi H) = \mathrm{d}\omega + i\xi(\iota_v\omega + \mathrm{d}H) = 0.$$

Here we have used closedness of ω and the definition of the Hamiltonian vector field.

Next, we define an equivariant Liouville form,

$$\mathcal{L}(\xi) = e^{\omega(\xi)} = e^{i\xi H} \sum_{k=1}^{m} \frac{\omega^k}{k!},$$

where the sum terminates because higher powers of ω vanish. The form $\mathcal{L}(\xi)$ is also equivariantly closed, and, moreover,

$$\int_M \mathcal{L}(\xi) = \int \mathcal{L}\ e^{i\xi H} = I(\xi).$$

Now we apply the localization formula (14) to the left hand side to obtain the Duistermaat-Heckman formula,

$$I(\xi) = \left(\frac{2\pi}{i\xi}\right)^m \sum_p \frac{e^{i\xi H(x_p)}}{\lambda_1^p \dots \lambda_m^p}. \tag{18}$$

Example: sphere S^2. Getting back to the example of the sphere S^2, we show that our observation on exact stationary phase is a particular case of the Duistermaat-Heckman formula.

Let us choose the area form on the sphere as the symplectic form. It is clearly closed, and non-degenerate. In the polar angles θ, ϕ the vector field generating rotations around z-axis is of the form $v = \partial/\partial\phi$. Then, the Hamiltonian of v is determined by equation,

$$\iota\left(\frac{\partial}{\partial\phi}\right) \mathrm{d}\cos(\theta)\mathrm{d}\phi + \mathrm{d}H = 0,$$

which implies $H = \cos(\theta) = z$ (up to a shift by a constant). Thus, the integral which we would like to compute,

$$I(\xi) = \int_{S^2} \mathrm{d}A\ e^{i\xi z}$$

is the Duistermaat-Heckman integral, and is given by the Duistermaat-Heckman formula (18).

There are two fixed points of the S^1-action on S^2, the North pole and the South pole. The values of the Hamiltonian are given by $z_N = 1, z_S = -1$, and the indices of the S^1-action are $\nu^N = 1$ and $\nu^S = -1$. Then, formula (18) yields

$$I(\xi) = \frac{2\pi}{i\xi}\left(\frac{e^{i\xi z_N}}{\nu^N} - \frac{e^{-i\xi z_S}}{\nu^S}\right) = 4\pi\ \frac{\sin(\xi)}{\xi}$$

confirming the results we obtained before.

3 Weil Model of Equivariant Cohomology

In this Section we develop technical tools for dealing with equivariant cohomology for any compact group G. We begin by introducing the Weil algebra, and the Weil model of equivariant cohomology. Then we establish the equivalence to the Cartan model which we used in the case of $G = S^1$. Finally, we give an expression for the equivariant Liouville form in the Weil model, and introduce equivariant cohomology with generalized coefficients.

For another physicist-oriented review of the subject see [CMR].

3.1 Weil algebra and Weil differential

Group actions on manifolds. In general, we shall study the situation when the group acting on M is not necessarily a circle S^1. Let G be a compact connected Lie group, and $\mathcal{G}$ be its Lie algebra. In many situations it will be convenient to choose a basis $\{e_a\} \subset \mathcal{G}$, and the dual basis $\{e^a\}$ in the space $\mathcal{G}^*$. We denote by f_{ab}^c the structure constants in this basis,

$$[e_a, e_b] = f_{ab}^c e_c.$$

If the group G acts on the manifold M, one can associate to each element $e \in \mathcal{G}$ a fundamental vector field on M which we denote by e_M. For instance, the vector fields corresponding to the basis elements e_a are $(e_a)_M$. The Lie derivatives and contractions corresponding to these fundamental vector fields act on the space of differential forms $\Omega(M)$. We denote them by L_a and ι_a, respectively. They satisfy the following relations,

$$\begin{aligned} [L_a, \iota_b] &= f_{ab}^c \iota_c, \\ [L_a, L_b] &= f_{ab}^c L_c, \\ [d, \iota_a] &= L_a. \end{aligned} \tag{19}$$

Here d is the de Rham differential, and $[,]$ stands for the super-commutator. For instance, $[d, \iota_a] = d\iota_a + \iota_a d$.

In a more abstract setting we can say that equations (19) define a superalgebra $\hat{\mathcal{G}}$ with generators L_a, ι_a, d. If M is a G-manifold, the space of forms $\Omega(M)$ carries a representation of $\hat{\mathcal{G}}$, where L_a are represented by Lie derivatives, ι_a by contractions, and d by the de Rham differential.

Weil algebra. In this section we construct a special representation of the algebra $\hat{\mathcal{G}}$ called the Weil algebra. It was suggested by H.Cartan in C1 as an 'algebraic model' of the space of forms on the classifying space EG.

By definition, the Weil algebra W_G is the product of the symmetric and exterior algebras of the dual space to the Lie algebra of G,

$$W_G := S\mathcal{G}^* \otimes \wedge\mathcal{G}^*. \tag{20}$$

The algebra $S\mathcal{G}^*$ is the algebra of polynomials on $\mathcal{G}$. It is convenient to introduce generators v^a of $S\mathcal{G}^*$ corresponding to the basis elements $e^a \in \mathcal{G}^*$. The generators v^a correspond to linear functions on $\mathcal{G}$, and naturally commute with each other,

$$v^a v^b - v^b v^a = 0.$$

We denote the generators of the exterior algebra $\wedge\mathcal{G}^*$ by y^a. They satisfy the anti-commutation relations,

$$y^a y^b + y^b y^a = 0.$$

One can introduce a grading on W_G by assigning degree 2 to v^a and degree 1 to y^a,

$$W_G^l = \oplus_{j+2k=l} S^k\mathcal{G}^* \otimes \wedge^j \mathcal{G}^*.$$

Following H.Cartan, one can view W_G as a model of the space of forms on EG, such that each y^a corresponds to a 1-form, and each v^a corresponds to a 2-form.

There is an action of $\hat{\mathcal{G}}$ on W_G defined as follows. Operators L_a are defined on generators,

$$L_a(v^c) = -f_{ab}^c v^b, \ L_a(y^b) = -f_{ab}^c y^c,$$

and extended by the Leibniz rule. In a similar fashion, one defines contractions ι_a,

$$\iota_a(v^b) = 0, \ \iota_a(y^b) = \delta_a^b.$$

Finally, the *Weil differential* d is defined by

$$d(y^a) = v^a - \frac{1}{2} f_{bc}^a y^b y^c, \ d(v^a) = -f_{bc}^a y^b v^c.$$

These formulas have a simple geometric meaning: if one interprets y^a as components of a connection on a principal G-bundle, then the first formula,

$$v^a = dy^a + \frac{1}{2} f_{bc}^a y^b y^c,$$

is the standard definition of the curvature. The second formula,

$$dv^a + f_{bc}^a y^b v^c = 0$$

gives the Bianchi identity.

Relation to BRST. The differential on W_G can be written in the form,

$$d = y^a(L_a \otimes 1) + (v^a - \frac{1}{2} f^a_{bc} y^b y^c)\iota_a,$$

where $(L_a \otimes 1)$ is the Lie derivative acting only on the elements of $S\mathcal{G}^*$.

It is often compared to the *BRST differential* which is defined as follows. Let V be a representation of the group G. Then, the BRST differential acts on the space $V \otimes \wedge\mathcal{G}^*$, and is given by formula,

$$d_{BRST} = y^a(L_a \otimes 1) - \frac{1}{2} f^a_{bc} y^b y^c \iota_a.$$

In the physical interpretation, y^a are called ghosts, and denoted by c^a. The dual contractions ι_a are called anti-ghosts, and denoted by b_a. The ghosts and anti-ghosts (generators of $\wedge\mathcal{G}^*$ and contractions) satisfy the anti-commutation relation,

$$c^k b_l + b_l c^k = \delta^k_l.$$

If we introduce a special notation for generators of the G-action on V, $T_a := (L_a \otimes 1)$, we get the standard formula for the BRST differential,

$$d_{BRST} = c^a T_a - \frac{1}{2} f^a_{bc} c^b c^c b_a.$$

The main difference between the Weil differential (and the equivariant differential) and the BRST differential is the extra term $v^a \iota_a$ in the Weil differential. One can interpret it as a BRST differential for the Abelian Lie algebra $\mathcal{G}^*$ with generators v^a and ghosts $b_a := \iota_a$. One can say that the Weil differential is a sum of two BRST differentials,

$$d = d^{\mathcal{G}}_{BRST} + d^{\mathcal{G}^*}_{BRST}.$$

3.2 Weil model of equivariant cohomology

In this section we define the Weil model of equivariant cohomology, and prove that it is equivalent to the Cartan model introduced before. Then, we extend our consideration to equivariant cohomology with generalized coefficients.

Definition of the Weil model. Let M be a G-manifold. It is our goal to define the space of equivariant forms in the Weil model, and the equivariant differential on this space.

Consider the product $\Omega(M) \otimes W_G$ of the space of differential forms on M and of the Weil algebra W_G. If one interprets W_G as the space of differential forms on EG, the product is naturally interpreted as $\Omega(M \times EG)$. Both $\Omega(M)$ and W_G carry representations of $\hat{\mathcal{G}}$. Hence, one can define the diagonal action on the tensor product. That is, L_a, ι_a and d are defined as operators on $(\Omega(M) \otimes S\mathcal{G}^*)$ by formulas,

$$L_a = L_a \otimes 1 + 1 \otimes L_a,$$
$$\iota_a = \iota_a \otimes 1 + 1 \otimes \iota_a,$$
$$d = d \otimes 1 + 1 \otimes d.$$

We define the space of equivariant forms on M as the *basic part* of $\Omega(M) \otimes W_G$,

$$\Omega_G(M) := \{\alpha \in \Omega(M) \otimes W_G | \, L_a\alpha = 0, \ \iota_a\alpha = 0\}.$$

In more geometric terms we are looking at the principal G-bundle

$$M \times EG \to (M \times EG)/G,$$

and define $\Omega_G(M)$ as the space of basic forms. These are forms which can be obtained as pull-backs of differential forms on the quotient space $(M \times EG)/G$.

The space of equivariant forms $\Omega_G(M)$ carries the action of the combined differential $(d \otimes 1 + 1 \otimes d)$. One defines the equivariant cohomology of M as

$$H_G(M) := H(\Omega_G(M), d \otimes 1 + 1 \otimes d).$$

In the next section we show that this definition is equivalent to the definition in the Cartan model which we used in the case of $G = S^1$.

Equivalence to the Cartan model. In Section 2 we used a simpler model for S^1-equivariant cohomology. This model does not use anti-commuting variables y^a, and is called the Cartan model [C2]. A simple transformation which establishes the relation between Weil and Cartan models was suggested by Kalkman [K].

Let us define an operator Φ on the space $\Omega(M) \otimes W_G$ by formula,

$$\Phi := \exp(-\iota_a \otimes y^a).$$

The key property of Φ is

$$\Phi(\iota_a \otimes 1 + 1 \otimes \iota_a)\Phi^{-1} = 1 \otimes \iota_a. \tag{21}$$

In order to prove this equality we use the formula,

$$\Phi X \Phi^{-1} = \sum_{j=0}^{\infty} \frac{1}{j!} \, \mathrm{ad}^{\mathrm{j}}(-\iota_{\mathrm{a}} \otimes \mathrm{y}^{\mathrm{a}})\mathrm{X}. \tag{22}$$

The calculation gives,

$$\mathrm{ad}(-\iota_{\mathrm{a}} \otimes \mathrm{y}^{\mathrm{a}})(\iota_{\mathrm{a}} \otimes 1 + 1 \otimes \iota_{\mathrm{a}}) = -\iota_{\mathrm{a}} \otimes 1,$$
$$\frac{1}{2!} \, \mathrm{ad}^2(-\iota_{\mathrm{a}} \otimes \mathrm{y}^{\mathrm{a}})(\iota_{\mathrm{a}} \otimes 1 + 1 \otimes \iota_{\mathrm{a}}) = 0.$$

After substitution to (22) we obtain equation (21).

The action of Φ maps the forms annihilated by $(\iota_a \otimes 1 + 1 \otimes \iota_a)$ to the forms annihilated by $1 \otimes \iota_a$. That is, it is mapped to $\Omega(M) \otimes S\mathcal{G}^*$. Taking into account that Φ commutes with the diagonal action of L_a, we conclude that the space of equivariant forms $\Omega_G(M)$ is mapped to

$$\Phi : \Omega_G(M) \to (\Omega(M) \otimes S\mathcal{G}^*)^G.$$

This is the new model of the space of equivariant forms called Cartan model. We already worked with it in the case of $G = S^1$.

Next, we compute the equivariant differential in the Cartan model. We apply formula (22) to the equivariant differential,

$$\begin{aligned} \mathrm{ad}(-\iota_a \otimes y^a)(d \otimes 1 + 1 \otimes d) &= L_a \otimes y^a - \iota_a \otimes (v^a - \frac{1}{2} f^a_{bc} y^b y^c), \\ \frac{1}{2!}\mathrm{ad}^2(-\iota_a \otimes y^a)(d \otimes 1 + 1 \otimes d) &= -\frac{1}{2} f^c_{ab} \iota_c \otimes y^a y^b, \\ \frac{1}{3!}\mathrm{ad}^2(-\iota_a \otimes y^a)(d \otimes 1 + 1 \otimes d) &= 0. \end{aligned}$$

We add all the terms and take into account that $(L_a \otimes 1 + 1 \otimes L_a)$ and $1 \otimes \iota_a$ annihilate the image of the space of equivariant forms. The final result for the differential in the Cartan model is quite simple,

$$d_G := d \otimes 1 - \iota_a \otimes v^a.$$

In the case of $G = S^1$ one should put $v = -i\xi$ to recover the expression (12) for d_{S^1}.

Equivariant Liouville form. As before, examples of equivariant classes are provided by symplectic geometry. Let (M, ω) be a symplectic manifold, and assume that the G-action on M is Hamiltonian. That is, there is a set of Hamiltonians, H_a such that [1]

$$\iota_a \omega = dH_a, \tag{23}$$

and

$$L_a H_b = f^c_{ab} H_c. \tag{24}$$

The collection of functions H_a can be viewed as the *moment map* $H : M \to \mathcal{G}^*$ with $H_a = \langle H, e_a \rangle$. The property (24) expresses equivariance of the map H with respect to the G-action on M and the co-adjoint action on $\mathcal{G}^*$.

Again, one can define the equivariant extension of the Liouville form in Cartan model,

$$\omega(v) := \omega + H_a v^a.$$

This form is equivariantly closed,

$$d_G \omega(v) = (d - v^a \iota_a)(\omega + v^b H_b) = d\omega + v^a (dH_a - \iota_a \omega) = 0.$$

[1] Note that we have changed the sign convention in comparison to the previous Section.

The same form in the Weil model is expressed by

$$\omega_W := \Phi^{-1}\omega(v) = \omega - y^a dH_a + H_a(v^a - \frac{1}{2} f^a_{bc} y^b y^c),$$

where we have used the property $\iota_a dH_b = f^c_{ab} H_c$ implied by (24).

Finally, we introduce the equivariant Liouville forms in Cartan and Weil models,

$$\mathcal{L}(v) := e^{\omega(v)} = e^{\omega}\, e^{v^a H_a},$$

and

$$\mathcal{L}_W := e^{\omega_W} = \exp(\omega)\, \exp(-y^a dH_a)\, \exp\left(-\frac{1}{2} H_a f^a_{bc} y^b y^c\right)\, \exp(v^a H_a). \quad (25)$$

In the next Section we shall use the expression for $\mathcal{L}_W$ to introduce the theory of 'group-valued' Hamiltonians.

Generalized coefficients. The last technical ingredient needed in the next Section is the notion of equivariant cohomology with generalized coefficients. It is sometimes convenient to make a Fourier-Laplace transform in the variables v^a such that they become distributions on the space $\mathcal{G}^*$ supported at the origin,

$$v^a = -\frac{\partial}{\partial \mu_a}\, \delta_0,$$

where μ_a are linear coordinates on $\mathcal{G}^*$, and δ_0 is the δ-function supported at the origin.

Then, it is natural to replace the space of polynomials $S\mathcal{G}^*$ in the definition of W_G by the space of all compactly supported distributions $\mathcal{E}'(\mathcal{G}^*)$. We denote the extended Weil algebra by

$$\hat{W}_G := \mathcal{E}'(\mathcal{G}^*) \otimes \wedge \mathcal{G}^*,$$

and the corresponding equivariant cohomology by $\hat{H}_G(M)$.

For instance, in the new notations the equivariant Liouville form contains the δ-function supported at the value H of the moment map,

$$\mathcal{L}_W = \exp(\omega)\, \exp(-y^a dH_a)\, \exp\left(-\frac{1}{2} H_a f^a_{bc} y^b y^c\right)\, \delta_H.$$

The distribution part of $\mathcal{L}_W$ is not supported at the origin, and defines a class in $\hat{H}_G(M)$.

4 Group-Valued Equivariant Localization

In this Section we explain how to *quantize* the Weil algebra [AM], and define the *group valued* equivariant cohomology. This leads to the new localization theorem [AMW2], and new moment map theory [AMM].

In contrast to the previous sections we only sketch the results. At this stage the proofs are too involved for these notes. So, refer the reader to the original papers.

4.1 Non-commutative Weil algebra

In this section we introduce the non-commutative counterpart of the Weil algebra. In the exposition we follow [AM].

Invariant inner product on $\mathcal{G}$. As before, we assume that G is a compact connected Lie group. In addition, we choose an invariant inner product on the Lie algebra $\mathcal{G}$, and suppose that G is a direct product of a compact simply-connected Lie group and a torus.

We denote the inner product on $\mathcal{G}$ by $(\cdot,\cdot)$. It induces a number of new structures. First, one can identify $\mathcal{G}$ with its dual space $\mathcal{G}^*$. The basis $\{e_a\}$ can be chosen orthonormal, $(e_a, e_b) = \delta_{ab}$. The corresponding structure constants $[e_a, e_b] = f_{abc}e_c$ are anti-symmetric with respect to the permutation of any two indices. Thus, one can define an element $\phi \in (\wedge^3\mathcal{G})^G$ by formula,

$$\phi := \frac{1}{6} f_{abc} e_a \otimes e_b \otimes e_c.$$

We define the left- and right-invariant vector fields e_a^L and e_a^R on the group G, and the dual left- and right-invariant 1-forms, θ_a^L and θ_a^R. They satisfy the Maurer-Cartan structure equations,

$$d\theta_a^L = -\frac{1}{2} f_{abc}\theta_b^L\theta_c^L, \; d\theta_a^R = \frac{1}{2} f_{abc}\theta_b^R\theta_c^R.$$

Using the identification $\wedge^3\mathcal{G} \cong \wedge^3\mathcal{G}^*$, one can define a bi-invariant 3-form on G,

$$\eta := \frac{1}{12} f_{abc}\theta_a^L\theta_b^L\theta_c^L = \frac{1}{12} f_{abc}\theta_a^R\theta_b^R\theta_c^R.$$

Finally, we introduce the distributions on G with support at the group unit corresponding to the vector fields e_a^L and e_a^R,

$$u_a^L := -e_a^L\delta_e, \; u_a^R := -e_a^R\delta_e.$$

Here δ_e is the δ-function supported at the group unit.

Definition of non-commutative Weil algebra. We recall that the Weil algebra W_G is a tensor product of symmetric and exterior algebras of the space $\mathcal{G}^*$. The non-commutative Weil algebra is a tensor product of the non-commutative counterparts of these algebras. The symmetric algebra is replaced by the *universal enveloping algebra* $U(\mathcal{G})$ with generators u_a and relations

$$u_a u_b - u_b u_a = f_{abc} u_c.$$

The exterior algebra is replaced by the Clifford algebra $\mathrm{Cl}(\mathcal{G})$ with generators x_a and relations

$$x_a x_b + x_b x_a = \delta_{ab}.$$

Here we use the fact that the basis $\{e_a\}$ is orthonormal. We denote the non-commutative Weil algebra by $\mathcal{W}_G$,

$$\mathcal{W}_G := U(\mathcal{G}) \otimes \mathrm{Cl}(\mathcal{G}).$$

Similar to W_G the algebra $\mathcal{W}_G$ carries a representation of $\hat{\mathcal{G}}$. The action of the Lie derivatives L_a is defined on generators,

$$L_a(u_b) = f_{abc}u_c, \; L_a(x_b) = f_{abc}x_c.$$

The contractions ι_a are given by formulas,

$$\iota_a(u_b) = 0, \; \iota_a(x_b) = \delta_{ab}.$$

Finally, the Weil differential has its analog on $\mathcal{W}_G$,

$$d(x_a) = u_a - \frac{1}{2} f_{abc}x_b x_c, \; d(u_a) = -f_{abc}x_b u_c.$$

These formulas are very similar to those which define the $\hat{\mathcal{G}}$-action on W_G. The important difference is that in the non-commutative algebra $\mathcal{W}_G$, the operators L_a, ι_a, d are inner derivations,

$$L_a = \mathrm{ad}(\mathrm{u_a} - \frac{1}{2}\mathrm{f_{abc}x_b x_c}), \; \iota_a = \mathrm{ad}(\mathrm{x_a}), \; \mathrm{d} = \mathrm{ad}(\mathrm{x_a u_a} - \frac{1}{6}\mathrm{f_{abc}x_a x_b x_c}).$$

As in the case of W_G, one can introduce the non-commutative Weil algebra with generalized coefficients,

$$\hat{\mathcal{W}}_G := \mathcal{E}'(G) \otimes \mathrm{Cl}(\mathcal{G}).$$

For any G-manifold M one can now define the space of 'group-valued' equivariant forms, $(\Omega(M) \otimes \mathcal{W}_G)_{basic}$ and $(\Omega(M) \otimes \hat{\mathcal{W}}_G)_{basic}$ and the 'group-valued' equivariant cohomology,

$$\mathcal{H}_G(M) := H((\Omega(M) \otimes \mathcal{W}_G)_{basic}, d \otimes 1 + 1 \otimes d),$$
$$\hat{\mathcal{H}}_G(M) := H((\Omega(M) \otimes \hat{\mathcal{W}}_G)_{basic}, d \otimes 1 + 1 \otimes d)$$

It is our main goal to present the localization formulas for classes in $\mathcal{H}_G(M)$ and in $\hat{\mathcal{H}}_G(M)$.

4.2 Group-valued moment maps

In this section we give examples of equivariant cocycles which give rise to classes in $\hat{\mathcal{H}}_G(M)$. We follow the paper [AMW1]. The idea is to find the counterpart of formula (25) for the equivariant Liouville form. We shall see that it naturally leads to moment maps with values in the Lie group rather than in the dual of the Lie algebra.

The right hand side of (25) is a product of four factors,

$$\exp(\omega)\ \exp(-y^a dH_a)\ \exp\left(-\frac{1}{2}H_a f_{abc} y^a y^b\right)\ \delta_H \in \Omega(M)\otimes \mathcal{E}'(\mathcal{G}^*)\otimes \wedge\mathcal{G}^*$$

where ω is a 2-form on M and H is the moment map $H : M \to \mathcal{G}^*$. In the group-valued case we still need a 2-form ω, but the moment map should take values in the group G, $\Phi : M \to G$. The reason is that instead of the space of distributions on the dual of the Lie algebra $\mathcal{E}'(\mathcal{G}^*)$ we now have the space of distributions on the group $\mathcal{E}'(G)$. Then, the first and the last terms in (25) have their counterparts, $\exp(\omega)$ and δ_Φ.

The third term is related to the *spinor representation* of G. In more detail, choose $H = H_a e_a \in g$ and define a map $\tau : G \to \mathrm{Cl}(\mathcal{G})$ by formula,

$$\tau(e^H) = \exp\left(-\frac{1}{2}H_a f_{abc} x_b x_c\right).$$

If G is a product of a compact simply-connected Lie group and a torus, the map τ is well-defined, and defines the representation of G (see *e.g.* [BGV]),

$$\tau(g_1 g_2) = \tau(g_1)\tau(g_2).$$

There are two possible candidates for the role of the second term, $\exp(-x_a \Phi^*\theta^L_a)$ and $\exp(-x_a\Phi^*\theta^R_a)$. We notice that

$$\exp(-x_a\Phi^*\theta^R_a)\tau(\Phi) = \tau(\Phi)\exp(-x_a\Phi^*\theta^L_a).$$

Thus, we can choose either left- or right-invariant Maurer-Cartan forms, but we should position them on the different sides of $\tau(\Phi)$.

Finally, our candidate for an group-valued equivariant Liouville form is,

$$\mathcal{L}_W = \exp(\omega)\ \exp(-x_a\Phi^*\theta^R_a)\tau(\Phi)\ \delta_\Phi. \tag{26}$$

The question is: under what conditions on ω and Φ, the form $\mathcal{L}_W$ is an equivariantly closed? According to [AMW1], there are 2 conditions to be satisfied: first, the differential of the 2-form ω is a pull-back of the bi-invariant 3-form η on G,

$$d\omega = \Phi^*\eta. \tag{27}$$

Second, there is an analog of the moment map condition,

$$\iota_a\omega = \frac{1}{2}\Phi^*(\theta^L_a + \theta^R_a). \tag{28}$$

We call a triple (M,ω,Φ) which satisfies these conditions a *group-valued Hamiltonian space.*

Examples of group-valued Hamiltonian spaces Our first example of a group-valued Hamiltonian G-space is the torus $T^2 = S^1 \times S^1$. We parametrize by it two angles, (ϕ, ψ), choose the S^1 action

$$\theta : (\phi, \psi) \mapsto (\phi, \psi + \theta)$$

and the two form,

$$\omega = d\psi \wedge d\phi.$$

Then,

$$\iota\left(\frac{\partial}{\partial \psi}\right) \omega = d\phi,$$

and one can define the moment map $\Phi : (\phi, \psi) \mapsto \phi$ with values in the group S^1. The 3-form η vanishes on S^1 for dimensional reasons, which is consistent with closedness of ω.

Our second example is a bit more complicated. We consider the group $G = SU(2)$, choose any element $f \in G$, and consider the corresponding conjugacy class,

$$C_f := \{gfg^{-1} |\ g \in G\}.$$

In other words, these are all unitary 2 by 2 matrices with the same eigenvalues as f. If f is e or $-e$, the corresponding conjugacy class f is a point. Otherwise, it is a 2-sphere. We define the moment map on C_f as its embedding into G. Then, the 2-form ω is uniquely determined by the conditions (27) and (28). Up to a scalar factor, ω coincides with the area form dA induced by the identification with the 2-sphere. If the eigenvalues of f are $\exp(i\lambda)$ and $\exp(-i\lambda)$, one gets [AMW1],

$$\omega = \sin(\lambda)\ dA.$$

Our last example is the product of two copies of $SU(2)$, $D := SU(2) \times SU(2)$. One can view it as a nonabelian counterpart of the torus T^2. We view D as an $SU(2) \times SU(2)$-manifold, with the action,

$$(g, h) : (a, b) \mapsto (gah^{-1}, gbh^{-1}),$$

and the moment map,

$$\Phi : (a, b) \mapsto (ab, a^{-1}b^{-1}).$$

The corresponding 2-form which satisfies conditions (27) and (28) is [AMW1],

$$\omega = \frac{1}{2}(a^*\theta^L_a b^*\theta^R_a + a^*\theta^R_a b^*\theta^L_a).$$

4.3 Group-valued localization

In this section we explain how the localization principle works for the classes in $\hat{\mathcal{H}}_G(M)$. We begin by recalling some standard fact from the theory of Lie groups. We assume that G is a direct product of a compact semi-simple Lie group and a compact torus.

Some facts about compact Lie groups. Let G be a product of a compact semi-simple Lie group and a torus. Let T be a maximal torus in G and $\mathcal{T}$ be its Lie algebra. The Lie algebra $\mathcal{T}$ contains the integral lattice $\Lambda \subset \mathcal{T}$,

$$\Lambda = \{x \in \mathcal{T} \mid \exp(x) = e\}.$$

The dual space $\mathcal{T}^*$ contains the dual lattice Λ^*. By choosing the set of positive roots R_+ we define the positive Weil chamber $\mathcal{T}^*_+ \subset \mathcal{T}^*$, and the set of dominant weights $\Lambda^* \cap \mathcal{T}^*_+$. A dominant weight λ defines a unique irreducible highest weight representation V_λ of G. The representation V_λ contains the highest weight vector v_λ which satisfies the following conditions,

$$e_\alpha \cdot v_\lambda = 0,$$

where e_α are the generators corresponding to positive roots, and

$$h_\alpha \cdot v_\lambda = (\alpha, \lambda) v_\lambda,$$

where h_α are the elements of $\mathcal{T}$ corresponding to the roots. All irreducible representations V_λ possess a Hermitian invariant scalar product. We normalize v_λ such that $(v_\lambda, v_\lambda) = 1$. Then, for each dominant weight λ one can define two functions on G, the character,

$$\chi_\lambda(g) = \mathrm{Tr}_{V_\lambda} g,$$

and the 'spherical harmonics',

$$\Delta_\lambda(g) = (v_\lambda, g \cdot v_\lambda).$$

A special weight is given by the half-sum of positive roots,

$$\rho = \frac{1}{2} \sum_{\alpha \in R_+} \alpha.$$

For example, for $G = SU(2)$ the representations are parametrized by the spin $j = 0, 1/2, 1 \ldots$. The weight ρ corresponds to $j = 1/2$. The corresponding representation is two-dimensional, and we obtain,

$$\chi_{\frac{1}{2}} \begin{pmatrix} a & b \\ c & d \end{pmatrix} = a + d, \; \Delta_{\frac{1}{2}} \begin{pmatrix} a & b \\ c & d \end{pmatrix} = a.$$

Usually, we identify $\mathcal{T}$ and $\mathcal{T}^*$ using the scalar product. Then, one can view the dominant weights as the elements of $\mathcal{T}$. Because the weights belong to a special lattice, the corresponding one-parameter subgroups $T_\lambda = \{\exp(s\lambda)\}$ (except for $\lambda = 0$) are circle subgroups of T. Note that a typical one-parameter subgroup of T is dense in T. So, the subgroups T_λ are very special, and this will play an important role in the localization theorem for $\hat{\mathcal{H}}_G(M)$.

Finally, we recall that the *classical r-matrix* is an element in $\wedge^2\mathcal{G}$ defined by formula,

$$r = \sum_{\alpha\in R_+} e_\alpha \wedge e_{-\alpha}.$$

It is convenient to represent r as $r = r_{ab}t_a \otimes t_b$. The r-matrix satisfies the *classical Yang-Baxter* equation,

$$\mathrm{Cycl_{abc}}(\mathrm{r_{as}f_{sbt}r_{tc}}) = \frac{1}{4}\mathrm{f_{abc}},$$

where $\mathrm{Cycl_{abc}}$ stands for summation over the cyclic permutations of the indices a, b, c.

The localization formula. Now we are ready to formulate the new localization formula. Let M be a compact G-manifold, and $\alpha \in (\Omega(M) \otimes \hat{\mathcal{W}}_G)_{basic}$ be an equivariant cocycle,

$$(d_M + d_{\mathcal{W}})\alpha = 0.$$

Then, one can define an integral of α over M,

$$\int_M \alpha \in (\hat{W}_G)_{basic}.$$

The elements of the space $(\hat{W}_G)_{basic}$ are annihilated by contractions, and, hence, belong to $\mathcal{E}'(G) \otimes 1$. By G-invariance, these distributions should be conjugation-invariant, $(\hat{W}_G)_{basic} \cong \mathcal{E}'(G)^G$. A conjugation-invariant distribution is completely characterized by its pairings with characters of irreducible representations of G,

$$\alpha_\lambda = \langle \int_M \alpha, \chi_\lambda \rangle.$$

It is easy to show that the numbers α_λ do not depend on the representative in the cohomology class.

The localization formula [AMW2] gives expressions for α_λ in terms of the fixed points of the action of $T_{\lambda+\rho}$ (note the shift by ρ). As usual, we assume that all these fixed points are isolated. Then, one obtains,

$$\alpha_\lambda = \left(\frac{2\pi}{i}\right)^m \dim V_\lambda \sum_p \frac{\langle exp(\frac{1}{2}\iota(r))\alpha, \Delta_\lambda\rangle(p)}{\nu_1^p \dots \nu_m^p}. \tag{29}$$

Here the dimension of M is $2m$, the dimension of the representation V_λ is $\dim V_\lambda$, $\iota(r)$ is defined as $r_{ab}\iota_a\iota_b$, and ν_i^p are the indices of the circle action of $T_{\lambda+\rho}$ at the point p.

Formula (29) simplifies if M is a Hamiltonian space with group valued moment map, and the cocycle is the equivariant Liouville form on M. In this case the localization formula reads [AMW2],

$$\mathcal{L}_\lambda = \left(\frac{2\pi}{i}\right)^m \dim V_\lambda \sum_p \frac{\Delta_{\lambda+\rho}(H(p))}{\nu_1^p \dots \nu_m^p}. \tag{30}$$

Note that if p is a fixed point for some $T_{\lambda+\rho}$, the value of the moment map $H(p)$ belongs to the maximal torus T. The spherical harmonics $\Delta_{\lambda+\rho}$ defines a character of T which generalizes the expression $\exp(i\xi H(p))$ in the Duistermaat-Heckman formula. (18).

Some simple application of the formula (30) can be found in [P]. In particular, certain integrals over the sphere S^4, and the space $SU(2) \times SU(2) \cong S^3 \times S^3$ can be computed using this technique. A more ambitious task is to show that formula (18) gives precise meaning to the path integrals of [W2].

References

[AB] Atiyah, M., Bott, R. (1984): The moment map and equivariant cohomology. Topology **23** no. 1, 1-28

[AMM] Alekseev, A., Malkin, A., Meinrenken E. (1998): Lie group valued moment maps. J. Differential Geom. **48** no. 3, 445-495

[AM] Alekseev, A., Meinrenken, E. (1999): The non-commutative Weil algebra. Preprint math.DG/990352, to be published in Inv. Math.

[AMW1] Alekseev, A., Meinrenken, E., Woodward, C. (1999): Duistermaat-Heckman distributions for group-valued moment maps. Preprint math.DG/9903087

[AMW2] Alekseev, A., Meinrenken, E., Woodward, C. (1999): Group-valued equivariant localization. Preprint math.DG/9905130

[BV] Berline, N., Vergne, M. (1983): Zéro d'un champ de vecteurs et classes caractéristiques équivariantes. Duke Math. J. **50**, 539-549

[BGV] Berline, N., Getzler, E., Vergne, M. (1992): Heat kernels and Dirac operators. Grundlehren der mathematischen Wissenschaften, vol. **298**, Springer-Verlag, Berlin-Heidelberg-New York

[BT] Blau, M., Thompson, G. (1995): Equivariant Kähler Geometry and Localization in G/G Model. Nucl. Phys. **B439**, 367-394

[C1] Cartan, H. (1950): La transgression dans un groupe de Lie et dans un fibré principal. In *Colloque de topologie (espaces fibrès)*, Bruxelles, 73-81

[C2] Cartan, H. (1950): Notions d'algèbre différentielle; application aux groupes de Lie et aux variétés où opère un groupe de Lie. In *Colloque de topologie (espaces fibrès)*, Bruxelles

[CMR] Cordes, S., Moore, G., Ramgoolam, S. (1995): Lectures on 2D Yang-Mills Theory, Equivariant Cohomology and Topological Field Theories. Nucl. Phys. Proc. Suppl. **41**, 184-244

[DH] Duistermaat, J.J., Heckman, G.J. (1982): One the variation in the cohomology of the symplectic form on the reduced phase space. Inv. Math. **69**, 259-268

[G] Gerasimov, A. (1993): Localization in GWZW and Verlinde formula. Preprint hep-th/9305090

[MNP] A., Morozov, A., Niemi, A., Palo, K. (1991): Supersymmetry and loop space geometry. Phys. Lett. **B 271** 365-371

[JK] Jeffrey, L., Kirwan, F. (1995): Localization for Nonabelian Group Actions. Topology **34**, 291-327

[K] Kalkman, J. (1993): A BRST model applied to symplectic geometry. Ph.D. thesis, Universiteit Utrecht.

[MNS] Moore, G., Nekrasov, N., Shatashvili, S. (1997): Integrating over Higgs branches. Preprint hep-th/9712241

[P] Plamenevskaya, O. (1999): A Residue Formula for $SU(2)$-valued Moment Maps. Preprint math.DG/9906093
[W1] Witten, E. (1982): Supersymmetry and Morse theory. J. Differential Geom. **17**, 661-692
[W2] Witten, E. (1992): Two-dimensional gauge theory revisited. J. Geom. Phys. **9**, 303-368

An Introduction to Spin Foam Models of BF Theory and Quantum Gravity

John C. Baez

Department of Mathematics, University of California, Riverside, California 92521, USA

Abstract. In loop quantum gravity we now have a clear picture of the quantum geometry of *space*, thanks in part to the theory of spin networks. The concept of 'spin foam' is intended to serve as a similar picture for the quantum geometry of *spacetime*. In general, a spin network is a graph with edges labeled by representations and vertices labeled by intertwining operators. Similarly, a spin foam is a 2-dimensional complex with faces labeled by representations and edges labeled by intertwining operators. In a 'spin foam model' we describe states as linear combinations of spin networks and compute transition amplitudes as sums over spin foams. This paper aims to provide a self-contained introduction to spin foam models of quantum gravity and a simpler field theory called BF theory.

1 Introduction

Spin networks were first introduced by Penrose as a radical, purely combinatorial description of the geometry of spacetime. In their original form, they are trivalent graphs with edges labelled by spins:

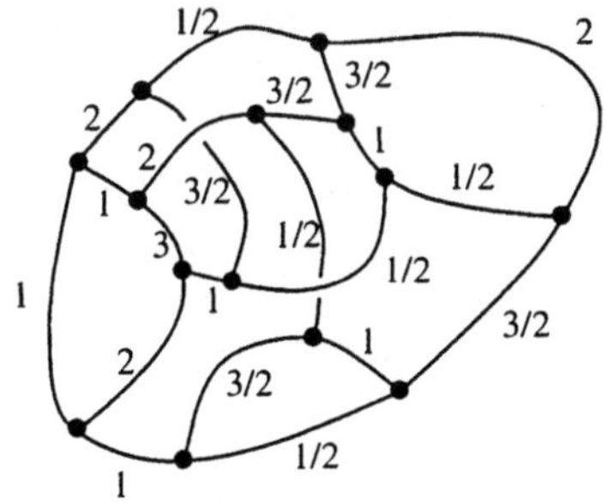

In developing the theory of spin networks, Penrose seems to have been motivated more by the quantum mechanics of angular momentum than by the details of general relativity. It thus came as a delightful surprise when Rovelli and Smolin discovered that spin networks can be used to describe states in loop quantum gravity.

Fundamentally, loop quantum gravity is a very conservative approach to quantum gravity. It starts with the equations of general relativity and attempts to apply the time-honored principles of quantization to obtain a Hilbert space of states. There are only two really new ideas in loop quantum gravity. The first is its insistence on a background-free approach. That is, unlike perturbative quantum gravity, it makes no use of a fixed 'background' metric on spacetime. The second is that it uses a formulation of Einstein's equations in which parallel transport, rather than the metric, plays the main role. It is very interesting that starting from such ideas one is naturally led to describe states using spin networks!

However, there is a problem. While Penrose originally intended for spin networks to describe the geometry of spacetime, they are really better for describing the geometry of *space*. In fact, this is how they are used in loop quantum gravity. Since loop quantum gravity is based on canonical quantization, states in this formalism describe the geometry of space at a fixed time. Dynamics enters the theory only in the form of a constraint called the Hamiltonian constraint. Unfortunately this constraint is still poorly understood. Thus until recently, we had almost no idea what loop quantum gravity might say about the geometry of *spacetime*.

To remedy this problem, it is natural to try to supplement loop quantum gravity with an appropriate path-integral formalism. In ordinary quantum field theory we calculate path integrals using Feynman diagrams. Copying this idea, in loop quantum gravity we may try to calculate path integrals using 'spin foams', which are a 2-dimensional analogue of Feynman diagrams. In general, spin networks are graphs with edges labeled by group representations and vertices labeled by intertwining operators. These reduce to Penrose's original spin networks when the group is SU(2) and the graph is trivalent. Similarly, a spin foam is a 2-dimensional complex built from vertices, edges and polygonal faces, with the faces labeled by group representations and the edges labeled by intertwining operators. When the group is SU(2) and three faces meet at each edge, this looks exactly like a bunch of soap suds with all the faces of the bubbles labeled by spins — hence the name 'spin foam'.

If we take a generic slice of a spin foam, we get a spin network. Thus we can think of a spin foam as describing the geometry of spacetime, and a slice of it as describing the geometry of space at a given time. Ultimately we would like a 'spin foam model' of quantum gravity, in which we compute transition amplitudes between states by summing over spin foams going from one spin network to another:

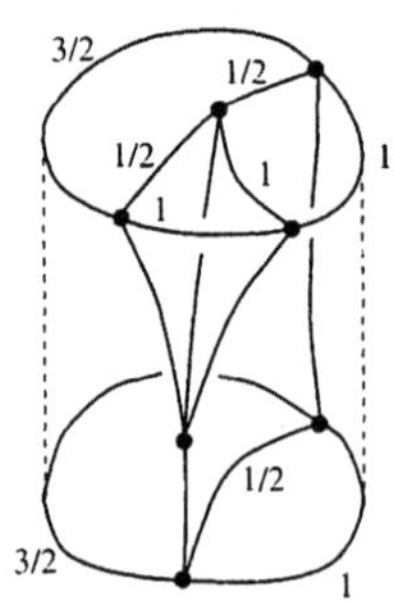

At present this goal has been only partially attained. For this reason it seems best to start by discussing spin foam models of a simpler theory, called BF theory. In a certain sense this the simplest possible gauge theory. It can be defined on spacetimes of any dimension. It is 'background-free', meaning that to formulate it we do not need a pre-existing metric or any other such geometrical structure on spacetime. At the classical level, the theory has no local degrees of freedom: all the interesting observables are global in nature. This remains true upon quantization. Thus BF theory serves as a simple starting-point for the study of background-free theories. In particular, general relativity in 3 dimensions is a special case of BF theory, while general relativity in 4 dimensions can be viewed as a BF theory with extra constraints. Most work on spin foam models of quantum gravity seeks to exploit this fact.

In what follows, we start by describing BF theory at the classical level. Next we canonically quantize the theory and show the space of gauge-invariant states is spanned by spin networks. Then we use the path-integral formalism to study the dynamics of the theory and show that the transition amplitude from one spin network state to another is given as a sum over spin foams. When the dimension of spacetime is above 2, this sum usually diverges. However, in dimensions 3 and 4, we can render it finite by adding an extra term to the Lagrangian of BF theory. In applications to gravity, this extra term corresponds to the presence of a cosmological constant. Finally, we discuss spin foam models of 4-dimensional quantum gravity.

At present, work on spin foam models is spread throughout a large number of technical papers in various fields of mathematics and physics. This has the unfortunate effect of making the subject seem more complicated and less beautiful than it really is. As an attempt to correct this situation, I have tried to make this paper as self-contained as possible. For the sake of smooth exposition, I have relegated all references to the Notes, which form a kind of annotated bibliography of the subject. The remarks at the end of each section contain information of a more technical nature that can safely be skipped.

2 BF Theory: Classical Field Equations

To set up BF theory, we take as our gauge group any Lie group G whose Lie algebra $\mathfrak{g}$ is equipped with an invariant nondegenerate bilinear form $\langle \cdot, \cdot \rangle$. We take as our spacetime any n-dimensional oriented smooth manifold M, and choose a principal G-bundle P over M. The basic fields in the theory are then:

- a connection A on P,
- an $\mathrm{ad}(P)$-valued $(n-2)$-form E on M.

Here $\mathrm{ad}(P)$ is the vector bundle associated to P via the adjoint action of G on its Lie algebra. The curvature of A is an $\mathrm{ad}(P)$-valued 2-form F on M. If we pick a local trivialization we can think of A as a $\mathfrak{g}$-valued 1-form on M, F as a $\mathfrak{g}$-valued 2-form, and E as a $\mathfrak{g}$-valued $(n-2)$-form.

The Lagrangian for BF theory is:

$$\mathcal{L} = \mathrm{tr}(E \wedge F).$$

Here $\mathrm{tr}(E \wedge F)$ is the n-form constructed by taking the wedge product of the differential form parts of E and F and using the bilinear form $\langle \cdot, \cdot \rangle$ to pair their $\mathfrak{g}$-valued parts. The notation 'tr' refers to the fact that when G is semisimple we can take this bilinear form to be the Killing form $\langle x, y \rangle = \mathrm{tr}(xy)$, where the trace is taken in the adjoint representation.

We obtain the field equations by setting the variation of the action to zero:

$$\begin{aligned} 0 &= \delta \int_M \mathcal{L} \\ &= \int_M \mathrm{tr}(\delta E \wedge F + E \wedge \delta F) \\ &= \int_M \mathrm{tr}(\delta E \wedge F + E \wedge d_A \delta A) \\ &= \int_M \mathrm{tr}(\delta E \wedge F + (-1)^{n-1} d_A E \wedge \delta A) \end{aligned}$$

where d_A stands for the exterior covariant derivative. Here in the second step we used the identity $\delta F = d_A \delta A$, while in the final step we did an integration by parts. We see that the variation of the action vanishes for all δE and δA if and only if the following field equations hold:

$$F = 0, \qquad d_A E = 0.$$

These equations are rather dull. But this is exactly what we want, since it suggests that BF theory is a topological field theory! In fact, all solutions of these equations look the same locally, so BF theory describes a world with no local degrees of freedom. To see this, first note that the equation $F = 0$ says the connection A is flat. Indeed, all flat connections are locally the same up to gauge transformations. The equation $d_A E = 0$ is a bit subtler. It is not true that all solutions of this are locally the same up to a gauge transformation in the usual sense. However, BF theory has another sort of symmetry. Suppose we define a transformation of the A and E fields by

$$A \mapsto A, \qquad E \mapsto E + d_A \eta$$

for some $\mathrm{ad}(P)$-valued $(n-3)$-form η. This transformation leaves the action unchanged:

$$\begin{aligned} \int_M \mathrm{tr}((E + d_A \eta) \wedge F) &= \int_M \mathrm{tr}(E \wedge F + d_A \eta \wedge F) \\ &= \int_M \mathrm{tr}(E \wedge F + (-1)^n \eta \wedge d_A F) \\ &= \int_M \mathrm{tr}(E \wedge F) \end{aligned}$$

where we used integration by parts and the Bianchi identity $d_A F = 0$. In the next section we shall see that this transformation is a 'gauge symmetry' of BF theory, in the more general sense of the term, meaning that two solutions differing by this transformation should be counted as physically equivalent. Moreover, when A is flat, any E field with $d_A E = 0$ can be written locally as $d_A \eta$ for some η; this is an easy consequence of the fact that locally all closed forms are exact. Thus locally, all solutions of the BF theory field equations are equal modulo gauge transformations and transformations of the above sort.

Why is general relativity in 3 dimensions a special case of BF theory? To see this, take $n = 3$, let $G = \mathrm{SO}(2,1)$, and let $\langle \cdot, \cdot \rangle$ be minus the Killing form. Suppose first that $E: TM \to \mathrm{ad}(P)$ is one-to-one. Then we can use it to define a Lorentzian metric on M as follows:

$$g(v,w) = \langle Ev, Ew \rangle$$

for any tangent vectors $v, w \in T_x M$. We can also use E to pull back the connection A to a metric-preserving connection Γ on the tangent bundle of M. The equation $d_A E = 0$ then says precisely that Γ is torsion-free, so that Γ is the Levi-Civita connection on M. Similarly, the equation $F = 0$ implies that Γ is flat. Thus the metric g is flat.

In 3 dimensional spacetime, the vacuum Einstein equations simply say that the metric is flat. Of course, many different A and E fields correspond to the same metric, but they all differ by gauge transformations. So in 3 dimensions, BF theory with gauge group $\mathrm{SO}(2,1)$ is really just an alternate formulation of Lorentzian general relativity without matter fields — at least when E is one-to-one. When E is not one-to-one, the metric g defined above will be degenerate, but the field equations of BF theory still make perfect sense. Thus 3d BF theory with gauge group $\mathrm{SO}(2,1)$ may be thought of as an extension of the vacuum Einstein equations to the case of degenerate metrics.

If instead we take $G = \mathrm{SO}(3)$, all these remarks still hold except that the metric g is Riemannian rather than Lorentzian when E is one-to-one. We call this theory 'Riemannian general relativity'. We study this theory extensively in what follows, because it is easier to quantize than 3-dimensional Lorentzian general relativity. However, it is really just a warmup exercise for the Lorentzian case — which in turn is a warmup for 4-dimensional Lorentzian quantum gravity.

We conclude with a word about double covers. We can also express general relativity in 3 dimensions as a BF theory by taking the double cover $\mathrm{Spin}(2,1) \cong \mathrm{SL}(2,\mathbb{R})$ or $\mathrm{Spin}(3) \cong \mathrm{SU}(2)$ as gauge group and letting P be the spin bundle. This does not affect the classical theory. As we shall see, it does affect the quantum theory. Nonetheless, it is very popular to take these groups as gauge groups in 3-dimensional quantum gravity. The question whether it is 'correct' to use these double covers as gauge groups seems to have no answer — until we couple quantum gravity to spinors, at which point the double cover is necessary.

Remarks 1. In these calculations we have been ignoring the boundary terms that arise when we integrate by parts on a manifold with boundary. They are valid if either M is compact or if M is compact with boundary and boundary conditions are imposed that make the boundary terms vanish. BF theory on manifolds with boundary is interesting both for its applications to black hole physics — where the event horizon may be treated as a boundary — and as an example of an 'extended topological field theory'.

3 Classical Phase Space

To determine the classical phase space of BF theory we assume spacetime has the form

$$M = \mathbb{R} \times S$$

where the real line $\mathbb{R}$ represents time and S is an oriented smooth $(n-1)$-dimensional manifold representing space. This is no real loss of generality, since any oriented hypersurface in any oriented n-dimensional manifold has a neighborhood of this form. We can thus use the results of canonical quantization to study the dynamics of BF theory on quite general spacetimes.

If we work in temporal gauge, where the time component of the connection A vanishes, we see the momentum canonically conjugate to A is

$$\frac{\partial \mathcal{L}}{\partial \dot{A}} = E.$$

This is reminiscent of the situation in electromagnetism, where the electric field is canonically conjugate to the vector potential. This is why we use the notation 'E'. Originally people used the notation 'B' for this field, hence the term 'BF theory', which has subsequently become ingrained. But to understand the physical meaning of the theory, it is better to call this field 'E' and think of it as analogous to the electric field. Of course, the analogy is best when $G = \mathrm{U}(1)$.

Let $P|_S$ be the restriction of the bundle P to the 'time-zero' slice $\{0\} \times S$, which we identify with S. Before we take into account the constraints imposed by the field equations, the configuration space of BF theory is the space $\mathcal{A}$ of connections on $P|_S$. The corresponding classical phase space, which we call the 'kinematical phase space', is the cotangent bundle $T^*\mathcal{A}$. A point in this phase space consists of a connection A on $P|_S$ and an $\mathrm{ad}(P|_S)$-valued $(n-2)$-form E on S. The symplectic structure on this phase space is given by

$$\omega((\delta A, \delta E), (\delta A', \delta E')) = \int_S \mathrm{tr}(\delta A \wedge \delta E' - \delta A' \wedge \delta E).$$

This reflects the fact that A and E are canonically conjugate variables. However, the field equations of BF theory put constraints on the initial data A and E:

$$B = 0, \qquad d_A E = 0$$

where B is the curvature of the connection $A \in \mathcal{A}$, analogous to the magnetic field in electromagnetism. To deal with these constraints, we should apply symplectic reduction to $T^*\mathcal{A}$ to obtain the physical phase space.

The constraint $d_A E = 0$, called the Gauss law, is analogous to the equation in vacuum electromagnetism saying that the divergence of the electric field vanishes. This constraint generates the action of gauge transformations on $T^*\mathcal{A}$. Doing symplectic reduction with respect to this constraint, we thus obtain the 'gauge-invariant phase space' $T^*(\mathcal{A}/\mathcal{G})$, where $\mathcal{G}$ is the group of gauge transformations of the bundle $P|_S$.

The constraint $B = 0$ is analogous to an equation requiring the magnetic field to vanish. Of course, no such equation exists in electromagnetism; this constraint is special to BF theory. It generates transformations of the form

$$A \mapsto A, \qquad E \mapsto E + d_A\eta,$$

so these transformations, discussed in the previous section, really are gauge symmetries as claimed. Doing symplectic reduction with respect to this constraint, we obtain the 'physical phase space' $T^*(\mathcal{A}_0/\mathcal{G})$, where $\mathcal{A}_0$ is the space of flat connections on $P|_S$. Points in this phase space correspond to physical states of classical BF theory.

Remarks 1. The space $\mathcal{A}$ is an infinite-dimensional vector space, and if we give it an appropriate topology, an open dense set of $\mathcal{A}/\mathcal{G}$ becomes an infinite-dimensional smooth manifold. The simplest way to precisely define $T^*(\mathcal{A}/\mathcal{G})$ is as the cotangent bundle of this open dense set. The remaining points correspond to connections with more symmetry than the rest under gauge transformations. These are called 'reducible' connections. A more careful definition of the physical phase space would have to take these points into account.
2. The space $\mathcal{A}_0/\mathcal{G}$ is called the 'moduli space of flat connections on $P|_S$'. We can understand it better as follows. Since the holonomy of a flat connection around a loop does not change when we apply a homotopy to the loop, a connection $A \in \mathcal{A}_0$ determines a homomorphism from the fundamental group $\pi_1(S)$ to G after we trivialize P at the basepoint $p \in S$ that we use to define the fundamental group. If we apply a gauge transformation to A, this homomorphism is conjugated by the value of this gauge transformation at p. This gives us a map from $\mathcal{A}_0/\mathcal{G}$ to $\hom(\pi_1(S), G)/G$, where $\hom(\pi_1(S), G)$ is the space of homomorphisms from $\pi_1(S)$ to G, and G acts on this space by conjugation. When S is connected this map is one-to-one, so we have

$$\mathcal{A}_0/\mathcal{G} \subseteq \hom(\pi_1(S), G)/G.$$

The space $\hom(\pi_1(S), G)/G$ is called the 'moduli space of flat G-bundles over S'. When $\pi_1(S)$ is finitely generated (e.g. when S is compact) this space is a real algebraic variety, and $\mathcal{A}_0/\mathcal{G}$ is a subvariety. Usually $\mathcal{A}_0/\mathcal{G}$ has singularities, but each component has an open dense set that is a smooth manifold. When we speak of $T^*(\mathcal{A}_0/\mathcal{G})$ above, we really mean the cotangent bundle of this open dense set, though again a more careful treatment would deal with the singularities.

We can describe $\mathcal{A}_0/\mathcal{G}$ much more explicitly in particular cases. For example, suppose that S is a compact oriented surface of genus n. Then the group $\pi_1(S)$ has a presentation with $2n$ generators $x_1, y_1, \ldots, x_n, y_n$ satisfying the relation

$$R(x_i, y_i) := (x_1 y_1 x_1^{-1} y_1^{-1}) \cdots (x_n y_n x_n^{-1} y_n^{-1}) = 1.$$

A point in $\hom(\pi_1(S), G)$ may thus be identified with a collection $g_1, h_1, \ldots, g_n, h_n$ of elements of G satisfying

$$R(g_i, h_i) = 1,$$

and a point in $\hom(\pi_1(S), G)/G$ is an equivalence class $[g_i, h_i]$ of such collections.

The cases $G = \mathrm{SU}(2)$ and $G = \mathrm{SO}(3)$ are particularly interesting for their applications to 3-dimensional Riemannian general relativity. When $G = \mathrm{SU}(2)$, all G-bundles over a compact oriented surface S are isomorphic, and $\mathcal{A}_0/\mathcal{G} = \hom(\pi_1(S), G)/G$. When $G = \mathrm{SO}(3)$, there are two isomorphism classes of G-bundles over S, distinguished by their second Stiefel-Whitney number $w_2 \in \mathbb{Z}_2$. For each of these bundles, the points $[g_i, h_i]$ that lie in $\mathcal{A}_0/\mathcal{G}$ can be described as follows. Choose representatives $g_i, h_i \in \mathrm{SO}(3)$ and choose elements $\tilde{g}_i, \tilde{h}_i$ that map down to these representatives via the double cover $\mathrm{SU}(2) \to \mathrm{SO}(3)$. Then $[g_i, h_i]$ lies in $\mathcal{A}_0/\mathcal{G}$ if and only if

$$(-1)^{w_2} = R(\tilde{g}_i, \tilde{h}_i).$$

For 3-dimensional Riemannian general relativity with gauge group $\mathrm{SO}(3)$, the relevant bundle is the frame bundle of S, which has $w_2 = 0$. For both $\mathrm{SU}(2)$ and $\mathrm{SO}(3)$, the space $\mathcal{A}_0/\mathcal{G}$ has dimension $6n - 6$ for $n \geq 2$. For the torus $\mathcal{A}_0/\mathcal{G}$ has dimension 2, and for the sphere it is a single point.

4 Canonical Quantization

In the previous section we described the kinematical, gauge-invariant and physical phase spaces for BF theory. All of these are cotangent bundles. Naively, quantizing any one of them should give the Hilbert space of square-integrable functions on the corresponding configuration space. We can summarize this hope with the following diagram:

$$\begin{array}{ccc}
T^*(\mathcal{A}) & \xrightarrow{\textit{quantize}} & L^2(\mathcal{A}) \\
\Big\downarrow \textit{constrain} & & \Big\downarrow \textit{constrain} \\
T^*(\mathcal{A}/\mathcal{G}) & \xrightarrow{\textit{quantize}} & L^2(\mathcal{A}/\mathcal{G}) \\
\Big\downarrow \textit{constrain} & & \Big\downarrow \textit{constrain} \\
T^*(\mathcal{A}_0/\mathcal{G}) & \xrightarrow{\textit{quantize}} & L^2(\mathcal{A}_0/\mathcal{G})
\end{array}$$

Traditionally it had been difficult to realize this hope with any degree of rigor because the spaces $\mathcal{A}$ and $\mathcal{A}/\mathcal{G}$ are typically infinite-dimensional, making it difficult to define $L^2(\mathcal{A})$ and $L^2(\mathcal{A}/\mathcal{G})$. The great achievement of loop quantum gravity is that it gives rigorous and background-free, hence diffeomorphism-invariant, definitions of these Hilbert spaces. It does so by breaking away from the traditional Fock space formalism and taking holonomies along paths as the basic variables to be quantized. The result is a picture in which the basic excitations are not 0-dimensional particles but 1-dimensional 'spin network edges'. As we shall see, this eventually leads us to a picture in which 1-dimensional Feynman diagrams are replaced by 2-dimensional 'spin foams'.

In what follows we shall assume that the gauge group G is compact and connected and the manifold S representing space is real-analytic. The case where S merely smooth is considerably more complicated, but people know how to handle it. The case where G is not connected would only require some slight modifications in our formalism. However, nobody really knows how to handle the case where G is noncompact! This is why, when we apply our results to quantum gravity, we consider the quantization of the vacuum Einstein equations for Riemannian rather than Lorentzian metrics: $\mathrm{SO}(n)$ is compact but $\mathrm{SO}(n,1)$ is not. The Lorentzian case is just beginning to receive the serious study that it deserves.

To define $L^2(\mathcal{A})$, we start with the algebra $\mathrm{Fun}(\mathcal{A})$ consisting of all functions on $\mathcal{A}$ of the form

$$\Psi(A) = f(Te^{\int_{\gamma_1} A}, \ldots, Te^{\int_{\gamma_n} A}).$$

Here γ_i is a real-analytic path in S, $Te^{\int_{\gamma_i} A}$ is the holonomy of A along this path, and f is a continuous complex-valued function of finitely many such holonomies. Then we define an inner product on $\mathrm{Fun}(\mathcal{A})$ and complete it to obtain the Hilbert space $L^2(\mathcal{A})$. To define this inner product, we need to think about graphs embedded in space:

Definition 1. *A finite collection of real-analytic paths $\gamma_i\colon [0,1] \to S$ form a* graph in S *if they are embedded and intersect, it at all, only at their endpoints. We then call them* edges *and call their endpoints* vertices. *Given a vertex v, we say an edge γ_i is* outgoing from v *if $\gamma_i(0) = v$, and we say γ_i is* incoming to v *if $\gamma_i(1) = v$.*

Suppose we fix a collection of paths $\gamma_1, \ldots, \gamma_n$ that form a graph in S. We can think of the holonomies along these paths as elements of G. Using this idea one can show that the functions of the form

$$\Psi(A) = f(Te^{\int_{\gamma_1} A}, \ldots, Te^{\int_{\gamma_n} A})$$

for these particular paths γ_i form a subalgebra of $\mathrm{Fun}(\mathcal{A})$ that is isomorphic to the algebra of all continuous complex-valued functions on G^n. Given two functions in this subalgebra, we can thus define their inner product by

$$\langle \Psi, \Phi \rangle = \int_{G^n} \overline{\Psi} \Phi$$

where the integral is done using normalized Haar measure on G^n. Moreover, given any functions $\Psi, \Phi \in \mathrm{Fun}(\mathcal{A})$ there is always some subalgebra of this form that contains them. Thus we can always define their inner product this way. Of course we have to check that this definition is independent of the choices involved, but this is not too hard. Completing the space $\mathrm{Fun}(\mathcal{A})$ in the norm associated to this inner product, we obtain the 'kinematical Hilbert space' $L^2(\mathcal{A})$.

Similarly, we may define $\mathrm{Fun}(\mathcal{A}/\mathcal{G})$ to be the space consisting of all functions in $\mathrm{Fun}(\mathcal{A})$ that are invariant under gauge transformations, and complete it in the above norm to obtain the 'gauge-invariant Hilbert space' $L^2(\mathcal{A}/\mathcal{G})$. This space can be described in a very concrete way: it is spanned by 'spin network states'.

Definition 2. *A* spin network *in S is a triple $\Psi = (\gamma, \rho, \iota)$ consisting of:*

1. *a graph γ in S,*
2. *for each edge e of γ, an irreducible representation ρ_e of G,*
3. *for each vertex v of γ, an intertwining operator*

$$\iota_v \colon \rho_{e_1} \otimes \cdots \otimes \rho_{e_n} \to \rho_{e'_1} \otimes \rho_{e'_m}$$

where $e_1, \ldots, e_n$ are the edges incoming to v and $e'_1, \ldots e'_m$ are the edges outgoing from v.

In what follows we call an intertwining operator an intertwiner.

There is an easy way to get a function in $\mathrm{Fun}(\mathcal{A}/\mathcal{G})$ from a spin network in S. To explain how it works, it is easiest to give an example. Suppose we have a spin network Ψ in S with three edges e_1, e_2, e_3 and two vertices v_1, v_2 as follows:

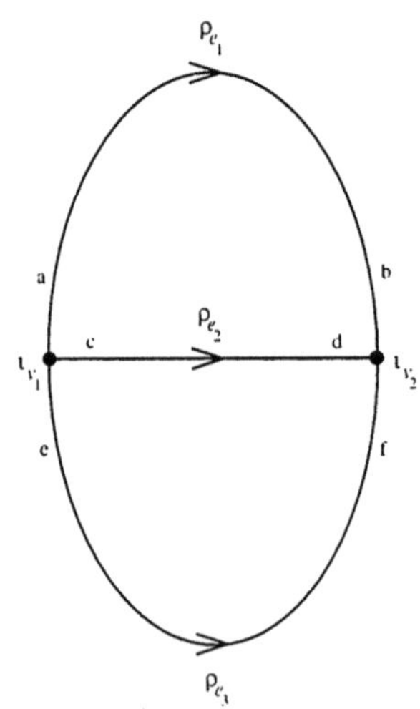

We draw arrows on the edges to indicate their orientation, and write little letters near the beginning and end of each edge. Then for any connection $A \in \mathcal{A}$ we define

$$\Psi(A) = \rho_{e_1}(Te^{\int_{e_1} A})^a_b \, \rho_{e_2}(Te^{\int_{e_2} A})^c_d \, \rho_{e_3}(Te^{\int_{e_3} A})^e_f \, (\iota_{v_1})_{ace} \, (\iota_{v_2})^{bdf}$$

In other words, we take the holonomy along each edge of Ψ, think of it as a group element, and put it into the representation labeling that edge. Picking a basis for this representation we think of the result as a matrix with one superscript and one subscript. We use the little letter near the beginning of the edge for the superscript and the little letter near the end of the edge for the subscript. In addition, we write the intertwining operator for each vertex as a tensor. This tensor has one superscript for each edge incoming to the vertex and one subscript for each edge outgoing from the vertex. Note that this recipe ensures that each letter appears once as a superscript and once as a subscript! Finally, using the Einstein summation convention we sum over all repeated indices and get a number, which of course depends on the connection A. This is $\Psi(A)$.

Since $\Psi: \mathcal{A} \to \mathbb{C}$ is a continuous function of finitely many holonomies, it lies in $\mathrm{Fun}(\mathcal{A})$. Using the fact that the ι_v are intertwiners, one can show that this function is gauge-invariant. We thus have $\Psi \in \mathrm{Fun}(\mathcal{A}/\mathcal{G})$. We call Ψ a 'spin network state'. The only hard part is to prove that spin network states span $L^2(\mathcal{A}/\mathcal{G})$. We give some references to the proof in the Notes.

The constraint $F = 0$ is a bit more troublesome. If we impose this constraint at the classical level, symplectic reduction takes us from $T^*(\mathcal{A}/\mathcal{G})$ to the physical phase space $T^*(\mathcal{A}_0/\mathcal{G})$. Heuristically, quantizing this should give the 'physical Hilbert space' $L^2(\mathcal{A}_0/\mathcal{G})$. However, for this to make sense, we need to choose a measure on $\mathcal{A}_0/\mathcal{G}$. This turns out to be problematic.

The space $\mathcal{A}_0/\mathcal{G}$ is called the 'moduli space of flat connections'. As explained in Remark 3 below, it has a natural measure when S is compact and of dimension 2 or less. It also has a natural measure when S is simply connected, since then it is a single point, and we can use the Dirac delta measure at that point. In these cases the physical Hilbert space is well-defined. In most other cases, there seems to be no natural measure on the moduli space of flat connections, so we cannot unambiguously define the physical Hilbert space.

If we are willing to settle for a mere vector space instead of a Hilbert space, there is something that works quite generally. Every function in $\mathrm{Fun}(\mathcal{A}/\mathcal{G})$ restricts to a gauge-invariant function on the space of flat connections, or in other words, a function on $\mathcal{A}_0/\mathcal{G}$. We denote the space of such functions by $\mathrm{Fun}(\mathcal{A}_0/\mathcal{G})$. In the cases listed above where there is a natural measure on $\mathcal{A}_0/\mathcal{G}$, the space $\mathrm{Fun}(\mathcal{A}_0/\mathcal{G})$ is dense in $L^2(\mathcal{A}_0/\mathcal{G})$. In what follows, we abuse language by calling elements of $\mathrm{Fun}(\mathcal{A}_0/\mathcal{G})$ 'physical states' even when there is no best measure on $\mathcal{A}_0/\mathcal{G}$. Of course, a space of physical states without an inner product is of limited use. Nonetheless the mathematics turns out to be very important for other things, so we proceed to study this space anyway.

We can understand $\mathrm{Fun}(\mathcal{A}_0/\mathcal{G})$ quite explicitly using the fact that every spin network in S gives a function in this space. In fact, if we give $\mathrm{Fun}(\mathcal{A}_0/\mathcal{G})$ a reasonable topology, like the sup norm topology, finite linear combinations of spin network states are dense in this space. Moreover, one can work out quite explicitly when two linear combinations of spin networks define the same physical state. For example, two spin networks in S differing by a homotopy define the same physical state, because the holonomy of a flat connection along a path does not change when we apply a homotopy to the path. There are also other relations, called 'skein relations', coming from the representation theory of the group G.

For example, suppose ρ is any irreducible representation of the group G. Then the following skein relation holds:

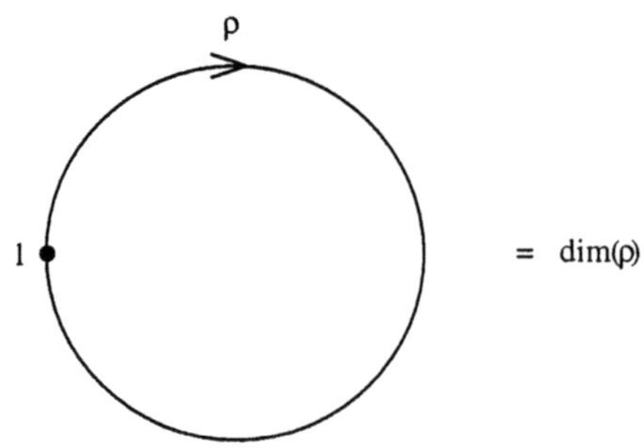

Here the left-hand side is a spin network with one edge e labeled by the representation ρ and one vertex labeled by the identity intertwiner. The edge is a contractible loop in S. The corresponding spin network state Ψ is given by

$$\Psi(A) = \mathrm{tr}(\rho(Te^{\oint_e A})).$$

The skein relation above means that $\Psi(A) = \dim(\rho)$ when A is flat. The reason is that the holonomy of a flat connection around a contractible loop is the identity, so its trace in the representation ρ is $\dim(\rho)$. As a consequence, whenever a spin network has a piece that looks like the above picture, if we eliminate that piece and multiply the remaining spin network state by $\dim(\rho)$, we obtain the same physical state.

People usually do not bother to draw vertices that are labeled by identity intertwiners. From now on we shall follow this custom. Thus instead of the above skein relation, we write:

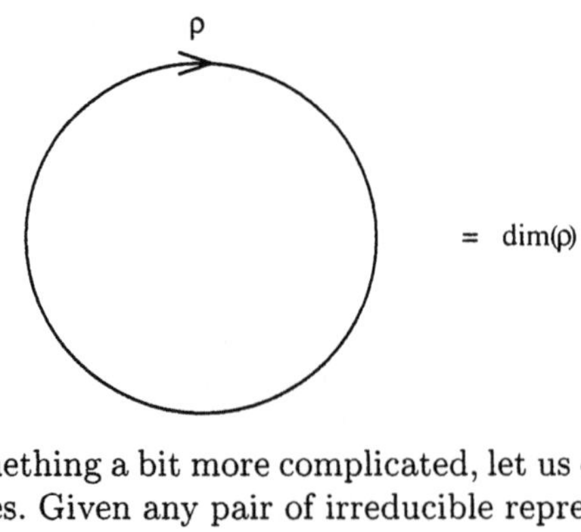

Moving on to something a bit more complicated, let us consider spin networks with trivalent vertices. Given any pair of irreducible representations ρ_1, ρ_2 of G, their tensor product can be written as a direct sum of irreducible representations. Picking one of these and calling it ρ_3, the projection from $\rho_1 \otimes \rho_2$ to ρ_3 is an intertwiner that we can use to label a trivalent vertex. However, it is convenient to multiply this projection by a constant so as to obtain an intertwiner $\iota: \rho_1 \otimes \rho_2 \to \rho_3$ with $\mathrm{tr}(\iota\iota^*) = 1$. We then have the skein relation

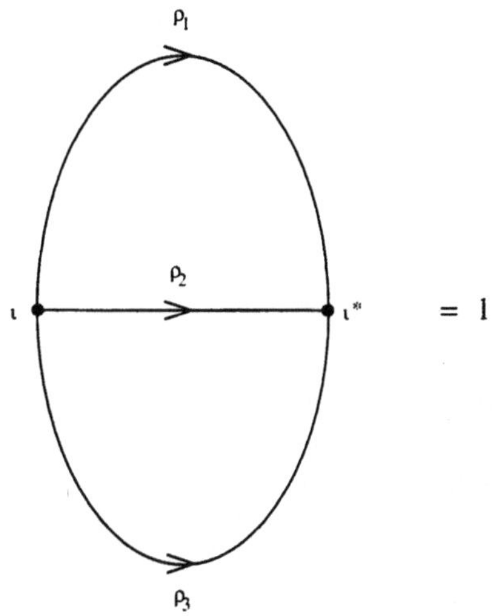

whenever this graph sits in S in a contractible way. Again, this skein relation means that the spin network on the left side of the equation defines a function $\Psi \in \mathrm{Fun}(\mathcal{A}/\mathcal{G})$ that equals 1 on all flat connections. Whenever a spin network in S has a piece that looks like this, we can eliminate that piece without changing the physical state it defines.

Of course, if the irreducible representation ρ_3 appears more than once in the direct sum decomposition of $\rho_1 \otimes \rho_2$ there will be more than one intertwiner of the above form. We can always pick a basis of such intertwiners such that $\iota_1 \iota_2^* = 0$ for any two distinct intertwiners ι_1, ι_2 in the basis. We then have the following skein relation:

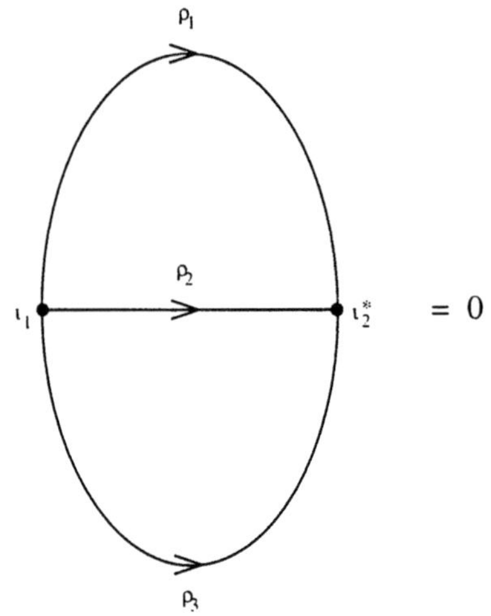

Let us pick such a basis of intertwiners for each triple of irreducible representations of G. To get enough states to span $\mathrm{Fun}(\mathcal{A}_0/\mathcal{G})$, it suffices to use these special intertwiners — appropriately dualized when necessary — to label trivalent vertices. What about vertices of higher valence? We can break any 4-valent vertex into two trivalent ones using the following sort of skein relation:

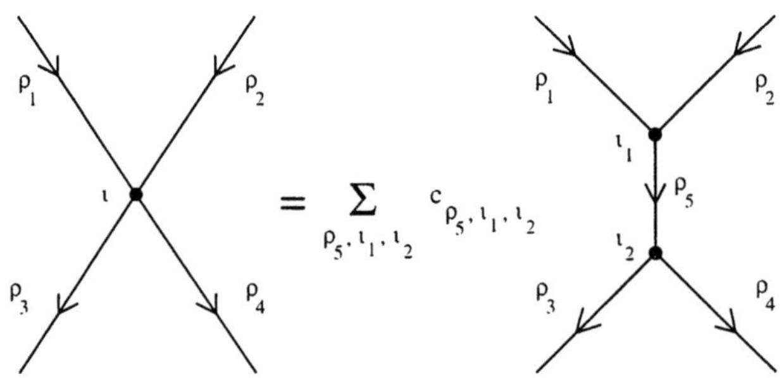

Here the sum is over irreducibles ρ_5 and intertwiners ι_1, ι_2 in the chosen bases. The coefficient depend on the details of the intertwiners in question. Both sides of this relation are to be interpreted as part of a larger spin network. The rest of the spin network, not shown in the figure, is arbitrary but the same for both sides. Similar skein relations hold for vertices of valence 5 or more. Using these skein relations and the tricks discussed in Remark 2 below, we can write any physical state as a linear combination of states coming from trivalent spin networks.

Philosophically, skein relations are intriguing because they can be interpreted in two different ways: either as facts about BF theory, or as facts about group representation theory. In the first interpretation, which we have emphasized here, the spin network edges represent actual curves embedded in space. In the second interpretation, they are merely an abstract notation for representations of G. The fact that both interpretations are possible shows that in some sense BF theory is nothing but a clever way to encode the representation theory of G in a quantum field theory. Ultimately, this is the real reason why BF theory is so interesting.

Remarks 1. The reason for assuming S is real-analytic is that given a finite collection of real-analytic paths γ_i in S, there is always some graph in S such that each path γ_i is a product of finitely many edges of this graph. This is not true in the smooth context: for example, two smoothly embedded paths can intersect in a Cantor set. One can generalize the construction of $L^2(\mathcal{A})$ and $L^2(\mathcal{A}/\mathcal{G})$ to the smooth context, but one needs a generalization of graphs known as 'webs'. The smooth and real-analytic categories are related as nicely as one could hope: a paracompact smooth manifold of any dimension admits a real-analytic structure, and this structure is unique up to a smooth diffeomorphism.
2. There are various ways to modify a spin network in S without changing the state it defines:

- We can reparametrize an edge by any orientation-preserving diffeomorphism of the unit interval.
- We can reverse the orientation of an edge while simultaneously replacing the representation labeling that edge by its dual and appropriately dualizing the intertwiners labeling the endpoints of that edge.
- We can subdivide an edge into two edges labeled with the same representation by inserting a vertex labeled with the identity intertwiner.
- We can eliminate an edge labeled by the trivial representation.

In fact, two spin networks in S define the same state in $L^2(\mathcal{A}/\mathcal{G})$ if and only if they differ by a sequence of these moves and their inverses. It is usually best to treat two such spin networks as 'the same'.
3. When S is a circle, $\mathcal{A}_0/\mathcal{G}$ is just the space of conjugacy classes of G. The normalized Haar measure on G can be pushed down to this space. We can easily extend this idea to put a measure on $\mathcal{A}_0/\mathcal{G}$ whenever S is a compact and 1-dimensional. When S is compact and 2-dimensional the space $\mathcal{A}_0/\mathcal{G}$ is an algebraic variety described as in Remark 2 of the previous section. There is a natural symplectic structure on the smooth part of this variety, given by

$$\omega([\delta A],[\delta A']) = \int_S \mathrm{tr}(\delta A \wedge \delta A')$$

where $\delta A, \delta' A$ are tangent vectors to $\mathcal{A}_0$, i.e., $\mathrm{ad}(P)$-valued 1-forms. Raising ω to a suitable power we obtain a volume form, and thus a measure, on $\mathcal{A}_0/\mathcal{G}$.
4. The theory of Reidemeister torsion helps to explain why there is typically a natural measure on the moduli space of flat connections only in dimensions 2 or less. The Reidemeister torsion is a natural section of a certain bundle on the moduli space of flat connections. In dimensions 2 or less we can think of this section as a volume form, but in most other cases we cannot.

5 Observables

The true physical observables in BF theory are self-adjoint operators on the physical Hilbert space, when this space is well-defined. Nonetheless it is interesting to consider operators on the gauge-invariant Hilbert space $L^2(\mathcal{A}/\mathcal{G})$. These are relevant not only to BF theory but also other gauge theories, such as 4-dimensional Lorentzian general relativity in terms of real Ashtekar variables, where the gauge group is SU(2). In what follows we shall use the term 'observables' to refer to operators on the gauge-invariant Hilbert space. We consider observables of two kinds: functions of A and functions of E.

Since A is analogous to the 'position' operator in elementary quantum mechanics while E is analogous to the 'momentum', we expect that functions of A act as multiplication operators while functions of E act by differentiation. As usual in quantum field theory, we need to smear these fields — i.e., integrate them over some region of space — to obtain operators instead of operator-valued distributions. Since A is like a 1-form, it is tempting to smear it by integrating it over a path. Similarly, since E is like an $(n-2)$-form, it is tempting to integrate it over an $(n-2)$-dimensional submanifold. This is essentially what we shall do. However, to obtain operators on the gauge-invariant Hilbert space $L^2(\mathcal{A}/\mathcal{G})$, we need to quantize gauge-invariant functions of A and E.

The simplest gauge-invariant function of the A field is a 'Wilson loop': a function of the form

$$\mathrm{tr}(\rho(Te^{\oint_\gamma A}))$$

for some loop γ in S and some representation ρ of G. In the simplest case, when $G = \mathrm{U}(1)$ and the loop γ bounds a disk, we can use Stokes' theorem to rewrite $\oint_\gamma A$ as the flux of the magnetic field through this disk. In general, a Wilson loop captures gauge-invariant information about the holonomy of the A field around the loop.

A Wilson loop is just a special case of a spin network, and we can get an operator on $L^2(\mathcal{A}/\mathcal{G})$ from any other spin network in a similar way. As we have seen, any spin network in S defines a function $\Psi \in \mathrm{Fun}(\mathcal{A}/\mathcal{G})$. Since $\mathrm{Fun}(\mathcal{A}/\mathcal{G})$ is an algebra, multiplication by Ψ defines an operator on $\mathrm{Fun}(\mathcal{A}/\mathcal{G})$. Since Ψ is a bounded function, this operator extends to a bounded operator on $L^2(\mathcal{A}/\mathcal{G})$. We call this operator a 'spin network observable'. Note that since $\mathrm{Fun}(\mathcal{A}/\mathcal{G})$ is an algebra, any product of Wilson loop observables can be written as a finite linear combination of spin network observables. Thus spin network observables give a way to measure correlations among the holonomies of A around a collection of loops.

When $G = \mathrm{U}(1)$ it is also easy to construct gauge-invariant functions of E. We simply take any compact oriented $(n-2)$-dimensional submanifold Σ in S, possibly with boundary, and do the integral

$$\int_\Sigma E.$$

This measures the flux of the electric field through Σ. Unfortunately, this integral is not gauge-invariant when G is nonabelian, so we need to modify the construction slightly to handle the nonabelian case. Write

$$E|_\Sigma = e\, d^{n-2}x$$

for some $\mathfrak{g}$-valued function e on Σ and some $(n-2)$-form $d^{n-2}x$ on Σ that is compatible with the orientation of Σ. Then

$$\int_\Sigma \sqrt{\langle e, e\rangle}\, d^{n-2}x$$

is a gauge-invariant function of E. One can check that it does not depend on how we write E as $e\, d^{n-2}x$. We can think of it as a precise way to define the quantity

$$\int_\Sigma |E|.$$

Recall that 3-dimensional BF theory with gauge group SU(2) or SO(3) is a formulation of Riemannian general relativity in 3 dimensions. In this case Σ is a curve, and the above quantity has a simple interpretation: it is the *length* of this curve. Similarly, in 4-dimensional BF theory with either of these gauge groups, Σ is a surface, and the above quantity can be interpreted as the *area* of this surface. The same is true for 4-dimensional Lorentzian general relativity formulated in terms of the real Ashtekar variables.

Quantizing the above function of E we obtain a self-adjoint operator $\mathcal{E}(\Sigma)$ on $L^2(\mathcal{A}/\mathcal{G})$, at least when Σ is real-analytically embedded in S. We shall not present the quantization procedure here, but only the final result. Suppose Ψ is a spin network in S. Generically, Ψ will intersect Σ transversely at finitely many points, and these points will not be vertices of Ψ:

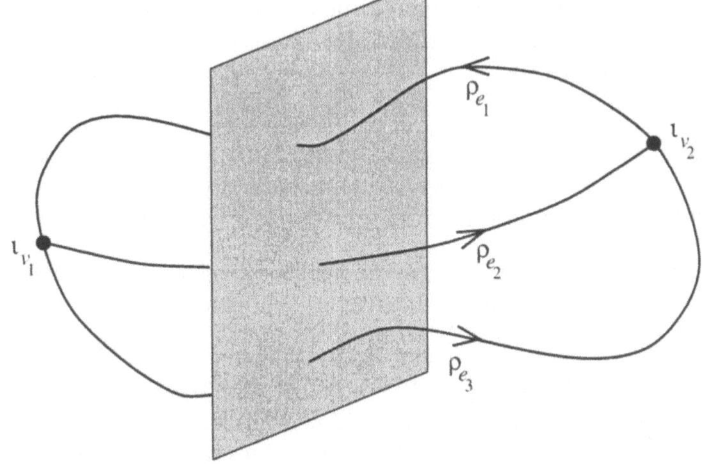

In this case we have

$$\mathcal{E}(\Sigma)\Psi = \big(\sum_i C(\rho_i)^{1/2}\big)\Psi$$

Here the sum is taken over all points p_i where an edge intersects the surface Σ, and $C(\rho_i)$ denotes the Casimir of the representation labeling that edge. Note that the same edge may intersect Σ in several points; if so, we count each point separately.

This result clarifies the physical significance of spin network edges: they represent *quantized flux lines of the E field.* In the case of 3-dimensional Riemannian quantum gravity they have a particularly simple geometrical meaning. Here the observable $\mathcal{E}(\Sigma)$ measures the length of the curve Σ. The irreducible representations of SU(2) correspond to spins $j = 0, \frac{1}{2}, 1\ldots$, and the Casimir equals $j(j+1)$ in the spin-j representation. Thus a spin network edge labeled by the spin j contributes a length $\sqrt{j(j+1)}$ to any curve it crosses transversely.

As an immediate consequence, we see that the length of a curve is not a continuously variable quantity in 3d Riemannian quantum gravity. Instead, it has a discrete spectrum of possible values! We also see here the difference between using SU(2) and SO(3) as our gauge group: only integer spins correspond to irreducible representations of SO(3), so the spectrum of allowed lengths for curves is sparser if we use SO(3). Of course, in a careful treatment we should also consider spin networks intersecting Σ nongenerically. As explained in Remark 1 below, these give the operator $\mathcal{E}(\Sigma)$ additional eigenvalues. However, our basic qualitative conclusions here remain unchanged.

Similar remarks apply to 4-dimensional BF theory with gauge group SU(2), as well as quantum gravity in the real Ashtekar formulation. Here $\mathcal{E}(\Sigma)$ measures the area of the surface Σ, area is quantized, and spin network edges give area to the surfaces they intersect! This is particularly intriguing given the Bekenstein-Hawking formula saying that the entropy of a black hole is proportional to its area. It it natural to try to explain this result by associating degrees of freedom of the event horizon to points where spin network edges intersect it. Attempts along these lines have been made, and the results look promising. Unfortunately, it is too much of a digression to describe these here, so we refer the reader to the Notes for more details.

Remarks 1. The formula for $\mathcal{E}(\Sigma)\Psi$ is slightly more complicated when the underlying graph γ of Ψ intersects Σ nongenerically. By subdividing its edges if necessary we may assume this graph has the following properties:

- If an edge of γ contains a segment lying in Σ, it lies entirely in Σ.
- Each isolated intersection point of γ and Σ is a vertex.
- Each edge of γ intersects Σ at most once.

For each vertex v of γ lying in Σ, we can divide the edges incident to v into three classes, which we call 'upwards', 'downwards', and 'horizontal'. The 'horizontal' edges are those lying in Σ; the other edges are separated into two classes according to which side of Σ they lie on; using the orientation of Σ we call these classes 'upwards' and 'downwards'. Reversing orientations of edges if necessary, we may assume all the upwards and downwards edges are incoming to v while the horizontal ones are outgoing. We can then write any intertwiner labeling v as a linear combination of intertwiners of the following special form:

$$\iota_v : \rho_v^u \otimes \rho_v^d \to \rho_v^h$$

where ρ_v^u (resp. ρ_v^d, ρ_v^h) is an irreducible summand of the tensor product of all the representations labeling upwards (resp. downwards, horizontal) edges. This lets us write any spin network state with γ as its underlying graph as a finite linear combination of spin network states with intertwiners of his special form. Now suppose Ψ is a spin network state with intertwiners of this form. Then we have

$$\mathcal{E}(\Sigma)\Psi = \frac{1}{2}\Big(\sum_v [2C(\rho_v^u) + 2C(\rho_v^d) - C(\rho_v^h)]^{1/2}\Big)\Psi$$

where the sum is over all vertices at which Σ intersects γ. In the generic case $C(\rho_v^u) = C(\rho_v^d)$ and $C(\rho_v^h) = 0$, so this formula reduces to the previous one.
2. When $G = \mathrm{U}(1)$ we can also quantize the observable $\int_\Sigma E$ when Σ is real-analytically embedded in S, obtaining an operator that measures the flux of the electric field through Σ. For any irreducible representation ρ of $\mathrm{U}(1)$ there is an integer $Q(\rho)$ such that

$$\rho(e^{i\theta}) = e^{iQ(\rho)\theta},$$

and using the notation of the previous remark this operator is given by

$$\Big(\int_\Sigma \hat{E}\Big)\Psi = \frac{1}{2}\Big(\sum_v Q(\rho_v^u) - Q(\rho_v^d)\Big)\Psi.$$

3. As noted, the true physical observables in BF theory are self-adjoint operators on the physical Hilbert space $L^2(\mathcal{A}_0/\mathcal{G})$. Examples include spin network observables: any spin network Ψ in S defines a bounded function on $\mathcal{A}_0/\mathcal{G}$, and multiplication by this function defines a bounded operator on the physical Hilbert space. Unlike the spin network observables on the gauge-invariant Hilbert space, these operators remain unchanged when we apply any homotopy to the underlying graph of the spin network, and they satisfy skein relations.

In the case of 3d BF theory with gauge group $\mathrm{SU}(2)$ or $\mathrm{SO}(3)$, a maximal commuting algebra of operators on $L^2(\mathcal{A}_0/\mathcal{G})$ is generated by Wilson loops corresponding to any set of generators of the fundamental group $\pi_1(S)$. For example, we can use the generators described in Remark 2 of Section 3. It suffices to use Wilson loops labeled by the fundamental representation of the gauge group.

6 Canonical Quantization via Triangulations

Starting from classical BF theory, canonical quantization has led us to a picture in which states are described using spin networks embedded in the manifold representing space. But our discussion of skein relations has shown that spin networks may also be regarded as abstract diagrams arising naturally from the representation theory of the gauge group G. This is very appealing to those who cherish the hope that someday quantum gravity will replace the differential-geometric conception of spacetime by something more algebraic or combinatorial in nature. If something like this is true, spin networks may ultimately be seen as more important than the manifold containing them! To study this possibility, we may isolate the following 'abstract' notion of spin network:

Definition 3. *A* spin network *is a triple* $\Psi = (\gamma, \rho, \iota)$ *consisting of:*

1. *a* graph γ*: i.e., a finite set* e *of edges, a finite set* $\mathcal{V}$ *of vertices, and* source *and* target *maps* $s, t\colon \mathrm{e} \to \mathcal{V}$ *assigning to each edge its two endpoints,*
2. *for each edge* e *of* γ*, an irreducible representation* ρ_e *of* G*,*
3. *for each vertex* v *of* γ*, an intertwiner*

$$\iota_v\colon \rho_{e_1} \otimes \cdots \otimes \rho_{e_n} \to \rho_{e'_1} \otimes \cdots \otimes \rho_{e'_m}$$

where $e_1, \ldots, e_n$ *are the edges incoming to* v *and* $e'_1, \ldots, e'_m$ *are the edges outgoing from* v.

Here we say an edge is incoming *to* v *if its target is* v*, and* outgoing *from* v *if its source is* v.

People have already begun formulating physical theories in which such abstract spin networks, not embedded in any manifold, describe the geometry of space. However it is still a bit difficult to relate such theories to more traditional physics. Thus it is useful to consider a kind of halfway house: namely, spin networks in the dual 1-skeleton of a triangulated manifold. While purely combinatorial, these objects still have a clear link to field theory as formulated on a pre-existing manifold.

In this case of BF theory this halfway house works as follows. As before, let us start with an $(n-1)$-dimensional real-analytic manifold S representing space. Given any triangulation of S we can choose a graph in S called the 'dual 1-skeleton', having one vertex at the center of each $(n-1)$-simplex and one edge intersecting each $(n-2)$-simplex. Using homotopies and skein relations, we can express any state in $\mathrm{Fun}(\mathcal{A}_0/\mathcal{G})$ as a linear combination of states coming from spin networks whose underlying graph is this dual 1-skeleton. So at least for BF theory, there is no loss in working with spin networks of this special form.

It turns out that the working with a triangulation this way sheds new light on the observables discussed in the previous section. Moreover, the dynamics of BF theory is easiest to describe using triangulations. Thus it pays to formalize the setup a bit more. To do so, we borrow some ideas from lattice gauge theory.

Given a graph γ, define a 'connection' on γ to be an assignment of an element of G to each edge of γ, and denote the space of such connections by $\mathcal{A}_\gamma$. As in lattice gauge theory, these group elements represent the holonomies along the edges of the graph. Similarly, define a 'gauge transformation' on γ to be an assignment of a group element to each vertex, and denote the group of gauge transformations by $\mathcal{G}_\gamma$. This group acts on $\mathcal{A}_\gamma$ in a natural way that mimics the usual action of gauge transformations on holonomies. Since $\mathcal{A}_\gamma$ is just a product of copies of G, we can use normalized Haar measure on G to put a measure on $\mathcal{A}_\gamma$, and this in turn pushes down to a measure on the quotient space $\mathcal{A}_\gamma/\mathcal{G}_\gamma$. Using these we can define Hilbert spaces $L^2(\mathcal{A}_\gamma)$ and $L^2(\mathcal{A}_\gamma/\mathcal{G}_\gamma)$.

In Section 4 we saw how to extract a gauge-invariant function on the space of connections from any spin network embedded in space. The same trick works in the present context: any spin network Ψ with γ as its underlying graph defines a function $\Psi \in L^2(\mathcal{A}_\gamma/\mathcal{G}_\gamma)$. For example, if Ψ is this spin network:

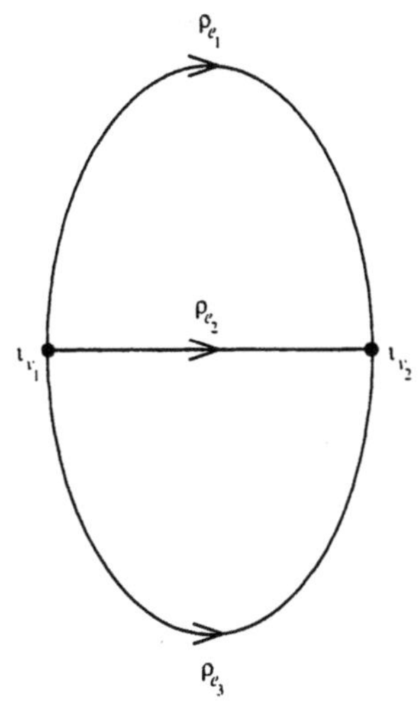

and the connection A assigns the group elements g_1, g_2, g_3 to the three edges of Ψ, we have

$$\Psi(A) = \rho_{e_1}(g_1)^b_a \; \rho_{e_2}(g_2)^c_d \; \rho_{e_3}(g_3)^e_f \; (\iota_{v_1})^f_{ac} \; (\iota_{v_2})^e_{db}.$$

We again call such functions 'spin network states'. Not only do these span $L^2(\mathcal{A}_\gamma/\mathcal{G}_\gamma)$, it is easy to choose an orthonormal basis of spin network states. Let Irrep(G) be a complete set of irreducible unitary representations of G. To obtain spin networks $\Psi = (\gamma, \rho, \iota)$ giving an orthonormal basis of $L^2(\mathcal{A}_\gamma/\mathcal{G}_\gamma)$, let ρ range over all labelings of the edges of γ by representations in Irrep(G), and for each ρ and each vertex v, let the intertwiners ι_v range over an orthonormal basis of the space of intertwiners

$$\iota \colon \rho_{e_1} \otimes \cdots \otimes \rho_{e_n} \to \rho_{e'_1} \otimes \cdots \otimes \rho_{e'_m}$$

where the e_i are incoming to v and the e'_i are outgoing from v.

How do these purely combinatorial constructions relate to our previous setup where space is described by a real-analytic manifold S equipped with a principal G-bundle? Quite simply: whenever γ is a graph in S, trivializing the bundle at the vertices of this graph gives a map from $\mathcal{A}$ onto $\mathcal{A}_\gamma$, and also a homomorphism from $\mathcal{G}$ onto $\mathcal{G}_\gamma$. Thus we have inclusions

$$L^2(\mathcal{A}_\gamma) \hookrightarrow L^2(\mathcal{A})$$

and

$$L^2(\mathcal{A}_\gamma/\mathcal{G}_\gamma) \hookrightarrow L^2(\mathcal{A}/\mathcal{G}).$$

These constructions are particularly nice when γ is the dual 1-skeleton of a triangulation of S. Consider 3-dimensional Riemannian quantum gravity, for example. In this case γ is always trivalent:

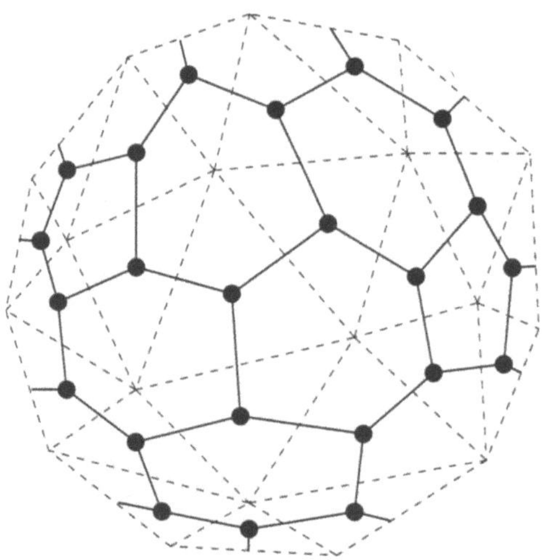

Since the representations of SU(2) satisfy

$$j_1 \otimes j_2 \cong |j_1 - j_2| \oplus \cdots \oplus (j_1 + j_2),$$

each basis of intertwiners $\iota: j_1 \otimes j_2 \to j_3$ contains at most one element. Thus we do not need to explicitly label the vertices of trivalent SU(2) spin networks with intertwiners; we only need to label the edges with spins. We can dually think of these spins as labeling the edges of the original triangulation. For example, the following spin network state:

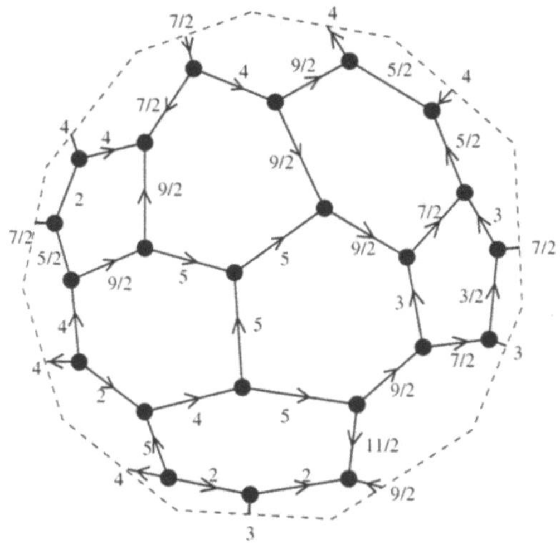

corresponds to a triangulation with edges labeled by spins as follows:

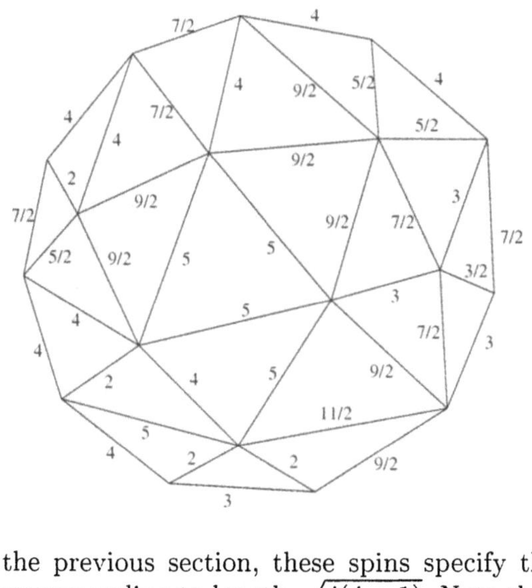

By the results of the previous section, these spins specify the *lengths* of the edges, with spin j corresponding to length $\sqrt{j(j+1)}$. Note that for there to be an intertwiner $\iota\colon j_1 \otimes j_2 \to j_3$, the spins j_1, j_2, j_3 labeling the three edges of a given triangle must satisfy two constraints. First, the triangle inequality must hold:

$$|j_1 - j_2| \leq j_3 \leq j_1 + j_2.$$

This has an obvious geometrical interpretation. Second, the spins must sum to an integer. This rather peculiar constraint would hold automatically if we had used the gauge group SO(3) instead of SU(2). If we consider all labelings satisfying these constraints, we obtain spin network states forming a basis of $L^2(\mathcal{A}_\gamma/\mathcal{G}_\gamma)$.

The situation is similar but a bit more complicated for 4-dimensional BF theory with gauge group SU(2). Let S be a triangulated 3-dimensional manifold and let γ be its dual 1-skeleton. Now γ is a 4-valent graph with one vertex in the center of each tetrahedron and one edge intersecting each triangle. To specify a spin network state in $L^2(\mathcal{A}_\gamma/\mathcal{G}_\gamma)$, we need to label each edge of γ with a spin and each vertex with an intertwiner:

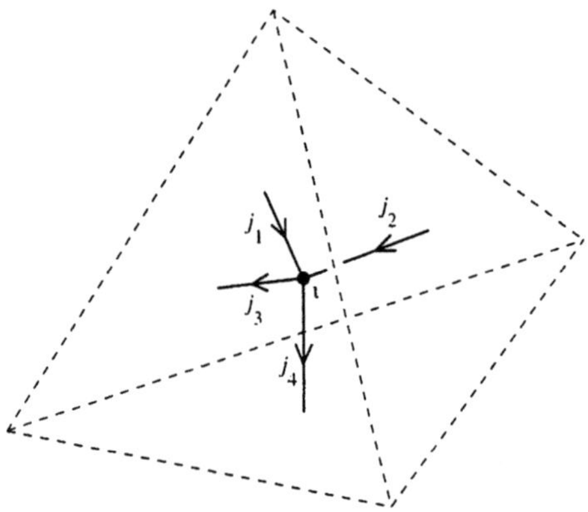

For each vertex there is a basis of intertwiners $\iota\colon j_1 \otimes j_2 \to j_3 \otimes j_4$ as described at the end of Section 4. We can draw such an intertwiner by formally 'splitting' the vertex into two trivalent ones and labeling the new edge with a spin j_5:

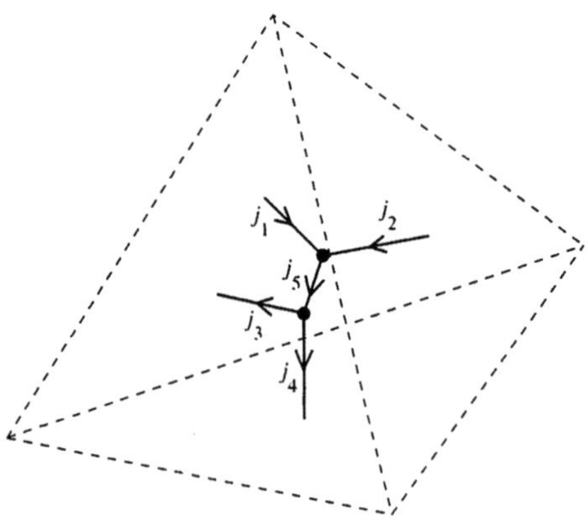

In the triangulation picture, this splitting corresponds to chopping the tetrahedron in half along a parallelogram:

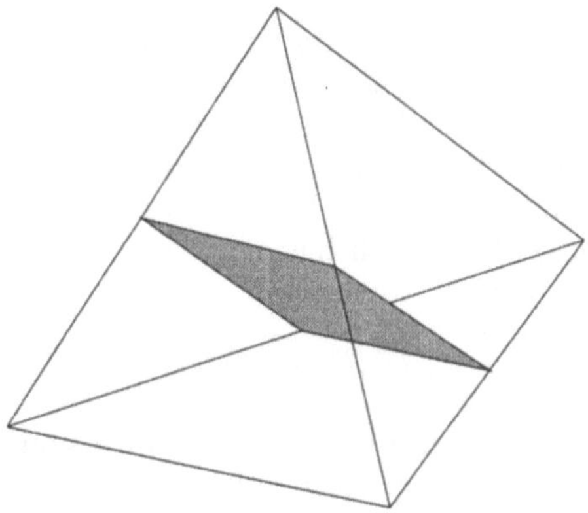

We can thus describe a spin network state in $L^2(\mathcal{A}_\gamma/\mathcal{G}_\gamma)$ by chopping each tetrahedron in half and labeling all the resulting parallelograms, along with all the triangles, by spins. These spins specify the *areas* of the parallelograms and triangles.

It may seem odd that in this picture the geometry of each tetrahedron is described by 5 spins, since classically it takes 6 numbers to specify the geometry of a tetrahedron. In fact, this is a consequence of the uncertainty principle. The area operators for surfaces do not commute when the surfaces intersect. There are three ways to chop a tetrahedron in half using a parallelogram, but we cannot simultaneously diagonalize the areas of these parallelograms, since they intersect. We can describe a basis of states for the quantum tetrahedron using 5 numbers: the areas of its 4 faces and any *one* of these parallelograms. Different ways of chopping tetrahedron in half gives us different bases of this sort, and the matrix relating these bases goes by the name of the '$6j$ symbols':

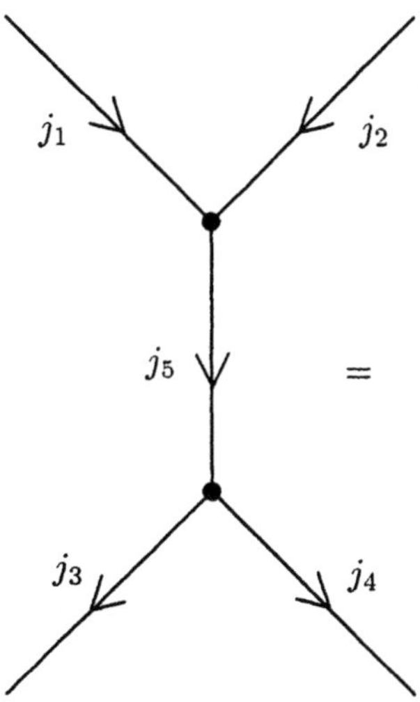

$$= \sum_{j_5} \begin{pmatrix} j_1 & j_2 & j_6 \\ j_4 & j_3 & j_5 \end{pmatrix}$$

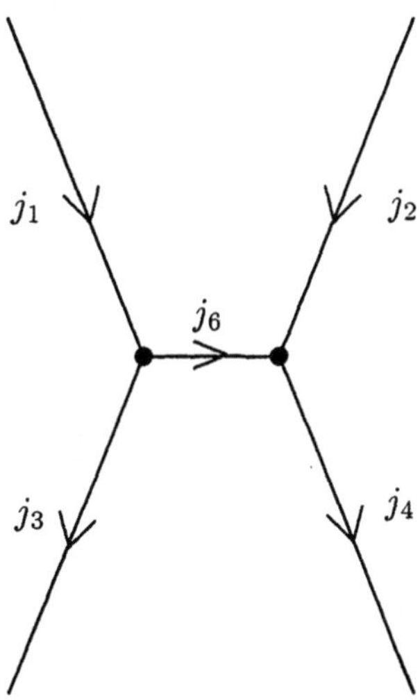

Remarks 1. For a deeper understanding of BF theory with gauge group SU(2), it is helpful to start with a classical phase space describing tetrahedron geometries and apply geometric quantization to obtain a Hilbert space of quantum states. We can describe a tetrahedron in $\mathbb{R}^3$ by specifying vectors $E_1, \ldots, E_4$ normal to its faces, with lengths equal to the faces' areas. We can think of these vectors as elements of $\mathfrak{so}(3)^*$, which has a Poisson structure familiar from the quantum mechanics of angular momentum:

$$\{J^a, J^b\} = \epsilon^{abc} J^c.$$

The space of 4-tuples $(E_1, \ldots, E_4)$ thus becomes a Poisson manifold. However, a 4-tuple coming from a tetrahedron must satisfy the constraint $E_1 + \cdots + E_4 = 0$. This constraint is the discrete analogue of the Gauss law $d_A E = 0$. In particular, it generates rotations, so if we take $(\mathfrak{so}(3)^*)^4$ and do Poisson reduction with respect to this constraint, we obtain a phase space whose points correspond to tetrahedron geometries modulo rotations. If we geometrically quantize this phase space, we obtain the 'Hilbert space of the quantum tetrahedron'.

We can describe this Hilbert space quite explicitly as follows. If we geometrically quantize $\mathfrak{so}(3)^*$, we obtain the direct sum of all the irreducible representations of SU(2):

$$\mathcal{H} \cong \bigoplus_{j=0,\frac{1}{2},1,\ldots} j.$$

Since this Hilbert space is a representation of SU(2), it has operators $\hat{J}^a$ on it satisfying the usual angular momentum commutation relations:

$$[\hat{J}^a, \hat{J}^b] = i\epsilon^{abc} \hat{J}^c.$$

We can think of $\mathcal{H}$ as the 'Hilbert space of a quantum vector' and the operators $\hat{J}^a$ as measuring the components of this vector. If we geometrically quantize $(\mathfrak{so}(3)^*)^{\otimes 4}$, we obtain $\mathcal{H}^{\otimes 4}$, which is the Hilbert space for 4 quantum vectors. There are operators on this Hilbert space corresponding to the components of these vectors:

$$\begin{aligned} \hat{E}_1^a &= \hat{J}^a \otimes 1 \otimes 1 \otimes 1 \\ \hat{E}_2^a &= 1 \otimes \hat{J}^a \otimes 1 \otimes 1 \\ \hat{E}_3^a &= 1 \otimes 1 \otimes \hat{J}^a \otimes 1 \\ \hat{E}_4^a &= 1 \otimes 1 \otimes 1 \otimes \hat{J}^a. \end{aligned}$$

One can show that the Hilbert space of the quantum tetrahedron is isomorphic to

$$\mathcal{T} = \{\psi \in \mathcal{H}^{\otimes 4} \colon\ (\hat{E}_1 + \hat{E}_2 + \hat{E}_3 + \hat{E}_4)\psi = 0\}.$$

On the Hilbert space of the quantum tetrahedron there are operators

$$\hat{A}_i = (\hat{E}_i \cdot \hat{E}_i)^{\frac{1}{2}}$$

corresponding to the areas of the 4 faces of the tetrahedron, and also operators

$$\hat{A}_{ij} = ((\hat{E}_i + \hat{E}_j) \cdot (\hat{E}_i + \hat{E}_j))^{\frac{1}{2}}$$

corresponding to the areas of the parallelograms. Since $\hat{A}_{ij} = \hat{A}_{kl}$ whenever $(ijkl)$ is some permutation of the numbers (1234), there are really just 3 different parallelogram area operators. The face area operators commute with each other and with the parallelogram area operators, but the parallelogram areas do not commute with each other. There is a basis of $\mathcal{T}$ consisting of states that are eigenvectors of all the face area operators together with any one of the parallelogram area operators. If for example we pick $\hat{A}_{12}$ as our preferred parallelogram area operator, any basis vector ψ is determined by 5 spins:

$$\hat{A}_i\psi = \sqrt{j_i(j_i+1)} \qquad 1 \le i \le 4,$$
$$\hat{A}_{12}\psi = \sqrt{j_5(j_5+1)}.$$

This basis vector corresponds to the intertwiner $\iota_j : j_1 \otimes j_2 \to j_3 \otimes j_4$ that factors through the representation j_5.

In 4d BF theory with gauge group SU(2), the Hilbert space $L^2(\mathcal{A}_\gamma/\mathcal{G}_\gamma)$ described by taking the tensor product of copies of $\mathcal{T}$, one for each tetrahedron in the 3-manifold S, and imposing constraints saying that when two tetrahedra share a face their face areas must agree. This gives a clearer picture of the 'quantum geometry of space' in this theory. For example, we can define observables corresponding to the volumes of tetrahedra. The results nicely match those of loop quantum gravity, where it has been shown that spin network vertices give volume to the regions of space in which they lie. In loop quantum gravity these results were derived not from BF theory, but from Lorentzian quantum gravity formulated in terms of the real Ashtekar variables. However, these theories differ only in their dynamics.

7 Dynamics

We now turn from the spin network description of the kinematics of BF theory to the spin foam description of its dynamics. Our experience with quantum field theory suggests that we can compute transition amplitudes in BF theory using path integrals. To keep life simple, consider the most basic example: the partition function of a closed manifold representing spacetime. Heuristically, if M is a compact oriented n-manifold we expect that

$$Z(M) = \int\int \mathcal{D}A\,\mathcal{D}E\; e^{i\int_M \mathrm{tr}(E\wedge F)}$$
$$= \int \mathcal{D}A\;\delta(F),$$

where formally integrating out the E field gives a Dirac delta measure on the space of flat connections on the G-bundle P over M. The final result should be the 'volume of the space of flat connections', but of course this is ill-defined without some choice of measure.

To try to make this calculation more precise, we can *discretize* it by choosing a triangulation for M and working, not with flat connections on P, but instead with flat connections on the dual 2-skeleton. By definition, the 'dual 2-skeleton' of a triangulation has one vertex in the center of each n-simplex, one edge intersecting each $(n-1)$-simplex, and one polygonal face intersecting each $(n-2)$-simplex. We call these 'dual vertices', 'dual edges', and 'dual faces', respectively. For example, when M is 3-dimensional, the intersection of the dual 2-skeleton with any tetrahedron looks like this:

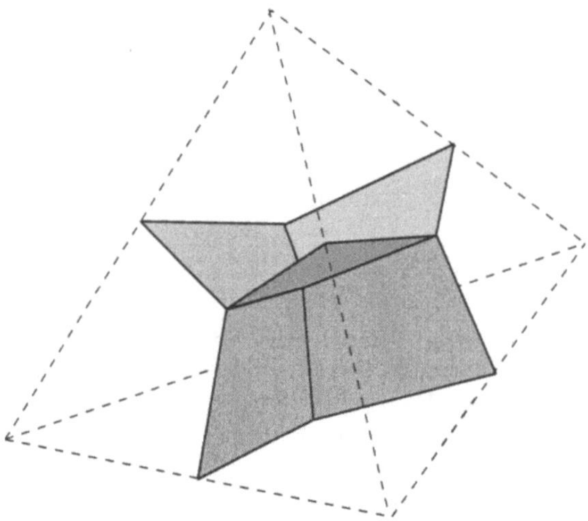

while a typical dual face looks like this:

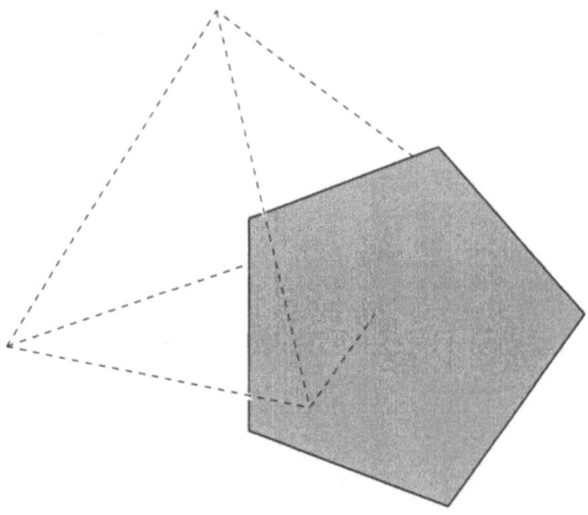

Note that the dual faces can have any number of edges. To keep track of these edges, we fix an orientation and distinguished vertex for each face f and call its edges $e_1 f, \ldots, e_N f$, taken in cyclic order starting from the distinguished vertex. Similarly, we call its vertices $v_1 f, \ldots, v_N f$:

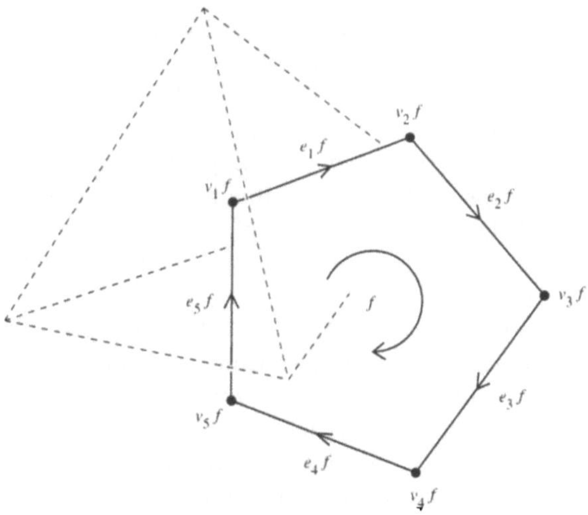

A 'connection' on the dual 2-skeleton is an object assigning a group element g_e to each dual edge e. For this to make sense we should fix an orientation for each dual edge. However, we can safely reverse our choice of the orientation as long as we remember to replace g_e by g_e^{-1} when we do so. We say that a connection on the dual 2-skeleton is 'flat' if that the holonomy around each dual face f is the identity:

$$g_{e_1 f} \cdots g_{e_N f} = 1$$

where we use the orientation of f to induce orientations of its edges.

To make sense of our earlier formula for the partition function of BF theory, we can try defining

$$Z(M) = \int \prod_{e \in \mathrm{e}} dg_e \prod_{f \in \mathcal{F}} \delta(g_{e_1 f} \cdots g_{e_N f}),$$

where $\mathcal{V}$ is the set of dual vertices, e is the set of dual edges, $\mathcal{F}$ is the set of dual vertices, and the integrals are done using normalized Haar measure on G. Of course, since we are taking a product of Dirac deltas here, we run the danger that this expression will not make sense. Nonetheless we proceed and see what happens!

We begin by using the identity

$$\delta(g) = \sum_{\rho \in \mathrm{Irrep(G)}} \dim(\rho) \mathrm{tr}(\rho(g)),$$

obtaining

$$Z(M) = \sum_{\rho: \mathcal{F} \to \mathrm{Irrep(G)}} \int \prod_{e \in \mathrm{e}} dg_e \prod_{f \in \mathcal{F}} \dim(\rho_f) \mathrm{tr}(\rho_f(g_{e_1 f} \cdots g_{e_N f})).$$

This formula is really a discretized version of

$$Z(M) = \int\int \mathcal{D}A\,\mathcal{D}E\; e^{i\int_M \mathrm{tr}(E\wedge F)}.$$

The analogue of A is the labeling of dual edges by group elements. The analogue of F is the labeling of dual faces by holonomies around these faces. These analogies make geometrical sense because A is like a 1-form and F is like a 2-form. What is the analogue of E? It is the labeling of dual faces by representations! Since each dual face intersects one $(n-2)$-simplex in the triangulation, we may dually think of these representations as labeling $(n-2)$-simplices. This is nice because E is an $(n-2)$-form. The analogue of the pairing $\mathrm{tr}(E\wedge F)$ is the pairing of a representation ρ_f and the holonomy around the face f to obtain the number $\dim(\rho_f)\mathrm{tr}(\rho_f(g_{e_1 f}\cdots g_{e_N f}))$.

Next we do the integrals over group elements in the formula for $Z(M)$. The details depend on the dimension of spacetime, and it is easiest to understand them with the aid of some graphical notation. In the previous section we saw how an abstract spin network Ψ together with a connection A on the underlying graph of Ψ give a number $\Psi(A)$. Since the connection A assigns a group element g_e to each edge of Ψ, our notation for the number $\Psi(A)$ will be a picture of Ψ together with a little circle containing the group element g_e on each edge e. When g_e is the identity we will not bother drawing it. Also, when two or more parallel edges share the same group element g we use one little circle for both edges. For example, we define:

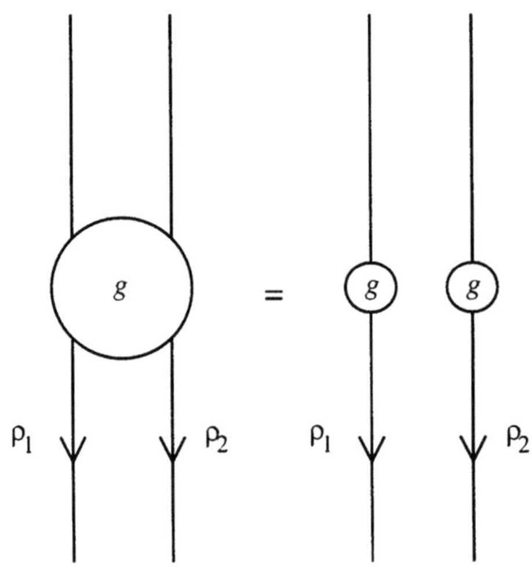

This is just the graphical analogue of the equation $(\rho_1\otimes\rho_2)(g) = \rho_1(g)\otimes\rho_2(g)$.

Now suppose M is 2-dimensional. Since each dual edge is the edge of two dual faces, each group element appears twice in the expression

$$\prod_{f\in\mathcal{F}} \mathrm{tr}(\rho_f(g_{e_1 f}\cdots g_{e_N f})).$$

In our graphical notation, this expression corresponds to a spin network with one loop running around each dual face:

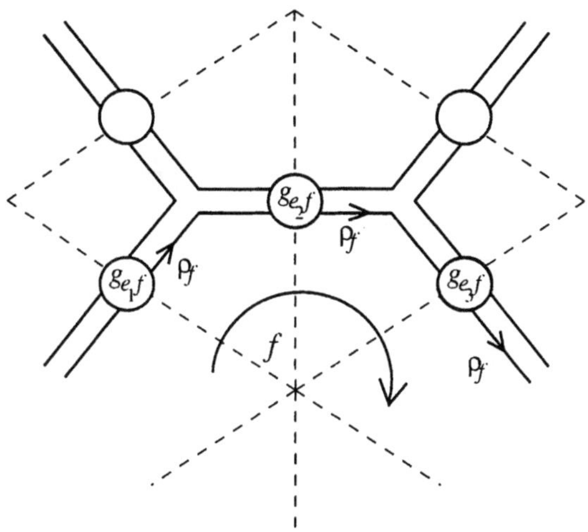

Here we have only drawn a small portion of the spin network. We can do the integral

$$\int \prod_{e \in \mathrm{e}} dg_e \prod_{f \in \mathcal{F}} \dim(\rho_f)\mathrm{tr}(\rho_f(g_{e_1 f} \cdots g_{e_N f}))$$

by repeatedly using the formula

$$\int dg \; \rho_1(g) \otimes \rho_2(g) = \begin{cases} \frac{\iota\iota^*}{\dim(\rho_1)} & \text{if } \rho_1 \cong \rho_2^* \\ 0 & \text{otherwise} \end{cases}$$

where $\iota\colon \rho_1 \otimes \rho_2 \to \mathbb{C}$ is the dual pairing when ρ_1 is the dual of ρ_2. This formula holds because both sides describe the projection from $\rho_1 \otimes \rho_2$ onto the subspace of vectors transforming in the trivial representation. Graphically, this formula can be written as the following skein relation:

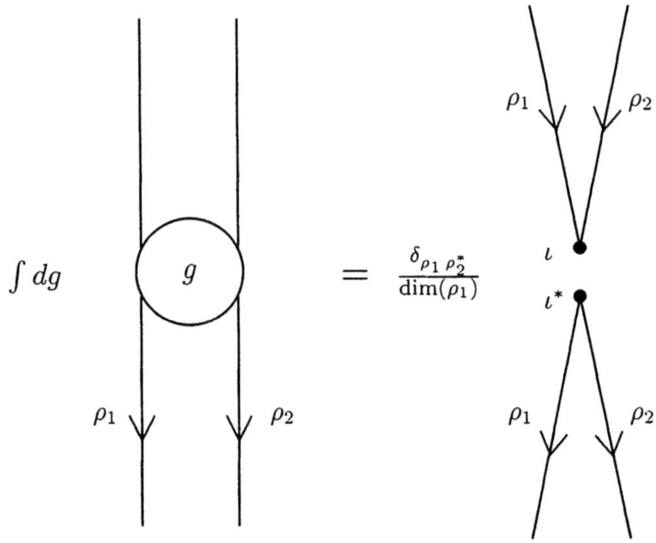

Applying this to every dual edge, we see that when M is connected the integral

$$\int \prod_{e \in \mathrm{e}} dg_e \prod_{f \in \mathcal{F}} \dim(\rho_f) \mathrm{tr}(\rho_f(g_{e_1 f} \cdots g_{e_N f}))$$

vanishes unless all the representations ρ_f are the same representation ρ, in which case it equals $\dim(\rho)^{|\mathcal{V}|-|\mathrm{e}|+|\mathcal{F}|}$. The quantity $|\mathcal{V}| - |\mathrm{e}| + |\mathcal{F}|$ is a topological invariant of M, namely the Euler characteristic $\chi(M)$. Summing over all labelings of dual faces, we thus obtain

$$Z(M) = \sum_{\rho \in \mathrm{Irrep(G)}} \dim(\rho)^{\chi(M)}$$

The Euler characteristic of a compact oriented surface of genus n is $2-2n$. When $\chi(M) < 0$, the sum converges for any compact Lie group G, and we see that the partition function of our discretized BF theory is well-defined and *independent of the triangulation!* This is precisely what we would expect in a topological quantum field theory. For $\chi(M) \geq 0$, that is, for the sphere and torus, the partition function typically does not converge.

In the 3-dimensional case each group element shows up in 3 factors of the product over dual faces, since 3 dual faces share each dual edge:

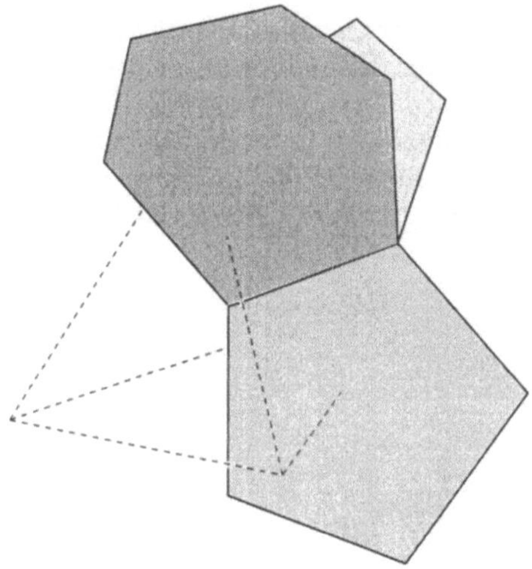

We can do the integral over each group element using the formula

$$\int dg\ \rho_1(g) \otimes \rho_2(g) \otimes \rho_3(g) = \sum_\iota \iota\iota^*$$

where the sum ranges over a basis of intertwiners $\iota\colon \rho_1 \otimes \rho_2 \otimes \rho_3 \to \mathbb{C}$, normalized as in Section 4, so that $\mathrm{tr}(\iota_1 \iota_2^*) = \delta_{\iota_1 \iota_2}$ for any two intertwiners ι_1, ι_2 in the basis. In our graphical notation this formula is written as:

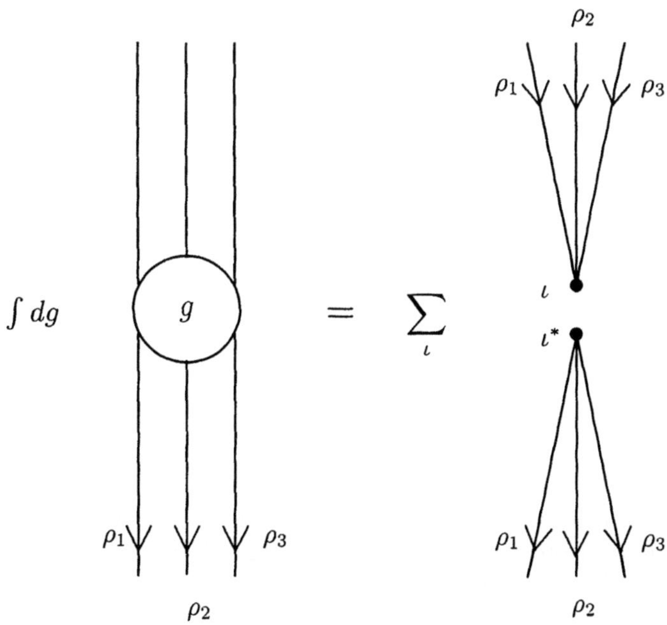

Both sides represent intertwiners from $\rho_1 \otimes \rho_2 \otimes \rho_3$ to itself. Again, the formula is true because both sides are different ways of describing the projection from $\rho_1 \otimes \rho_2 \otimes \rho_3$ onto the subspace of vectors that transform trivially under G. Using this formula once for each dual edge — or equivalently, once for each triangle in the triangulation — we can integrate out all the group elements g_e. Graphically, each time we do this, an integral over expressions like this:

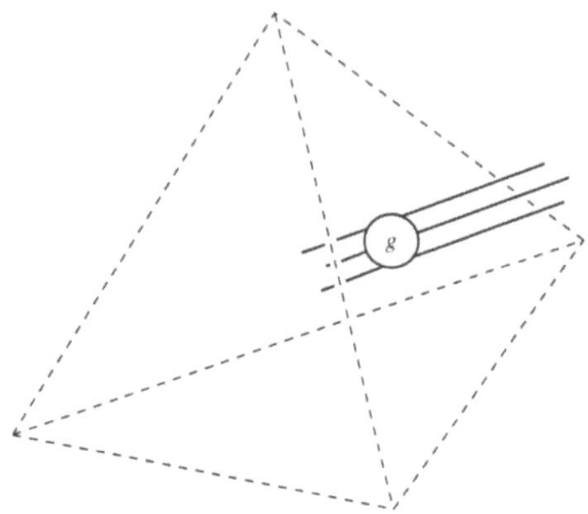

is replaced by a sum of expressions like this:

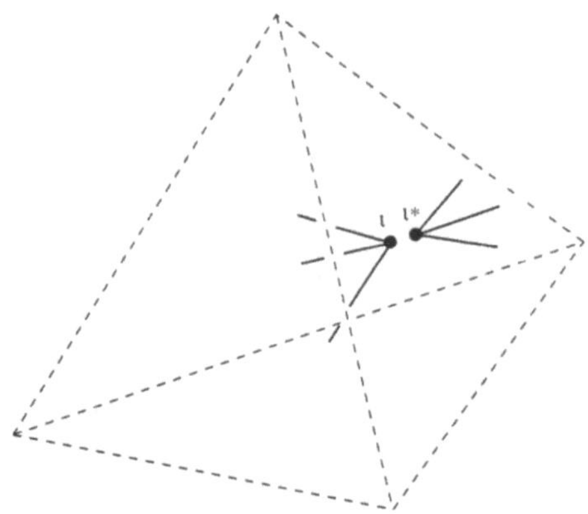

(We have not bothered to show the orientation of the edges in these pictures, since they depend on how we orient the edges of the dual 2-skeleton.) When we do this for all the triangular faces of a given tetrahedron, we obtain a little tetrahedral spin network like this:

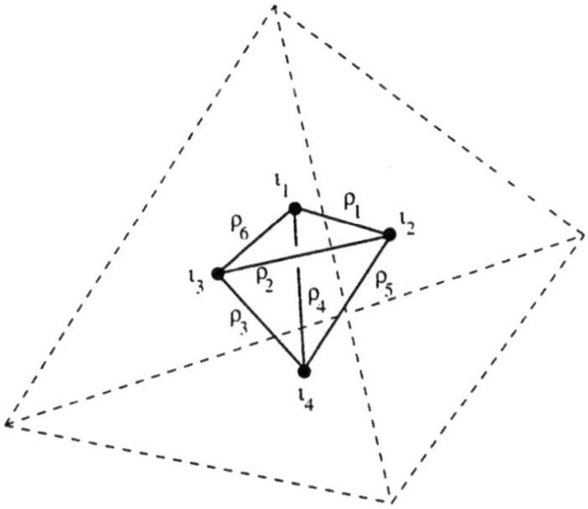

which we can evaluate in the usual way. This tetrahedral spin network is 'dual' to the original tetrahedron in the triangulation of M: its vertices (resp. edges, faces) correspond to faces (resp. edges, vertices) of the original tetrahedron.

We thus obtain the following formula for the partition function in 3-dimensional BF theory:

$$Z(M) = \sum_{\rho: \mathcal{F} \to \mathrm{Irrep(G)}} \sum_{\iota} \prod_{f \in \mathcal{F}} \dim(\rho_f) \prod_{v \in \mathcal{V}}$$

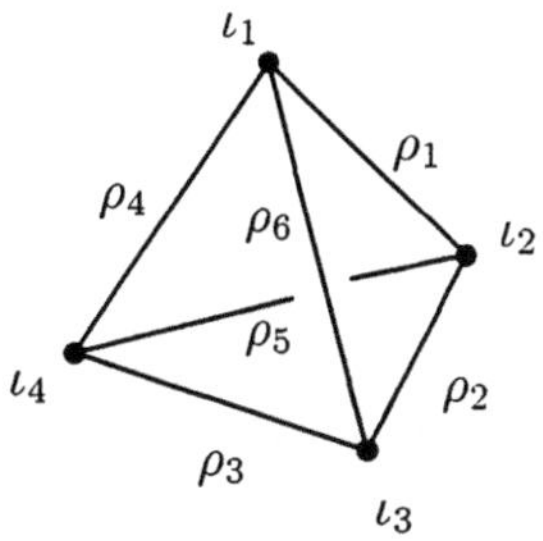

Here for each labeling $\rho: \mathcal{F} \to \mathrm{Irrep(G)}$, we take a sum over labelings ι of dual edges by intertwiners taken from the appropriate bases. For each dual vertex v, the tetrahedral spin network shown above is built using the representations ρ_i labeling the 6 dual faces incident to v and the intertwiners ι_i labeling the 4 dual edges incident to v. When $G = \mathrm{SU}(2)$ or $\mathrm{SO}(3)$, the labeling by intertwiners is trivial, so the tetrahedral spin network depends only on 6 spins. Using our graphical notation, it is not hard to express the value of this spin network in terms of the $6j$ symbols described in the previous section. We leave this as an exercise for the reader.

The calculation in 4 dimensions is similar, but now 4 dual faces share each dual edge, so we need to use the formula

$$\int dg \; \rho_1(g) \otimes \rho_2(g) \otimes \rho_3(g) \otimes \rho_4(g) = \sum_{\iota} \iota \iota^*$$

where now the sum ranges over a basis of intertwiners $\iota: \rho_1 \otimes \rho_2 \otimes \rho_3 \otimes \rho_4 \to \mathbb{C}$, normalized so that $\mathrm{tr}(\iota_1 \iota_2^*) = \delta_{\iota_1 \iota_2}$ for any intertwiners ι_1, ι_2 in the basis. Again both sides describe the projection on the subspace of vectors that transform in the trivial representation, and again we can write the formula as a generalized skein relation:

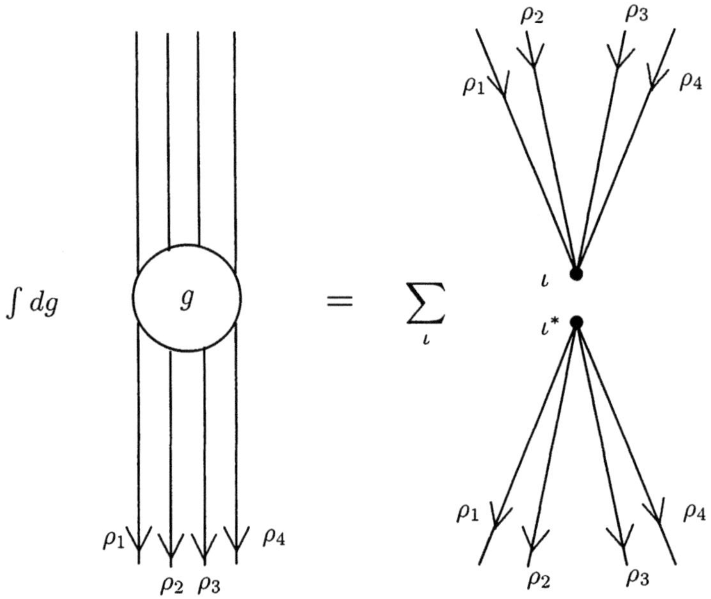

We use this formula once for each dual edge — or equivalently, once for each tetrahedron in the triangulation — to do the integral over all group elements in the partition function. Each time we do so, we introduce an intertwiner labeling the dual edge in question. We obtain

$$Z(M) = \sum_{\rho:\mathcal{F}\to\mathrm{Irrep(G)}} \sum_{\iota} \prod_{f\in\mathcal{F}} \dim(\rho_f) \prod_{v\in\mathcal{V}}$$

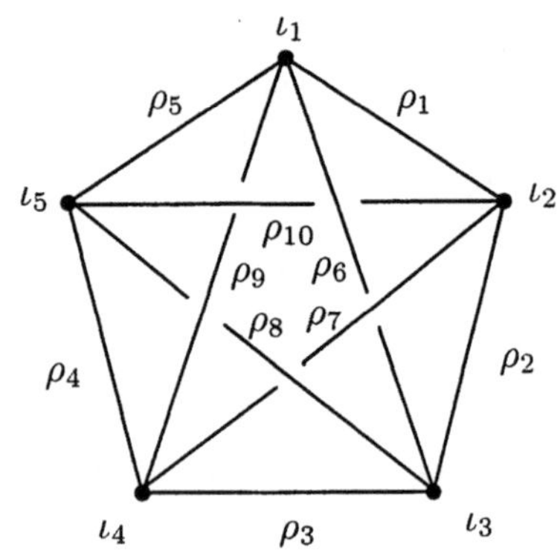

The 4-simplex in this formula is dual to the 4-simplex in the original triangulation that contains $v \in \mathcal{V}$. Its edges are labeled by the representations labeling the 10 dual faces incident to v, and its vertices are labeled by the intertwiners labeling the 5 dual edges incident to v.

People often rewrite this formula for the partition function by splitting each 4-valent vertex into two trivalent vertices using the skein relations described in Section 4. The resulting equation involves a trivalent spin network with 15 edges. In the SU(2) case this trivalent spin network is called a '$15j$ symbol', since it depends on 15 spins.

Having computed the BF theory partition function in 2, 3, and 4 dimensions, it should be clear that the same basic idea works in all higher dimensions, too. We always get a formula for the partition function as a sum over ways of labeling dual faces by representations and dual edges by intertwiners. There is, however, a problem. The sum usually diverges! The only cases I know where it converges are when G is a finite group (see Remark 2 below), when M is 0- or 1-dimensional, or when M is 2-dimensional with $\chi(M) < 0$. Not surprisingly, these are a subset of the cases when the moduli space of flat connections on M has a natural measure. In other cases, it seems there are too many delta functions in the expression

$$Z(M) = \int \prod_{e \in \mathsf{e}} dg_e \prod_{f \in \mathcal{F}} \delta(g_{e_1 f} \cdots g_{e_N f})$$

to extract a meaningful answer. We discuss this problem further in Section 9.

Of course, there is more to dynamics than the partition function. For example, we also want to compute vacuum expectation values of observables, and transition amplitudes between states. It is not hard to generalize the formulas above to handle these more complicated calculations. However, at this point it helps to explicitly introduce the concept of a 'spin foam'.

Remarks 1. Ponzano and Regge gave a formula for a discretized version of the action in 3-dimensional Riemannian general relativity. In their approach the spacetime manifold M is triangulated and each edge is assigned a length. The Ponzano-Regge action is the sum over all tetrahedra of the quantity:

$$S = \sum_e \ell_e \theta_e$$

where the sum is taken over all 6 edges, l_e is the length of the edge e, and θ_e is the dihedral angle of the edge e, that is, the angle between the outward normals of the two faces incident to this edge. One can show that in a certain precise their action is an approximation to the integral of the Ricci scalar curvature. In the limit of large spins, the value of the tetrahedral spin network described above is asymptotic to

$$\sqrt{\frac{2}{3\pi V}} \cos(S + \frac{\pi}{4}).$$

where the lengths ℓ_e are related to the spins j_e labeling the tetrahedron's edges by $\ell = j + 1/2$, and V is the volume of the tetrahedron. Naively one might have hoped to get $\exp(iS)$. That one gets a cosine instead can be traced back to the fact that the lengths of the edges of a tetrahedron only determine its

geometry modulo rotation *and reflection*. The phase $\frac{\pi}{4}$ shows up because calculating the asymptotics of the tetrahedral spin network involves a stationary phase approximation.
2. Ever since Section 4 we have been assuming that G is connected. The main reason for this is that it ensures the map from $\mathcal{A}$ to $\mathcal{A}_\gamma$ is onto for any graph γ in S, so that we have inclusions $L^2(\mathcal{A}_\gamma) \hookrightarrow L^2(\mathcal{A})$ and $L^2(\mathcal{A}_\gamma/\mathcal{G}_\gamma) \hookrightarrow L^2(\mathcal{A}/\mathcal{G})$. When G is not connected, these maps are usually not one-to-one.

Requiring that G be connected rules out all nontrivial finite groups. However, our formula for the BF theory partition function makes equally good sense for groups that are not connected. In fact, when G is finite, the partition function is convergent regardless of the dimension of M, and when a suitable normalization factor is included it becomes triangulation-independent. This is a special case of the 'Dijkgraaf-Witten model.'

In this model, the path integral is not an integral over flat connections on a fixed G-bundle over M, but rather a sum over isomorphism classes of G-bundles. In fact, our discretized formula for the path integral in BF theory always implicitly includes a sum over isomorphism classes of G-bundles, because it corresponds to an integral over the whole moduli space of flat G-bundles over M, rather than the moduli space of flat connections on a fixed G-bundle. (For the relation between these spaces, see Remark 2 in Section 3.) When G is a finite group, the moduli space of flat G-bundles is discrete, with one point for each isomorphism class of G-bundle.

8 Spin Foams

We have seen that in BF theory the partition function can be computed by triangulating spacetime and considering all ways of labeling dual faces by irreducible representations and dual edges by intertwiners. For each such labeling, we compute an 'amplitude' as a product of amplitudes for dual faces, dual edges, and dual vertices. (By cleverly normalizing our intertwiners we were able to make the edge amplitudes equal 1, rendering them invisible, but this was really just a cheap trick.) We then take a sum over all labelings to obtain the partition function.

To formalize this idea we introduce the concept of a 'spin foam'. A spin foam is the 2-dimensional analog of a spin network. Just as a spin network is a graph with edges labeled by irreducible representations and vertices labeled by intertwiners, a spin foam is a 2-dimensional complex with faces labeled by irreducible representations and edges labeled by intertwiners. Of course, to make this precise we need a formal definition of '2-dimensional complex'. Loosely, such a thing should consist of vertices, edges, and polygonal faces. There is some flexibility about the details. However, we certainly want the dual 2-skeleton of a triangulated manifold to qualify. Since topologists have already studied such things, this suggests that we take a 2-dimensional complex to be what they call a '2-dimensional piecewise linear cell complex'.

The precise definition of this concept is somewhat technical, so we banish it to the Appendix and only state what we need here. A 2-dimensional complex has a finite set $\mathcal{V}$ of vertices, a finite set e of edges, and a finite set $\mathcal{F}_N$ of N-sided faces for each $N \geq 3$, with only finitely many $\mathcal{F}_N$ being nonempty. In fact, we shall work with 'oriented' 2-dimensional complexes, where each edge and each face has an orientation. The orientations of the edges give maps

$$s, t\colon \mathrm{e} \to \mathcal{V}$$

assigning to each edge its source and target. The orientation of each face gives a cyclic ordering to its edges and vertices. Suppose we arbitrarily choose a distinguished vertex for each face $f \in \mathcal{F}_N$. Then we may number all its vertices and edges from 1 to N. If we think of these numbers as lying in $\mathbb{Z}_N$, we obtain maps

$$e_i\colon \mathcal{F}_N \to \mathrm{e}, \quad v_i\colon \mathcal{F}_N \to \mathcal{V} \qquad\qquad i \in \mathbb{Z}_N.$$

We say f is 'incoming' to e when the orientation of e agrees with the orientation it inherits from f, and 'outgoing' when these orientations do not agree:

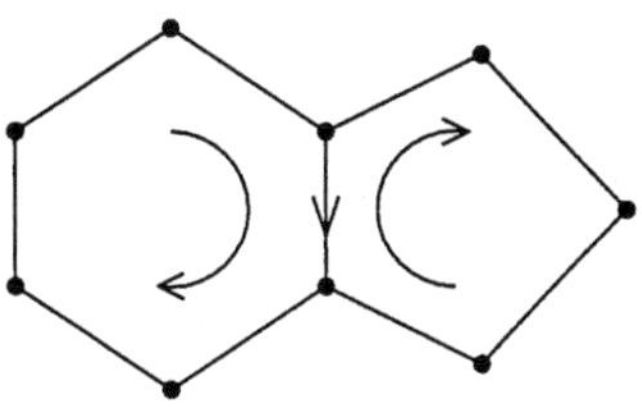

With this business taken care of, we can define spin foams. The simplest kind is a 'closed' spin foam. This is the sort we sum over when computing partition functions in BF theory.

Definition 4. *A* closed spin foam F *is a triple* (κ, ρ, ι) *consisting of:*

1. *a 2-dimensional oriented complex* κ*,*
2. *a labeling* ρ *of each face* f *of* κ *by an irreducible representation* ρ_f *of* G*,*
3. *a labeling* ι *of each edge* e *of* κ *by an intertwiner*

$$\iota_e\colon \rho_{f_1} \otimes \cdots \otimes \rho_{f_n} \to \rho_{f'_1} \otimes \cdots \otimes \rho_{f'_m}$$

where $f_1, \ldots, f_n$ *are the faces incoming to* e *and* $f'_1, \ldots, f'_m$ *are the faces outgoing from* e*.*

Note that this definition is exactly like that of a spin network, but with everything one dimension higher! This is why a generic slice of a spin foam is a spin network. We can formalize this using the notion of a spin foam $F\colon \Psi \to \Psi'$ going from a spin network Ψ to a spin network Ψ':

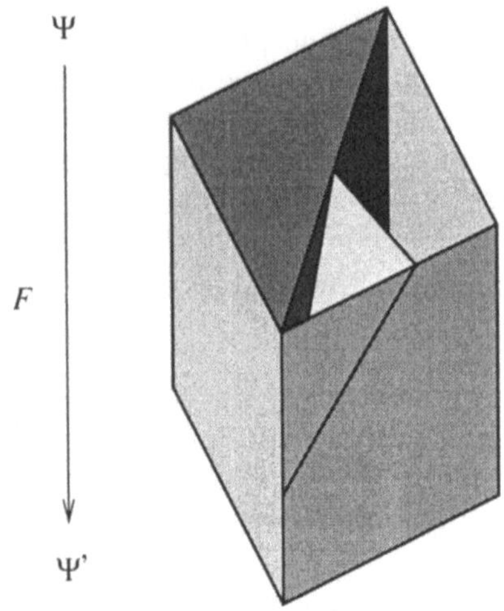

This is the sort we sum over when computing transition amplitudes in BF theory. (To reduce clutter, we have not drawn the labelings of edges and faces in this spin foam.) In this sort of spin foam, the edges that lie in Ψ and Ψ' are not labeled by intertwiners. Also, the edges ending at spin network vertices must be labeled by intertwiners that match those labeling the spin network vertices. These extra requirements are lacking for closed spin foams, because a closed spin foam is just one of the form $F\colon \emptyset \to \emptyset$, where $\emptyset$ is the 'empty spin network': the spin network with no vertices and no edges.

To make this more precise, we need to define what it means for a graph γ to 'border' a 2-dimensional oriented complex κ. The reader can find this definition in Appendix A. What matters here is that if γ borders κ, then each vertex v of γ is the source or target of a unique edge $\tilde{v}$ of κ, and each edge e of γ is the edge of a unique face $\tilde{e}$ of κ. Using these ideas, we first define spin foams of the form $F\colon \emptyset \to \Psi$:

Definition 5. *Suppose that* $\Psi = (\gamma, \rho, \iota)$ *is a spin network. A* spin foam $F\colon \emptyset \to \Psi$ *is a triple* $(\kappa, \tilde{\rho}, \tilde{\iota})$ *consisting of:*

1. *a 2-dimensional oriented complex* κ *such that* γ *borders* κ,
2. *a labeling* $\tilde{\rho}$ *of each face* f *of* κ *by an irreducible representation* $\tilde{\rho}_f$ *of* G,
3. *a labeling* $\tilde{\iota}$ *of each edge* e *of* κ *not lying in* γ *by an intertwiner*

$$\tilde{\iota}_e\colon \rho_{f_1} \otimes \cdots \otimes \rho_{f_n} \to \rho_{f'_1} \otimes \cdots \otimes \rho_{f'_m}$$

   where $f_1, \ldots, f_n$ *are the faces incoming to* e *and* $f'_1, \ldots, f'_m$ *are the faces outgoing from* e,

such that the following hold:

1. *For any edge* e *of* γ, $\tilde{\rho}_{\tilde{e}} = \rho_e$ *if* $\tilde{e}$ *is incoming to* e, *while* $\tilde{\rho}_{\tilde{e}} = (\rho_e)^*$ *if* $\tilde{e}$ *is outgoing from* e.

2. For any vertex v of γ, $\tilde{\iota}_{\tilde{e}}$ equals ι_e after appropriate dualizations.

Finally, to define general spin foams, we need the notions of 'dual' and 'tensor product' for spin networks. The dual of a spin network $\Psi = (\gamma, \rho, \iota)$ is the spin network Ψ^* with the same underlying graph, but with each edge e labelled by the dual representation ρ_e^*, and each vertex v labelled by the appropriately dualized form of the intertwining operator ι_v. Given spin networks $\Psi = (\gamma, \rho, \iota)$ and $\Psi' = (\gamma', \rho', \iota')$, their tensor product $\Psi \otimes \Psi'$ is defined to be the spin network whose underlying graph is the disjoint union of γ and γ', with edges and vertices labelled by representations and intertwiners using ρ, ρ' and ι, ι'. As usual, duality allows us to think of an input as an output:

Definition 6. *Given spin networks Ψ and Ψ', a* spin foam *$F: \Psi \to \Psi'$ is defined to be a spin foam $F: \emptyset \to \Psi^* \otimes \Psi'$.*

Here is how we compute transition amplitudes in BF theory as a sum over spin foams. Suppose spacetime is given by a compact oriented cobordism $M: S \to S'$, where S and S' are compact oriented manifolds of dimension $n-1$:

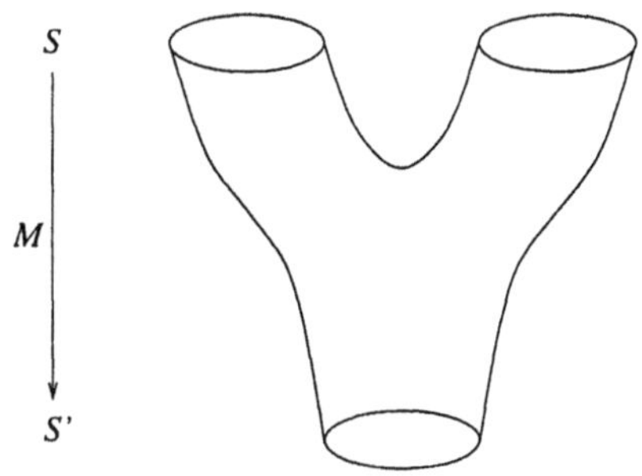

Choose a triangulation of M. This induces triangulations of S and S' with dual 1-skeletons γ and γ', respectively. As described in Section 6, in this triangulated context we can use $L^2(\mathcal{A}_\gamma/\mathcal{G}_\gamma)$ as the gauge-invariant Hilbert space for S. This Hilbert space has a basis given by spin networks whose underlying graph is the dual 1-skeleton of S. Similarly, we use $L^2(\mathcal{A}_{\gamma'}/\mathcal{G}_{\gamma'})$ as the space of gauge-invariant states on S'. We describe time evolution as an operator

$$Z(M): L^2(\mathcal{A}_\gamma/\mathcal{G}_\gamma) \to L^2(\mathcal{A}_{\gamma'}/\mathcal{G}_{\gamma'}).$$

To specify this operator, it suffices to describe the transition amplitudes $\langle \Psi', Z(M)\Psi \rangle$ when Ψ, Ψ' are spin network states. We write this transition amplitude as a sum over spin foams going from Ψ to Ψ':

$$\langle \Psi', Z(M)\Psi \rangle = \sum_{F: \Psi \to \Psi'} Z(F)$$

Since we are working with a fixed triangulation of M, we restrict the sum to spin foams whose underlying complex is the dual 2-skeleton of M. The crucial thing is the formula for the amplitude $Z(F)$ of a given spin foam F.

We have already given a formula for the amplitude of a closed spin foam in the previous section: it is computed as a product of amplitudes for spin foam faces, edges and vertices. A similar formula works for any spin foam $F: \Psi \to \Psi'$, but we need to make a few adjustments. First, when we take the product over faces, edges and vertices, we exclude edges and vertices that lie in Ψ and Ψ'. Second, we use the square root of the usual edge amplitude for edges of the form $\tilde{v}$, where v is a vertex of Ψ or Ψ'. Third, we use the square root of the usual face amplitudes for faces of the form $\tilde{e}$, where e is an edge of Ψ or Ψ'. The reason for these adjustments is that we want to have

$$Z(M')Z(M) = Z(M'M)$$

when $M: S \to S'$ and $M': S' \to S''$ are composable cobordisms and $M'M: S \to S''$ is their composite:

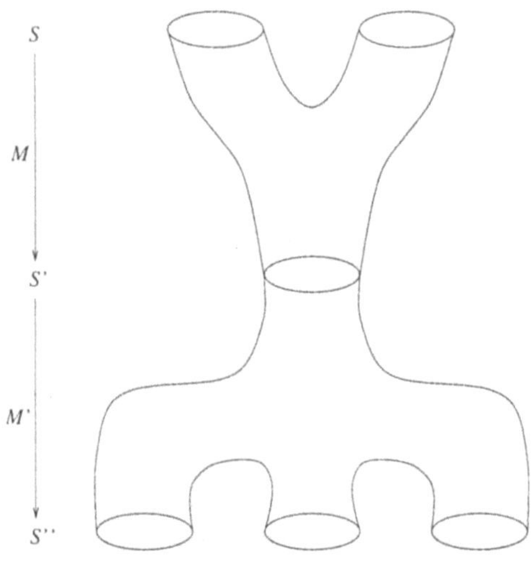

For this to hold, we want

$$Z(F')Z(F) = Z(F'F)$$

whenever $F'F: \Psi \to \Psi''$ is the spin foam formed by gluing together $F: \Psi \to \Psi'$ and $F': \Psi' \to \Psi''$ along their common border Ψ' and erasing the vertices and edges that lie in Ψ'. The adjustments described above make this equation true. Of course, the argument that $Z(F')Z(F) = Z(F'F)$ implies $Z(M')Z(M) = Z(M'M)$ is merely formal unless the sums over spin foams used to define these time evolution operators converge in a sufficiently nice way.

Let us conclude with some general remarks on the meaning of the spin foam formalism. Just as spin networks are designed to merge the concepts of *quantum state* and the *geometry of space*, spin foams are designed to merge the concepts of *quantum history* and the *geometry of spacetime*. However, the concept of 'quantum history' is a bit less familiar than the concept of 'quantum state', so it deserves some comment. Perhaps the most familiar example of a quantum history is a Feynman diagram. A Feynman diagram determines an operator on Fock space, but there is more information in the diagram than this operator, since besides telling us transition amplitudes between states, the diagram also tells a story of 'how the transition happened'. In other words, the internal edges and vertices of the diagram describe a 'quantum history' in which various virtual particles are created and annihilated.

Similarly, spin foams can be used to describe operators, but they contain extra information. If Ψ and Ψ' are spin networks with underlying graphs γ and γ', respectively, then any spin foam $F: \Psi \to \Psi'$ determines an operator from $L^2(\mathcal{A}_\gamma/\mathcal{G}_\gamma)$ to $L^2(\mathcal{A}_{\gamma'}/\mathcal{G}_{\gamma'})$, which we also denote by F, such that

$$\langle \Phi', F\Phi \rangle = \langle \Phi', \Psi' \rangle \langle \Psi, \Phi \rangle$$

for any states Φ, Φ'. The time evolution operator $Z(M)$ is a linear combination of these operators weighted by the amplitudes $Z(F)$. But a spin foam contains more information than the operator it determines, since the operator depends only on the initial state Ψ and the final state Ψ', not on the details of the spin foam at intermediate times. This extra information is what we call a 'quantum history'.

How exactly does a spin foam describe the geometry of spacetime? In part, this follows from how spin networks describe the geometry of space. Consider, for example, 4d BF theory with gauge group SU(2). Spin network edges give area to surfaces they puncture, while spin network vertices give volume to regions of space in which they lie. But a spin network edge is really just a slice of a spin foam face, and a spin network vertex is a slice of a spin foam edge. Thus in the spacetime context, spin foam faces give area to surfaces they intersect, while spin foam edges give 3-volume to 3-dimensional submanifolds they intersect. Continuing the pattern, one expects that spin foam vertices give 4-volume to regions of spacetime in which they lie. However, calculations have not yet been done to confirm this, in part because a thorough picture of the metric geometry of spacetime in 4 dimensions requires that one impose constraints on the E field. We discuss this a bit more in Section 10.

A similar story holds for 3d BF theory with gauge group SU(2), or in other words, Riemannian quantum gravity in 3 dimensions. In this case, spin foam faces give length to curves they intersect and spin foam edges give area to surfaces they intersect. We expect that spin foam vertices give volume to regions of spacetime in which they lie, but so far the calculations remain a bit problematic.

Remarks 1. The notation $F: \Psi \to \Psi'$ is meant to suggest that there is a category with spin networks as objects and spin foams as morphisms. For this, we should be able to compose spin foams $F: \Psi \to \Psi'$ and $F': \Psi' \to \Psi''$ and obtain a spin foam $F'F: \Psi \to \Psi''$. This composition should be associative, and for each spin network Ψ we want a spin foam $1_\Psi: \Psi \to \Psi$ serving as a left and right unit for composition.

To get this to work, we actually need to take certain equivalence classes of spin foams as morphisms. In my previous paper on this subject, the equivalence relation described was actually not coarse enough to prove associativity and the left and right unit laws. The quickest way to fix this problem is to simply impose extra equivalence relations of the form $F(GH) \sim (FG)H$ and $1_\Psi F \sim F \sim 1_{\Psi'}$, to ensure that these laws hold.

2. The physical meaning of the time evolution operators

$$Z(M)\colon L^2(\mathcal{A}_\gamma/\mathcal{G}_\gamma) \to L^2(\mathcal{A}_{\gamma'}/\mathcal{G}_{\gamma'})$$

is somewhat subtle in a background-independent theory. For example, when $M = S\times[0,1]$ is a cylinder cobordism from S to itself, we should have $Z(M)^2 = Z(M)$. In this case $Z(M)$ should represent the projection from the gauge-invariant Hilbert space to the space of physical states.

9 q-Deformation and the Cosmological Constant

As we have seen, BF theory leads to a beautiful interplay between representation theory and geometry, in which the distinction between the two subjects gradually fades away. In the end, spin networks serve simultaneously as a tool for calculations in representation theory and as a description of the quantum geometry of space. Spin foams extend this idea to the geometry of spacetime. This is exactly the sort of thing one would hope for in a theory of quantum gravity, since quantum mechanics is largely based on representation theory, while general relativity is founded on differential geometry.

But so far, our treatment has been plagued by a serious technical problem. Mathematically, the problem is that the moduli space of flat connections only has a natural measure in dimensions 2 or less. We need this measure to define the physical Hilbert space, so canonical quantization only works when the dimension of *space* is at most 2. But we also need this measure to do path integrals in BF theory, so transition amplitudes between states are only well-defined when the dimension of *spacetime* is at most 2. Physically, the problem is the presence of infrared divergences. For example, in 3-dimensional Riemannian quantum gravity, spin networks describe the geometry of space, while spin foams describe the geometry of spacetime. When we compute a transition amplitude from one spin network to another, we sum over spin foams going between them. The transition amplitude diverges because we are summing over spin foams with faces labelled by arbitrarily high spins. These correspond to arbitrarily large spacetime geometries.

In quantum field theory, one can often learn to live with infrared divergences by restricting the set of questions one expects the theory to answer. Crudely speaking, the idea is that we can ignore the behavior of a theory on length scales greatly exceeding the characteristic length scale of the experiment whose outcome we are seeking to predict. For example, certain infrared divergences in quantum electrodynamics can be ignored if we assume our apparatus is unable to detect 'soft photons', i.e., those with very long wavelengths. Similarly, one can argue that the possibility of arbitrarily large spacetime geometries should not affect the outcome of an experiment that occurs within a bounded patch of spacetime. Thus it is quite possible that with a little cleverness we can learn to extract extra sensible physics from spin foam models with infrared divergences.

Luckily, when it comes to BF theory, we have another option: we can completely *eliminate* the infrared divergences by adding an extra term to the Lagrangian of our theory, built using only the E field. This trick only works when spacetime has dimension 3 or 4. In dimension 3, the modified Lagrangian is

$$\mathcal{L} = \mathrm{tr}(E \wedge F + \frac{\Lambda}{6} E \wedge E \wedge E),$$

while in dimension 4 it is

$$\mathcal{L} = \mathrm{tr}(E \wedge F + \frac{\Lambda}{12} E \wedge E).$$

For reasons that will become clear, the coupling constant Λ is called the 'cosmological constant'. We only consider the case $\Lambda > 0$.

Adding this 'cosmological term' has a profound effect on BF theory: it changes all our calculations involving the representation theory of the gauge group into analogous calculations involving the representation theory of the corresponding quantum group. This gives us a well-defined and *finite-dimensional* physical Hilbert space, and turns the divergent sum over spin foams into a *finite* sum for the transition amplitudes between states. This process is known as 'q-deformation', because the quantum group depends on a parameter q, and reduces to the original group at $q = 1$. Often people think of q as a function of $\hbar$, but for us it is a function of Λ, and we have $q = 1$ when $\Lambda = 0$. Thus, at least in the present context, quantum groups should really be called 'cosmological groups'!

To understand how quantum groups are related to BF theory with a cosmological term, we need to exploit its ties to Chern-Simons theory. This is a background-free gauge theory in 3 dimensions whose action depends only the connection A:

$$S_{CS}(A) = \frac{k}{4\pi} \int_M \mathrm{tr}(A \wedge dA + \frac{2}{3} A \wedge A \wedge A)$$

This formula only makes sense after we have chosen a trivialization of P. Luckily, if we assume G is simply connected, every G-bundle over a 3-manifold admits a trivialization. The Chern-Simons action is not invariant under large gauge transformations. However, if we also assume that G is semisimple and 'tr' is defined using the Killing form, then the Chern-Simons action changes by an integer multiple of $2\pi k$ when we do a large gauge transformation. This implies that $\exp(iS_{CS}(A))$ is gauge-invariant when the quantity k, called the 'level', is an integer. Since this exponential of the action is what actually appears in the path integral, one might hope that Chern-Simons theory admits a reasonable quantization in this case. And indeed this is so — at least when G is compact. Unfortunately, Chern-Simons theory with noncompact gauge group is still poorly understood.

The vacuum expectation values of spin network observables are very interesting in Chern-Simons theory. Suppose Ψ is a spin network in M. We can try to compute the vacuum expectation value

$$\langle \Psi \rangle = \frac{\int \Psi(A)\, e^{iS_{CS}(A)}\, \mathcal{D}A}{\int e^{iS_{CS}(A)}\, \mathcal{D}A}.$$

Naively, we would expect from the diffeomorphism-invariance of the Chern-Simons action that $\langle \Psi \rangle$ remains unchanged when we apply a diffeomorphism to Ψ. In fact, this expectation value is ill-defined until we smear Ψ by equipping it with a 'framing'. Roughly, this means that we thicken each edge of Ψ into a ribbon, put a small disc at each vertex, and demand that the ribbons merge with the discs smoothly at each vertex to form an orientable surface with boundary. The expectation values of these framed spin networks are diffeomorphism-invariant, and they satisfy skein relations which allow one to calculate them in a completely combinatorial way.

The reader will recall that in BF theory without cosmological term, spin network observables also satisfied skein relations. In that case, the skein relations encoded the representation theory of G. That is what allowed us to give a purely combinatorial, or algebraic, description of the theory. Marvelously, a similar thing is true in Chern-Simons theory! In Chern-Simons theory, however, the skein relations encode the representation theory of the quantum group $U_q\mathfrak{g}$. This is an algebraic gadget depending on a parameter q which is related to k by the formula

$$q = \exp(2\pi i/(k+h))$$

where h is the value of the Casimir in the adjoint representation of $\mathfrak{g}$. Alas, it would vastly expand the size of this paper to really explain what quantum groups are, and how they arise from Chern-Simons theory. To learn these things, the reader must turn to the references in the Notes. For our purposes, the most important thing is that the representation theory of $U_q\mathfrak{g}$ closely resembles that of G. In particular, each representation of G gives a representation of $U_q\mathfrak{g}$. This lets us think of spin network edges as labelled by representations of the quantum group rather than the group. However, only finitely many irreducible representations of the group give irreducible representations of the quantum group with nice algebraic properties. We shall call these 'good' representations. For example, when $G = \mathrm{SU}(2)$, only the representations of spin $j = 0, \frac{1}{2}, \ldots, \frac{k}{2}$ give good representations of $U_q\mathfrak{g}$. It turns out that Chern-Simons theory admits an algebraic formulation involving only the good representations of $U_q\mathfrak{g}$.

With this information in hand, let us turn to 3-dimensional BF theory with cosmological term. Starting from the action one can derive the classical field equations:

$$F + \frac{\Lambda}{2} E \wedge E = 0, \qquad d_A E = 0.$$

For $G = \mathrm{SO}(2,1)$, these are equivalent to the vacuum Einstein equations *with a cosmological constant* when E is one-to-one. One can show this using the same sort of argument we gave in Section 2 for the case $\Lambda = 0$. This reason this works is that $\mathrm{tr}(E \wedge E \wedge E)$ is proportional to the volume form coming from the metric defined by E. Up to a constant factor, it is therefore just a rewriting of the usual cosmological term in the action for general relativity. Similar remarks apply to $G = \mathrm{SO}(3)$, which gives us Riemannian general relativity with cosmological constant. We can also use the double covers of these gauge groups without affecting the classical theory.

The relation between 3d BF theory with cosmological term and Chern-Simons theory is as follows. Starting from the A and E fields in BF theory, we can define two new connections $A_\pm$ as follows:

$$A_\pm = A \pm \sqrt{\Lambda} E.$$

Ignoring boundary terms, we then have

$$\int_M \mathrm{tr}(E \wedge F + \frac{\Lambda}{6} E \wedge E \wedge E) = S_{CS}(A_+) - S_{CS}(A_-)$$

where

$$k = \frac{4\pi}{\sqrt{\Lambda}}.$$

In short, the action for 3d BF theory with cosmological term is a difference of two Chern-Simons actions. Thus we can quantize this BF theory whenever we can quantize Chern-Simons theory at levels k and $-k$, and we obtain a theory equivalent to two independent copies of Chern-Simons theory with these two opposite values of k. The physical Hilbert space is thus the tensor product of Hilbert spaces for two copies of Chern-Simons theory with opposite values of k, and a similar factorization holds for the time evolution operators associated to cobordisms. Actually, we can simplify this description using the fact that the Hilbert space for Chern-Simons theory at level $-k$ is naturally the dual of the Hilbert space at level k. This let us describe 3d BF theory with cosmological constant Λ completely in terms of Chern-Simons theory at level k.

Using this description together with the formulation of Chern-Simons theory in terms of quantum groups, one can derive a formula for the partition function 3d BF theory with cosmological term. This formula is almost identical to the one given in Section 7 for 3d BF theory with $\Lambda = 0$. The main difference is that now the quantum group $U_q\mathfrak{g}$ takes over the role of the group G. In other words, we now label dual faces by good representations of $U_q\mathfrak{g}$ and label dual edges by intertwiners between tensor products of these representations. A similar formula holds for transition amplitudes. In short, we have a spin foam model of a generalized sort, based on the representation theory of a quantum group instead of a group. The wonderful thing about this spin foam model is that the sums involved are finite, since there are only finitely many good representations of $U_q\mathfrak{g}$. With the infrared divergences eliminated, the partition function and transition amplitudes are truly well-defined. Even better, one can check that they are triangulation-independent!

The first example of this sort of spin foam model is due to Turaev and Viro, who considered the case $G = \mathrm{SU}(2)$. As we have seen, this model corresponds to 3-dimensional Riemannian gravity with cosmological constant Λ. In this case only spins $j \leq \frac{k}{2}$ correspond to good representations of $U_q\mathfrak{g}$. This constraint on the spins labeling dual faces corresponds to an upper bound on the lengths of the edges of the original triangulation. We thus have, not only a minimum length due to nonzero Planck's constant, but also a maximum length due to nonzero cosmological constant! As $\Lambda \to 0$, this maximum length goes to infinity.

Now let us turn to 4-dimensional BF theory with cosmological term. Here the classical field equations are

$$F + \frac{\Lambda}{6}E = 0, \qquad d_A E = 0.$$

If we canonically quantize the theory, we discover something interesting: for any compact oriented 3-manifold S representing space, the space of physical states is 1-dimensional. To see this, note first that 'kinematical' states should be functions on $\mathcal{A}$, just as we saw in Section 4 for the case $\Lambda = 0$. Physical states are solutions of the constraints

$$B + \frac{\Lambda}{6}E = 0, \qquad d_A E = 0,$$

where B is the curvature of $A \in \mathcal{A}$. As before, the constraint $d_A E = 0$ generates gauge transformations, so imposing this constraint should restrict us to gauge-invariant functions on $\mathcal{A}$. But the other constraint has a very different character when $\Lambda \neq 0$ than it did for $\Lambda = 0$. If we naively replace A and E by operators following the usual rules of canonical quantization, we see that states satisfying this constraint should be functions $\Psi\colon \mathcal{A} \to \mathbb{C}$ with

$$(B^a_{ij} + \frac{\Lambda}{6i}\epsilon_{ijk}\frac{\delta}{\delta A_{ka}})\psi = 0.$$

For $\Lambda \neq 0$ this equation has just one solution, the so-called 'Chern-Simons state':

$$\psi(A) = e^{-\frac{3i}{\Lambda}\int_S \mathrm{tr}(A\wedge dA + \frac{2}{3}A\wedge A\wedge A)}.$$

By our previous remarks, if G is simple, connected and simply-connected and 'tr' is defined using the Killing form, the Chern-Simons state is gauge-invariant exactly when the quantity

$$k = \frac{12\pi}{\Lambda}$$

is an integer. If in addition G is compact, we can go further: we can compute expectation values of framed spin networks in the Chern-Simons state using skein relations.

How do we describe dynamics in 4-dimensional BF theory with cosmological term? Unlike the other cases we have discussed, there is not yet a plausible 'derivation' of a spin foam model for this theory. At present, about the best one can do is note the following facts. There is a quantum group analog of the spin foam model for 4d BF theory discussed in Section 7, and this theory has finite and triangulation-independent partition function and transition amplitudes. One can show that this theory has a 1-dimensional physical Hilbert space for any compact oriented 3-manifold S. Moreover, one can compute the expectation values of framed spin networks in this theory, and one gets the same answers as in the Chern-Simons state. Thus it seems plausible that this theory is the correct spin foam model for 4d BF theory with cosmological term. However, this subject deserves further investigation.

10 4-Dimensional Quantum Gravity

We finally turn to theory that really motivates the interest in spin foam models: quantum gravity in 4 dimensions. Various competing spin foam models have been proposed for 4-dimensional quantum gravity — mainly in the Riemannian case so far. While some of these models are very elegant, their physical meaning has not really been unraveled, and some basic problems remain unsolved. The main reason is that, unlike BF theory, general relativity in 4 dimensions has local degrees of freedom. In short, the situation is full of that curious mix of promise and threat so typical of quantum gravity. In what follows we do not attempt a full description of the state of the art, since it would soon be outdated anyway. Instead, we merely give the reader a taste of the subject. For more details, see the Notes!

We begin by describing the Palatini formulation of general relativity in 4 dimensions. Let spacetime be given by a 4-dimensional oriented smooth manifold M. We choose a bundle $\mathcal{T}$ over M that is isomorphic to the tangent bundle, but not in any canonical way. This bundle, or any of its fibers, is called the 'internal space'. We equip it with an orientation and a metric η, either Lorentzian or Riemannian. Let P denote the oriented orthonormal frame bundle of M. This is a principal G-bundle, where G is either $\mathrm{SO}(3,1)$ or $\mathrm{SO}(4)$ depending on the signature of η. The basic fields in the Palatini formalism are:

- a connection A on P,
- a $\mathcal{T}$-valued 1-form e on M.

The curvature of A is an $\mathrm{ad}(P)$-valued 2-form which, as usual, we call F. Note however that the bundle $\mathrm{ad}(P)$ is isomorphic to the second exterior power $\Lambda^2\mathcal{T}$. Thus we are free to switch between thinking of F as an $\mathrm{ad}(P)$-valued 2-form and a $\Lambda^2\mathcal{T}$-valued 2-form. The same is true for the field $e \wedge e$.

The Lagrangian of the theory is

$$\mathcal{L} = \mathrm{tr}(e \wedge e \wedge F).$$

Here we first take the wedge products of the differential form parts of $e \wedge e$ and F while simultaneously taking the wedge products of their 'internal' parts, obtaining the $\Lambda^4\mathcal{T}$-valued 4-form $e \wedge e \wedge F$. The metric and orientation on $\mathcal{T}$

give us an 'internal volume form', that is, a nowhere vanishing section of $\Lambda^4 \mathcal{T}$. We can write $e \wedge e \wedge F$ as this volume form times an ordinary 4-form, which we call $\mathrm{tr}(e \wedge e \wedge F)$.

To obtain the field equations, we set the variation of the action to zero:

$$\begin{aligned} 0 &= \delta \int_M \mathcal{L} \\ &= \int_M \mathrm{tr}(\delta e \wedge e \wedge F + e \wedge \delta e \wedge F + e \wedge e \wedge \delta F) \\ &= \int_M \mathrm{tr}(2\delta e \wedge e \wedge F + e \wedge e \wedge d_A \delta A) \\ &= \int_M \mathrm{tr}(2\delta e \wedge e \wedge F - d_A(e \wedge e) \wedge \delta A). \end{aligned}$$

The field equations are thus

$$e \wedge F = 0, \qquad d_A(e \wedge e) = 0.$$

These equations are really just an extension of the vacuum Einstein equation to the case of degenerate metrics. To see this, first define a metric g on M by

$$g(v, w) = \eta(ev, ew).$$

When $e\colon TM \to \mathcal{T}$ is one-to-one, g is nondegenerate, with the same signature as η. The equation $d_A(e \wedge e) = 0$ is equivalent to $e \wedge d_A e = 0$, and when e is one-to-one this implies $d_A e = 0$. If we use e to pull back A to a metric-preserving connection Γ on the tangent bundle, the equation $d_A e = 0$ says that Γ is torsion-free, so Γ is the Levi-Civita connection of g. This lets us rewrite $e \wedge F$ in terms of the Riemann tensor. In fact, $e \wedge F$ is proportional to the Einstein tensor, so $e \wedge F = 0$ is equivalent to the vacuum Einstein equation.

There are a number of important variants of the Palatini formulation which give the same classical physics (at least for nondegenerate metrics) but suggest different approaches to quantization. Most simply, we can pick a spin structure on M and use the double cover $\mathrm{Spin}(3,1) \cong \mathrm{SL}(2,\mathbb{C})$ or $\mathrm{Spin}(4) \cong \mathrm{SU}(2) \times \mathrm{SU}(2)$ as gauge group. A subtler trick is to work with the 'self-dual' or 'left-handed' part of the spin connection. In the Riemannian case this amounts to using only one of the $\mathrm{SU}(2)$ factors of $\mathrm{Spin}(4)$ as gauge group; in the Lorentzian case we need to complexify $\mathrm{Spin}(3,1)$ first, obtaining $\mathrm{SL}(2,\mathbb{C}) \times \mathrm{SL}(2,\mathbb{C})$, and then use one of these $\mathrm{SL}(2,\mathbb{C})$ factors. It it not immediately obvious that one can formulate general relativity using only the left-handed part of the connection, but the great discovery of Plebanski and Ashtekar is that one can. A further refinement of this trick allows one to formulate the canonical quantization of Lorentzian general relativity in terms of the e field and an $\mathrm{SU}(2)$ connection. These so-called 'real Ashtekar variables' play a crucial role in most work on loop quantum gravity. Indeed, much of the spin network technology described in this paper was first developed for use with the real Ashtekar variables. However, to keep the discussion focused, we only discuss the Palatini formulation in what follows.

The Palatini formulation of general relativity brings out its similarity to BF theory. In fact, if we set $E = e \wedge e$, the Palatini Lagrangian looks exactly like the BF Lagrangian. The big difference, of course, is that not every ad(P)-valued 2-form E is of the form $e \wedge e$. This restricts the allowed variations of the E field when we compute the variation of the action in general relativity. As a result, the equations of general relativity in 4 dimensions:

$$e \wedge F = 0, \qquad d_A E = 0$$

are weaker than the BF theory equations:

$$F = 0, \qquad d_A E = 0.$$

Another, subtler difference is that, even when E is of the form $e \wedge e$, we cannot uniquely recover e from E. In the nondegenerate case there is only a sign ambiguity: both e and $-e$ give the same E. Luckily, changing the sign of e does not affect the metric. In the degenerate case the ambiguity is greater, but we need not be unduly concerned about it, since we do not really know the 'correct' generalization of Einstein's equation to degenerate metrics.

The relation between the Palatini formalism and BF theory suggests that one develop a spin foam model of quantum gravity by taking the spin foam model for BF theory and imposing extra constraints: quantum analogues of the constraint that E be of the form $e \wedge e$. However, there are some obstacles to doing this. First, BF theory is only well-understood when the gauge group is compact. If we work with a compact gauge group, we are limited to Riemannian quantum gravity. Of course, this simply means that we should work harder and try to understand BF theory with noncompact gauge group. Work on this is currently underway, but the picture is still rather murky, and a fair amount of new mathematics will need to be developed before it clears up. For this reason, we only consider the Riemannian quantum gravity in what follows.

Second, when computing transition amplitudes in BF theory, we only summed over spin foams living in the dual 2-skeleton of a fixed triangulation of spacetime. This was acceptable because we could later show triangulation-independence. But triangulation-independence is closely related to the fact that BF theory lacks local degrees of freedom: if we study BF theory on a triangulated manifold, subdividing the triangulation changes the gauge-invariant Hilbert space, but it does not increase the number of physical degrees of freedom. There is no particular reason to expect something like this to hold in 4d quantum gravity, since general relativity in 4 dimensions *does* have local degrees of freedom. So what should we do? Nobody knows! This problem requires careful thought and perhaps some really new ideas. In what follows, we simply ignore it and restrict attention to spin foams lying in the dual 2-skeleton of a fixed triangulation, for no particular good reason.

We begin by considering at the classical level the constraints that must hold for the E field to be of the form $e \wedge e$. We pick a spin structure for spacetime and take the double cover Spin(4) as our gauge group. Locally we may think of the E field as taking values in the Lie algebra $\mathfrak{so}(4)$, but the splitting

$$\mathfrak{so}(4) \cong \mathfrak{so}(3) \oplus \mathfrak{so}(3)$$

lets us write E as the sum of left-handed and right-handed parts $E^{\pm}$ taking values in $\mathfrak{so}(3)$. If $E = e \wedge e$, the following constraint holds for all vector fields v, w on M:

$$|E^+(v,w)| = |E^-(v,w)|$$

where $|\cdot|$ is the norm on $\mathfrak{so}(3)$ coming from the Killing form. In fact, this constraint is almost sufficient to guarantee that E is of the form $e \wedge e$. Unfortunately, in addition to solutions of the desired form, there are also solutions of the form $-e \wedge e$, $*(e \wedge e)$, and $-*(e \wedge e)$, where $*$ is the Hodge star operator on $\Lambda^2\mathcal{T}$.

If we momentarily ignore this problem and work with the constraint as described, we must next decide how to impose this constraint in a spin foam model. First recall some facts about 4d BF theory with gauge group SU(2). In this theory, a spin foam in the dual 2-skeleton of a triangulated 4-manifold is given by labeling each dual face with a spin and each dual edge with an intertwiner. This is equivalent to labeling each triangle with a spin and each tetrahedron with an intertwiner. We can describe these intertwiners by chopping each tetrahedra in half with a parallelogram and labeling all these parallelograms with spins. Then all the data is expressed in terms of spins labeling surfaces, and each spin describes the integral of $|E|$ over the surface it labels.

Now we are trying to describe 4-dimensional Riemannian quantum gravity as a BF theory with extra constraints, but now the gauge group is Spin(4). Since Spin(4) is isomorphic to SU(2) × SU(2), irreducible representation of this group are of the form $j^+ \otimes j^-$ for arbitrary spins j^+, j^-. Thus, before we take the constraints into account, a spin foam with gauge group Spin(4) can be given by labeling each triangle and parallelogram with a *pair* of spins. These spins describe the integrals of $|E^+|$ and $|E^-|$, respectively, over the surface in question. Thus, to impose the constraint

$$|E^+(v,w)| = |E^-(v,w)|$$

at the quantum level, it is natural to restrict ourselves to labelings for which these spins are equal. This amounts to labeling each triangle with a representation of the form $j \otimes j$ and each tetrahedron with an intertwiner of the form $\iota_j \otimes \iota_j$, where $\iota_j: j_1 \otimes j_2 \to j_3 \otimes j_4$ is given in our graphical notation by:

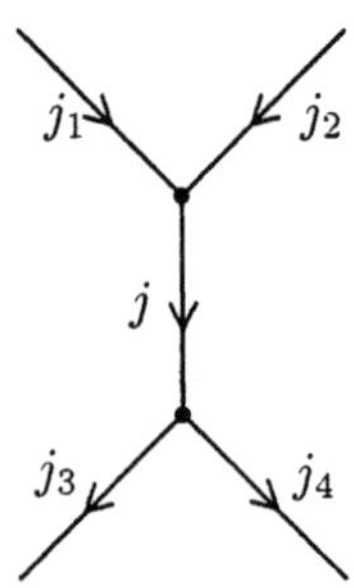

and $j_1, \dots, j_4$ are the spins labeling the 4 triangular faces of the tetrahedron. More generally, we can label the tetrahedron by any intertwiner of the form $\sum_j c_j(\iota_j \otimes \iota_j)$.

However, there is a subtlety. There are three ways to split a tetrahedron in half with a parallelogram P, and we really want the constraint

$$\int_P |E^+| = \int_P |E^-|$$

to hold for all three. To achieve this, we must label tetrahedra with intertwiners of the form $\sum_j c_j(\iota_j \otimes \iota_j)$ that *remain* of this form when we switch to a different splitting using the $6j$ symbols. Barrett and Crane found an intertwiner with this property:

$$\iota = \sum_j (2j+1)(\iota_j \otimes \iota_j).$$

Later, Reisenberger proved that this was the unique solution. Thus, in this spin foam model for 4-dimensional Riemannian quantum gravity, we take the partition function to be:

$$Z(M) = \sum_{j:\mathcal{F}\to\{0,\frac{1}{2},1,\dots\}} \prod_{f\in\mathcal{F}} (2j_f+1) \prod_{v\in\mathcal{V}}$$

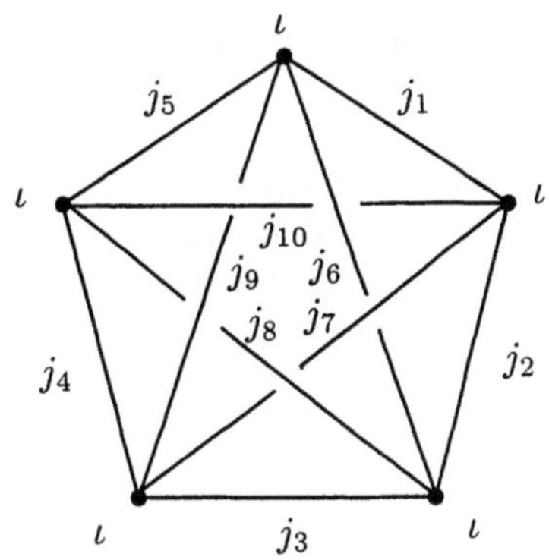

Here $j_1, \dots, j_{10}$ are the spins labeling the dual faces meeting at the dual vertex in question, and ι is the Barrett-Crane intertwiner. One can also write down a similar formula for transition amplitudes.

The sums in these formulas probably diverge, but there is a q-deformed version where they become finite. This q-deformed version appears *not* to be triangulation-independent. We expect that it is related to general relativity with a nonzero cosmological constant. As a piece of evidence for this, note that adding a cosmological term to general relativity in 4 dimensions changes the Lagrangian to

$$\mathcal{L} = \mathrm{tr}(e \wedge e \wedge F + \frac{\Lambda}{12} e \wedge e \wedge e \wedge e).$$

We can think of this as the BF Lagrangian with cosmological term together with a constraint saying that $E = e \wedge e$.

So, where do we stand? We have a specific proposal for a spin foam model of quantum gravity. In this theory, a quantum state of the geometry of space is described by a linear combination of spin networks. Areas and volumes take on a discrete spectrum of quantized values. Transition amplitudes between states are computed as sums over spin foams. In the q-deformed version of the theory these sums are finite and explicitly computable.

This sounds very nice, but there are severe problems as well. The theory is actually a theory of Riemannian rather than Lorentzian quantum gravity. It depends for its formulation on a fixed triangulation of spacetime. Even worse, our ability to do computations with the theory is too poor to really tell if it reduces to classical Riemannian general relativity in the large-scale limit, i.e. the limit of distances much larger than the Planck length. We thus face the following tasks:

- Develop spin foam models of Lorentzian quantum gravity.
- Determine what role, if any, triangulations or related structures should play in spin foam models with local degrees of freedom.
- Develop computational techniques for studying the large-scale limit of spin foam models.

Luckily, work on these tasks is already underway.

Remarks 1. Regge gave a formula for a discretized version of the action in 4-dimensional Riemannian general relativity. In his approach, spacetime is triangulated and each edge is assigned a length. The Regge action is the sum over all 4-simplices of:

$$S = \sum_t A_t \theta_t$$

where the sum is taken over the 10 triangular faces t, A_t is the area of the face t, and θ_t is the dihedral angle of t, that is, the angle between the outward normals of the two tetrahedra incident to this edge. Calculations suggest that the spin foam vertex amplitudes in the Barrett-Crane theory are related to the Regge action by a formula very much like the one relating vertex amplitudes in 3d Riemannian quantum gravity to the Ponzano-Regge action (see Remark 1 of Section 7).

2. Our heuristic explanation of the Barrett-Crane model may make it seem more ad hoc than it actually is. For a more thorough treatment one should see the references in the Notes. At present our best understanding of this model comes from a 4-dimensional analogue of the theory of the quantum tetrahedron discussed in Remark 1 of Section 6. In particular, this approach allows a careful study of the 'spurious solutions' to the constraint $|E^+(v,w)| = |E^-(v,w)|$. It appears that at the quantum level, use of the Barrett-Crane intertwiner automatically excludes solutions of the form $E = \pm *(e \wedge e)$, but does not exclude solutions of the form $E = -e \wedge e$. The physical significance of this is still not clear.

Appendix: Piecewise Linear Cell Complexes

Here we give the precise definition of 'piecewise linear cell complex'. A subset $X \subseteq \mathbb{R}^n$ is said to be a 'polyhedron' if every point $x \in X$ has a neighborhood in X of the form

$$\{\alpha x + \beta y \colon \alpha, \beta \geq 0,\ \alpha + \beta = 1,\ y \in Y\}$$

where $Y \subseteq X$ is compact. A compact convex polyhedron X for which the smallest affine space containing X is of dimension k is called a 'k-cell'. The term 'polyhedron' may be somewhat misleading to the uninitiated; for example, $\mathbb{R}^n$ is a polyhedron, and any open subset of a polyhedron is a polyhedron. Cells, on the other hand, are more special. For example, every 0-cell is a point, every 1-cell is a compact interval affinely embedded in $\mathbb{R}^n$, and every 2-cell is a convex compact polygon affinely embedded in $\mathbb{R}^n$.

The 'vertices' and 'faces' of a cell X are defined as follows. Given a point $x \in X$, let $\langle x, X\rangle$ be the union of lines L through x such that $L \cap X$ is an interval with x in its interior. If there are no such lines, we define $\langle x, X\rangle$ to be $\{x\}$ and call x a 'vertex' of X. One can show that $\langle x, X\rangle \cap X$ is a cell, and such a cell is called a 'face' of X. (In the body of this paper we use the words 'vertex', 'edge' and 'face' to stand for 0-cells, 1-cells and 2-cells, respectively. This should not be confused with the present use of these terms.)

One can show that any cell X has finitely many vertices v_i and that X is the convex hull of these vertices, meaning that:

$$X = \{\sum \alpha_i v_i \colon \alpha_i \geq 0, \sum \alpha_i = 1\}.$$

Similarly, any face of X is the convex hull of some subset of the vertices of X. However, not every subset of the vertices of X has a face of X as its convex hull. If the cell Y is a face of X we write $Y \leq X$. This relation is transitive, and if $Y, Y' \leq X$ we have $Y \cap Y' \leq X$.

Finally, one defines a 'piecewise linear cell complex', or 'complex' for short, to be a collection κ of cells in some $\mathbb{R}^n$ such that:

1. If $X \in \kappa$ and $Y \leq X$ then $Y \in \kappa$.
2. If $X, Y \in \kappa$ then $X \cap Y \leq X, Y$.

In this paper we restrict our attention to complexes with finitely many cells.

A complex is 'k-dimensional' if it has cells of dimension k but no higher. A complex is 'oriented' if every cell is equipped with an orientation, with all 0-cells being equipped with the positive orientation. The union of the cells of a complex κ is a polyhedron which we denote by $|\kappa|$.

When discussing spin foams we should really work with spin networks whose underlying graph is a 1-dimensional oriented complex. Suppose γ is a 1-dimensional oriented complex and κ is a 2-dimensional oriented complex. Note that the product $\gamma \times [0,1]$ becomes a 2-dimensional oriented complex in a natural way. We say γ 'borders' κ if there is a one-to-one affine map $c\colon |\gamma| \times [0,1] \to |\kappa|$ mapping each cell of $\gamma \times [0,1]$ onto a unique cell of κ in an orientation-preserving way, such that c maps $\gamma \times [0,1)$ onto an open subset of $|\kappa|$. Note that in this case, c lets us regard each k-cell of γ as the face of a unique $(k+1)$-cell of κ.

Notes

While long-winded, this bibliography has no pretensions to completeness. In particular, as a mathematician by training, my selection of references inevitably has an emphasis on mathematically rigorous work. This gives a somewhat slanted view of the subject, which is bound to make some people unhappy. I apologize for this in advance, and urge the reader to look at some of the references written by physicists to get a more balanced picture.

2 *BF* Theory: Classical Field Equations

For all aspects of BF theory, the following papers are invaluable:

A. S. Schwartz, The partition function of degenerate quadratic functionals and Ray-Singer invariants, *Lett. Math. Phys.* **2** (1978), 247-252.
G. Horowitz, Exactly soluble diffeomorphism-invariant theories, *Comm. Math. Phys.* **125** (1989) 417-437.
D. Birmingham, M. Blau, M. Rakowski and G. Thompson, Topological field theories, *Phys. Rep.* **209** (1991), 129-340.
M. Blau and G. Thompson, Topological gauge theories of antisymmetric tensor fields, *Ann. Phys.* **205** (1991), 130-172.

For BF theory on manifolds with boundary, see:

V. Husain and S. Major, Gravity and BF theory defined in bounded regions, *Nucl. Phys.* **B500** (1997), 381-401.
A. Momen, Edge dynamics for BF theories and gravity, *Phys. Lett.* **B394** (1997), 269-274.

3 Classical Phase Space

The space $\mathcal{A}/\mathcal{G}$ and its cotangent bundle have mainly been studied in the context of Yang-Mills theory:

V. Moncrief, Reduction of the Yang-Mills equations, in *Differential Geometrical Methods in Mathematical Physics*, eds. P. Garcia, A. Pérez-Rendón, and J. Souriau, Lecture Notes in Mathematics 836, Springer-Verlag, New York, 1980, pp. 276-291.
P. K. Mitter, Geometry of the space of gauge orbits and Yang-Mills dynamical system, in *Recent developments in Gauge Theories*, eds. G. 't Hooft et al., Plenum Press, New York, 1980, pp. 265-292.

The moduli space of flat G-bundles and the moduli space of flat connections on any particular G-bundle have been extensively studied when the base manifold is a Riemann surface. See for example:

M. Narasimhan and C. Seshadri, Stable and unitary vector bundles on a compact Riemann surface, *Ann. Math.* **82** (1965) 540-567.

Later, Goldman and others studied these spaces when the base space is a compact 2-dimensional smooth manifold, without any complex structure:

W. Goldman, The symplectic nature of fundamental groups of surfaces, *Adv. Math.* **54** (1984) 200-225.
W. Goldman, Invariant functions on Lie groups and Hamiltonian flows of surface group representations, *Invent. Math.* **83** (1986) 263-302.
W. Goldman, Topological components of spaces of representations, *Invent. Math.* **93** (1988) 557-607.
A. Alekseev, A. Malkin, Symplectic structure of the moduli space of flat connections on a Riemann surface, *Commun. Math. Phys.* **169** (1995), 99-120.

4 Canonical Quantization

The idea of taking functions of holonomies as the basic observables or states in a quantized gauge theory has a long history. The earliest work dealt with Yang-Mills theory and used Wilson loops; later the idea was applied to gravity, and the importance of spin networks became clear still later. Some good books and review articles include:

R. Gambini and J. Pullin, *Loops, Knots, Gauge Theories, and Quantum Gravity*, Cambridge U. Press, Cambridge, 1996.
R. Loll, Chromodynamics and gravity as theories on loop space, preprint available as hep-th/9309056.
C. Rovelli, Loop quantum gravity, *Living Reviews in Relativity* (1998), available online at ⟨http://www.livingreviews.org⟩.

The first really systematic attempt to formulate quantum gravity in terms of Wilson loops is due to Rovelli and Smolin:

C. Rovelli and L. Smolin, Loop representation for quantum general relativity, *Nucl. Phys.* **B331** (1990), 80-152.

An important step towards a rigorous description of the space of states in loop quantum gravity was made by Ashtekar and Isham:

A. Ashtekar and C. J. Isham, Representations of the holonomy algebra of gravity and non-abelian gauge theories, *Class. Quan. Grav.* **9** (1992), 1069-1100.

This work used piecewise smooth loops, which turn out to be technically difficult to handle, so these authors were unable to construct $L^2(\mathcal{A}/\mathcal{G})$ except when G is abelian. Later, Ashtekar and Lewandowski used piecewise real-analytic loops to give a rigorous construction of $L^2(\mathcal{A}/\mathcal{G})$ for $G = \mathrm{SU}(2)$:

A. Ashtekar and J. Lewandowski, Representation theory of analytic holonomy C*-algebras, in *Knots and Quantum Gravity*, ed. J. Baez, Oxford, Oxford U. Press, 1994.

Then graphs with real-analytic edges were introduced, and used to construct $L^2(\mathcal{A}/\mathcal{G})$ for more general groups:

J. Baez, Diffeomorphism-invariant generalized measures on the space of connections modulo gauge transformations, in *Proceedings of the Conference on Quantum Topology*, ed. D. Yetter, World Scientific, Singapore, 1994.

Later graphs were used to construct the space $L^2(\mathcal{A})$:

J. Baez, Generalized measures in gauge theory, *Lett. Math. Phys.* **31** (1994), 213-223.

The use of graphs for integral and differential calculus on $\mathcal{A}$ and $\mathcal{A}/\mathcal{G}$ is systematically developed in the following papers:

A. Ashtekar and J. Lewandowski, Projective techniques and functional integration, *Jour. Math. Phys.* **36** (1995), 2170-2191.
A. Ashtekar and J. Lewandowski, Differential geometry for spaces of connections via graphs and projective limits, *Jour. Geom. Phys.* **17** (1995), 191-230.

The history of spin networks is rather complicated and I cannot do justice to it here. For a good introduction see:

L. Smolin, The future of spin networks, preprint available as gr-qc/9702030.

Briefly, spin networks were first invented by Penrose:

R. Penrose, Angular momentum: an approach to combinatorial space-time, in *Quantum Theory and Beyond*, ed. T. Bastin, Cambridge U. Press, Cambridge, 1971, pp. 151-180.
R. Penrose, Applications of negative dimensional tensors, in *Combinatorial Mathematics and its Applications*, ed. D. Welsh, Academic Press, New York, 1971, pp. 221-244.
R. Penrose, On the nature of quantum geometry, in *Magic Without Magic*, ed. J. Klauder, Freeman, San Francisco, 1972, pp. 333-354.
R. Penrose, Combinatorial quantum theory and quantized directions, in *Advances in Twistor Theory*, eds. L. Hughston and R. Ward, Pitman Advanced Publishing Program, San Francisco, 1979, pp. 301-317.

Penrose considered trivalent graphs labelled by spins. He wanted to use these as the basis for a purely combinatorial approach to spacetime. The following thesis is still invaluable for anyone interested in these ideas:

J. Moussouris, Quantum models of space-time based on recoupling theory, Ph.D. thesis, Department of Mathematics, Oxford University, 1983.

Later, as part of an attempt to understand the Jones polynomial and related knot invariants, the notion of spin network was generalized to include arbitrary graphs labelled by representations of any quantum group:

N. Reshetikhin and V. Turaev, Ribbon graphs and their invariants derived from quantum groups, *Comm. Math. Phys.* **127** (1990), 1-26.

In this more general context a framing of the graph is required, hence the term 'ribbon graph'. Spin networks were introduced into loop quantum gravity by Rovelli and Smolin:

C. Rovelli and L. Smolin, Spin networks in quantum gravity, *Phys. Rev.* **D52** (1995), 5743-5759.

The fact that spin network states span $L^2(\mathcal{A}/\mathcal{G})$ was shown in:

J. Baez, Spin networks in gauge theory, *Adv. Math.* **117** (1996), 253-272.

For an expository account of this proof and a general introduction to quantum gravity, try:

J. Baez, Spin networks in nonperturbative quantum gravity, in *The Interface of Knots and Physics*, ed. L. Kauffman, American Mathematical Society, Providence, Rhode Island, 1996.

For a rigorous approach to the canonical quantization of diffeomorphism-invariant gauge theories using spin networks, see:

A. Ashtekar, J. Lewandowski, D. Marolf, J. Mourão, and T. Thiemann, Quantization of diffeomorphism invariant theories of connections with local degrees of freedom, *Jour. Math. Phys.* **36** (1995), 6456-6493.

For the theory of $L^2(\mathcal{A})$ and $L^2(\mathcal{A}/\mathcal{G})$ in the smooth context, which involves the notion of 'webs', see:

J. Baez and S. Sawin, Functional integration on spaces of connections, *Jour. Funct. Anal.* **150** (1997), 1-27.
J. Baez and S. Sawin, Diffeomorphism-invariant spin network states, *Jour. Funct. Anal.* **158** (1998), 253-266.

J. Lewandowski and T. Thiemann, Diffeomorphism invariant quantum field theories of connections in terms of webs, preprint available as gr-qc/9901015.

For the canonical quantization of 3-dimensional general relativity, see:

E. Witten, 2+1 dimensional gravity as an exactly soluble system, *Nucl. Phys.* **B311** (1988), 46-78.
A. Ashtekar, V. Husain, C. Rovelli, J. Samuel and L. Smolin, 2+1 gravity as a toy model for the 3+1 theory, *Class. Quant. Grav.* **6** (1989), L185-L193.
A. Ashtekar, Lessons from (2+1)-dimensional quantum gravity, *Strings 90*, World Scientific, Singapore, 1990, pp. 71-88.
A. Ashtekar, R. Loll, New loop representations for 2+1 gravity, *Class. Quant. Grav.* **11** (1994), 2417-2434.
S. Carlip, *Quantum Gravity in 2+1 Dimensions,* Cambridge U. Press, Cambridge, 1998.

For a discussion of torsion and BF theory, see:

M. Blau and G. Thompson, A new class of topological field theories and the Ray-Singer torsion, *Phys. Lett.* **B228** (1989), 64-68.

5 Observables

The first calculation of area and volume operators in loop quantum gravity was by Rovelli and Smolin:

C. Rovelli and L. Smolin, Discreteness of area and volume in quantum gravity, *Nucl. Phys.* **B442** (1995), 593-622. Erratum, *ibid.* **B456** (1995), 753.

A rigorous construction and analysis of area and volume operators on $L^2(\mathcal{A}/\mathcal{G})$, using a somewhat different quantization scheme, was given in the following series of papers:

A. Ashtekar and J. Lewandowski, Quantum theory of geometry I: area operators, *Class. Quantum Grav.* **14** (1997), A55-A81.
A. Ashtekar and J. Lewandowski, Quantum theory of geometry II: volume operators, *Adv. Theor. Math. Phys.* **1** (1998), 388-429.
A. Ashtekar, A. Corichi and J. Zapata, Quantum theory of geometry III: noncommutativity of Riemannian structures, *Class. Quantum Grav.* **15** (1998), 2955-2972.

The area operator considered in these papers is the same as the operator $\mathcal{E}(\Sigma)$ in the special case when space is 3-dimensional and the gauge group is SU(2); however, the generalization to other dimensions and gauge groups is straightforward. For a simplified derivation of the area operator, see:

C. Rovelli and P. Upadhya, Loop quantum gravity and quanta of space: a primer, preprint available as gr-qc/9806079.

For attempts to compute the entropy of black holes in loop quantum gravity, see:

L. Smolin, Linking topological quantum field theory and nonperturbative quantum gravity, *Jour. Math. Phys.* **36** (1995) 6417-6455.
C. Rovelli, Loop quantum gravity and black hole physics, *Helv. Phys. Acta* **69** (1996), 582-611.
K. Krasnov, Counting surface states in loop quantum gravity, *Phys. Rev.* **D55** (1997), 3505-3513.
K. Krasnov, On quantum statistical mechanics of a Schwarzschild black hole, *Gen. Rel. Grav.* **30** (1998), 53-68.
A. Ashtekar, J. Baez, A. Corichi and K. Krasnov, Quantum geometry and black hole entropy, *Phys. Rev. Lett.* **80** (1998), 904-907.
A. Ashtekar, A. Corichi and K. Krasnov, Isolated black holes: the classical phase space, to appear.
A. Ashtekar, J. Baez, and K. Krasnov, Quantum geometry of black hole horizons, to appear.

6 Canonical Quantization via Triangulations

The relation between canonical quantum gravity on a triangulated manifold and other simplicial approaches to quantum gravity was noted by Rovelli:

C. Rovelli, The basis of the Ponzano-Regge-Turaev-Viro-Ooguri model is the loop representation basis, *Phys. Rev.* **D48** (1993), 2702-2707.

In a series of papers, Loll developed a version of loop quantum gravity on a cubical lattice:

R. Loll, Non-perturbative solutions for lattice quantum gravity, *Nucl. Phys.* **B444** (1995), 619-640.
R. Loll, The volume operator in discretized quantum gravity, *Phys. Rev. Lett.* **75** (1995) 3048-3051.
R. Loll, Spectrum of the volume operator in quantum gravity, *Nucl. Phys.* **B460** (1996) 143-154.
R. Loll, Further results on geometric operators in quantum gravity, *Class. Quantum Grav.* **14** (1997), 1725-1741.
R. Loll, Imposing $\det E > 0$ in discrete quantum gravity, *Phys. Lett.* **B399** (1997), 227-232.

For a definition of $L^2(\mathcal{A})$ and $L^2(\mathcal{A}/\mathcal{G})$ in the piecewise-linear context, see:

J. A. Zapata, A combinatorial approach to diffeomorphism invariant quantum gauge theories, *Jour. Math. Phys.* **38** (1997), 5663-5681.
J. A. Zapata, Combinatorial space from loop quantum gravity, *Gen. Rel. Grav.* **30** (1998), 1229-1245.

The study of the quantum tetrahedron was initiated by Barbieri:

A. Barbieri, Quantum tetrahedra and simplicial spin networks, *Nucl. Phys.* **B518** (1998) 714-728.

For a treatment of the quantum tetrahedron using geometric quantization, see:

J. Baez and J. Barrett, The quantum tetrahedron in 3 and 4 dimensions, preprint available as gr-qc/9903060.

7 Dynamics

The formulation of 3d Riemannian quantum gravity as a sum over labelings of the edges of a triangulated 3-manifold by spins was first given by Ponzano and Regge:

G. Ponzano and T. Regge, Semiclassical limit of Racah coefficients, in *Spectroscopic and Group Theoretical Methods in Physics*, ed. F. Bloch, North-Holland, New York, 1968.

The relation to Penrose's spin networks was noted by Hasslacher and Perry:

B. Hasslacher and M. Perry, Spin networks are simplicial quantum gravity, *Phys. Lett.* **B103** (1981), 21-24.

We can now see the work of Ponzano and Regge as providing a formula for the partition function of 3d BF theory with gauge group SU(2). Much later, Witten gave a similar formula in the 2-dimensional case:

E. Witten, On quantum gauge theories in two dimensions, *Comm. Math. Phys.* **141** (1991) 153-209.

and Ooguri gave a similar formula in the 4-dimensional case:

H. Ooguri, Topological lattice models in four dimensions, *Mod. Phys. Lett.* **A7** (1992) 2799-2810.

For the Dijkgraaf-Witten model see:

R. Dijkgraaf and E. Witten, Topological gauge theories and group cohomology, *Commun. Math. Phys.* **129** (1990) 393-429.

D. Freed and F. Quinn, Chern-Simons theory with finite gauge group, *Commun. Math. Phys.* **156** (1993), 435-472.

Ponzano and Regge's original argument relating the asymptotics of $6j$ symbols to their discretized action for 3-dimensional Riemannian general relativity turned out to be surprisingly hard to make precise. A rigorous proof was recently given by Roberts:

J. Roberts, Classical $6j$-symbols and the tetrahedron, *Geometry and Topology* **3** (1999), 21-66.

8 Spin Foams

The idea that transition amplitudes in 4d quantum gravity should be expressed as a sum over surfaces was proposed in the following paper:

J. Baez, Strings, loops, knots and gauge fields, in *Knots and Quantum Gravity*, ed. J. Baez, Oxford U. Press, Oxford, 1994.

This idea was developed by Iwasaki and Reisenberger, who stressed the importance of summing over 2-dimensional complexes, as opposed to 2-manifolds:

J. Iwasaki, A definition of the Ponzano-Regge quantum gravity model in terms of surfaces, *Jour. Math. Phys.* **36** (1995), 6288-6298.
M. Reisenberger, Worldsheet formulations of gauge theories and gravity, preprint available as gr-qc/9412035.

Later, Reisenberger and Rovelli showed how to derive such a 'sum over surfaces' formulation from a formula for the Hamiltonian constraint in quantum gravity:

M. Reisenberger and C. Rovelli, "Sum over surfaces" form of loop quantum gravity, *Phys. Rev.* **D56** (1997), 3490-3508.

The relation between spin network evolution and triangulated spacetime manifolds was clarified by Markopoulou:

F. Markopoulou, Dual formulation of spin network evolution, preprint available as gr-qc/9704013.

The general notion of a spin foam was defined in the following paper:

J. Baez, Spin foam models, *Class. Quant. Grav.* **15** (1998) 1827-1858.

For an attempt to systematically derive spin foam models from the Lagrangians for BF theory and related theories, see:

L. Freidel and K. Krasnov, Spin foam models and the classical action principle, *Adv. Theor. Phys.* **2** (1998), 1221-1285.

For a discussion of the mathematical and philosophical underpinnings of the spin foam approach, see:

J. Baez, Higher-dimensional algebra and Planck-scale physics, to appear in *Physics Meets Philosophy at the Planck Scale,* eds. C. Callender and N. Huggett, Cambridge U. Press, preprint available as gr-qc/9902017.

For a study of volume in 3-dimensional quantum gravity, see:

L. Freidel and K. Krasnov, Discrete space-time volume for 3-dimensional BF theory and quantum gravity, *Class. Quant. Grav.* **16** (1999), 351-362.

9 q-Deformation and the Cosmological Constant

The relation between Chern-Simons theory and the Jones polynomial was first glimpsed in Witten's seminal paper:

E. Witten, Quantum field theory and the Jones polynomial, *Comm. Math. Phys.* **121** (1989), 351-399.

The relation to quantum groups was clarified by Reshetikhin and Turaev:

N. Reshetikhin and V. Turaev, Invariants of 3-manifolds via link polynomials and quantum groups, *Invent. Math.* **103** (1991), 547-597.

By now the subject has grown to enormous proportions, and we can scarcely begin to list all the relevant refereces here. Instead, we merely direct the reader to the following textbooks:

M. Atiyah, *The Geometry and Physics of Knots,* Cambridge U. Press, Cambridge, 1990.
V. Chari and A. Pressley, *A Guide to Quantum Groups,* Cambridge U. Press, Cambridge, 1994.
J. Fuchs, *Affine Lie Algebra and Quantum Groups,* Cambridge U. Press, Cambridge, 1992.
C. Kassel, *Quantum Groups,* Springer-Verlag, New York, 1995.
L. Kauffman, *Knots and Physics,* World Scientific Press, Singapore, 1993.
L. Kauffman and S. Lins, *Temperley-Lieb Recoupling Theory and Invariants of 3-Manifolds,* Princeton U. Press, Princeton, New Jersey, 1994.
V. Turaev, *Quantum Invariants of Knots and 3-Manifolds,* de Gruyter, New York, 1994.

The book by Kauffman and Lins is especially handy whenever one needs a compendium of skein relations for $U_q\mathfrak{su}(2)$. For an overview of the relations between

BF theory and Chern-Simons theory, see:

A. Cattaneo, P. Cotta-Ramusino, J. Fröhlich and M. Martellini, Topological BF theories in 3 and 4 dimensions, *Jour. Math. Phys.* **36** (1995), 6137-6160.

The q-deformed version of the Ponzano-Regge model was discovered by Turaev and Viro:

V. Turaev and O. Viro, State sum invariants of 3-manifolds and quantum $6j$ symbols, *Topology* **31** (1992), 865-902.

They formulated the theory both in terms of a triangulation of the 3-manifold and, dually, in terms of a 2-dimensional complex embedded in the manifold. We may now see their theory as a spin foam model for 3d Riemannian quantum gravity with nonzero cosmological constant. Their construction was soon generalized by isolating the properties of the $6j$ symbols that make it work, and tracing these back to the properties of certain categories of representations. One can read about these generalizations in the book by Turaev, and also in the following papers:

B. Durhuus, H. Jakobsen and R. Nest, Topological quantum field theories from generalized $6j$-symbols, *Rev. Math. Phys.* **5** (1993), 1-67.
J. Barrett and B. Westbury, Invariants of piecewise-linear 3-manifolds, *Trans. Amer. Math. Soc.* **348** (1996), 3997-4022.
D. Yetter, State-sum invariants of 3-manifolds associated to Artinian semisimple tortile categories, *Topology and its Applications* **58** (1994), 47-80.

Turaev also described a related model in 4 dimensions, formulated in terms of a 2-dimensional complex embedded in the manifold:

V. Turaev, Quantum invariants of 3-manifolds and a glimpse of shadow topology, in *Quantum Groups*, Springer Lecture Notes in Mathematics 1510, Springer-Verlag, New York, 1992, pp. 363-366.

This model is also discussed in Turaev's book. Crane and Yetter developed an isomorphic theory, formulated in terms of a triangulation, by q-deforming Ooguri's formula for the partition function of 4d BF theory with gauge group SU(2):

L. Crane and D. Yetter, A categorical construction of 4d TQFTs, in *Quantum Topology*, eds. L. Kauffman and R. Baadhio, World Scientific, Singapore, 1993, pp. 120-130.

The isomorphism between Turaev's theory and the Crane-Yetter model was worked out by Roberts:

J. Roberts, Skein theory and Turaev-Viro invariants, *Topology* **34** (1995), 771-787.

The generalization of this theory to other quantum groups was later worked out by Turaev (see his book above) and in the following paper:

L. Crane, L. Kauffman and D. Yetter, State-sum invariants of 4-manifolds, *J. Knot Theory & Ramifications* **6** (1997), 177-234.

For an argument that this theory is really a spin foam model of BF theory with cosmological term, see:

J. Baez, Four-dimensional BF theory as a topological quantum field theory, *Lett. Math. Phys.* **38** (1996), 129-143.

Along closely related lines, there is also some interesting work on the canonical quantization of Chern-Simons theory and 3d BF theory in the piecewise-linear context:

A. Alekseev, H. Grosse and V. Schomerus, Combinatorial quantization of the Hamiltonian Chern-Simons theory I, *Commun. Math. Phys.* **172** (1995), 317-358.
A. Alekseev, H. Grosse and V. Schomerus, Combinatorial quantization of the Hamiltonian Chern-Simons theory II, *Commun. Math. Phys.* **174** (1995), 561-604.
D. Bullock, C. Frohman, and J. Kania-Bartoszynska, Topological interpretations of lattice gauge field theory, *Commun. Math. Phys.* **198** (1998), 47-81.
D. Bullock, C. Frohman, and J. Kania-Bartoszynska, Skein modules and lattice gauge field theory, preprint available as math.GT/9802023.

10 4-Dimensional Quantum Gravity

For a tour of various formulations of Einstein's equation, see:

P. Peldan, Actions for gravity, with generalizations: a review, *Class. Quant. Grav.* **11** (1994), 1087-1132.

For an introduction to canonical quantum gravity, try the following books:

A. Ashtekar and invited contributors, *New Perspectives in Canonical Gravity*, Bibliopolis, Napoli, Italy, 1988. (Available through the American Institute of Physics; errata available from the Center for Gravitational Physics and Geometry at Pennsylvania State University.)
A. Ashtekar, *Lectures on Non-perturbative Canonical Quantum Gravity*, World Scientific, Singapore, 1991.

The spin foam model of 4-dimensional Riemannian quantum gravity which we discuss here was invented by Barrett and Crane:

J. Barrett and L. Crane, Relativistic spin networks and quantum gravity, *Jour. Math. Phys.* **39** (1998), 3296-3302.

A detailed discussion of their model appears in my first paper on spin foam models (see the Notes for Section 7). A more detailed treatment of general relativity as a constrained Spin(4) BF theory can be found in the following papers:

M. Reisenberger, Classical Euclidean general relativity from 'lefthanded area = righthanded area', preprint available as gr-qc/9804061.
R. De Pietri and L. Freidel, $\mathfrak{so}(4)$ Plebanski action and relativistic spin foam model, preprint available as gr-qc/9804071.

A heuristic argument for the uniqueness of the Barrett-Crane intertwiner was given by Barbieri:

A. Barbieri, Space of the vertices of relativistic spin networks, preprint available as gr-qc/9709076.

Later, Reisenberger gave a rigorous proof:

M. Reisenberger, On relativistic spin network vertices, preprint available as gr-qc/9809067.

An explanation of the uniqueness of the Barrett-Crane intertwiner in terms of geometric quantization was given in my paper with Barrett on the quantum tetrahedron (see the Notes for Section 5.) Similar intertwiners for vertices of higher valence have been constructed by Yetter:

D. Yetter, Generalized Barrett-Crane vertices and invariants of embedded graphs, preprint available as math.QA/9801131.

Barrett found an integral formula for the Barrett-Crane intertwiner:

J. Barrett, the classical evaluation of relativistic spin networks, preprint available as math.QA/9803063.

Later, he and Williams used this to give a heuristic argument relating the asymptotics of the amplitudes in the Barrett-Crane model to the Regge action:

J. Barrett and R. Williams, The asymptotics of an amplitude for the 4-simplex, preprint available as gr-qc/9809032.

For the Regge action, see:

T. Regge, General relativity without coordinates, *Nuovo Cimento* **19** (1961), 558-571.

Reisenberger and Iwasaki have proposed alternative spin foam models of 4-dimensional Riemannian quantum gravity. As with the Barrett-Crane model, the basic idea behind these models is to treat general relativity as a constrained BF theory. However, the models of Reisenberger and Iwasaki involve only the left-handed part of the spin connection, so the gauge group is SU(2):

M. Reisenberger, A lattice worldsheet sum for 4-d Euclidean general relativity, preprint available as gr-qc/9711052.

J. Iwasaki, A surface theoretic model of quantum gravity, preprint available as gr-qc/9903112.

Freidel and Krasnov have constructed spin foam models of Riemannian quantum gravity in higher dimensions by treating the theory as a constrained BF theory with gauge group SO(n):

L. Freidel, K. Krasnov, and R. Puzio, *BF* description of higher-dimensional gravity theories, preprint available as hep-th/9901069.

Barrett and Crane have also begun work on a Lorentzian version of their theory, but so far their formula for the amplitude of a spin foam vertex remains formal, because the evaluation of spin networks typically diverges when the gauge group is noncompact, apparently even after q-deformation:

J. Barrett and L. Crane, A Lorentzian signature model for quantum general relativity, preprint available as gr-qc/9904025.

In a different but related line of development, Markopoulou and Smolin have considered a class of local, causal rules for the time evolution of spin networks. Rules in this class are the same as spin foam models.

F. Markopoulou and L. Smolin, Quantum geometry with intrinsic local causality, *Phys. Rev.* **D58**:084032 (1998).

Smolin has suggested a relationship between these models and string theory, and proposed a specific model of this type as a candidate for a background-free formulation of M-theory. Ling and Smolin have begun to develop the supersymmetric analogue of the theory of spin networks:

L. Smolin, Strings as perturbations of evolving spin networks, preprint available as hep-th/9801022.

L. Smolin, Towards a background-independent approach to M theory, preprint available as hep-th/9808192.

Y. Ling, L. Smolin, Supersymmetric spin networks and quantum supergravity, preprint available as hep-th/9904016.

Appendix

For more details on piecewise linear cell complexes, try:

C. Rourke and B. Sanderson, *Introduction to Piecewise-Linear Topology*, Springer Verlag, Berlin, 1972.

The 2-dimensional case is explored more deeply in:

C. Hog-Angeloni, W. Metzler, and A. Sieradski, *Two-dimensional Homotopy and Combinatorial Group Theory*, London Mathematical Society Lecture Note Series 197, Cambridge U. Press, Cambridge, 1993.

T-Duality and the Gravitational Description of Gauge Theories

Cesar Gomez[♣][◇] and Pedro Silva[♠][♭]

[♣] *Instituto de Fisica Teorica, C-XVI, Universidad Autonoma de Madrid E-28049-Madrid, Spain*[**]
[◇] *I.M.A.F.F., C.S.I.C., Calle de Serrano 113 E-28006-Madrid, Spain*
[♠] *Physics department, University of Newcastle Upon Tyne NE1 7RU, UK*
[♭] *Depto de Fisica Teorica, C-XI, Universidad Autonoma de Madrid, E-28049-Madrid, Spain*

Abstract. This is a review of some basic features on the relation between supergravity and pure gauge theories with special emphasis on the relation between T-duality and supersymmetry. Some new results concerning the interplay between T-duality and near horizon geometries are presented.

1 Introduction

String theory [1], [2], [3] is defined by the two dimensional non linear sigma-model

$$S = \frac{1}{4\pi\alpha'} \int d^2\xi \left\{ \left[\sqrt{-g} g^{ij} G_{\mu\nu} + \varepsilon^{ij} B_{\mu\nu}\right] \partial_i X^\mu \partial_j X^\nu + \alpha' \Phi \sqrt{-\gamma} R^{(2)} \right\}, \quad (1)$$

provided the following space-time interpretations (fig. 1):

1. X are the space-time coordinates where the string is embed.
2. G, Φ, B are the external background fields called the metric, the dilaton and the torsion, respectively.
3. g is the world sheet metric.
4. ξ are the world sheet coordinates.

The governing principle of string theory is the world sheet Weyl invariance of (1), understood as a two dimensional field theory. Interpreting G, B and Φ as coupling constraints the requirement of Weyl invariance becomes equivalent to the following equations:

$$\begin{aligned}
\beta^G_{\alpha\beta} &= R_{\alpha\beta} - \frac{1}{4} H^2_{\alpha\beta} + 2\nabla_\alpha \nabla_\beta \Phi, \\
\beta^B_{\alpha\beta} &= \frac{1}{2} \nabla^\gamma H_{\gamma\alpha\beta} - \nabla^\gamma \Phi H_{\gamma\alpha\beta} H, \\
\beta^\Phi &= \frac{D}{6} + \frac{\alpha'}{2}[-R + \frac{H^2}{12} + 4(\nabla\Phi)^2 - 4\nabla^2\Phi], \quad (2)
\end{aligned}$$

[**] Unidad de Investigacion asociada al centrode Fisica Miguel Catalan (C.S.I.C.)

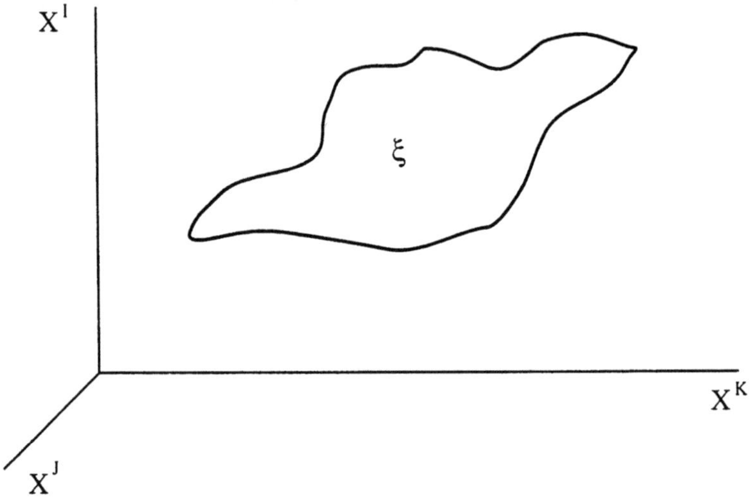

Fig. 1. Space-Time

where the β's in (2) are the different beta functions for the Lagrangian of (1) and R,H are the strength fields of G,B.

One of the most interesting aspects of string theory can be already obtained by direct inspection of the first equation of (2) which present a strong similarity with general relativity Einstein's equations. More precisely we can define associated with the string theory (1) an effective Lagrangian on the background fields $\mathcal{L}(G, B, \Phi)$ such that the corresponding equations of motions

$$\frac{\delta \mathcal{L}}{\delta \Psi_i} = 0 \quad , \quad \Psi_i = G, B, \Phi, \tag{3}$$

coincide precisely with the beta function equation (2). This effective Lagrangian is given by:

$$I = \frac{1}{2\kappa^2} \int dx^{26} \sqrt{-G} e^{2\Phi} \left[R + 4\nabla_\mu \phi \nabla^\mu \Phi - \frac{1}{12} H^2 \right] \tag{4}$$

Note that this lagrangian is not defined in the canonical form of Einstein general relativity. To make contact with the more familiar Einstein formalism we perform a rescaling on the metric absorbing the dilaton into the new metric. The resulting metric is known as the Einstein metric or the canonical metric, while the original metric is called string metric. The precise transformation is given by:

$$G^e_{\alpha\beta} = e^{-\Phi/2} G^s_{\alpha\beta}. \tag{5}$$

The previous discussion correspond to the closed bosonic string. The physical spectrum of this string contains a tachyonic mode with mass square $\frac{-2}{\alpha'}$, a

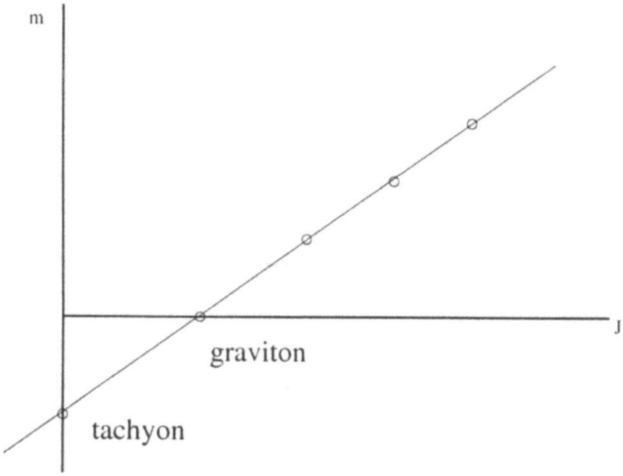

Fig. 2. Regge Trajectory

massless state of spin two containing the graviton, antisymmetric tensor and the trace part that is identify with the dilaton. On the top of this, a tower of massive states with masses proportional to $\frac{-1}{\alpha'}$ (fig.2).

Graviton and tachyon scattering amplitudes for this string theory are defined in terms of vertex operators:

$$V_{tachyon} = e^{ikx},$$
$$V_{graviton} = \eta_{\mu\nu}\partial X^{\mu}\partial X^{\nu}e^{ikx}, \tag{6}$$

At tree level these amplitudes are consistently defined for $D \leq 26$, for D the space-time dimension. At one loop level the requirement of unitarity implies that D should be equal to the critical dimension $D = 26$.

A slightly different type of strings are the open string. In this case the Regge trajectory contains a massless spin one state that we can try to identify with some sort of gauge boson. The definition of open strings requires to specify precise boundary conditions at the end points of the string. If we want to preserved target space-time Lorenz invariance we should choose Newman boundary conditions.

$$\partial_{\sigma}X = 0 \,|_{\text{end points}} \tag{7}$$

with the parameterization of the world sheet as indicated on (fig.3).

In addition the above type of strings allow us to decorate its ends with additional information, the so called Chan-Paton factors, that we can heuristically imagine as pairs of "quark-antiquark" transforming in the fundamental representation of some gauge group G. This makes for $G = U(N)$ that the open string

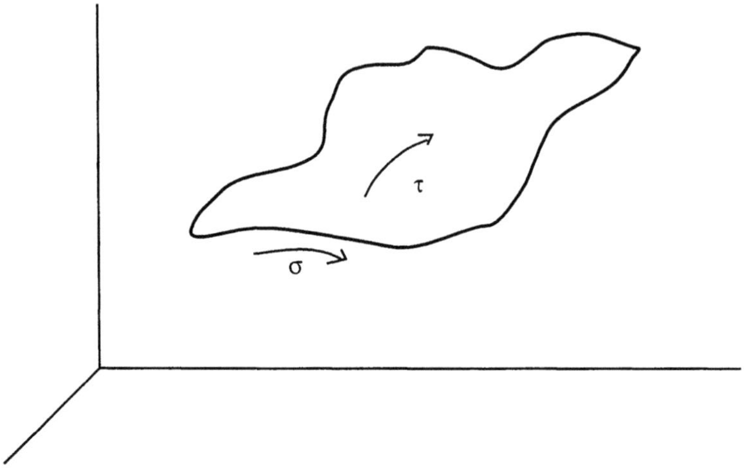

Fig. 3. World-sheet parametrisation

states will transform in the adjoin representation as it should be for a gauge boson. In case we consider orthogonal groups $G = SO(N)$ the requirement for the gauge boson to transform in the adjoint representation implies to introduce an orientation to the open string. Gauge boson associated with exceptional algebras can not be included on this way. This was one of the main motivations for the discovery of the "heterotic string", that surprisely enough are closed strings that contains in its spectrum massless gauge bosons for the group $E_8 \otimes E_8$. Gauge boson amplitudes for the open bosonic string can be easily computed using the following vertex operators:

$$V = \xi \partial X e^{ikx}. \tag{8}$$

As in the case of the closed bosonic string, the open strings contains a tower on massive states, with masses of order $1/\alpha'$, and a tachyon with negative square mass. In figure (4) we have depicted some open string one loop amplitudes. Examples like (a) are planar, i.e. they can be draw in a plane, while examples as (b) can not, and are called non-planar.

One of the deepest aspects of open string theory can be already discussed from direct inspection of diagram (b). Namely, from the standard scattering theory point of view, what we are seeing is a scattering of open string with closed string states contributing to the internal channel. In other words closed strings appears naturally as interaction products of open strings. This simple and basic fact of string theory should immediately ring a bell of any quantum field theorist. In fact we can always work out string theory in the infinite tension limit ($\alpha' \to 0$) where we decouple all the tower of massive states. From the open string point of view the result should be a pure gauge theory, while from the closed string point of view should land in pure gravity. We may wonder then, if there is any residual effect of the string open-closed relation that survives at the limit $\alpha' \to 0$?. This is a basic question that will allow us to enter into the very recent important developments connecting Yang Mill theories and gravity but before that we need to discuss another aspect of the open-closed interplay namely the D-branes and T-duality.

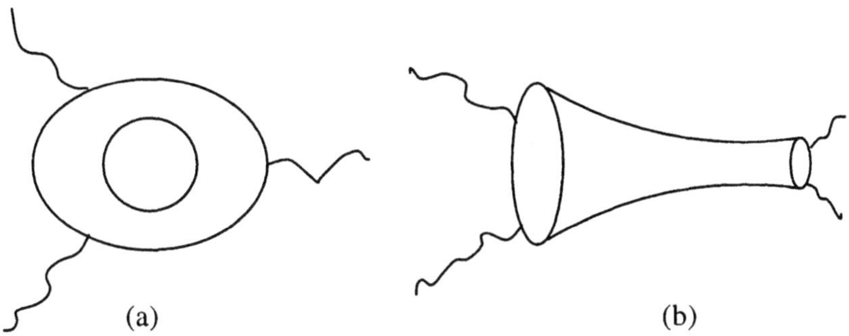

Fig. 4. One loop amplitudes

2 D-Branes and T-Duality

Let us star considering a closed string in a space time of the type $\Re^d \otimes S^1_R$, with R the radius of the S^1. The extended nature of the closed string allowed us to define a new quantum number, namely the "winding number" of the closed string around the circle. Let us called it m. Now consider the mass formula for the closed string states after compactification on the circle S^1 (i.e. in our example masses from the point of view of the observers in $\Re^d$):

$$M^2 = -p^\mu p_\mu = \frac{2}{\alpha'}(\alpha_0^{25})^2 + \frac{4}{\alpha'}(N-1) \tag{9}$$

and

$$\alpha_0^{25} = \left(\frac{n}{R} + \frac{mR}{\alpha}\right)\sqrt{\frac{\alpha'}{2}}, \tag{10}$$

where μ runs only over the non-compact dimensions, N is the total level of the left-moving excitations.

This mass formula posses a very interesting and amazing symmetry defined by:

$$n \leftrightarrow m \quad , \quad R \leftrightarrow \frac{\alpha'}{R}, \tag{11}$$

This symmetry is known as T-duality [4].

As it should be clear from the effective Lagrangian (4) if we require not only invariance of the spectrum but also invariance for the amplitude we should change the dilaton field as

$$\tilde{\Phi} = \Phi - ln(\frac{R}{\alpha'^{1/2}}). \tag{12}$$

The above example of T-duality can be generalized to generic backgrounds with a Killing vector. The T-duality transformation with respect to this isometry is given by the so called Buscher transformations [13]:

$$\begin{aligned}
\tilde{G}_{kk} &= \frac{1}{G_{kk}}, \\
\tilde{G}_{k\alpha} &= \frac{B_{k\alpha}}{G_{kk}}, \qquad \tilde{B}_{k\alpha} = \frac{G_{k\alpha}}{G_{kk}}, \\
\tilde{G}_{\alpha\beta} &= G_{\alpha\beta} - \frac{G_{k\alpha}G_{k\beta} - B_{k\alpha}B_{k\beta}}{G_{kk}}, \\
\tilde{B}_{\alpha\beta} &= B_{\alpha\beta} - \frac{G_{k\alpha}B_{k\beta} - G_{k\beta}B_{k\alpha}}{G_{kk}},
\end{aligned} \tag{13}$$

where the letter k stands for the direction of the isometry. Obviously the above transformations are generalisable to more than one isometry. Coming back to our initial example, we note that for the bosonic string $R \to 0$ and $R \to \infty$ are in all aspects equivalents provided we interchange the winding by the momenta. In practice what happen is that for instance in the limit $R = 0$ an effective "extra dimension" appears due to the generation of massless winding modes.

For a while nobody ask, concerning this strange T-duality symmetry for closed strings, the most natural question, namely what is the interplay between T-duality and the already mentioned closed-open string relation?.

Intuitively the problem we face is quite clear. In fact for the open string there is no winding number, so once we go to the limit $R \to 0$ we most expect to end up with a d-dimensional theory (as usual in a quantum field theory), however open strings in interactions will produce closed strings and as we have seen in this case we get a new effective dimension, on the above limit. So what is really happening?

The first answer to this puzzle is D-branes [5]. As we will see in a moment open strings live in a $d+1$-dimensional space time, but with the end points attached to a d-dimensional region that we identify with the D-brane, more over the closed string in the $d+1$-dimensional space time will induce a gravitational life for the D-branes that will appears as a real source of gravity. Let us see all this in more detail. As we mentioned before open strings are characterized by the boundary conditions at the ends points of the string. The T-duality transformation of (11) is geometrically understand as mirror symmetry on one sector of the string mode expansion, this can be better seen by inspectioning the

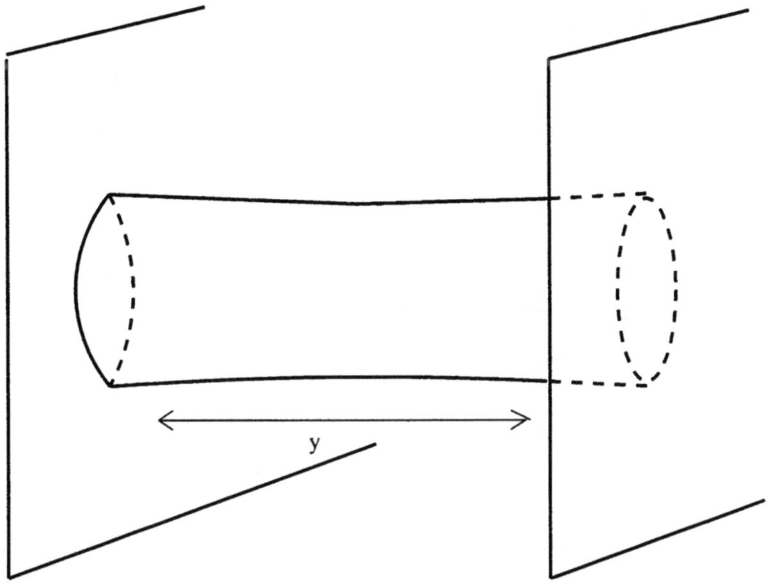

Fig. 5. D-brane interaction

closed string expansion that solve the field equation,

$$X^\mu = x^\mu - i\sqrt{\frac{\alpha'}{2}}(\alpha_0^\mu + \bar{\alpha}_0^\mu)\tau + \sqrt{\frac{\alpha'}{2}}(\alpha_0^\mu - \bar{\alpha}_0^\mu)\sigma$$
$$+ i\sqrt{\frac{\alpha'}{2}}\sum_{m\neq 0}\left(\frac{\alpha_m^\mu}{m}z^{-m} + \frac{\bar{\alpha}_m^\mu}{m}\bar{z}^{-m}\right), \tag{14}$$

where $z = e^{\tau - i\sigma}$ and $\bar{z} = e^{\tau + i\sigma}$. This expansion can be rewritten in terms of homomorphic ad anti-homomorphic functions as,

$$X^\mu = X^\mu(z) + \bar{X}^\mu(\bar{z}), \tag{15}$$

then the T-duality transformation translates into a parity transformation for the anti-holomorphic function on the above expansion,

$$\tilde{X}^{25}(z,\bar{z}) = X^{25}(z) - \bar{X}^{25}(\bar{z}). \tag{16}$$

for the open string solution we have,

$$X^\mu = x^\mu - i\alpha' p^\mu \ln z\bar{z} + i\sqrt{\frac{\alpha'}{2}}\sum_{m\neq 0}\frac{\alpha_m^\mu}{m}(z^{-m} + \bar{z}^{-m}). \tag{17}$$

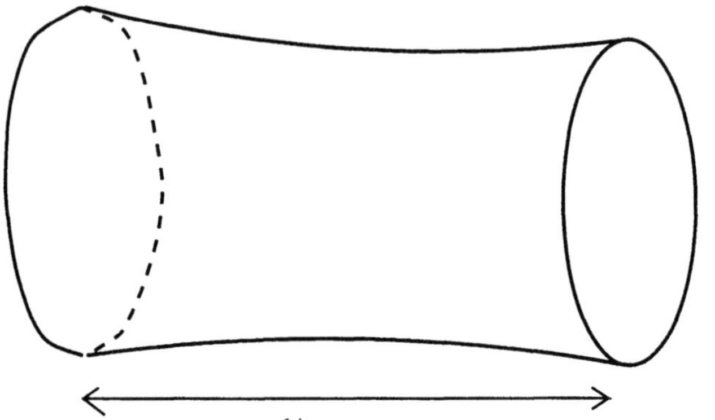

Fig. 6. Modular parameter

Therefore using the same transformation as in the closed string sector we get the expansion,

$$\tilde{X}^{25} = x^{25} - i\alpha' p^{25} \ln \frac{z}{\bar{z}} + i\sqrt{\frac{\alpha'}{2}} \sum_{m \neq 0} \frac{\alpha_m^{25}}{m} (z^{-m} - \bar{z}^{-m}). \tag{18}$$

Therefore the boundary condition $\partial_\sigma X^{25} = 0$ is transformed into $\partial_\tau \tilde{X}^{25} = 0$, that means Dirichlet conditions at the ends of the open string. We can see that T-duality exchange Newman boundary conditions into Dirichlet boundary conditions. This implied that the ends of the open string are constraint to move on a given hypersurface, that we called D(irichlet)-branes.

At this point we can come back to our previous discussion concerning the filed theory limit $\alpha' \to 0$. In fact if in this limit closed strings are still surviving them we should expect that T-duality in the field theory limit $\alpha' \to 0$ of an open string will present somehow the phenomena of quantum generation of extra dimension. We will come back to this issue in chapter (8).

As a warming up exercise in D-brane dynamics let us consider the D-brane amplitude represented in (fig.5).

Up to numerical actors that will not be relevant for us at this point the amplitude is given by:

$$A(y) = (\alpha')^{-d/2} \int \frac{dt}{t} t^{-d/2} \eta(it)^{(2-D)} e^{y^2 t/\alpha'}, \tag{19}$$

where η is the Dedekind function, y the space-time distance between the Dp-brane, $d = p + 1$ and t the modular parameter of the cylinder (see fig.6).

We can adopt two different points of view to interpret the amplitude in (fig.5). From the open string point of view we have the open string stretched between the Dp-brane and a "time" evolution along the loop with the value of

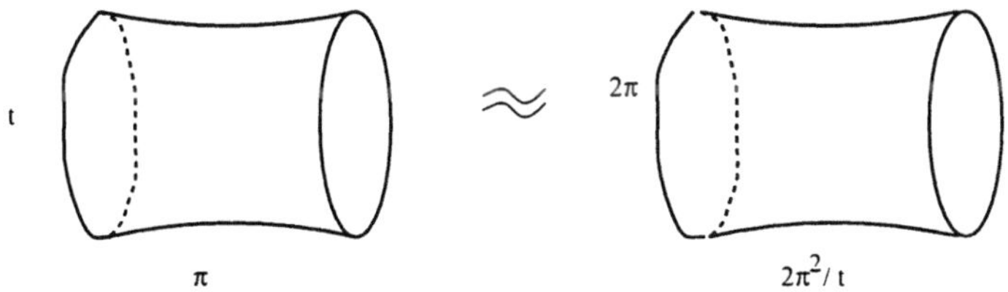

Fig. 7. Conformal mapping

the time equal to t. From the closed string point of view we have the emission-absorption of closed string states with the "time" of the process given by $1/t$. Both pictures are related by the conformal mapping described in (fig.7). We will be first interested in computing (19) in the limit $t \to 0$ that is the regime (see fig.6) dominated by the contribution of light closed string states.

Using the transformation properties of the Dedeking function:

$$\eta(-\frac{1}{it}) = t^{-d/2}\eta(it) \tag{20}$$

and the expansion of η for $-1/it \to \infty$ we get:

$$A(y) = \left[(\alpha')^{-d/2} \int \frac{dt}{t} t^{-d/2} t^{(2-D)/2} e^{y^2 t/\alpha'}\right] (D-2), \tag{21}$$

where D is the dimension of the target space time and where we have avoided the tachyon contribution in the expansion of the expansion of η-function [1]

In critical dimension $D = 26$ we get from the well known expression

$$A(y) \approx \Gamma\left(\frac{23-p}{2}\right) |y|^{p-23} (\alpha')^{11-p}. \tag{23}$$

We can compare equation (23) with the effective Lagrangian computation. In fact (23) is a long distance contribution where we have keep only massless dilatons and gravitons, therefore we should compare (23) with the tree level Feynman diagrams for the effective Lagrangian in the Einstein frame and a D-brane coupled i.e. the Feynman diagram in (fig.8). The coupling in the vertices in (fig.8) will depend on the gravitational constant κ and on the Dp-brane tension τ_p that we want to discover by identifying the amplitude in (fig.8) and the amplitude in (fig.4). In order to do this we need the Lagrangian describing the gravitational interaction of a Dp-brane with the target space-time metric, the simplest ansatz is the p-dimensional generalization of the Nambu-Goto action

[1] Recalled that:

$$\eta(is)|_{is\to\infty} = (e^{1/24}(1 + e^{-25} + \ldots \tag{22}$$

with the factor 1 representing the tachyon contribution.

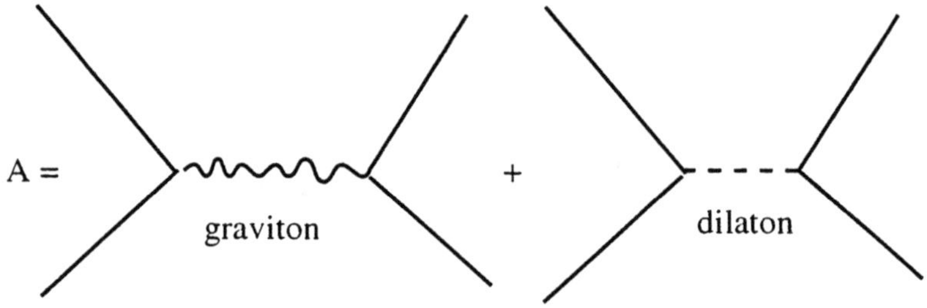

Fig. 8. Feynman diagrams

$$S = \tau_p \int d\xi^{p+1} \sqrt{-detG^e}. \tag{24}$$

Using this amplitude and identifying the string amplitude (fig.4) with the gravitational field theory amplitude (fig.8) we get the Dp-brane tension formulae

$$\tau_p = \frac{\pi(4\pi\alpha')^{11-p}}{256\kappa^2}. \tag{25}$$

In principle nothing prevent us from doing the computation of the amplitude in th limit $t \to 0$. Using again the amplitude for the Dedekind function what we get in this case is

$$A(y) \simeq (\alpha')^{-d/2}(D-2) \int \frac{dt}{t} t^{-d/2} e^{y^2 t/\alpha'}. \tag{26}$$

After performing the integration we get,

$$A(y) \simeq (D-2) \frac{1}{(\alpha')^d} \Gamma(-d/2) |y|^d. \tag{27}$$

We should notice a few things concerning (27). First of all the only dependence on the target space-time dimension D is in the irrelevant front factor $(D-2)$. This indicates that (27) reflects only the dynamics on the world volume of the Dp-brane. Secondly contrary to the case (23) the amplitude (27) is singular.

Concerning the amplitude (27) we can take the near D-brane field theory limit

$$y \to 0 \,, \ \alpha' \to 0 \,, \ u \equiv \frac{y}{\alpha'}. \tag{28}$$

In this case we get:

$$A(u) \simeq (D-2)\Gamma(-d/2) |u|^d. \tag{29}$$

The limit $\alpha' \to 0$ of (23) can be nicely taken for the special case $p = 11$ that correspond to the half dimensional brane in 26-dimensions. Moreover for $p = 11$ it is easy to see that the dilaton exchange in the (fig.8) vanishes.

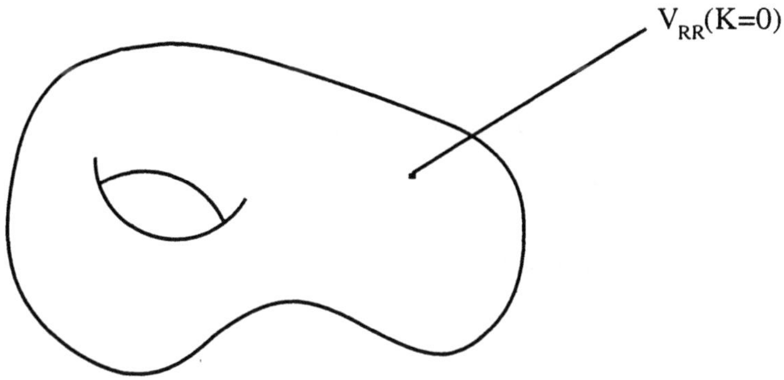

Fig. 9. RR vertex operator on strigs

3 R.R Charged D-Branes

Perhaps the most interesting dynamics of the Dp-branes appears in the case of superstrings (for a good review of superstrings see [1]).

The Hilbert space of superstring contains different sectors depending on the world sheet fermion boundary conditions. For periodic boundary conditions, both in the left and right components we have the so called $R \otimes R$ sector. For antiperiodic boundary conditions we have the $NS \otimes NS$ sector. The two sectors correspond to space-time bosons. In the $NS \otimes NS$ sector we have the standard gravity multiplet containing the dilaton graviton and antisymmetric tensor. In the $R \otimes R$ sector we have also space-time bosons but this time corresponding to the factors appearing in the decomposition of the product of two "Ramond"-vacua,

$$\begin{aligned} &\text{type A} \;\; 8x\bar{8} = [0] + [2] + [4], \\ &\text{type A} \;\; 8x8 =, [1] + [3] + [5] \end{aligned} \tag{30}$$

with 8 and $\bar{8}$ representing the different chiralities. Space-time fermions and gravitons are in the $NS \otimes R$ and $R \otimes NS$ sectors.

One loop modular invariance determines the different GSO projections. At this point we get four different types of superstring theories. For type (A) we get type II with two gravities and no tachyons, also type 0 where we have no gravitons and closed string tachyons. The same for the type B strings.

What is the string meaning of the $R \otimes R$ forms appearing in (30)?. This is a difficult question from the world sheet point of view. The reason for that comes from the fact that strings are not sources of the $R \otimes R$ fields. The simplest way to see this is to consider a string diagram as the one represented in (fig.9), where we have a $R \otimes R$ vertex operator inserted in a string amplitude at zero momentum. From the form of the vertex operator we easily see that the amplitude contains

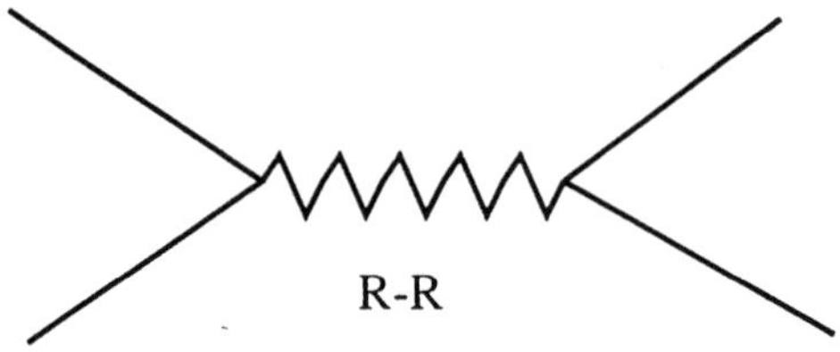

Fig. 10. RR vertex operator on strigs

a factor of the form

$$\mathbf{k}F_{\mu_1 \dots \mu_n}, \tag{31}$$

for F the $R \otimes R$ stress tension and k the momentum, hence vanishing for $k = 0$.

If strings are not sources of $R \otimes R$ fields is not clear what can be the stringy meaning of defining backgrounds with non vanishing vacuum expectation values for $R \otimes R$ tensors, moreover it is not clear at all what can act as a source of those fields in string theory. The answer to this puzzle comes again from the Dp-branes. In fact Dp-branes are natural sources of $R \otimes R$ charges, the amplitude in (fig.11) where we represent the amplitude for the Dp-brane emission of a $R \otimes R$ quanta is now non vanishing mainly due to the fact that the world sheet entering into the game is a disc, due to the Dirichlet boundary condition on the Dp-brane world volume. In order to prove that the amplitude in (fig.11) is now non vanishing we need to invoke a "picture changing" manipulation [6] to cancel the k in (31).

Once we know that Dp-branes are -or superstrings- sources of $R \otimes R$ fields we can compute the type of interaction between parallel Dp-branes mediated by the interchange of $R \otimes R$ quanta. This amplitude will be exactly the type of amplitude depicted in (fig.4), but this time we will consider the contribution to the cylinder of $R \otimes R$ states in the spectrum. If we choose a GSO projection implying space-time supersymmetry we will get exactly the same amplitude with the reverse sing and with $D = 10$, the critical dimension for superstrings,

$$A(y) \sim -(\alpha')^{3-p} \Gamma \left(\frac{7-p}{2} \right) |y|^{p-7}. \tag{32}$$

Now we would like to compare this amplitude with the corresponding Feynman diagram in (fig.10). In order to do that we need again to defined a quantum field theory coupling between the Dp-brane and the $R \otimes R$ $(p+1)$-form

$$\mu_p \int d\xi^{p+1} A. \tag{33}$$

with the kinetic term for the $R \otimes R$ stress tension $F = dA$ given by

$$\int d^{10}xF. \tag{34}$$

By comparing the amplitude in (fig.11) with space-time (32) we get the value of the $R \otimes R$ charge density μ_p for a Dp-brane:

$$\mu_p = e^{\Phi} \tau_p \ , \quad \tau_p = \frac{\pi (4\pi\alpha')^{3-p}}{\kappa^2} \tag{35}$$

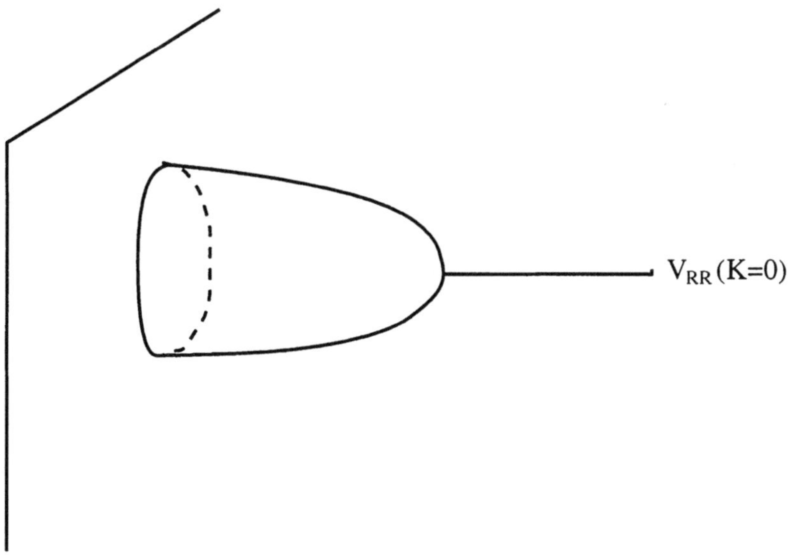

Fig. 11. RR vertex operator and D-branes

Probing the BPS nature of the Dp-brane. The above forms for the interaction of the brane with the $R \otimes R$ field can be deduce from the equation of motion that the open string on the D-brane background must satisfy [7]. The resulting action is given by,

$$S = T_p \int d^{p+1}\xi \ e^{-\Phi} \sqrt{-\det(g_{\alpha\beta} + b_{\alpha\beta} + 2\pi\alpha' f_{\alpha\beta})}, \tag{36}$$

where g, b and Φ are the pull backs of the 10D metric, antisymmetric tensor and dilaton to the D-brane world volume, while f is the field strength of the world volume $U(1)$ gauge field A_α and $T_p = \tau e^{<\Phi>}$.

For the supersymmetric string theory we must extend the action to a supersymmetric Born Infield type action, that includes Chern-Simons type terms that couple the Dp-brane to the $R \otimes R$ fields. Of course this last part of the action, in the leading term correspond to the coupling of equation (33).

To simplify the above action we can consider the background space-time to be flat, the Dp-brane almost flat and in the static gauge, hence giving the expansion

$$g_{\alpha\beta} \approx \eta_{\alpha\beta} + \partial_\alpha X^a \partial_\beta X^a + \mathcal{O}\left((\partial X)^4\right). \tag{37}$$

On the top of this we can consider vanishing antisymmetric field b and that $2\pi\alpha' F_{\alpha\beta}$ and $\partial_\alpha X^a$ are small and of the same order. The resulting low energy action is:

$$S = \tau_p \int d\xi^{p+1}\sqrt{-det\hat{G}} +$$
$$\frac{1}{4g_{\mathrm{YM}}^2}\int d^{p+1}\xi\left(F_{\alpha\beta}F^{\alpha\beta} + \frac{2}{(2\pi\alpha')^2}\partial_\alpha X^a \partial^\alpha X^a\right), \tag{38}$$

where the Yang-Mills coupling is given by,

$$g_{YM}^2 = \frac{1}{4\pi^2\alpha'^2\tau_p} = \frac{g}{\sqrt{\alpha'}}(2\pi\sqrt{\alpha'})^{p-2}. \tag{39}$$

The second term correspond to the dimensional reduction of the ten dimension SYM action. Once the fermions are include, the resulting action is the dimensional reduction of supersymmetric $N = 1$, $U(1)$ Yang-Mills theory in ten dimension.

$$S = \frac{1}{g_{YM}^2}\int d^{10}\xi\left(-\frac{1}{4}F_{\mu\nu}F^{\mu\nu} + \frac{i}{2}\bar{\psi}\Gamma^\mu\partial_\mu\psi\right). \tag{40}$$

4 Microscopic Dp-Brane String Amplitudes and Metrics

Before going into the supergravity effective Lagrangian associated with superstring, let us motivate from the previous discussion on amplitudes the form of the metric generated by a Dp-brane in the limit $t \to 0$ i.e.

$$A(y) \sim (\alpha')^{3-p}\Gamma\left(\frac{7-p}{2}\right)|y|^{p-7}. \tag{41}$$

In the post-Newtonian approximation we can think of (41) as a gravitational correction to the flat metric. Moreover we can invoke the T-duality relation between different Dp-branes to look for an ansatz for the metric of the type

$$ds^2 = \left(1 - \frac{g(\alpha')^{7-p/2}\Gamma\left(\frac{7-p}{2}\right)}{y^{7-p}}\right)dx_{||}^2 + \left(1 + \frac{g(\alpha')^{7-p/2}\Gamma\left(\frac{7-p}{2}\right)}{y^{7-p}}\right)dx_{\perp}^2. \tag{42}$$

In the next section we will see that (42) is at first order in the string coupling constant g a solution to the supergravity equations.

5 Supergravity as an Effective Theory

In this section we are going to talk about the supergravity description of Dp-branes. This kind of description involves the effective theory describing the low energy massless states of the Type II superstring theories, where these massless states are the graviton, the dilaton the NS two-form and the corresponding $R \otimes R$ forms. An important point to consider about this effective theories, is that their specific form is completely determined by requiring supersymmetry on the massless spectrum just mentioned. Therefore if the superstring theory has a given symmetry, the effective theory should also be invariant under this symmetry (supersymmetry acts as a protecting shield to quantum corrections). Actually the action itself doesn't have to be invariant under this symmetry, is good enough to have invariant field equations. Recall that what we really obtain from the beta function in the non-linear supersymmetric σ-model is the field equation for some of the massless modes. The other important point that we should keep in mind is that because the Dp-brane are BPS states, some of their characteristics are also protected from quantum effects, for example the mass, charge and even the degeneracy of their spectrum, as we move on the moduli space of the full superstring theory. Therefore the low energy description of Dp-branes and BPS objects in general is very important. In fact these solutions are one of the most powerful ways to study the new dualities within the different string theories.

The effective theories (for a good review see [9]) obtained from the type II superstrings are the type IIA and type IIB supergravity. Type IIA has for spectrum the graviyton G, dilaton Φ, the NS two-form B, and in the Ramod sector the one-form potential $A_{[1]}$ and the three-form potential $A_{[3]}$. The action in the string frame is given by

$$\begin{aligned} S_{IIA} &= \tfrac{1}{2\kappa^2}\int d^{10}x \Big\{ \sqrt{-G}\, e^{-2\Phi}\left[R + 4|d\Phi|^2 - \tfrac{1}{12}|H|^2\right] \\ &\quad -\sqrt{-G}\left[\tfrac{1}{4}|F_{[2]}|^2 + \tfrac{1}{48}|F_{[4]}|^2\right]\Big\} + \tfrac{1}{4\kappa^2}\int F_{[4]} \wedge F_{[4]} \wedge B\,, \end{aligned} \tag{43}$$

where $F_{[4]} = dA_{[3]} + 12B \wedge F_{[2]}$ is the non-linear version of the 4-form field strength.

The type IIB spectrum is given by the graviton G, dilaton Φ, the NS two-form B and in the Ramond sector we get a pseudoscalar potential $A_{[0]}$, a two-form potential $A_{[2]}$ and a four-form potential $A_{[4]}$ whose field strength is self-dual. The action in the string frame is given by

$$\begin{aligned} S_{IIB} = \tfrac{1}{2\kappa^2}\int d^{10}x\sqrt{-G}\Big\{ & e^{-2\Phi}\left[R + 4|d\Phi|^2 - \tfrac{1}{12}|H|^2\right] - 2|d\ell|^2 \\ & -\tfrac{1}{3}|F_{[3]} - \ell H|^2 - \tfrac{1}{60}|F_{[5]}|^2\Big\} - \tfrac{1}{4\kappa^2}\int A_{[4]} \wedge H \wedge F_{[3]}, \end{aligned} \tag{44}$$

where we have ignore the self-duality condition of $F_{[5]}$ to write this action[2]. The full non-linear Bianchi identity satisfied by $F_{[5]}$ is now $dF_{[5]} = H \wedge F_{[3]}$. By combining this 'modified' Bianchi identity with the self-duality condition on $F_{[5]}$ we deduce that $d \star F_{[5]} = H \wedge F_{[3]}$, which is just the $A_{[4]}$ field equation. Thus, the modification of the Bianchi identity is needed for consistency with the self-duality condition.

To study the Dp-brane solutions of the above actions it is convenient to consider truncations on the Lagrangian that simplify and clarify the task of finding out the right ansatz. Basically we can always set to zero all but one of the field strength, living us we the following type of action (written in the Einstein frame for simplification)

$$I = \int dx^{10}\sqrt{-G}\Big[R - \tfrac{1}{2}|d\Phi|^2 - \tfrac{1}{2(d+1)!}e^{a\Phi}|F|^2_{[d+1]}\Big]. \tag{45}$$

where $a^2 = (d - 4/2)^2$.

With this Lagrangian we are ready to look for solutions that are sensible to be interpreted as Dp-branes. The ansatz we have to consider in the most simplified case is:

- Asymptotic flat space time.
- Broken Lorentz invariance from $SO(1,9)$ to $SO(1,p) \otimes SO(9-p)$.
- A bosonic supersymmetric solution.

This conditions can be understand as follows: The first item allows us to define the meaningful quantities that characterize a BPS state like the mass, charge, and supersimetric killing spinors. The second item assumes we are considering the so called static gauge on the coordinates used i.e. the world-volume coordinates of the corresponding Dp-brane are exactly the same as the first $d = p + 1$ space-time coordinates x^μ. Also implies that none of the configuration depends on the above coordinates and that, in the simplest assumption all the possible non-trivial dependence of the involved fields goes in term of a radial coordinate defined on the perpendicular space to the brane, with coordinates y^m. The resulting form of the ansatz is given by

$$ds^2 = e^{2A(r)}dx^\mu dx^\nu \eta_{\mu\nu} + e^{2B(r)}dy^m dy^n \delta_{mn} \ , \tag{46}$$

where $r = \sqrt{y^m y^m}$. For the dilaton we have only radial dependence $\Phi = \Phi(r)$ and for the RR field we have two possibilities, one related to the electric solution and the other to the magnetic solution, the form of the ansatz for the electric case is

$$F_{d+1} = dA_d \ ; \quad A_{\mu_1 \dots\dots \mu_d} = \epsilon_{\mu_1 \dots\dots \mu_d} e^{C(r)}. \tag{47}$$

[2] The self-duality condition for the $F_{[5]}$ makes very difficult to write an action for type IIB supergravity. In any case we can always introduce this condition as a constraint at the level of field equations

For the magnetic ansatz we can only give an expression in terms a o the strength filed, as there is no global definition for the associated potential,

$$F^{(\mathrm{mag})}_{m_1\cdots m_n} = g_{\bar{d}}\epsilon_{m_1\cdots m_n p}\frac{y^p}{r^{n+1}}\,, \qquad \text{others zero.} \tag{48}$$

By now the only condition we haven't used is the supersymmetry character of the solutions. In forthcoming sections we will be dealing more carefully with the supersymmetric equations, but for our actual proposes, it is sufficient to know that the constraints imposed by supersymmetry restrict the number of independent functions from four (A, B, Φ, C) to one, let say C. Therefore using the field equations obtained varying the action (45), we get the electrically charged Dp-brane solution (in the Einstein frame),

$$\begin{aligned} ds^2 &= H^{\frac{-\bar{d}}{d+\bar{d}}} dx^\mu dx_\mu + H^{\frac{d}{d+\bar{d}}} dy^m dy_m \\ e^{\Phi} &= H^{\frac{a}{2}} \\ H(r) &= \begin{cases} 1 + \frac{K_d}{r^{\bar{d}}} & \text{if } \bar{d} > 0 \\ C_o + K_o ln(r) & \text{if } \bar{d} = 0 \end{cases} \end{aligned} \tag{49}$$

where $a = -\sqrt{4 - \frac{2d\bar{d}}{d+\bar{d}}}$.

Note that the constant integration K_d, K_o, C_o and the parameter $g_{\bar{d}}$ are not fixed by this ansatz. Also we have introduce a new constant $\bar{d}$, which satisfies the equation $d + \bar{d} = 8$, being the worldvolume dimension of the magnetic solution associated to a strength field $F_{[d+1]}$. To get the solitonic solutions, one replaces d by $\bar{d}$ and set $a(\bar{d}) = -a(d)$. This will give a $\bar{d}$-brane magnetically charged.

We can define the mass, electric charge and magnetic charge of the above solutions, by the expressions,

$$\begin{aligned} m &= \frac{1}{2\kappa}\int_{S^{\bar{d}+1}} d^{\bar{d}}\Sigma^m(\partial^n h_{mn} - \partial_m h^n_n), \\ e_d &= \frac{1}{\sqrt{2}\kappa}\int_{S^{\bar{d}+1}} e^{-a(d)\phi}F^*, \\ g_{\bar{d}} &= \frac{1}{\sqrt{2}\kappa}\int_{S^{d+1}} F, \end{aligned} \tag{50}$$

where we expand the metric G as $G = \eta + h$. When we solve the mass, and charges for the above metrics we find that this solutions have the characteristic relation between the mass and charge. Also the electrically charge solution and the magnetically charge solution satisfy the Dirac quantization condition

$$e_d g_{\bar{d}} = 2\pi n, \tag{51}$$

with n an integer.

A very peculiar fact about the electrically charge Dp-branes is that they solve the supergravity field equations with sources at the origin, where the source is the action of the elementary Dp-brane (24) plus the coupling with the RR potential.

This fact allow us to define the value of $K_, e_d, g_{\bar{d}}$ in terms of the string coupling constant and the string length α', giving $K_d = 2^{6-d}\pi^{(10-d)/2}\Gamma(6-d/2)gl_p^{d-4}$. For future discussions it is important to rewrite the Dp-brane metrics in the string frame, here we show the result:

For the elementary case, we have

$$\begin{aligned} ds^2 &= H^{-1/2}dx^\mu dx_\mu + H^{1/2}dy^m dy_m \\ e^\Phi &= H^{1-d/4} \\ H(r) &= \begin{cases} 1+\frac{K_d}{r^{\bar{d}}} & \text{if } \bar{d}>0 \\ C_o + K_o ln(r) & \text{if } \bar{d}=0 \end{cases} \\ F_{\mu_1 \dots\dots \mu_d m} &= \epsilon_{\mu_1 \dots\dots \mu_d}\partial_m H^{-1} \end{aligned} \tag{52}$$

where $\mu = 0, ..., d-1, \quad m = d, ..., 9$.

For the the solitonic case,

$$\begin{aligned} ds^2 &= H^{-1/2}dx^\mu dx_\mu + H^{1/2}dy^m dy_m \\ e^\Phi &= H^{d/4-1} \\ H(r) &= \begin{cases} 1+\frac{K_{\bar{d}}}{r^{d}} & \text{if } d>0 \\ C_o + K_o ln(r) & \text{if } d=0 \end{cases} \\ F_{m_1 \dots\dots m_{d+1}} &= g_{\bar{d}}\epsilon_{m_1 \dots\dots m_{d+1} l}\frac{y^l}{r^{d+2}}. \end{aligned} \tag{53}$$

where $\mu = 0, ...\bar{d}-1, \quad m = \bar{d}, ..., 9$.

The Lagrangian of equation (45), as we just showed above, contains the class of solutions to its field equations, relevant for the Dp-brane studies. This type of Lagrangian is not restricted to the ten dimensional bosonic sector of a truncated type II supergravity, in fact we found the same structure in the bosonic sector of eleventh dimensional $N = 1$ supergravity theory. Where the bosonic part of the Lagrangian is given by

$$S = \frac{1}{2\kappa^2}\int d^{11}x\sqrt{-\hat{G}}[R - \frac{1}{48}|\hat{F}|^2] + \frac{1}{12\kappa^2}\int \hat{F}\wedge\hat{F}\wedge\hat{A}. \tag{54}$$

This time the field potential is a three-form $\hat{A}_{[3]}$, and we put hats on each field to difierenciate from ten dimensional variables. Therefore the associated p-branes solutions are the electrically charged M3-brane and the magnetic M5-brane. Their names comes from the idea that $D = 11$, $N = 1$ supergravity is the low energy effective theory of the famous M-theory. In fact $D = 11$, $N = 1$ supergravity compactified on a small circle gives type IIA supergravity in ten dimensions, plus the relevant RR field strength, coming as Kalusa Klein modes on the dimensional reduction. This is part of the conjectured relation of type IIA superstring theory and M-theory. In any case the M-branes solutions of $D = 11$, $N = 1$ supergravity are given below. For the electric M3-brane we have,

$$\begin{aligned} ds^2 &= \left(1+\frac{K_3}{r^6}\right)^{-2/3} dx^\mu dx_\mu + \left(1+\frac{K_3}{r^6}\right)^{-1/3} dy^m dy_m, \\ A_{\mu\nu\rho} &= \epsilon_{\mu\nu\rho}\left(1+\frac{K_3}{r^6}\right)^{-1}, \end{aligned} \tag{55}$$

where $\mu = 0, ..., 2,\ \ m = 3, ..., 10$. For the magnetic M5-brane we get

$$\begin{aligned} ds^3 &= \left(1+\frac{K_6}{r^3}\right)^{-1/3} dx^\mu dx_\mu + \left(1+\frac{K_6}{r^3}\right)^{2/3} dy^m dy_m, \\ F_{mnop} &= 3K_6 \epsilon_{mnopq}\frac{y^q}{r^5}, \end{aligned} \tag{56}$$

where $\mu = 0, ..., 5,\ \ m = 6, ..., 10$.

As in the case of the Dp-branes the constant of integration can be related to the relevant structure constant of M-theory, obtaining that $K_6 = \pi l_p^3$ and $K_3 = 2^2\pi^2 l_p^6$.

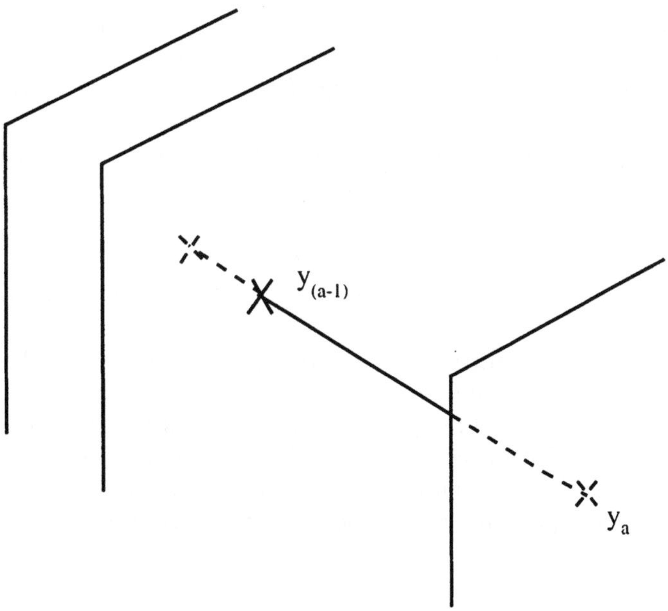

Fig. 12. Multicentre D-branes

Before finishing the discussion of Dp-branes as solutions of the effective low energy theories of superstrings theories, we should recall that there is a very simple generalization to the above solutions which will be used extensible on future discussions. Basically we can consider what is called a multiple centre branes solution i.e. a solutions representing a given number of branes, say N in a parallel configuration. Because the branes are all oriented in the same way, the supersymmetry conserved is the same as in the previous cases. If we prepare the solution such that each brane is located at $\mathbf{y}_a$ (see fig. 12) then the only difference in the brane solution comes in the form of the harmonic function H

$$H = 1 + \sum_{a=1}^{N} \frac{K_{(d,a)}}{|\mathbf{y} - \mathbf{y}_a|^{\tilde{d}}}. \tag{57}$$

The form of the electric field ansatz is unchanged , but the magnetic field needs the following change,

$$F_{m_1 \ldots\ldots m_{d+1}} = \frac{-1}{\tilde{d}} \epsilon_{m_1 \ldots\ldots m_{d+1}} \partial_p \sum_{a=1}^{N} \frac{g_{(\tilde{d},a)}}{|\mathbf{y} - \mathbf{y}_a|^{\tilde{d}}}. \tag{58}$$

This solutions are relevant when we consider Dp-brane configurations with open string sectors including Chan-Paton factors. In principle this factors produces a family of Dp-branes, rather than a single brane. In particular strings stretching between different branes correspond to massive states of the gauge theory defined on the world volume of the Dp-branes. If we start with N different Dp-branes all at the same place, we have a $SU(N)$ ($U(N)$) gauge theory living on the world volume of the brane. By pulling out one of these branes, we break the group $SU(N)$ to $SU(N) \otimes U(1)$, and the mass of the associated W-vector boson is given in terms of the distance between the bunch of $N-1$ Dp-branes and the pulled out Dp-brane, r and the squared string length

$$|mass| = \frac{r}{\alpha'}. \tag{59}$$

Therefore we have a very nice geometrical picture of symmetry breaking on gauge theories.

As a last remark we introduce the notion of T-duality for the RR-fields. From the point of view of the string theory the T-duality transformation is (as we said on section 2) a hybrid parity operation, this parity restricted to the anti-holomorphic worlsheet sector is realized in the spinor space as the operator $-i\Gamma^9\Gamma_{11}$ (where x^9 is the compact direction). As the RR-fields are bispinors defined by the corresponding vertex operator, its transformation is defided as,

$$F = -iF\Gamma^9\Gamma_{11}. \tag{60}$$

therefore we get the final relation in terms of the index of F as,

$$\begin{aligned} F_{\mu_1 \ldots \mu_n} &= -F_{9\mu_1 \ldots \mu_n} \\ F_{9\mu_1 \ldots \mu_n} &= -F_{\mu_1 \ldots \mu_n} \end{aligned} \tag{61}$$

The general form of the transformation in the case of non trivial background metric NSNS-antisimmetric field and dilaton can be found on [10].

6 Near Horizon Geometry

Given a Dp-brane metric

$$ds^2 = H^{-1/2} dx^\mu dx_\mu + H^{1/2} dy^m dy_m \tag{62}$$

with

$$H = 1 + \frac{(4\pi)^{(5-p)/2} (\alpha')^{(7-p)/2} g N \Gamma((7-p)/2)}{r^{7-p}} \tag{63}$$

where $r = \sqrt{y^m y_m}$, the near horizon geometry is defined by the double limit [12]

$$\begin{aligned} \alpha' \to 0 \quad &, \quad r \to 0 \\ \frac{r}{\alpha'} &= U \end{aligned} \tag{64}$$

with U arbitrary. Using (64) we can rewrite the harmonic funtion H as

$$H = 1 + \frac{(4/pi)^{(5-p)/2} g N \Gamma((7-p)/2)}{U^{7-p} (\alpha')^{(7-p)/2}} \approx \frac{(4/pi)^{(5-p)/2} g N \Gamma((7-p)/2)}{U^{7-p(\alpha')^{(7-p)/2}}} \tag{65}$$

Geometrically we can think of (63) as a function defined on the (α', r) plane. In these conditions (64) defines a blow up of the point $(0,0)$ generating a divisor with coordinate U. Using (65) we get for the D3-brane the following near horizon geometry,

$$\begin{aligned} ds_3^2 &= \alpha' \left[\frac{U^2}{(gN4\pi)^{1/2}} dx_{||}^2 + \frac{(gN4\pi)^{1/2}}{U^2} dU^2 + (gN4\pi)^{1/2} d\Omega_5) \right] \\ e^\phi &= g. \end{aligned} \tag{66}$$

which is the metric of $AdS_5 \otimes S^5$.

For a generic Dp-brane we can write the metric in a similar form provided we use instead of g -the string coupling constant- the Yang Mills coupling constant on the Dp-brane which is given by (39).

$$g_{YM}^2 = \frac{1}{4\pi^2 \alpha'^2 \tau_p} = \frac{g}{\sqrt{\alpha'}} (2\pi\sqrt{\alpha'})^{p-2}. \tag{67}$$

Note that the Yang-Mills coupling constant is also defined as the result of a limiting procedure where g is sent to infinite or zero, depending on the value of p. Using the above equation, the near horizon form of the harmonic function H becomes

$$H = \frac{d_p g_{YM}^2 N}{U^{7-p} (\alpha')^{-2}} \tag{68}$$

where $d_p = 2^{7-2p} \pi^{(9-3p)/2} \Gamma((7-p)/2)$. Therefore we get the following near horizon metric,

$$\begin{aligned} ds^2 &= \alpha' \left[\frac{U^{(7-p)/2}}{g_{YM} (N d_p)^{1/2}} dx_{||}^2 + \frac{g_{YM} (N d_p)^{1/2}}{U^{(7-p)/2}} dU^2 + g_{YM} (N d_p)^{1/2} U^{p-3} d\Omega_5) \right] \\ e^\phi &= 2\pi^{2-p} g_{YM}^2 \left(\frac{g_{YM}^2 N d_p}{U^{7-p}} \right)^{(3-p)/4} . \end{aligned} \tag{69}$$

How should we interpret the near horizon geometries? (To find a more complete list of reference and a deeper introduction see [8]). Since we are performing the $\alpha' \to 0$ limit we can think of these geometries as related somehow to the pure Yang-Mills theory living on the Dp-brane. We can try to establish this relation from 3 different points of view.

1. **Gauge Singlets**: The main idea underlying this approach will consist in looking for an isomorphism between gauge singlets of the gauge theory on the Dp-brane and the spectrum of supergravity defined on the near horizon geometry. Since supergravity is not a complete theory we should look for its string ancestor on the near horizon geometry background. Supergravity approximation will be good if the curvature of the near horizon geometry is large in string units, which in general will implies to take a large N number of branes.
   Notice that the standard open-closed relation in string theory always allows to think of the closed gravitational sector in terms of gauge singlets of the open string theory. The problem is to find a limit where the closed string spectrum precisely describes the physics of the gauge singlets (glue-balls) of the field theory limit $\alpha' \to 0$ of the open string sector.
2. **Renormalization Group**: In this approach we start by identifying the blow up variable U with the renormalization group scale of the gauge theory. The direct connection between string theory on the near horizon geometry and the pure Yang-Mills theory on the Dp-brane in the $\alpha' \to 0$ limit, will be identifying the renormalization group equation of the field theory with the dilaton behaviour dictated by the Dp-brane near horizon geometry through the string beta function equations.
3. **Matching of Symmetries**: A more concrete procedure to find a relation between the near horizon geometry and the pure Yang-Mills theory is by matching the symmetries of the Yang-Mills theory with the isometries of the near horizon geometry. the relevant data for the Yang-Mills theory are their supersymmetries and global R-symmetries. From the supergravity point of view we should consider the corresponding superalgabra.

Let us consider the simplest case of the D3-brane. The corresponding super-Yang-Mills theory is $D = 4, N = 4$. This is a theory with 16 supersymmetries and conformal invariance that is described by the superconformal $N = 4$ superalgebra. As we will see in the next section, the matching with the near horizon symmetries comes from the fact that in $AdS_5 \otimes S^5$ we get an enhancement of supersymmetry.

From the point of view of the renormalization group the matching is in this case trivial since on one side we have $\beta = 0$ and on the other a constant dilaton.

The most difficult aspect is of course the correspondence between gauge singlest and supergravity fields. Here the geometry of AdS space time is specially important. In fact we will look for a correspondence between supergravity fields ψ_i and gauge singlets observable θ_i of super-Yang-Mills theory is $D = 4, N = 4$

in such a way that ψ_i at the boundary of AdS_5 can act as a source for the operators θ_i through an interaction term of the type,

$$\int dx^4\ \theta_i(x)\psi_i(x) \tag{70}$$

with $\psi_i(x, U)$ the boundary value of $\psi_i(x, U)$.

The operator θ_i can be characterized by their conformal weight Δ_i hence we need ψ_i to have dimensions of the type $[lenght]^{\Delta_i - 4}$. This condition fixes the mass of the field ψ_i to be determined by the following relation,

$$\Delta_i = 2 + \sqrt{1 + R^2 m^2} \tag{71}$$

for R the AdS_5 radius. Once we have this correspondence between AdS_5 fields ψ_i and the observable θ_i of super-Yang-Mills theory is $D = 4, N = 4$ the recipe for computation of the amplitudes is given by,

$$\langle e^{\int dx^4\ \theta_i(x)\hat{\psi}_i(x)} \rangle_{SSYM} = e^{S_{sugra}(\psi_i(x,U))} \tag{72}$$

with $\psi_i(x, U)\ |_{U=0} = \hat{\psi}_i$. Relation (72) is of course valid as long as the AdS_5 radius is large enough. It is conjectured that string corrections to the r.h.s. of equation (72) still reproduce SSYM dynamics, but for the time being this conjecture has not been proved.

In Summary the correspondence between $AdS_5 \otimes S^5$ supergravity and super-Yang-Mills theory is $D = 4, N = 4$ is based on two facts: i) the isomorphism of superalgebras for $N = 4$ SSYM and $N = 8$ AdS_5 supergravity, isomorphism that actually depends on the enhancement of supersymmetry in AdS_5 and ii) the special structure of the conformal infinity of AdS_5, which is isomorphic to four dimensional Minkowski space time. How to extend this picture to non-conformal cases is an interesting open problem. Most likely the starting point in trying to solve this problem will consist in a systematic string reinterpretation of quantum field theory renormalization group equations.

7 Supersymmetry

M-branes and Dp-branes are known to be BPS states. That is to say that saturate a Bogoumoly inequality relating their mass and charge. In turns this imply that the supersymmetry preserved by this objects is a fraction of the maximal supersymmetry appearing on the theory. In this case we have only 16 real supercharges or 1/2 of the maximal number of real supercharges that is 32 for $D = 11$, $N = 1$ supergravity and $D = 10$, $N = 2$ supergravity. A nice way to see why we have this relation between BPS states and fractional number of the maximal supersymmetry conserved, is to look on the superalgebra of the above theories at infinity (where we can defined mass and charge, associated with the Poincare group). The general form of this superalgebra is, in the presence of branes given skematically by

$$\{Q, Q\} = \Gamma^M P_m + \Gamma^{M_0 \dots M_p} Z_{M_0 \dots M_p}, \tag{73}$$

where P is the momentum and Z is form defining the charge carried by the brane. The BPS status of the brane tell us that the mass and the charge are equal, therefore in static configurations we get

$$\{Q,Q\} = m(1 \pm \Gamma^{0\dots p}), \tag{74}$$

clearly the eigenspinors of $\{Q,Q\}$ satisfy

$$\Gamma^{0\dots p}\epsilon_0^{\pm} = \pm\epsilon_0^{\mp}. \tag{75}$$

The operator $\Gamma^{0\dots p}$ squares to the identity and is trassless. Therefore the number of independent eigenvalues is one half of the maximum possible, in other words 16.

We can see the above mechanism in the supergravity solutions of the Branes that we study before. In the type II supergravity case the supersymmetry transformation acting on the fermionic field gives the following equations,

$$\delta\psi_M = \partial_M\epsilon - \frac{1}{4}\omega_M^{AB}\gamma_{AB}\epsilon + \frac{(-1)^p}{8(p+2)!}e^{\phi}F_{M_1\dots Mp+2}\gamma^{M_1\dots M_{p+2}}\gamma_M\epsilon'_{(p)} \tag{76}$$

$$\delta\lambda = \gamma^M(\partial_M\phi)\epsilon + \frac{3-p}{4(p+2)!}e^{\phi}F_{M_1\dots M_{p+2}}\gamma^{M_1\dots M_{p+2}}\epsilon'_{(p)} \tag{77}$$

$$\epsilon'_{(0,4.8)} = \epsilon \qquad \epsilon'_{(2,6)} = \gamma_{11}\epsilon \qquad \epsilon'_{(-1,3,7)} = \imath\epsilon \qquad \epsilon'_{(1,5)} = \imath\epsilon^* \tag{78}$$

where ϵ is a 32-component spinor, and ω is the spin connection After solving for the Dp-brane solutions we found the equations

$$\delta\lambda = H^{-1/4}\gamma^r\partial_r\phi\epsilon + \frac{(3-p)e^{\phi}(\partial_r H)\gamma^r}{4H^{\frac{8-p}{4}}}\gamma_0\dots\gamma_p\epsilon'$$

$$\delta\psi_\alpha = \partial_\alpha\epsilon + \frac{(\partial_r H)}{8H^{\frac{3}{2}}}\gamma^r\gamma_\alpha\epsilon + \frac{e^{\phi}(\partial_r H)}{8H^{\frac{9-p}{4}}}\gamma^r\gamma_\alpha\gamma_0\dots\gamma_p\epsilon'$$

$$\delta\psi_r = \partial_r\epsilon - \frac{e^{\phi}(\partial_r H)}{8H^{\frac{7-p}{4}}}\gamma_0\dots\gamma_p\epsilon'$$

$$\epsilon'_{(0,4.8)} = \epsilon \qquad \epsilon'_{(2,6)} = \gamma_{11}\epsilon \qquad \epsilon'_{(-1,3,7)} = \imath\epsilon \qquad \epsilon'_{(1,5)} = \imath\epsilon^* \tag{79}$$

Where we have used the split $M = (\alpha, r, \theta)$ where (r,θ) are perpendicular coordinates to the brane, also ϵ is a 32-component spinor, and ω is the spin connection [16]. Note that we have on propose leave the dilaton unspecified in terms of H.

The first thing to note about this system of equation as is that for the D3-brane case the dilatino equation is satisfied independently of the type of spinor consider, therefore we are left with only the gravitino constraints [3]. This is a consequence of the fact that in this case the dilaton is constant. For the other Dp-branes the dilaton equation is present and the solutions are dilatonic.

[3] This case is similar to the M2-brane and the M5-branes, where the supersymmetry equations are written in terms of the gravitino constraints only (in $D = 11$ there is no dilatino)

The dilatino equation is up to multiplicative factors the projector operator appearing on equation (74), therefore its presence implies the breaking of 1/2 of supersymmetry. Also note that the gravitino equation on the worldvolume coordinates is proportional to the same projector up to a additive factor of the form $\partial_\alpha \epsilon$. The last equation referring to the gravitino components on the perpendicular space doesn't correspond to to the full projector but just the part showed on equation (75) plus a partial derivative on the perpendicular direction. The solution of this equations is obtained by assuming no dependence on the worldvolume coordinates plus using the eigenspinors defined on equation (75), giving the result

$$\epsilon = H^{-1/8}\epsilon_0. \tag{80}$$

where ϵ_0 satisfied that $\epsilon_0 = -\gamma_0 ... \gamma_p \epsilon^{'}$. Hence we get only the expected 16 real supercharges.

Let's next consider the case for the near horizon geometries obtained from the Dp-branes solutions. At first this question could be consider a bit trivial, after all we are talking about Dp-brane solutions in certain regions, and we already know the BPS structure of this type of solutions. Nevertheless note that the near horizon limit change the asymptotic structure of the metrics. In those cases we lost the Minkowski structure and the relation with the Poincare invariants like the mass. The argument showed at the beginning of this section simply doesn't applied to this situation.

To study the supersymmetry properties of these near horizon geometries we start with the equation (79). Our goal is to define the near horizon limit of this set of equations. For the gravitino equations, the limiting recepee is quite simple due to the fact that the gravitino supersymmetry variations are components of a one-form, therefore we have an geometrical object where to define the near horizon limit, basically

$$\delta\hat{\psi} \equiv lim_{NH}\left(\delta\psi\right) = lim_{NH}\left(dx^M \delta\psi_M\right) = lim_{NH}\left(dx^M\right) lim_{NH}\left(\delta\psi_M\right), \tag{81}$$

hence we get

$$\begin{aligned} \delta\psi_\alpha &= \partial_\alpha \epsilon + \frac{(\partial_u h)}{8h^{\frac{3}{2}}}\gamma^u \gamma_\alpha [\epsilon + \gamma_0 ... \gamma_p \epsilon^{'}], \\ \delta\psi_u &= \partial_u \epsilon - \left(\frac{\partial_u h}{8h}\right)\gamma_0 ... \gamma_p \epsilon^{'}, \end{aligned} \tag{82}$$

where $h = g_{YM}^2 / U^{7-p}$. On the other hand the dilatino equation is a bit more subtle, as it is a scalar from the point of view of ten dimensional supergravity. We can define its near horizon limit by consider the dilatino variation as part of the relevant gravitino in a higher dimensional theory like eleventh or twelve dimensional dimensional supergravity. Once this is done the resulting near horizon limit is,

$$\delta\lambda = \alpha'^{1/6}\left[\frac{(3-p)(\partial_u h)\gamma^u}{4h^{\frac{5}{4}}}[\epsilon + \gamma_0 ... \gamma_p \epsilon']\right]. \tag{83}$$

In this case the dilatino equation goes to zero in the near horizon limit, giving no constraints. Nevertheless the consistency conditions corresponding to the gravitino equation, contains the dilatino equation among others. After the usual decomposition of the killing spinors, defined by the canonical projector on this ansatz, we find that one half of the supersymmetry is always preserved if the dilatino constraint is satisfied , but the other half is conserved only if the function h, behaves as

$$h \propto \frac{1}{U^4}, \tag{84}$$

therefore we found the possibility of enhancement for $p = 3$ only. The killing spinors equation found in this case corresponds to the AdS equation

$$\partial_\alpha \epsilon - \frac{u}{g_{YM}^{1/2}}\gamma_\alpha(1-\gamma_u)\epsilon = 0,$$
$$\partial_u \epsilon - \frac{1}{2u}\gamma_u \epsilon = 0, \tag{85}$$

with the well known solution.

$$\epsilon_1 = u^{1/2}\epsilon_0^+,$$
$$\epsilon_2 = \left(u^{-1/2} + \frac{u^{1/2}}{2g_{YM}^{1/2}}x^\alpha \gamma_\alpha\right)\epsilon_0^-. \tag{86}$$

The M-theoretical cases go along the same line of reasoning as before, here we show in detail the M2-brane only. The supersymmetry transformation for the gravitino in eleventh dimensions give,

$$\delta\hat{\psi}_M = \partial_M \hat{\epsilon} - \frac{1}{4}\hat{\omega}_M^{AB}\hat{\gamma}_{AB}\hat{\epsilon} - \frac{1}{288}\left(\hat{\gamma}_M^{BCDE} - 8\hat{e}_M^B \hat{\gamma}^{CDE}\right)\hat{F}_{BCDE}, \tag{87}$$

where M, N, O refers to Einstein index and A, B, C refers to Lorenz index, $\hat{e}$ is the elfvien and $\hat{\omega}_M^{AB}$ is the spin connection. Also we have hatted the variables to avoid confusion with ten dimensional variables. After solving for the M2-brane ansatz we get,

$$\delta\hat{\psi}_\alpha = \partial_\alpha \hat{\epsilon} - \frac{(\partial_r H)}{12H^{\frac{3}{2}}}\hat{\gamma}^r \hat{\gamma}_\alpha (1 + \hat{\gamma}_0\hat{\gamma}_1\hat{\gamma}_2)\hat{\epsilon},$$
$$\delta\hat{\psi}_r = \partial_r \hat{\epsilon} - \frac{(\partial_r H)}{6H}\hat{\gamma}_0\hat{\gamma}_1\hat{\gamma}_2\hat{\epsilon}. \tag{88}$$

The solution for this system of equations is again of the form,

$$\epsilon = H^{-1/6}\epsilon_0. \tag{89}$$

where ϵ_0 is a constant spinor with only the expected 16 real degrees of freedom.

To study the the supersymmetries in the near horizon limit for the eleventh dimensional case, we consider the relevant near horizon limit, giving as a result on the killing spinor equation

$$\partial_\alpha \hat{\epsilon} - \frac{u}{k^{1/2}} \hat{\gamma}_\alpha (1 - \hat{\gamma}_u) \hat{\epsilon} = 0,$$
$$\partial_u \hat{\epsilon} - \frac{1}{2u} \hat{\gamma}_u \hat{\epsilon} = 0, \tag{90}$$

which tell us that also in this case we get enhancement of supersymmetry.

8 T-Duality and Near Horizon Geometries

One of the questions that naturally arise, once we have a duality between QFT and string theory on given backgrounds, is the meaning of pure stringy symmetries from the point of view of the QFT. For example in the celebrated AdS/CFT correspondence the $SL(2,Z)$ S-duality of the Type IIB-string can be interpreted as the string version of the well known Monton Olive $SL(2,Z)$ duality of $N = 4$ SYM [11]. The quantum field theory meaning of T-duality is however a bit less clear since T-duality interpolates different D-branes and therefore tends to define maps between Yang Mills theories in different dimensions. Moreover T-duality transformation generally produce explicit breaking of supersymmetry.

The simplest possible way to address this questions on T-duality in a purely quantum field theoretical framework is of course by defining the quantum field theory using the near horizon limit of D-branes metrics and working out in this limit T-duality transformations.

To begin with, let us start considering a D3-brane living in a ten dimensional space time with one of the orthogonal coordinates compactified on a circle S^1 of radius R. The corresponding metric is given by

$$ds_3^2 = H(r,R)^{-1/2}(dx_{||}^2) + H(r,R)^{1/2}(d(\theta R)^2 + dr^2 + r^2 d\Omega_4)$$
$$e^\phi = g, \tag{91}$$

with $x_{||}$ standing for world volume coordinates and the harmonic function H given by

$$H(r,R) = 1 + g l_s^4 \sum_{n=1}^{N} \sum_{n_i} \left(\frac{Q_n}{[(y_9 - y_{n_9} + n_9 R)^2 + \mid r - r_n \mid^2]^{4/2}} \right), \tag{92}$$

where g is the string coupling constant, l_s is the string length, Q_n is the charge of the D3-brane, R is the radius of the circle S^1 and r is the radius on spherical coordinates for the $\Re^5$ space time. From now on we will ignore any constant and will work with meaningful variables on the discussion, and always at large N.

At distances much longer than R, we can Poisson resume the expression (92) obtaining

$$H(r,R) = 1 + \frac{l_s^4 g}{Rr^3}. \tag{93}$$

This is a good solution as far as R is small enough.

I we are interested in the near horizon limit [12] of this metric we would be forced to defined this limit by performing a double "blow up", namely

$$\left(\alpha' \to 0 \quad , \quad U \equiv \frac{r}{\alpha'} = constant \quad , \quad v \equiv \frac{R}{\alpha'} = constant\right). \tag{94}$$

Notice that the new variable v correspond properly speaking to a blow up of the point $R = 0$ in the moduli of the target space time metric (91) with the harmonic function of (93). After performing the blow up (94) we get the metric

$$\begin{aligned} ds_3^2 &= \alpha' \left[\frac{U^{3/2}v^{1/2}}{g^{1/2}}(dx_{||}^2) + \frac{g^{1/2}}{U^{3/2}v^{1/2}}dU^2 + \frac{g^{1/2}U^{1/2}}{v^{1/2}}d\Omega_4 + \frac{g^{1/2}v^{3/2}}{U^{3/2}}d\theta^2\right] \\ e^{\phi} &= g. \end{aligned} \tag{95}$$

Note that the dilaton field for this solution is constant. The topology of the space time (95) is that o a fibration of a circle S^1 and a sphere S^4 on the space defined on the coordinates (x^α, U) with the corresponding radius

$$\begin{aligned} R_{S^1} &= (\frac{gv^3}{U^3})^{1/4}, \\ R_{S^4} &= (\frac{gU}{v})^{1/4}, \end{aligned} \tag{96}$$

which depend on the moduli v. It is easy to see that this space time admits only 16 real supercharges, the simples way to understand this is observing that $\frac{1}{2}$ of the Killing spinors for the near horizon geometry of the D3-brane are projected out once we compactify a transverse direction (we will come back to this point later on).

From the QFT point of view we should expect the metric (95) to be related to a SYM in $3+1$ with 16 real supercharges and with a peculiar R-symmetry given by the isometries of $S^4 \otimes S^1$. The type of strings living on the space time (95) is Type IIB.

Consider next the candidate for a T-dual geometry, namely the D4-brane compactified on a S^1 of radius R, on its world volume. This solutions is given by

$$\begin{aligned} ds_4^2 &= H^{(-1/2)}(dx_{||}^2 + d(\theta R)^2) + H^{(1/2)}(dr^2 + r^2 d\Omega_4) \\ e^{\phi} &= gH^{-1/4}, \end{aligned} \tag{97}$$

with $x_{||}$ expanding the non-compact four dimensional part of the world volume of the D4-brane, and the corresponding harmonic function

$$H = 1 + \frac{gl_s^3}{r^3}. \tag{98}$$

Taking the near horizon limit, defined by again a double blow up

$$\left(\alpha' \to 0 \quad , \quad U \equiv \frac{r}{\alpha'} = constant \quad , \quad g_5^2 \equiv g\alpha'^{1/2} = constant\right), \tag{99}$$

we get the metric

$$\begin{aligned} ds_4^2 &= \alpha' \left[\frac{U^{3/2}}{g_5}(dx_{||}^2 + (d\theta R)^2) + \frac{g_5}{U^{3/2}} dU^2 + g_5 U^{1/2} d\Omega_4\right] \\ e^{\phi} &= g_5^{3/2} U^{3/4}. \end{aligned} \tag{100}$$

Note that this time the dilaton is not constant. The number of supersymmetries in this brane is again 16 real supercharges, however the QFT interpretation is a bit different, In this case we have a $D = 5$ SYM theory living on $\Re^4 \otimes S^1$. The corresponding radius of the four-sphere and the circle are

$$\begin{aligned} R_{S^4} &= (g_5^2 U)^{1/4}, \\ R_{S^1} &= (\frac{U^3 R^4}{g_5^2})^{1/4}. \end{aligned} \tag{101}$$

Before proceeding any further, let us clarify the picture we have (see fig. 13).

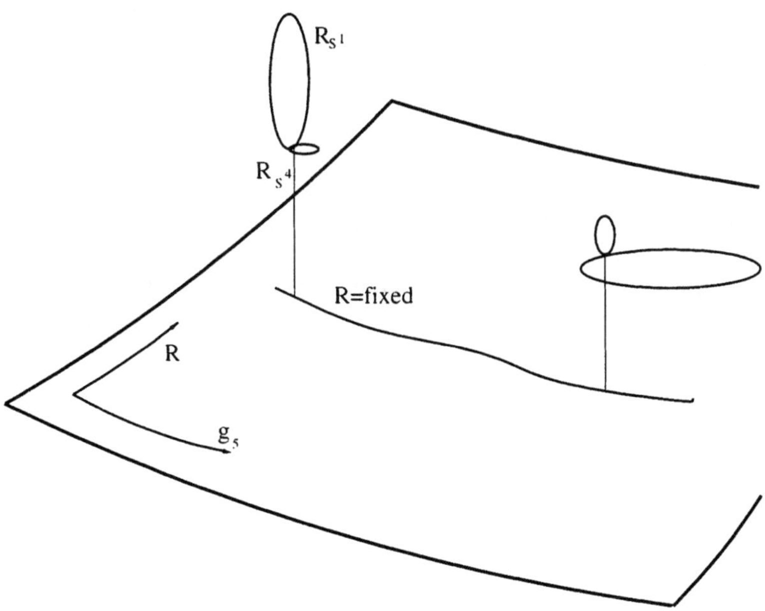

Fig. 13. Moduli space for near horizon D4-brane.

In the D3-brane case we obtained the near horizon geometry as the result of a limit where it was left one of the moduli g constant but we allowed R to varied such that, at $R = 0$ we create a divisor v. This two variables define our moduli (v, g). The resulting geometry is that of a base space expanded by the coordinates (x, U) and fibers S^4 and S^1 with the radius of equation (96). In the D4-brane, we obtained the near horizon geometry as the result a another limit where one of the moduli R is maintained constant while the other g varies such that at infinite point we create a new divisor g_5. This two variables define the new moduli (R, g_5). Again the geometry obtained is that of a base manifold expanded by the coordinates (x, U) and fibers S^4 and S^1 with the radius of equation (101). In order to identify both metrics (94,100) by T-duality we most require the following relation between the different moduli,

$$g = \frac{g_5^2}{R},$$
$$v = \frac{1}{R}. \tag{102}$$

Provided that this relation holds, we can perform the T-duality transformation following the normal Buscher rules [13]. In principle we could run into difficulties if there appears to many singular point on the fibration, so that T-duality could loose its natural meaning. This T-dual map is well defined all over the base manifold, Actually we only have singularities at $U = 0$ and $U = \infty$ but both points are related rather to wrong coordinate patches than real singularities. Therefore this T-dual map is a very trivial example of fiber-wise T-duality [14]. The map described above is defined by T-duality, and effectively acts between the two moduli.

By now we have a neat relation between the bare coupling constants of both gauge theories in the two near horizon metrics. On the other hand it is usually associated to the dilaton behavior, the value of the corresponding running coupling constant i.e. the effective gauge coupling constant. Therefore to obtain the effective coupling of the compactified gauge theory on the world volume of the D4-brane, we considered the ratio of the effective coupling constant of the five dimensional gauge theory g_5 squared, with the effective radius of compactification namely the radius of the S^1 given on equation (101), hence we get

$$g_{4_{eff}}^2 \equiv \frac{g_{5_{eff}}^2}{R_{eff}}. \tag{103}$$

Then after solving for the moduli variables (v, g) we obtain

$$g_{4_{eff}}^2 = g. \tag{104}$$

Therefore the effective coupling constant of the gauge theory we are studying from the point of view of the near horizon D4-brane has the same running as the the gauge theory on the near horizon D3-brane, as it should be expected invoking its duality relation. Note that the equation (103) was obtained by plausibly physical relations, however this equation is nothing more than the changing rule for the dilaton under T-duality!.

On the other hand the super Yang Mills theory on the D4 brane is not renormalizable, we can trust it only at low energies. This aspect of the gauge theory can be seen from the gravitational point of view. Note that for this geometry (100) the dilaton grows for large U. Actually, we can trust on this solution as long as $U \geq 1/g_5^2$, after this point we should think in terms of M-theory.

The other possibility we have is to consider the D3-brane wrapped on a circle S^1 of radius R. This time the near horizon geometry is defined by the limit

$$\left(\alpha' \to 0 \quad , \quad U \equiv \frac{r}{\alpha'} = constant\right), \tag{105}$$

where (g, R) are kept constant on the process. Note that this time we don't have the double blow up of the above cases. The resulting metric is given by

$$\begin{aligned} ds_3^2 &= \alpha' \left[\frac{U^2}{g^{1/2}}(d(R\theta)^2 + dx_{||}^2) + \frac{g^{1/2}}{U^2}dU^2 + g^{1/2}d\Omega_5)\right] \\ e^{\phi} &= g. \end{aligned} \tag{106}$$

Again the dilaton is constant, and the topology is that of a fibration of a circle S^1 and the sphere S^5, on the space defined by the coordinates (x^α, U), with the corresponding radius

$$\begin{aligned} R_{S^1} &= \frac{UR}{g^{1/4}}, \\ R_{S^5} &= g^{1/4}. \end{aligned} \tag{107}$$

This configuration also only admits 16 real supercharges. The situation on the field theory should be that of a $D = 4$ SYM theory living on $\Re^3 \otimes S^1$ with 16 real supercharges and R-symmetry contained on $SO(6)$.

The T-dual near horizon background is the result of first, a Poisson resume of the D2-brane solution with a transverse direction compactified on a small circle S^1 of radius R, second its near horizon limit defined by the triple blow up, here showed

$$\begin{aligned} \alpha' \to 0 \quad &, \quad U \equiv \frac{r}{\alpha'} = constant, \\ v \equiv \frac{R}{\alpha'} \quad &, \quad g_3^2 \equiv g\alpha'^{-1/2} = constant. \end{aligned} \tag{108}$$

The resulting metric and dilaton are given by

$$\begin{aligned} ds_2^2 &= \alpha' \left[\frac{U^2 v^{1/2}}{g_3}(dx_{||}^2) + \frac{g_3}{U^2 v^{1/2}}dU^2 + \frac{g_3}{v^{1/2}}d\Omega_5 + \frac{g_3 v^{3/2}}{U^2}d\theta^2\right] \\ e^{\phi} &= \frac{g_3^{5/2}}{Uv^{1/4}}. \end{aligned} \tag{109}$$

This time we have space time with 16 real supercharges, the metric defines a fibration on a circle S^1 and the sphere S^5 on the base space expanded by the coordinates (x^α, U) and the field theory point of view should be that of a SYM theory on 2+1 dimensions, with 16 real supercharges, and R-symmetry contained in $SO(6) \otimes U(1)$. The corresponding radius of the fibers are

$$\begin{aligned} R_{S^1} &= \frac{g_3^{1/2} v^{3/4}}{U^{1/2}}, \\ R_{S^5} &= \frac{g_3^{1/2}}{v^{1/4}}. \end{aligned} \tag{110}$$

To perform the T-duality map between this two metrics, we require to identify the moduli as follows,

$$\begin{aligned} g &= \frac{g_3^2}{v}, \\ R &= \frac{1}{v}. \end{aligned} \tag{111}$$

Similar remarks about the validity of T-duality of the D3-brane and the D4-brane are applicable to this case.

In general, we can consider the above type of compactification of the D3-brane on a T^{3-s}, which take us to the near horizon D(s)-brane on a T^{3-s} in the perpendicular coordinates. The metric of these geometries looks like $AdS_{s+2} \otimes S^5 \otimes T^{3-s}$. When the holography map between the gauge theories and the bulk geometries is defined [4], the radius of the torus is very small. Also we found that when $R \to \infty$ we recover on these D(s)-branes, the full range of running for U $(0, \infty)$, while for small R we are forced to stay at big values of the holographic variable U.

It is well known that T-duality breaks supersymmetry in some cases [15], our previous metrics are not exceptions to this phenomenon. Note that we are relating theories with 16 real supercharges, but we alredy showed that the near horizon D3-brane shows enhancement of supersymmetry. The matching condition for the number of supersymmetries is given by the fact that the compact direction on the D3-brane (on both cases) eliminates the possibility of that enhancement, while the other Dp-branes don't show the enhancement at all.

It is important to notice that the supersymmetries which are broken by T-duality correspond to those which get enhanced in the particular case of the D3-brane, i.e. the bilinears associated with the broken Killing spinors, are the conformal Killing vectors.

In our previous analysis we have consider the two dual pairs (D3/D4) and (D2/D3). Both star with the D3-brane with the only difference that the compactified dimension is or not on the transversal direction. Let us study a bit more carefully both pairs. In the (D3/D4) case the D4-metric is characterized by two

[4] Recall that the holographic conjecture of Maldacena only applies within a range of validity of U for the case p different from 3

different moduli (g_5, R). After the work in Matrix theory [17], it is natural to interpret g_5 as related to the eleventh dimension of M-theory and therefore to consider this metrics as coming from a compactification of M-theory on a T^2, with sizes determined by the two moduli g_5 and R. Also it is well known [18] that in the limit where the volume of the two torus goes to zero, we should recover type IIB theory. This mechanism implies the dynamical generation of a "quantum" dimension with the corresponding Kaluza Klein modes associated with the menbrane wrapped on the two torus. When we apply this mechanism to our case it is natural to expect to get in the limit of zero volume for the two torus ($Vol(T^2) \to 0$) the type IIB D3-brane metric of equation (95). Using the relations (96) and (102) we observe that the limit $Vol(T^2) \to 0$ corresponds to a ten dimensional type IIB theory, where the extra "quantum" dimension is the S^1 circle in the limit $v \to \infty$, with the radius given by (96). The up lifted of the D4-brane to M-theory give us a M5-brane wrapped on a circle determined by the value of g_5. In addition to this, in our case we wrap the M5-brane on another circle defining the two torus characterized by the two moduli (g_5, R). In the limit $Vol(T^2) \to 0$ what we get is the M5-brane wrapped on a two torus of zero volume, that produce a D3-brane with the extra dimension defining the transversional circle in the metric (95).

Hence the theory on the D4-brane gets embedded in the six dimensional (2,0) theory on the M5-brane. As it is well known for the M5-brane we get enhancement of supersymmetry and therefore we can say that the D4-brane theory will flow to a conformal point in strong coupling. *In other words what we observed is that once we break the superconformal generators by T-duality the resulting theory naturally flows to recover the supersymmetry by up lifting to M-theory.*

Let us now consider the T-dual pair (D2/D3). This is very similar to the previous case. In the D2-brane metric the moduli is characterized by (g_3, v), which again should by interpreted as M-theory compactified on a two torus of size g_3 and v. In the limit when the volume of the two torus goes to zero, we should recover the type IIB-picture by exactly the same mechanism described above. The D2-brane is now up lifted to a M2-brane but contrary to what happens in the (D3/D4) case, the extra "quantum" dimension becomes now part of the world volume dimension of the T-dual D3-brane. More precisely what we observed is that the compact "world volume" dimension in the metric (106) is the extra "quantum" dimension in the type IIB that we get when we compactify M-theory on the two torus characterized by (g_3, v). *In other words what we observed, is that the T-dual description in (106) of the uplifted D2-brane is a three dimensional theory becoming four dimensional for "strong coupling" $v \to 0$ in equation (111).*

As before with the (D3/D4) pair we also observe here that the theories flow to reach superconformal invariance. Notice that this two dual saturate the known examples of superconformal theories namely the D3-brane, M2-brane and M5-brane. The (D3/D4) pair is related to the M5-brane and the (D2/D3) pair to the M2-brane. The previous conjecture is in contrast to the mechanism suggested in [19] for solving the cosmological constant problem. In that we can

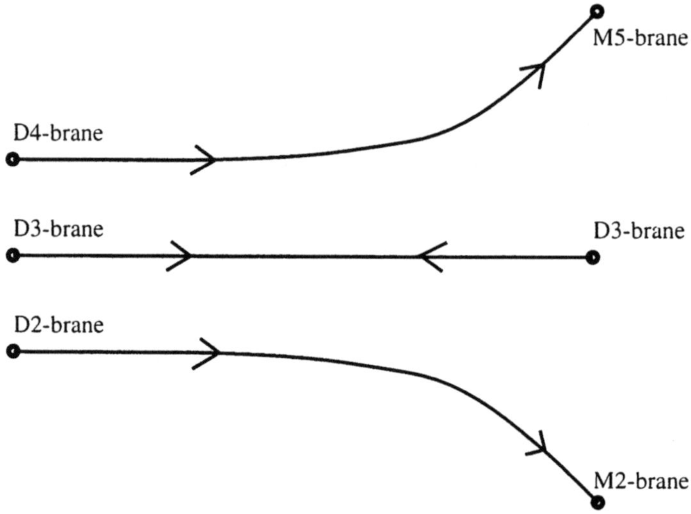

Fig. 14. Flow to the superconformal QFT.

start with a three-dimensional theory that is expected to flow in strong coupling to a four-dimensional theory. Massive particles in three dimensions are associated with conical geometries, when some amount of supersymmetry is broken. The suggested solution to the cosmological constant problem, is based on the assumption that these supersymmetries are not restore in the strong coupling four-dimensional limit. In our case, we have simply studied supersymmetry generators associated with conformal transformations that are the ones naturally broken by the action of T-duality, and we find they are restored in the up lifted "M-theory" limit.

The general picture emerging from the previous discussion is that once we start with a superconformal theory, T-duality generally breaks the supersymmetries associated with the superconformal transformations, however the T-dual theory tends to flow to recover these supersymmetries broken by T-duality up lifting to M-theory.

To be more precise, starting with a D3-brane with the world volume compactified on a circle of radius R, we break for finite radius the supersymmetries associated with those Killing spinors depending on world volume coordinates. Those are associated with the enhanced supersymmetry. In order to decide if T-duality breaks or not supersymmetry, we perform a T-duality to a D2-brane. Once we have done that, we send the radius R to infinite. In this limit we recover for the D3-brane the whole superconformal algebra, then if T-duality is not braking supersymmetry we should find that the T-dual of the $R \to \infty$ limit possesses enhanced superconformal invariance. In fact this is what happened. By relation (111), when $R \to \infty$ then $v \to 0$ and $g_3^2 \to 0$ for finite g, but g_3^2 can be interpreted as $1/\Delta$ for Δ the size of the eleventh dimension. Thus the D2-brane becomes uplifted to M2-brane recovering the superconformal transormations. In a certain sense M-theory is there to work out the breaking of supersymmetry induced by T-duality.

Acknowledgements

We are grateful for stimulating discusions with E. Alvarez and T. Ortin. This work has been partially supported by European Union TMR program FMRX-CT96-0012 *Integravility, Non-pertubative Effects, and Symmetry in Quantum Field Theory* and by the spanish grant AEN96-1655. The work of P.Silva was also suported by the Venezuelan goverment.

References

[1] M.B. Green, J.H. Schwarz and E. Witten, *Superstring Theory*(Cambridge U. Press 1987).
[2] J. Polchinski *String theory* Vol. I and Vol. II (Cambridge U. Press 1998).
[3] E. Kiritsis, *Introduction to Superstring Theory.* Leuven: Leuven University, to be publish, hep-th/9709062.
[4] A. Giveon, M. Porrati and E.Ravinivici, *Phys. Rep. 244*,77 (1994).
[5] J. Polchinski, *Phys. Rev. Lett.75*,4724 (1995).
[6] D. Friedam, E. Martinec and S. Shenker, *Nucl. Phy. B271*,93 (1986).
[7] J. Polchinski, In *Fields, Strings and Duality. TASI 1996* eds. C. Efthimiou and B. Greene, 293.
[8] O. Aharony, S. S. Gubser, J. Maldacena, H. Ooguri, Y. Oz, hep-th/9905111.
[9] K. S. Stelle, hep-th/9803116.
[10] E. Bergshoeff, C. M. Hull and T. Ortin *Nucl.Phys. B451.* 547 (1995).
[11] L. Susskind, hep-th/9611164
[12] J. Maldacena, hep-th/9711200
N. Itzhaki et all, hep-th/9802042
[13] T.H. Buscher,*Phys. Lett. B194* 59 (1987).
T.H. Buscher, *Phys. Lett. B201* 466 (1988)
E. Alvarez, L. Alarez-Gaume and Y. Lozano, hep-th/9406206
[14] P.Aspinwall and R. Plesser, hep-th/9905036
[15] E. Alvarez, L. Alvarez-Gaume and I. Bakas, hep-th/9507112
E. Alvarez, L. Alvarez-Gaume and I. Bakas,hep-th/9510028
[16] E. Bergshoeff, hep-th/9611099
[17] N. Seiberg, hep-th/9705117
M. Berkooz, M. Rozali an N. Seiberg, hep-th/9704089
M. Rozali, hep-th/9702136
[18] S. Sethi and L. Susskind, hep-th/9702101
[19] E. Witten, hep-th/9409111
E. Witten, hep-th/9506101

Noncommutative Geometry and Basic Physics

Daniel KASTLER

Centre de Physique Théorique
CNRS - Luminy, Case 907
F-13288, MARSEILLE CEDEX 09 (FRANCE)

Contents

Foreword

Alain Connes' noncommutative geometry, started in 1982 [0], widely developed in 1994 as expounded in his book at this date [0] (it has grown meanwhile) is a systematic quantization of mathematics parallel to the quantization of physics effected in the twenties.This theory widens the scope of mathematics in a manner congenial to physics, reorganizes the existing ("classical") mathematics of which it produces an hitherto unsuspected unification, and provides basic physics (the synthesis of elementary particles and gravitation) with a programme of renewal which has thus far achieved a clarification of the classical (tree-level) aspects of a new synthesis of the (Euclidean) standard model with gravitation [32],[33]: this is the subject of the present lectures - with the inherent tentative prediction of the Higgs mass.

The hinge of the programme is the notion of *spectral triple* (see [**1.1**] below and [[0], Chapt.IV]) - the central concept of Connes' "metric noncommutative geometry", a wide generalization of the notion of (Riemannian) Dirac operator recognized in the classical (space-time) case as encoding the geometry of the spin manifold (including the Yang-Mills action of classical electrodynamics now obtained by means of a "quantum Yang-Mills formalism") - but with the crucial capacity of playing this role in a considerably wider frame covering amongst others the Lagrangian aspect of the standard model. The first version of the theory (Connes-Lott model [3],[9],[16]) successfully obtained the bosonic action of the (Euclidean) standard model by applying the quantum Yang-Mills formalism to a combined spectral triple, tensor product of the space-time (electrodynamics) spectral triple by an "inner spectral triple" featuring the inner degrees of freedom governed by the gauge group $U(1) \times SU(2) \times SU(3)$ - accordingly based on the tensor product algebra , $(\mathbb{C} \oplus \mathbb{H}) \otimes (\mathbb{C} \oplus \mathbf{M}_3(\mathbb{C}))$, since the respective groups of unitaries of $\mathbb{C}$, $\mathbb{H}$ and $\mathbf{M}_3(\mathbb{C})$ are $U(1)$, $SU(2)$, and $U(3)$, the latter to be turned into $SU(3)$ by a "modular correction" (the supplementary direct-summand $\mathbb{C}$ in the chromodynamics tensorial factor affords maneuverability for this). The same combined spectral triple also yields an elegant procedure for obtaining the fermion action.

The exposition of the subject (and the consultation of its literature) is now complicated by the fact that the original (quantum Yang-Mills) Connes-Lott model has undergone two major structural improvements causing modifications of doctrine:

– the passage [13] from "dual metric pairs" to "S_0-real spectral triples" (specifically: the replacement of the above inner spectral triple featuring a tensor product of an electroweak and a strong inner algebra by a (manifestly charge-conjugation-symmetric, Poincaré self-dual) S_0-real spectral triple based on the direct-sum algebra $\mathbb{C} \oplus \mathbb{H} \otimes \mathbf{M}_3(\mathbb{C})$ - now used as before within the quantum Yang-Mills scheme for producing the bosonic and fermionic actions;

– a subsequent change [32], [33] of the Lagrangian-creating paradigm for the bosonic action (with maintenance of the S_0-real spectral triple). In spite of its great success the quantum Yang-Mills formalism is now replaced by a new paradigm (*spectral action*) technically based on the heat-kernel expansion and yielding a unified standard model-gravitation Lagrangian (with parallel unification of the diffeomorphism and gauge groups). This Lagrangian has to be interpreted as yielding a theory of the "primal matter", and thus requires for the obtention of realistic previsions (e.g. that of the Higgs mass at accelerator energy) a renormalization group treatment.

Given this historical development the problem of exposition is the following: the aim is to provide the reader with a working knowledge of the modern state of affairs: the heat-kernel expansion of the spectral-action - which utilizes as we mentioned the same S_0-real spectral triple as the second version of the Yang-Mills theory. Now, in want of an a-priori derivation of the latter object (as one hopefully expects from a quantum group - but we still await this) one needs to motivate its construction which, if dogmatically expounded, would appear strange to the reader. Our task is thus to describe the way in which this concept was evolved within the now abandoned Yang-Mills frame without burdening our exposition with the intricacies of the Yang-Mills computation nowadays presumably only of historical interest. We attempt to sketch this "historical path" in our section [**2**], hopefully providing the reader with enough motivation to make him accept the subsequent technical sections. In fact the large measure of success of the Yang-Mills theory (yielding for the Higgs mass, with the right choice of scalar product on Ω_D, results very similar to those of the spectral action theory) poses the problem of a better understanding of the (at present mysterious) relationship between the two formalisms.

Here is a word of advice on how to read this text diagonally. The elements required for a working knowledge of the present description of the (bosonic) action of the standard model + gravitation are the following:

- (a): a general idea about Alain Connes metric noncommutative geometry (spectral triples, i.e. generalized Dirac operators);[1]
- (b): construction of the covariant Dirac operator of the standard model;
- (c): asymptotic evaluation of the spectral action (defined in terms of the letter). These items correspond respectively to our sections [**1**], [**5**], [**6**][2] Now (b) is obtained from the S_0-real spectral triple resulting from a previous metric dual pair through a process described in section [**4**]. And motivation for the definition of the metric dual pair comes from the historical development of the quantum Yang-Mills standard model as sketched in section [**3**]-itself motivated by the two first examples [**1.8**] (classical electrodynamics in quantum guise), and [**1.9**] (embryonal Higgs).

The reader can accordingly, after reading [**1**] and [**2.1**] trough [**2.3**], browse

[1] including the quantum Yang-Mills which is part of it as the classical Yang-Mills is part of classical differential geometry.

[2] especially [**6.6**] and [**6.7**] which the reader might like to try and prove ab initio as an exercise: this yielding then in combination with [**7**] the fastest (if somewhat immotivated) access to the final results. Geodesic: understand (2.35), [**5.6**],[**6.5**] through [**6.7**], then [**7**].

quickly trough the rest of [**2**] and [**3**], then go to [**4**], [**5.3**] and [**7**] where the actual story begins, apart from motivation. The principle of the quantum Yang-Mills formalism (including cumbersome division through $d\mathcal{K}^1$"- elimination of the junk" as Thomas Schücker calls it) is illustrated by the simple examples [**1.8**], [**1.9**] (and the somewhat more subtle inner electroweak action [**2.5**]). For the standard model the junk elimination becomes somewhat of a - now probably obsolete - nightmare.

For a shorter, more descriptive account of the subject of these lectures, the reader might like to consult the Portugal school text [39].We recommend also Thomas Schücker's Portugal school lectures [38] (less technical details, more physical flavour than the present text) and the writings of our Costa-Rica friends [6],[23]. We refer to [13],[21],[32] for high-flying surveys, and to [34] for a physicist's point of view. These lectures describe the present status of the classical (lagrangian=tree-level) stage of the theory. Last year Alain Connes and Dirk Kreimer started a programme of investigation of field quantization from the point of view of noncommutative geometry. The reader interested in this quest of renewal of a deeper level of the theory should consult [46],[47].

We express our indebtedness to our colleague Galina Erochenkova for her kind and efficient dressing in $\LaTeX$ my self-made Word5-text to render it gesellschafts fähig. Thanks are due to Springer-Verlag for tolerating microdeviations from their (constraining) macros. Thomas Schücker also deserves warm thanks for his scientific advice and for rescuing me from typographic ship wreck in the last hour. Last but not least, I thank H.Gausterer, H.Grosse and L.Pittner for their invitation to the pleasant and informative Schladming winter school.

1 Spectral Triples as Generalized Dirac Operators. Sketch of the Quantum Yang-Mills Algorithm. Electrodynamics and the Two-Point Functions as the First Examples

One of the essential contributions of Alain Connes was the recognition, opening the way to "metric differential geometry", of the essential role of the Dirac operator (suitably generalized, as *spectral triple*) which then becomes a universal object in mathematics – both in his new noncommutative geometry and (for its unifying role) in classical mathematics. This section aims at introducing to this circle of ideas (proofs are sketched).

1.1 The Dirac operator of a Riemannian spin manifold

Let $\mathbf{M}$ be a Riemannian spin manifold of even dimension d= $2m$, $m \in \mathbb{N}$, with $\mathcal{A} = C^\infty(\mathbf{M}, \mathbb{C})$ (or $\mathcal{A} = C^\infty(\mathbf{M}, \mathbb{R})$) and with spin bundle $\mathcal{S}_\mathbf{M}$. The Hilbert space $\mathbf{H} = L^2(\mathcal{S}_\mathbf{M})$ of square-integrable spinors is acted upon by the Dirac operator $\widetilde{D} = i\gamma^\mu \widetilde{\nabla}_\mu = i\gamma^\mu(\partial_\mu + \sigma_\mu)$, σ_μ the Levi-Civita spin connection with covariant derivative $\widetilde{\nabla}$ (for items pertaining to the spin structure of $\mathbf{M}$ we use the symbols $\widetilde{D}$ and $\widetilde{\nabla}$, reserving the letters D and ∇ for the generic items below). We look for constitutive properties of this setting : we have at hand :

- a unital *-algebra $\mathcal{A}$ over $\mathbb{C}$ (over $\mathbb{R}$) with unit $\mathbb{I}$;
- a $\mathbb{Z}/2$-graded complex Hilbert space $\mathbf{H} = \mathbf{H}^0 \oplus \mathbf{H}^1$ (with grading involution γ^{2m+1});
- a $\mathbb{C}$-linear ($\mathbb{R}$-linear) *-representation $\mathcal{A} \ni a \to \underline{a} \in B(\mathbf{H})$ of $\mathcal{A}$ by even bounded operators ($\underline{a}$ is the continuous extension of the pointwise multiplication by $a \in \mathcal{A}$ in $\mathcal{S}_{\mathbf{M}}$);
- an odd self-adjoint operator $\widetilde{D}$ of $\mathbf{H}$ such that all $[\widetilde{D}, \underline{a}]$, $a \in \mathcal{A}$, are bounded, $\widetilde{D}^{-1}$ is a compact operator and $\widetilde{D}^{\text{-d}}$ has discrete eigenvalues μ_n such that [3]

$$\sum_{n=1}^{N} \mu_n = O(\ln N) \ . \tag{1.1}$$

The two following remarks are now crucial:

(i): the geodesic distance between two points $p, q \in \mathbf{M}$ is given as follows in terms of the Dirac operator: with $\|\ \|$ denoting the operator-norm of $B(\mathbf{H})$, one has:

$$\delta(p,q) = \sup\{|\, a(p) - a(q)|\, , \text{ a Lipschitzian s.t. } \|\, [\widetilde{D}, \underline{a}]\, \| \leq 1\} \ , \tag{1.2}$$

implying that the whole information concerning the manifold (metric topology, differential geometry) is encoded in the Dirac operator;

(ii): the constitutive properties recorded above do not refer to the fact that the *-algebra $\mathcal{A}$ is Abelian nor continuous, and thus make sense for *-algebras at large.

Given a *-algebra $\mathcal{A}$, one is accordingly led to *decree* that one endows $\mathcal{A}$ with a "noncommutative metric geometry" by requiring the existence of a d-dimensional spectral triple in the following sense:

1.2 Spectral triples

With $\mathcal{A}$ a unital *-algebra over $\mathbb{C}$ (over $\mathbb{R}$) an **even spectral triple** $(\mathcal{A}, \mathbf{H}, D)$ is the data of:

- a $\mathbb{Z}/2$-graded complex Hilbert space $\mathbf{H} = \mathbf{H}^0 \oplus \mathbf{H}^1$ (with grading involution χ);
- a $\mathbb{C}$-linear ($\mathbb{R}$-linear) *-representation $\mathcal{A} \ni a \to \underline{a} \in B(\mathbf{H})$ of $\mathcal{A}$ by even bounded operators (if the representation $a \to \underline{a}$ is faithful we can simply write a for $\underline{a}$);
- an odd self-adjoint operator D of $\mathbf{H}$ such that all $[D, \underline{a}]$, $\underline{a} \in \mathcal{A}$, are bounded, and D^{-1} is a compact operator.[1]

[3] Take the resolvent of the Dirac operator instead of its inverse if it has zero modes.

Note that this definition has two versions, one with $\mathcal{A}$ complex (customary) and one with $\mathcal{A}$ real (relevant to our physical applications). The spectral triple $(\mathcal{A}, \mathbf{H}, D)$ is called d$^+$- **summable** (or d-**dimensional**), $\mathrm{d} = 2m$, $m \in \mathbb{N}$, whenever the operator $D^{-\mathrm{d}}$ has its discrete eigenvalues μ_n such that $\sum_{n=1}^{N} \mu_n = O(\ln N)$.[4] Note that this condition rules out the algebras which are not "small" in the sense that their thereby defined "cohomological dimension" equals d. [5] Terminological remark: the phrase (d-dimensional) "spectral triple" $(\mathcal{A}, \mathbf{H}, D)$ has replaced the earlier term (d-dimensional) **K-cycle** $(\mathbf{H}, D)$ **of** $\mathcal{A}$, cf. [0], which we shall occasionally still use if we want to emphasize the operator D. Note also that Alain Connes also defines "odd spectral triples" for which the Hilbert space is not graded - we do not discuss these objects which we shall not need in physics. The expectation that these data endow $\mathcal{A}$ with a "metric geometry" is now comforted by the fact that they will successively yield the notions of formal forms, quantum forms, connections and their curvatures, and quantum volume form. We briefly review these items, which transpose to the "noncommutative frame" the essential features of the "**classical case**", i.e. that of the algebra $\mathcal{A} = C^\infty(\mathbf{M})$ of smooth functions on a compact spin Riemannian manifold $\mathbf{M}$ of even dimension d, equipped with its Dirac spectral triple

$$(\mathcal{A}, L^2(\mathcal{S}_{\mathbf{M}}), \widetilde{D}),\ \widetilde{D} = i\gamma^\mu \widetilde{\nabla}_\mu = i\gamma^\mu(\partial_\mu + \sigma_\mu))\ .$$

The phrase "noncommutative frame" covers all kind of generalizations: noncommutative, discrete, etc.

1.3 Formal forms

The **formal forms** are the elements of the **unital differential envelope** $\Omega\mathcal{A}$ **of** $\mathcal{A}$, a $\mathbb{N}$-graded differential algebra $(\Omega\mathcal{A}, \mathrm{d})$ defined as a $\mathbb{N}$-graded algebra through symbols:

$$\begin{cases} a \in \mathcal{A}\,. \\ da,\ a \in \mathcal{A}\,.\ , \end{cases} \tag{1.3}$$

and relations (where operations in the free algebra are indicated by a dot):

$$\begin{cases} \lambda \cdot a \dot{+} \mu \cdot b \dot{-} (\lambda a + \mu b) = 0 \\ a \cdot b \dot{-} (ab) = 0 \\ \lambda \cdot da \dot{+} \mu \cdot db \dot{-} d(\lambda a + \mu b) = 0 \\ da \cdot b \dot{+} a \cdot db \dot{-} d(ab) = 0 \\ d\mathbb{1} = 0\ . \end{cases} \tag{1.4}$$

consequently linearly spanned by elements of the form:

$$\omega = a_0 da_1 da_2 \ldots da_n, \qquad a_0, a_1, a_2, \ldots, a_n \in \mathcal{A}\ , \tag{1.5}$$

---

[4] To be quite exact: one requires the somewhat more stringent property: $D^{-\mathrm{d}} \in \mathcal{L}^{1+}(\mathbf{H})$, the definition ideal of the Dixmier trace Tr_ω (see below).

[5] "Large" algebras like loop-space algebras or the local algebras of quantum field theory are thereby excluded.

spanning the n-grade part $\Omega\mathcal{A}^n$ of $\Omega\mathcal{A}$. The differential d is specified as follows:

$$d\omega = \mathbb{1} da_0 da_1 da_2 \dots da_n \ . \tag{1.6}$$

With the algebraic *-operation specified by the requirement of reducing to the * of $\mathcal{A}$ on $\Omega\mathcal{A}^0$ and anticommuting with the differential d, $\Omega\mathcal{A}$ becomes a *-algebra.

Amongst the requirements (1.4) the significant ones are those of the lines four and five (essentially expressing the Leibniz rule in grade 0 and 1). The fourth line allows the reordering of arbitrary products which leads to (1.5), and moreover entails the facts that (1.6) defines a differential, and that (1.7) below defines a representation π_D of $\Omega\mathcal{A}$.

1.4 Quantum DeRham complex

The **quantum forms** are the elements of the **quantum DeRham complex** obtained as the quotient $\Omega_D\mathcal{A} = \Omega\mathcal{A}/\mathcal{I}$ of the $\mathbb{N}$-graded differential algebra $\Omega\mathcal{A}$ through the differential graded *-ideal $\mathcal{I} = \oplus_{n\in\mathbb{N}} (\mathcal{K}^n + d\mathcal{K}^{n-1})$, $\mathcal{K}^n = \mathrm{Ker}\, \pi_D \cap \Omega\mathcal{A}^n$, and $\mathrm{Ker}\, \pi_D$ is the kernel of the following bounded *-representation π_D of $\Omega\mathcal{A}$:

$$\pi_D(a_0 da_1 da_2 \dots da_n) = (-i)^n \underline{a}_0 [D, \underline{a}_1][D, \underline{a}_2] \dots [D, \underline{a}_n] \ , \qquad (a_0, a_1, a_2, \dots, a_n \in \mathcal{A}) \ . \tag{1.7}$$

Note that π_D is a *-representation of $\Omega\mathcal{A}$ *as a *-algebra*, [6] thus $\mathrm{Ker}\, \pi_D$ is an algebraic (not a differential) *-ideal, $\oplus_{n\in\mathbb{N}} \mathcal{K}^n$ is a graded *-ideal, and $\mathcal{I}$ is a differential graded *-ideal. The indicated procedure is the general one for generating differential ideals of differential algebras starting from algebraic (not differential) ideals. One has then in grade n:

$$(\Omega_D\mathcal{A})^n = \frac{\pi_D((\Omega\mathcal{A})^n)}{\pi_D(d\mathcal{K}^{n-1})} \left(= \frac{(\Omega\mathcal{A})^n / \mathcal{K}^n}{(\mathcal{K}^n + d\mathcal{K}^{n-1})/\mathcal{K}^n} = \frac{(\Omega\mathcal{A})^n / \mathcal{K}^n}{d\mathcal{K}^{n-1}/\mathcal{K}^n} \right) \tag{1.8}$$

in particular since $\mathcal{K}^0$ vanishes

$$\Omega_D\mathcal{A}^1 = \pi_D(\Omega\mathcal{A})^1 , \quad \Omega_D\mathcal{A}^1 = \frac{\pi_D(\Omega\mathcal{A})^2}{\pi_D(d\mathcal{K}^1)} \ . \tag{1.9}$$

The name " quantum DeRham complex" is justified by the fact that in the classical case one has an isomorphism of $\mathbb{N}$-graded differential *-algebras $(\Omega_{\tilde{D}}\mathcal{A}, d) \approx (\Omega(\mathbf{M}), \mathbf{d})$, $\Omega(\mathbf{M})$ the $\mathbb{N}$-graded *-algebra of differential forms on $\mathbf{M}$ with product its exterior product $\wedge$, and differential the exterior derivative $\mathbf{d}$ (specifically the isomorphism i: $\Omega_D\mathcal{A} \to \Omega(\mathbf{M})$ maps $a_0 da_1 \dots da_n$, $a_0, a_1, \dots, a_n \in \mathcal{A}$,

[6] It is intuitive that π_D is multiplicative: indeed the product in $\Omega\mathcal{A}$ essentially results for the Leibniz rule postulated in the fourth line (1.4), whilst the operation $[D, \underline{\cdot}]$ of bracketing by D also fulfills the Leibniz rule (without however being of vanishing square – in contradistinction from $[F, \underline{\cdot}]$, F the sign of D (with $F^2 = \mathbf{1}$).

into the classical form $\frac{1}{n!}a_0\mathbf{d}a_1\wedge\ldots\wedge\mathbf{d}a_n$ - this result in all grades requires non trivial arguing). In the representation one has then

$$\pi_{\tilde{D}}(a_0da_1\ldots da_n)=a_0\gamma^\mu\partial\mu a_1\ldots\gamma^\mu\partial\mu a_n=\gamma(a_0\mathbf{d}a_1\otimes\ldots\otimes\mathbf{d}a_n)\ , \qquad (1.10)$$

with γ the canonical map from covariant tensors to Clifford elements, ranging in the **Clifford module** $\mathbb{C}1(\mathrm{M})$. Note that the latter is economically obtained as the bijective image

$$\mathbb{C}1(\mathrm{M})=\gamma\left(\Omega(\mathrm{M})\right)=\oplus_{n=1,\ldots,\mathrm{d}}\,\gamma\left(\Omega(\mathrm{M})^n\right) \qquad (1.11)$$

of the DeRham complex, this yielding a canonical grading of $\mathbb{C}1(\mathrm{M})$ as a vector space (not as an algebra!).
The isomorphism i applied to $\psi=\pi_{\tilde{D}}(\omega)\operatorname{mod}\pi_{\tilde{D}}\left(d\mathcal{K}^{n-1}\right)=\gamma(T)$ then yields the component of maximum grade of T.

1.5 Quantum volume form

With Tr_ω the **Dixmier trace** the **volume-form** is the positive trace: [7]

$$\tau_D(\omega)=\mathrm{Tr}_\omega\left\{D^{\text{-d}}\pi_D(\omega)\right\}\ , \qquad (1.12)$$

and will be used
– for "integrating the curvature" in the next paragraph;
– for handling $\Omega_D\mathcal{A}$ concretely as follows: consider the positive semi-definite scalar product of $\pi_D(\Omega\mathcal{A})^n$:

$$(S,T)_n=\begin{cases}\mathrm{Tr}_\omega\left\{D^{\text{-d}}S^*T\right\}, & S,T\in\pi_D\left(\Omega\mathcal{A}\right)^n\ ,\\ \mathrm{Re}\,\mathrm{Tr}_\omega\left\{D^{\text{-d}}S^*T\right\}, & \text{if } \mathcal{A} \text{ is over the reals}\ ,\end{cases} \qquad (1.13)$$

yielding the Hilbert space completion $\mathbf{H}_n$. We obtain concrete objects representing the quantum n-forms by projecting in $\mathbf{H}_n$ onto the orthogonal complement of $\pi_D\left(d\mathcal{K}^{n-1}\right)$. With P_n the corresponding projection, we shall call P_nT the **concrete representative of the class of** $T\in\pi_D(\Omega\mathcal{A})^n$ in $\Omega_D\mathcal{A}^n$ (slight abuse of notation: stricto sensu this representative is the image under P_n of T/N, N the null-space of the scalar product (1.13). Note that $N=0$ in the classical case, as follows from (1.14), since the Clifford trace Tr is a faithful trace).

The linear form τ_D is a trace of $\pi_D(\Omega\mathcal{A})$ because the Dixmier trace Tr_ω vanishes on the commutators $\left[D^{\text{-d}},\underline{a}\right]$ which are trace-class. Positivity stems

[7] The Dixmier trace Tr_ω is a trace defined on the ideal $\mathcal{L}^{1+}(\mathbf{H})$ of $B(\mathbf{H})$ generated by the positive compact operators whose sum of the N first eigenvalues (taken in decreasing order) is $O(\ln N)$. The expression (1.12) makes sense, since $D^{\text{-d}}$ lies by assumption in the ideal $\mathcal{L}^{1+}(\mathbf{H})$, and $\pi_D(\omega)$ is bounded. Note that Tr_ω vanishes on trace - class operators.

from the fact that the trace Tr_ω is positive on products of two positive operators. τ_D is a faithful trace in all our applications. The name " volume-form" is justified by the fact that in the classical case $\tau_{\tilde{D}}$ is given as follows:

$$\tau_{\tilde{D}}(a_0 da_1 \ldots da_n) = \frac{1}{2^m m! \pi^m} \int \mathrm{Tr}\left\{a_0 \gamma(\mathbf{d}a_1) \ldots \gamma(\mathbf{d}a_n)\right\} dv \ , \qquad (1.14)$$
$$a_0, a_1, \ldots, a_n \in \mathcal{A} \ ,$$

where Tr denotes the normalized canonical trace of the Clifford algebra, dv denotes the Riemannian volume element of $\mathbf{M}$, and we use the shorthand $\gamma(\mathbf{d}a) = \gamma^\mu \partial_\mu a$, $a \in C^\infty(\mathbf{M})$ (coherent with our denoting by γ the canonical map from covariant tensors to Clifford elements, and using a bold $\mathbf{d}$ for the exterior derivative). Note that $\tau_{\tilde{D}}$ is now a faithful trace of $\pi_{\tilde{D}}(\Omega\mathcal{A})$ owing to faithfulness of the Clifford trace, thus all scalar products $(\cdot, \cdot)_n$ are positive-definite. Furthermore the expression (1.13) for $S \in \gamma((\Omega\mathcal{A})^i)$, $T \in \gamma((\Omega\mathcal{A})^j)$, vanishes if $i \neq j$.
The general context of (1.14) is the fact that for a pseudodifferential operator P of order -d one has the following expression of the Dixmier trace of its Sobolev extension:

$$\mathrm{Tr}_\omega(P) = \frac{1}{\mathrm{d}(2\pi)^{\mathrm{d}}} \iint dx d\xi \, \mathrm{Tr}\left\{\sigma^P(x, \xi)\right\} (|\xi|^2 - 1) \ , \qquad (1.15)$$

where Tr denotes the trace on the fiber and σ^P is the principal symbol of P. Formula (1.14) is the hinge of the computation of τ_D in our physical applications.

1.6 Quantum connections and curvature

We first recall the description of the classical connections and curvature under the form lending itself (by a simple periphrase) to noncommutative generalization. Let $\mathcal{E}$ be a smooth bundle of finite rank over $\mathbf{M}$. Replacing $\mathcal{E}$ by the equivalent data of its $C^\infty(\mathbf{M})$-module $\mathbf{E}$ of smooth sections, one completely algebraïzed the concept: $\mathbf{E}$ is indeed a projective-finite $C^\infty(\mathbf{M})$-module,[8] all such objects being obtained from smooth bundles over $\mathbf{M}$ in this fashion (*Serre-Swann theorem*). Let now ∇ be a connection of $\mathcal{E}$, usually considered as yielding linear maps $\nabla_\xi : \mathbf{E} \to \mathbf{E}$ indexed by tangent vectors ξ from which they depend $C^\infty(\mathbf{M})$-linearly. Writing $\nabla_\xi \eta = \nabla(\eta, \xi)$ we get a map ∇: $\mathbf{E} \to \mathbf{E} \otimes_{C^\infty(\mathbf{M})} \Omega(\mathbf{M})^1$ which extends uniquely as an odd $\mathbf{d}$-derivation ∇ of the right $\Omega(\mathbf{M})$-module $\mathbf{E}_\Omega = \mathbf{E} \otimes_{C^\infty(\mathbf{M})} \Omega(\mathbf{M})$: the **exterior covariant derivative:** we have namely the module-derivation property:

$$\nabla(\eta\omega) = (\nabla\eta)\omega + (-1)^{\partial\eta}\eta\mathbf{d}\omega, \quad \eta \in \mathbf{E}, \ \omega \in \Omega(\mathbf{M}) \ , \qquad (1.16)$$

the corresponding curvature being then the $\Omega(\mathbf{M})$-module endomorphism ∇^2.

[8] The projective-finite $\mathcal{A}$-modules $\mathbf{E}$ are characterizable as possessing a dual basis (e_i, ε^i), $e_i \in \mathbf{E}$, $\varepsilon^i \in \mathbf{E}^*$, fulfilling the completeness condition $\sum_i e_i \varepsilon^i = id_{\mathbf{E}}$, where $(e_i, \varepsilon^i)(\xi) = e_i \varepsilon^i(\xi)$, $\xi \in \mathbf{E}$.

In order to get the noncommutative generalization we need only paraphrase the foregoing:
(i): general case (the reader can forego this and pass to the special case $\mathbf{E} = \mathcal{A}$ in (ii) below solely needed in our applications): with $\mathbf{E}$ a projective-finite module over $\mathcal{A}$ the **quantum connections of E** are the grade-one graded d-derivations ∇ of the $\mathbb{N}$-graded right $\Omega_D\mathcal{A}$-module $\mathbf{E}_\Omega = \mathbf{E}\otimes_{\mathcal{A}}\Omega_D\mathcal{A}$: one has $\nabla\mathbf{E}_{\Omega^n} \subset \mathbf{E}_{\Omega^{n+1}}$, and

$$\nabla(\eta\omega) = (\nabla\eta)\omega + (-1)^{\partial\eta}\eta d\omega, \quad \eta \in \mathbf{E}, \quad \omega \in \Omega_D(\mathcal{A}) \ , \tag{1.17}$$

with corresponding **curvature** the $\Omega_D(\mathcal{A})$-module-endomorphism ∇^2: one has:

$$\nabla^2(\eta\omega) = (\nabla\eta)\omega, \quad \eta \in \mathbf{E},\ \omega \in \Omega_D(\mathcal{A}) \ . \tag{1.18}$$

Given a coordinatization $\{e_i, \varepsilon^i\}_{i=1,\dots,n}$ of $\mathbf{E}$ inducing a coordinatization of $\mathbf{E}_\Omega$, the latter are parametrized as follows: we have

$$\nabla = d + \rho, \quad \rho = (\rho^i_k) \in \mathrm{End}_{\Omega_D\mathcal{A}}(\mathbf{E}_\Omega)^1$$

(identified with its left action on $\mathbf{E}_\Omega$) the matrix-valued **connection one-form** of ∇, and $\nabla^2 = \theta = (\theta^i_k) = d\rho + \rho^2 \in \mathrm{End}_{\Omega_D\mathcal{A}}(\mathbf{E}_\Omega)^2$, with d acting coordinate-wise and ρ^i_k and θ^i_k acting by multiplication from the left. One gets a trace $\underline{\tau}_D$ of $\mathrm{End}_{\Omega_D\mathcal{A}}\mathbf{E}_\Omega$ by specifying the latter on "dyads" as follows:

$$\underline{\tau}_D(X\Phi) = \tau_D(\Phi X), \quad X \in \mathbf{E}_\Omega,\ \Phi \in \mathbf{E}^*_\Omega \ , \tag{1.19}$$

where $X\Phi \in \mathrm{End}_{\Omega_D\mathcal{A}}\mathbf{E}_\Omega$ is given by $(X\Phi)Y = X(\Phi Y)$, $Y \in \mathbf{E}_\Omega, \Phi X$, ΦY the values of Φ on X, resp. Y.
(ii): With $\mathcal{A}$ considered as a (free) right module over itself, the **quantum connections (of $\mathcal{A}$)** are the grade-one graded d-derivations ∇ of the (free) right $\Omega_D\mathcal{A}$-module $\Omega_D\mathcal{A}$ ($= \mathcal{A}\otimes_{\mathcal{A}}\Omega_D\mathcal{A}$): one has $\nabla\mathbf{E}_{\Omega^n} \subset \mathbf{E}_{\Omega^{n+1}}$, and

$$\nabla(\eta\omega) = (\nabla\eta)\omega + (-1)^{\partial\eta}\eta d\omega, \quad \eta \in \mathbf{E},\ \omega \in \Omega_D(\mathcal{A}) \ , \tag{1.20}$$

with corresponding **curvatures** the $\Omega_D(\mathcal{A})$-module-endomorphisms ∇^2: one has:

$$\nabla^2(\eta\omega) = (\nabla\eta)\omega, \quad \eta \in \mathbf{E},\ \omega \in \Omega_D(\mathcal{A}) \ . \tag{1.21}$$

As follows immediately from the fact that the difference of two d-derivations is an endomorphism, these objects are parametrized as follows: we have $\nabla = d + \rho$, $\rho \in \Omega_D\mathcal{A}^1 \cong \pi_D(\Omega_D\mathcal{A}^1)$ the **connection form of** ∇ with **curvature** $\nabla^2 = \theta = d\rho + \rho^2 \in \Omega_D\mathcal{A}^2 = \pi_D(\Omega\mathcal{A})^2/\pi_D(d\mathcal{K}^1)$, both acting on $\Omega_D\mathcal{A}$ by multiplication from the left. Note that in the latter formula $d\rho + \rho^2$ is supposed to be computed in $\Omega_D\mathcal{A}$, with ρ belonging to $\Omega_D\mathcal{A}^1$. Concretely this consists in computing $\theta' = d\rho' + \rho'^2$ in $\Omega\mathcal{A}$ for $\rho' \in \Omega\mathcal{A}^1$ such that $\rho = \pi_D(\rho')$, and taking $\theta = \pi_D(\theta') \bmod \pi_D(d\mathcal{K}^1)$.

The connection ρ is **compatible** (or **Euclidean**) iff $\rho^* = -\rho$, implying $\theta = \theta^*$, the self-adjoint $i\rho$ is called the (quantum) **vector potential**. The **gauge group** G of $\mathcal{A}$ is its group of unitaries:

$$G = \{u \in \mathcal{A};\ uu^* = u^*u = \mathbb{1}\} \ , \tag{1.22}$$

with the gauge transformation $u \in G$ acting on $\Omega_D\mathcal{A}$ by multiplication from the left. Using the same notation u for this action, the action of $u \in G$ is as follows: we have on connections:

$$\nabla = d + \rho \to {}^u\nabla = u^* \circ \nabla \circ u = d + {}^u\rho \quad \text{with} \quad {}^u\rho = u^*du + u^*\rho u \ , \tag{1.23}$$

and on curvatures:

$$\nabla^2 = \theta \to {}^u\theta = {}^u\nabla^2 = u^*\theta^2 u \ , \tag{1.24}$$

where the operations in the r.h.s. of these formulae are within $\Omega_D\mathcal{A}$. Defining analogously the action of G on $\Omega\mathcal{A}$, the canonical map: $\Omega\mathcal{A} \to \Omega_D\mathcal{A}$ intertwines the two actions.

1.7 The quantum Yang-Mills algorithm (for a given even d-dimensional spectral triple $(\mathcal{A}, \mathbf{H}, D)$)

We now dispose of all we need to transcribe Yang-Mills to the noncommutative frame. We shall do this in the previous special case where the module is the algebra itself (case at hand for our applications of interest: two-point functions, electrodynamics, the standard model - the case of a general projective-finite right module $\mathbf{E}$ could be treated analogously with $\underline{\tau}_D$ instead of τ_D). Our task consists in paraphrasing the definition of the familiar Yang-Mills action density as the "integral of the square of the curvature". The "integral" is furnished by the trace τ_D defined on $\pi_D(\Omega\mathcal{A})$, whilst ρ and θ belongs to $\Omega_D\mathcal{A}$. We thus specify the quantum Yang-Mills action in two steps: first define the **primary Yang-Mills action** as the following functional of $\rho' \in \Omega\mathcal{A}^1$ such that $\rho = \pi_D(\rho')$

$$YM^0(\rho') = \mathrm{Tr}_\omega\{D^{-4}\pi_D(\theta')^2\} \quad \text{with} \quad \theta' = d\rho' + \rho'^2 \in \Omega\mathcal{A}^2 \ . \tag{1.25}$$

(Observe that as ρ' ranges through the anti-image of ρ for $\pi_D, \pi_D(d\rho' + \rho'^2)$ ranges through the class modulo $\pi_D(d\mathcal{K}^1)$). Then define the **Yang-Mills action proper** by minimizing over the $\Omega_D\mathcal{A}$-class

$$YM(\rho) = \inf\{YM^0(\rho');\ \pi_D(\rho') = \rho\} \ . \tag{1.26}$$

This recipe is then such that:
(i): YM is gauge-invariant: $YM({}^u\rho) = YM(\rho)$, $u \in G$. The reason for this is as follows: gauge-invariance is manifest if one replaces D by the covariant generalized Dirac operator $D_\rho = D + \rho$ such that $D_{u_\rho} = \pi_D(u^*)D_\rho\pi_D(u)$, $u \in G$. However, this does not alter those expressions since D_ρ^{-4} and D^4 differ by a trace-class operator vanishing under the Dixmier trace.

(ii): One has the following expression in terms of the concrete representative of θ:

$$YM(\rho) = \begin{cases} (\pi_D(\theta), P_2\pi_D(\theta))_2 = \mathrm{Tr}_\omega\{D^{-4}P_2\pi_D(\theta^2)\}; \\ \text{for a real algebra take instead} \\ (\mathrm{Re}\,(\pi_D(\theta), P_2\pi_D(\theta))_2 = \mathrm{Re}\ \mathrm{Tr}_\omega\left\{D^{-4}P_2\pi_D(\theta^2)\right\} \ . \end{cases} \tag{1.27}$$

1.8 The electrodynamics case

Mathematically this is the "classical case" with d=4. We recall that, for $a \in \mathcal{A} = C^\infty(\mathbf{M})$, $[\widetilde{D}, \underline{a}] =$ (multiplication by) $\gamma^\mu \partial_\mu a = \gamma(\mathbf{d}a)$. For $\rho' = \sum_j a_0^j da_1^j \in \Omega\mathcal{A}$, selfadjoint, $a_0^j, a_1^j \in \mathcal{A}$, yielding the quantum potential:

$$\pi_{\widetilde{D}}(\rho') = \sum_j a_0^j \gamma^\mu \partial_\mu a_1^j = \gamma(A) \ , \tag{1.28}$$

with $A = \sum_j a_0^j \mathbf{d} a_1^j$ the classical potential. Then:

$$\begin{aligned} \pi_{\widetilde{D}}(d\rho' + \rho'^2) &= \pi_{\widetilde{D}}\Big(\sum_j da_0^j da_1^j\Big) + \gamma(A)\gamma(A) \\ &= \sum_j \gamma\left(\mathbf{d}a_0^j\right)\gamma\left(\mathbf{d}a_1^j\right) + \gamma(A)\gamma(A) = \gamma\Big(\sum_j \mathbf{d}a_0^j \otimes \mathbf{d}a_1^j\Big) + \gamma\left(A \otimes A\right) \\ &= \sum_j \left[\gamma\left(\frac{1}{2}\mathbf{d}a_0^j \wedge \mathbf{d}a_1^j + \frac{1}{2}\mathbf{d}a_0^j \vee \mathbf{d}a_1^j\right)\right] + (A, A) \\ &= \frac{1}{2}\gamma(\mathbf{d}A) + \sum_j \left(\mathbf{d}a_0^j, \mathbf{d}a_1^j\right) + (A, A) \ . \end{aligned} \tag{1.29}$$

We have to project this on the orthogonal complement of $\pi_{\widetilde{D}}(d\mathcal{K}^1)$. Since $\sigma' = \sum_j b_0^j db_1^j \in \Omega\mathcal{A}^1$ belongs to $\mathcal{K}^1$ whenever $B = \sum_j b_0^j \mathbf{d} b_1^j$ vanishes, we then have:

$$\pi_{\widetilde{D}}\left(d\sigma'\right) = \pi_{\widetilde{D}}\Big(d\sum_j b_0^j db_1^j\Big) = \frac{1}{2}\gamma\left(\mathbf{d}B\right) + \sum_j \left(\mathbf{d}b_0^j, \mathbf{d}b_1^j\right) = \sum_j \left(\mathbf{d}b_0^j, \mathbf{d}b_1^j\right) \ , \tag{1.30}$$

which ranges through $C^\infty(\mathbf{M}, \mathbb{R})$ as σ' ranges through $\mathcal{K}^1$:

$$\pi_{\widetilde{D}}(d\mathcal{K}^1) = C^\infty(\mathbf{M}, \mathbb{R}) \ . \tag{1.31}$$

Hence $P_2\pi_{\widetilde{D}}(d\rho' + \rho'^2)$ arises by asking $\frac{1}{2}\gamma(\mathbf{d}A) + \sum_j(\mathbf{d}a_0^j, \mathbf{d}a_1^j) + (A, A)$ to be orthogonal to the zero-grade part $C^\infty(\mathbf{M}, \mathbb{R})$ of the Clifford bundle: since the latter is orthogonal to the one-grade part, this requires vanishing of $(A, A)_1 + \sum_j(\mathbf{d}a_0^j, \mathbf{d}a_1^j)$, hence leads to $P_2\pi_{\widetilde{D}}(d\rho' + \rho'^2) = \frac{1}{2}\gamma(\mathbf{d}A)$. Plugging this into (1.27) then yields the familiar classical action of electrodynamics.

1.9 The two-point algebra $\mathbb{C} \oplus \mathbb{C}$ (embryonal Higgs)

We now look at the simplest possible non trivial example

$$\mathcal{A} = \mathbb{C} \oplus \mathbb{C} = \{a = (f, g);\ f, g \in \mathbb{C}\} . \tag{1.32}$$

For the discrete algebra $\mathcal{A}$ we define the differential geometry by the following even spectral triple $(\mathcal{A}, \mathbf{H}\, D, \chi)$: $\mathbf{H} = \mathbb{C}^N \oplus \mathbb{C}^N$, (where the integer N prefigures the number of fermion families). Linear endomorphisms of $\mathbf{H}$ are described by 2×2 matrices with entries in $\mathbf{M}_2(\mathbb{C}) \otimes \mathbf{M}_N(\mathbb{C})$: in this sense we set:

$$\chi = \begin{pmatrix} \mathbb{I}_N & 0 \\ 0 & -\mathbb{I}_N \end{pmatrix} , \tag{1.33}$$

$$\underline{a} = \begin{pmatrix} f\mathbb{I}_N & 0 \\ 0 & g\mathbb{I}_N \end{pmatrix}, \qquad a = (f, g) \in \mathcal{A}_2 , \tag{1.34}$$

and

$$D = \begin{pmatrix} 0 & M^* \\ M & 0 \end{pmatrix} , \tag{1.35}$$

where the invertible $N \times N$ complex matrix M with adjoint M^* prefigures the fermion mass matrix. For

$$\omega = \sum_{i=1}^{n} a_0^i da_1^i, \quad a_0^i = (f_0^i, g_0^i),\ a_1^i = (f_1^i, g_1^i) \in \mathcal{A} , \tag{1.36}$$

one easily computes

$$\pi_D(\omega) = \sum_{i=1}^{n} \begin{pmatrix} 0 & -f_0^i\,(f_1^i, g_1^i)\, M \\ g_0^i\,(f_1^i, g_1^i)\, M & 0 \end{pmatrix} , \tag{1.37}$$

and

$$\pi_D(\omega) = \sum_{i=1}^{n} (f_0^i, g_0^i)\,(f_1^i, g_1^i) \begin{pmatrix} M^*M & 0 \\ 0 & MM^* \end{pmatrix} . \tag{1.38}$$

This implies that $\mathcal{K}^1$ consists of the ω for which

$$\sum_{i=1}^{n} f_0^i\,(f_1^i, g_1^i) = \sum_{i=1}^{n} g_0^i\,(f_1^i, g_1^i) = 0 . \tag{1.39}$$

For those ω the value of $\pi_D(d\omega)$ vanishes, hence $\pi_D(d\mathcal{K}^1)$ vanishes. Since $\mathcal{K}^0$ vanishes, one has thus $\Omega_D\mathcal{A}^1 = \pi_D(\Omega_D\mathcal{A}^1)$ and $\Omega_D\mathcal{A}^2 = \pi_D(\Omega_D\mathcal{A}^2)$. The one-form of the compatible connections are the antihermitean elements of $\pi_D(\Omega_D\mathcal{A}^1)$, thus of the form

$$\rho = \begin{pmatrix} 0 & -hM^* \\ hM^* & 0 \end{pmatrix}, \qquad h \in \mathbb{C} , \tag{1.40}$$

with

$$d\rho = -(h + \bar{h}) \begin{pmatrix} M^*M & 0 \\ 0 & MM^* \end{pmatrix} , \tag{1.41}$$

and

$$\rho^2 = -\bar{h}h \begin{pmatrix} M^*M & 0 \\ 0 & MM^* \end{pmatrix} , \tag{1.42}$$

thus with curvature

$$\theta = -(\bar{h}h + h + \bar{h}) \begin{pmatrix} M^*M & 0 \\ 0 & MM^* \end{pmatrix} . \tag{1.43}$$

Setting $h+1=\phi$ one has $\bar{h}h+h+\bar{h}=|\phi|^2-1$. In this finite case the Dixmier trace reduces to the usual trace Tr. The Yang-Mills action thus turns to be

$$\mathrm{Tr}(\theta^2) = 2\left(|\phi|^2-1\right)^2 \mathrm{Tr}\left[(M^*M)^2\right] \tag{1.44}$$

the characteristic form of a Higgs potential. We obtain the "embryonal Higgs "!

Morale of our findings: we have up to now described two spectral triples : that of the (Euclidean) Dirac operator, relevant to (Euclidean, compact) space-time, leading to (the Euclidean version of) electrodynamics; and the two-point function spectral triple, remindful of the Higgs particle, thus tentatively interpretable as describing the 2N "inner degrees of freedom" of a (simplified) world with N generations of one single species of fermions. For constructing a physical model we need a spectral triple combination of the two items. At this point mathematics plays into our hands with the following thoroughly canonical notion (in fact coïnciding with the (exterior, easy) product of K-cycles in Kasparov's KK -theory):

1.10 Definition-lemma

With $(\mathcal{A}',\mathbf{H}',D')$ and $(\mathcal{A}'',\mathbf{H}'',D'')$ two even spectral triples with respective grading involution χ' and χ'', let

$$\begin{cases} \mathcal{A} = \mathcal{A}' \otimes \mathcal{A}'' \\ \mathbf{H} = \mathbf{H}' \otimes \mathbf{H}'' \\ \chi = \chi' \otimes \chi'' \\ \underline{a' \otimes a''} = \underline{a}' \otimes \underline{a}'' ,\ a' \in \mathcal{A}',\ a'' \in \mathcal{A}'' \\ D = D' \otimes \mathbb{1}'' + \chi' \otimes D'' \end{cases} , \tag{1.45}$$

then $(\mathcal{A},\mathbf{H},D)$ is an even spectral triple with grading involutions χ, called the **tensor product of** $(\mathcal{A}',\mathbf{H}',D')$ **and** $(\mathcal{A}'',\mathbf{H}'',D'')$. *If furthermore $(\mathcal{A}',\mathbf{H}',D')$ and $(\mathcal{A}'',\mathbf{H}'',D'')$ are d'^{+}-,respectively d''^{+}-summable, $(\mathcal{A},\mathbf{H},D)$ is $(\mathrm{d}'+\mathrm{d}'')^{+}$-summable.*

These claims are easily established. Note that the procedure for defining D resembles the tensor product of graded derivation (infinitesimal operators).

At this point we could calculate the Yang-Mills action of the tensor product of the two above spectral triples and investigate the thus obtained model of a world with a simplified fermion structure. Since we are interested in the real world, we shall instead construct right away in the next Chapter **2** the "inner spectral triple" describing the "inner degrees of freedom" of the electroweak sector of our world featuring N generation of 7 fermions (thus far the experimental evidence is that N=3, but the construction is the same for general N).

1.11 Remark

Of course, with $(\mathcal{A}',\mathbf{H}',D')$ and $(\mathcal{A}'',\mathbf{H}'',D'')$ two even spectral triples with respective grading involutions χ' and χ'', and respective left-module actions $a \to a'$ on $\mathbf{H}'$, and $a \to a''$ on $\mathbf{H}''$, setting

$$\begin{cases} \mathbf{H} = \mathbf{H}' \oplus \mathbf{H}'' \\ \chi = \chi' \oplus \chi'' \\ \underline{a' \oplus a''} = \underline{a}' \oplus \underline{a}'' , \; a' \in \mathcal{A}', a'' \in \mathcal{A}'' \\ D = D' \oplus D'' \end{cases} , \tag{1.46}$$

yields an is an even spectral triple $(\mathcal{A},\mathbf{H},D)$ with grading involution χ, called the **direct sum of** $(\mathcal{A}',\mathbf{H}',D')$ **and** $(\mathcal{A}'',\mathbf{H}'',D'')$. *If furthermore* $(\mathcal{A}',\mathbf{H}',D')$ *and* $(\mathcal{A}'',\mathbf{H}'',D'')$ *are* d^+*-summable, so is* $(\mathcal{A},\mathbf{H},D)$.

2 The Electroweak Inner Spectral Triple

We met in [**1.9**] above (two-point algebra) a spectral triple of zero cohomological dimension whose Yang-Mills action has the form of a Higgs potential - which thus seems related with the inner degrees of freedom of elementary particles.We shall now construct a more realistic such object actually related to the inner degrees of freedom of the electroweak sector of standard model: a spectral triple acting on a fermion inner space of 21 dimension (3 families of 3 leptons and 4 quarks which for the moment we take colourless, concentrating on electroweak aspects). Our construction rests on our phenomenological knowledge of the electroweak group $U(1) \times SU(2)$ which, postulated to be the group of unitaries of the inner space algebra, induces us to take for the latter the direct sum $\mathbb{C} \oplus \mathbb{H}$ of the complex numbers and the quaternions.

We recall that the real vector space

$$\mathbb{H}=\left\{\begin{pmatrix} a & b \\ -\bar{b} & \bar{a}\end{pmatrix}, \; a,b \in \mathbb{C}\right\} \text{ or } \mathbb{H}=\left\{\begin{pmatrix} H_2 & H^1 \\ -H_1 & H^2\end{pmatrix}, \; H^1=\overline{H}_1, \; H^2=\overline{H}_2 \in \mathbb{C}\right\}^{9} \tag{2.1}$$

(consisting of sums of hermitean scalar and antihermitean traceless matrices) is a real subalgebra of $M_2(\mathbb{C})$ stable under the hermitean conjugation * and intersecting the unitaries along $SU(2)$. As a real vector space, $\mathbb{H}$ is spanned by the unit matrix $\mathbb{I}$ and the 3 quaternions:

$$I = \begin{pmatrix} 0 & i \\ i & 0\end{pmatrix}, \quad J = \begin{pmatrix} 0 & 1 \\ -1 & 0\end{pmatrix}, \quad K = \begin{pmatrix} i & 0 \\ 0 & -i\end{pmatrix} \tag{2.2}$$

[9] This alternative notation for quaternions will be used for the "Higgs doublet".

fulfilling $JK = -KJ = -I = I^* = I^{-1}$, $KI = -IK = -J = J^* = J^{-1}$, $IJ = -JI = -K = K^* = K^{-1}$, thus orthogonal for the Euclidean scalar product with corresponding norm $|\ |$:[10]

$$\left(q,q'\right)=\frac{1}{2}\mathrm{Tr}\left(q^*q'\right)=\frac{1}{2}\mathrm{Tr}\left(q'^*q\right)\ \text{ s.t. } \frac{1}{2}\left(q^*q'+q'^*q\right)=\left(q,q'\right)\mathbb{1},\ q,\,q'\in\mathbb{H},$$

$$|q|^2=(q,q)=\frac{1}{2}\mathrm{Tr}\,(q^*q)=\mathrm{Det}(q)=\mathrm{Det}(q^*)\ \left(\text{ thus } q^*q=qq^*=|q|^2\mathbb{1}\right)\ . \tag{2.3}$$

2.1 Basic *-algebra. Gauge group

The basic *-algebra is the real algebra

$$\mathbf{A}_{ew} = \mathbb{C} \oplus \mathbb{H} = \{(p,q);\ \ p \in \mathbb{C},\ \ q \in \mathbb{H}\} \tag{2.4}$$

provided with the *-operation ($\mathbb{R}$-linear, product-reversing involution):

$$(p,q)^* = (\bar{p},q^*),\ \ p \in \mathbb{C},\ q \in \mathbb{H}\ , \tag{2.5}$$

direct sum of the complex conjugation of complex numbers and hermitean conjugation of quaternions.

The gauge group of $\mathbf{A}_{ew}$ is then

$$\left\{\mathbf{u}=(u,v)\in\mathbf{A}_{ew},\ u^*u=uu^*=\mathbb{1},\ v^*v=vv^*=\mathbb{1}\right\} = U(1)\times SU(2)\ , \tag{2.6}$$

i.e. the gauge group of the electroweak theory. Accordingly, $\mathbf{A}_{ew}$ will also serve (taken as a right module over itself) as the module serving to define connections. Note that the bijection:

$$\mathbb{C} \ni p \leftrightarrow p_{\mathrm{diag}} = \begin{pmatrix}\bar{p} & 0\\ 0 & p\end{pmatrix} \in \mathbb{H}_{\mathrm{diag}}\ (\mathbb{H}_{\mathrm{diag}} \text{ the set of diagonal quaternions}) \tag{2.7}$$

is an isomorphism of real algebras, significant in the sequel inasmuch as the elements of $\mathbb{C}$ often behave as diagonal quaternions. We occasionally write p for p_{diag} when the context is clear.

2.2 The spectral triple $(\mathbf{A}_{ew}, H_f, D_f)$

The spectral triple ($\mathbf{A}_{ew}$,H_f,D_f) is the direct sum $(\mathbf{A}_{ew}, H_l, D_l)\oplus(\mathbf{A}_{ew},H_q,D_q)$ of a leptonic spectral triple ($\mathbf{A}_{ew}$,H_l,D_l) and a quarkonic spectral triple ($\mathbf{A}_{ew}$,H_q,D_q) The leptonic spectral triple ($\mathbf{A}_{ew}$,H_l,D_l) is as follows: the Hilbert space is:

$$H_l = [\underset{e_R}{\mathbb{C}^1_R} \oplus \underset{\nu_L\, e_L}{\mathbb{C}^2_L}] \otimes \mathbb{C}^N \tag{2.8}$$

The respective indices f, l and q stand for "fermion", "lepton" and "quark". The symbols e (electron), ν (neutrino), u (upper quark), d (lower quark), affected

[10] Note that quaternions have real traces, since $\mathrm{Tr}\,1 = 2$, and $\mathrm{Tr}\,I = \mathrm{Tr}\,J = \mathrm{Tr}\,K = 0$.

by subscripts L (left handed) or R (right-handed) indicate the type of fermions. Tensoring by $\mathbb{C}^N$ corresponds to the existence of N fermion families (N=3 in nature). (We note in anticipation that subsequent introduction of the gluons will leave the leptonic spectral triple unchanged, whilst endowing the quarkonic spectral triple with a threefold color multiplicity, with the color degrees of freedom neither acted upon by the electroweak algebra nor by the Dirac operator (the latter " entirely on the electroweak side"). For the moment we plan to describe merely the electroweak sector. Writing the endomorphisms of H_l as 3×3 matrices with entries in $M_N(\mathbb{C})$, and denoting by $\mathbb{I}_N$ the identity of $\mathbb{C}^N$, one defines respectively as follows the representation π_l of $\mathbf{A}_{ew}$ on H_l, the "Dirac operator" D_l, and the "chirality" χ_l:

$$\pi_l((p,q)) = \begin{pmatrix} p\mathbb{I}_N & 0 & 0 \\ 0 & a\mathbb{I}_N & b\mathbb{I}_N \\ 0 & -\bar{b}\mathbb{I}_N & \bar{a}\mathbb{I}_N \end{pmatrix} \quad \left(p,\ q = \begin{pmatrix} a & b \\ -\bar{b} & \bar{a} \end{pmatrix}\right) \in \mathbf{A}_{ew} , \tag{2.9}$$

(columns labelled e_R, ν_L, e_L)

$$D_l = \begin{pmatrix} 0 & 0 & M_e^* \\ 0 & 0 & 0 \\ M_e & 0 & 0 \end{pmatrix} , \tag{2.10}$$

(columns labelled e_R, ν_L, e_L)

$$\chi_l = \begin{pmatrix} \mathbb{I}_N & 0 & 0 \\ 0 & -\mathbb{I}_N & 0 \\ 0 & 0 & -\mathbb{I}_N \end{pmatrix} . \tag{2.11}$$

(columns labelled e_R, ν_L, e_L)

The quarkonic spectral triple ($\mathbf{A}_{ew}$,H_q,D_q) is as follows: the Hilbert space is:

$$H_q = [\underset{u_R\, d_R}{\mathbb{C}_R^2} \oplus \underset{u_L\, d_L}{\mathbb{C}_L^2}] \otimes \mathbb{C}^N . \tag{2.12}$$

Writing the endomorphisms of H_l as 4×4 matrices with entries in $M_N(\mathbb{C})$, one has now:

$$\pi_q((p,q)) = \begin{pmatrix} \bar{p}\mathbb{I}_N & 0 & 0 & 0 \\ 0 & p\mathbb{I}_N & 0 & 0 \\ 0 & 0 & a\mathbb{I}_N & b\mathbb{I}_N \\ 0 & 0 & -\bar{b}\mathbb{I}_N & \bar{a}\mathbb{I}_N \end{pmatrix} , \quad \left(p,\ q = \begin{pmatrix} a & b \\ -\bar{b} & \bar{a} \end{pmatrix}\right) \in \mathbf{A}_{ew} , \tag{2.13}$$

(columns labelled u_R, d_R, u_L, d_L)

$$D_q = \begin{pmatrix} 0 & 0 & M_u^* & 0 \\ 0 & 0 & 0 & M_d^* \\ M_u & 0 & 0 & 0 \\ 0 & M_d & 0 & 0 \end{pmatrix} , \tag{2.14}$$

(columns labelled u_R, d_R, u_L, d_L)

$$\chi_q = \begin{pmatrix} \mathbb{1}_N & 0 & 0 & 0 \\ 0 & \mathbb{1}_N & 0 & 0 \\ 0 & 0 & -\mathbb{1}_N & 0 \\ 0 & 0 & 0 & -\mathbb{1}_N \end{pmatrix} \quad (\text{columns } u_R \; d_R \; u_L \; d_L) . \tag{2.15}$$

Here M_e, M_d, M_u, are the $N \times N$ complex **mass matrices** of the respective electron, lower quark, and upper quark families, assumed to be invertible. Since one passes from the relations (2.13)-(2.15) to the relations (2.9)-(2.11), trough the changes $M_u \to 0, M_d \to M_e$ followed by restriction to the right-lower corner 3×3 matrix, these changes (for which we coin the phrase **"leptonic reduction"**) yield the description of the leptonic sector in terms of that of the quark sector, on which we now concentrate.

For our calculations it will be convenient, using the previous convention $p_{\text{diag}} = \begin{pmatrix} \bar{p} & 0 \\ 0 & p \end{pmatrix}$ and the notation:

$$\begin{gathered} \mathbb{M} = E \otimes M_u + F \otimes M_d = \mathbb{1} \otimes \sigma + iK \otimes \delta \ , \\ \sigma = \frac{1}{2}(M_u + M_d) \ , \quad \delta = \frac{1}{2}(M_u - M_d) \ , \\ \text{with } E = \begin{pmatrix} 1 & 0 \\ 0 & 0 \end{pmatrix} , \ F = \begin{pmatrix} 0 & 0 \\ 0 & 1 \end{pmatrix} \text{ and } K = \begin{pmatrix} i & 0 \\ 0 & -i \end{pmatrix} \in \mathbb{H} \ , \end{gathered} \tag{2.16}$$

to handle the 4×4 matrices (2.13) through (2.15) as the 2×2 matrices (2.18) through (2.21) below with entries in $M_2(\mathbb{C}) \times M_N(\mathbb{C})$, corresponding to the decomposition:

$$H_q = H_{q_R} \oplus H_{q_L} \quad \text{with} \quad \begin{cases} H_{q_R} = \mathbb{C}_R^2 \otimes \mathbb{C}^N \\ H_{q_L} = \mathbb{C}_L^2 \otimes \mathbb{C}^N \end{cases} , \tag{2.17}$$

(whereby the 4×4 matrix reading $\begin{pmatrix} S_1^1 & S_2^1 \\ S_1^2 & S_2^2 \end{pmatrix}$ in terms of 2×2 blocks $S_k^i \in M_2(M_N(\mathbb{C}))$ becomes the 2×2 matrix $\begin{pmatrix} T_1^1 & T_2^1 \\ T_1^2 & T_2^2 \end{pmatrix}$ with $T_k^i = \sum_{i,k} e_i^k \otimes S_k^i$, e_i^k the $M_2(\mathbb{C})$-matrix units):

$$\pi_q((p,q)) = \begin{pmatrix} \mathbb{p} & 0 \\ 0 & \mathbb{q} \end{pmatrix} \quad (\text{columns } R \; L) \tag{2.18}$$

with

$$\begin{cases} \mathbb{p} = p_{\text{diag}} \otimes \mathbb{1}_N \\ \mathbb{q} = q \otimes \mathbb{1}_N \end{cases} , \tag{2.19}$$

$$D_q = \begin{pmatrix} 0 & \mathbb{M}^* \\ \mathbb{M} & 0 \end{pmatrix} \quad (\text{columns } R \; L) , \tag{2.20}$$

$$\chi_q = \begin{pmatrix} \mathbb{I}\otimes\mathbb{I}_N & 0 \\ 0 & -\mathbb{I}\otimes\mathbb{I}_N \end{pmatrix} \quad (R,\ L) \ . \tag{2.21}$$

We gather some useful formulae. Note the important fact that $\mathbb{p}$ and $\mathbb{M}$ commute. From (2.13), (2.14) we compute, for $(p,q)\in \mathbf{A}_{ew}$:

$$D_q\pi_q((p,q)) = \begin{pmatrix} 0 & \mathbb{M}^* \\ \mathbb{M} & 0 \end{pmatrix}\begin{pmatrix} \mathbb{p} & 0 \\ 0 & \mathbb{q} \end{pmatrix} = \begin{pmatrix} 0 & \mathbb{M}^*\mathbb{q} \\ \mathbb{M}\mathbb{p} & 0 \end{pmatrix} = \begin{pmatrix} 0 & \mathbb{M}^*\mathbb{q} \\ \mathbb{p}\mathbb{M} & 0 \end{pmatrix} \ , \tag{2.22}$$

$$\pi_q((p,q))D_q = \begin{pmatrix} \mathbb{p} & 0 \\ 0 & \mathbb{q} \end{pmatrix}\begin{pmatrix} 0 & \mathbb{M}^* \\ \mathbb{M} & 0 \end{pmatrix} = \begin{pmatrix} 0 & \mathbb{p}\mathbb{M}^* \\ \mathbb{q}\mathbb{M} & 0 \end{pmatrix} = \begin{pmatrix} 0 & \mathbb{M}^*\mathbb{p} \\ \mathbb{q}\mathbb{M} & 0 \end{pmatrix} \ , \tag{2.23}$$

hence:

$$i\pi_q(d(p,q)) = [D_q,\pi_q((p,q))] = \begin{pmatrix} 0 & \mathbb{M}^*(\mathbb{q}-\mathbb{p}) \\ (\mathbb{p}-\mathbb{q})\mathbb{M} & 0 \end{pmatrix} \ . \tag{2.24}$$

We then have, for $(p,q),(\pi,\chi),(s,r)\in\mathbf{A}_{ew}$, using the noted fact that $\mathbb{p}$ commutes with $\mathbb{M}$ and $\mathbb{M}^*$:

$$i\pi_q((p,q)d(\pi,\chi)) = \begin{pmatrix} 0 & \mathbb{M}^*\mathbb{p}(\boldsymbol{\chi}-\boldsymbol{\pi}) \\ \mathbb{q}(\boldsymbol{\pi}-\boldsymbol{\chi})\mathbb{M} & 0 \end{pmatrix} \ , \tag{2.25}$$

further

$$\begin{aligned} &\pi_q(d(p,q)d(\pi,\chi)) \\ &= \begin{pmatrix} \mathbb{M}^*(\mathbb{p}-\mathbb{q})(\boldsymbol{\pi}-\boldsymbol{\chi})\mathbb{M} & 0 \\ 0 & (\mathbb{p}-\mathbb{q})\mathbb{M}\mathbb{M}^*(\boldsymbol{\pi}-\boldsymbol{\chi}) \end{pmatrix} \ , \end{aligned} \tag{2.26}$$

$$\begin{aligned} &\pi_q((s,r)d(p,q)d(\pi,\chi)) \\ &= \begin{pmatrix} \mathbb{M}^*\mathbb{s}(\mathbb{p}-\mathbb{q})(\boldsymbol{\chi}-\boldsymbol{\pi})\mathbb{M} & 0 \\ 0 & \mathbb{r}(\mathbb{p}-\mathbb{q})\mathbb{M}\mathbb{M}^*(\boldsymbol{\pi}-\boldsymbol{\chi}) \end{pmatrix} \ . \end{aligned} \tag{2.27}$$

2.3 The $\mathbf{A}_{ew}$-bimodule $\Omega_{D_q}A^1_{ew}$ of D_q-quantum one-forms

We recall that $\Omega_{D_q}A^1_{ew} = \pi_q\left(\Omega\left(\mathbf{A}_{ew}\right)^1\right)$. Therefore $\Omega_{D_q}A^1_{ew}$ is obtained by linearity from (2.22): we have:

$$i\pi_q(\rho) = \begin{pmatrix} 0 & \mathbb{M}^*\mathbb{Q}' \\ \mathbb{Q}\mathbb{M} & 0 \end{pmatrix}, \quad \rho = \sum_{j=1}^{k}(p_j,q_j)d(\pi_j,\chi_j) \ , \tag{2.28}$$

with

$$\begin{cases} \mathbb{Q} = \sum_{j=1}^{k}\mathbb{q}_j(\boldsymbol{\pi}_j-\boldsymbol{\chi}_j) = Q\otimes\mathbb{I}_N \\ \mathbb{Q}' = \sum_{j=1}^{k}\mathbb{p}_j(\boldsymbol{\chi}_j-\boldsymbol{\pi}_j) = Q'\otimes\mathbb{I}_N \end{cases} \begin{cases} Q = Q(\rho) = \sum_{j=1}^{k} q_j(\pi_j-\chi_j) \\ Q' = Q'(\rho) = \sum_{j=1}^{k} p_j(\chi_j-\pi_j) \end{cases} \ . \tag{2.29}$$

Since $\mathbb{M}$ has been assumed invertible, the specification of $i\pi_q(\rho)$ is equivalent to that of the pair (Q, Q'), obviously ranging through $\mathbb{H} \times \mathbb{H}$, as ρ ranges through $i\Omega(\mathbf{A}_{ew})^1$. The bimodule $\Omega_{D_q} A^1_{ew}$ can thus be considered as consisting of pairs $(Q, Q') \in \mathbb{H} \oplus \mathbb{H}$, suggestively noted

$$\begin{pmatrix} & Q' \\ Q & \end{pmatrix} = \begin{pmatrix} 0 & \mathbb{M}^*\mathbb{Q}' \\ \mathbb{Q}\mathbb{M} & 0 \end{pmatrix}, \quad \left(\mathbb{Q} = Q \otimes \mathbb{1}_N,\ \mathbb{Q}' = Q' \otimes \mathbb{1}_N\right) \ . \tag{2.30}$$

From (2.30) it is obvious that $\begin{pmatrix} & Q' \\ Q & \end{pmatrix}^* = \begin{pmatrix} & Q^* \\ Q'^* & \end{pmatrix}$; and we have the bimodule multiplications:

$$\begin{pmatrix} \mathbb{p} & 0 \\ 0 & \mathbb{q} \end{pmatrix} \begin{pmatrix} 0 & \mathbb{M}^*\mathbb{Q}' \\ \mathbb{Q}\mathbb{M} & 0 \end{pmatrix} = \begin{pmatrix} 0 & \mathbb{M}^*\, \mathbb{p}\,\mathbb{Q}' \\ \mathbb{q}\,\mathbb{Q}\mathbb{M} & 0 \end{pmatrix} , \tag{2.31}$$

$$\begin{pmatrix} 0 & \mathbb{M}^*\mathbb{Q}' \\ \mathbb{Q}\mathbb{M} & 0 \end{pmatrix} \begin{pmatrix} \mathbb{p} & 0 \\ 0 & \mathbb{q} \end{pmatrix} = \begin{pmatrix} 0 & \mathbb{M}^*\, \mathbb{Q}'\,\mathbb{q} \\ \mathbb{Q}\,\mathbb{p}\,\mathbb{M} & 0 \end{pmatrix} , \tag{2.32}$$

whence (2.38) below; and the scalar product (for real C^*-algebras):

$$\begin{aligned}
&\left(\begin{pmatrix} & Q_1' \\ Q_1 & \end{pmatrix}, \begin{pmatrix} & Q_2' \\ Q_2 & \end{pmatrix} \right)_{D_q} = \\
&= \mathrm{Re\ Tr} \left\{ \begin{pmatrix} 0 & \mathbb{M}^*\mathbb{Q}_1' \\ \mathbb{Q}_1\mathbb{M} & 0 \end{pmatrix}^* \begin{pmatrix} 0 & \mathbb{M}^*\mathbb{Q}_2' \\ \mathbb{Q}_2\mathbb{M} & 0 \end{pmatrix} \right\} \\
&= \mathrm{Re\ Tr} \left\{ \begin{pmatrix} 0 & \mathbb{M}^*\mathbb{Q}_1^* \\ \mathbb{Q}'^*_1\mathbb{M} & 0 \end{pmatrix} \begin{pmatrix} 0 & \mathbb{M}^*\mathbb{Q}_2' \\ \mathbb{Q}_2\mathbb{M} & 0 \end{pmatrix} \right\} \\
&= \mathrm{Re\ Tr} \left\{ \begin{pmatrix} \mathbb{M}^*\mathbb{Q}_1^*\mathbb{Q}_2\mathbb{M} & 0 \\ 0 & \mathbb{Q}'^*_1\,\mathbb{M}\mathbb{M}^*\mathbb{Q}_2' \end{pmatrix} \right\} \\
&= \mathrm{Re\ Tr} \left\{ \mathbb{M}\mathbb{M}^* \left[\mathbb{Q}_1^*\mathbb{Q}_2 + \mathbb{Q}_2'\mathbb{Q}'^*_1 \right] \right\} \\
&= \mathrm{Re\ Tr} \left\{ [\mathbb{1} \otimes \Sigma - iK \otimes \Delta] \left[\left(Q_1^*Q_2 + Q_2'Q_1'^* \right) \otimes \mathbb{1}_N \right] \right\} \\
&= \mathrm{Re} \left\{ \mathrm{Tr}\,\Sigma\, \mathrm{Tr} \left[Q_1^*Q_2 + Q_2'Q_1'^* \right] - i \mathrm{Tr}\,\Delta\, \mathrm{Tr} \left[K \left(Q_1^*Q_2 + Q_2'Q_1'^* \right) \right] \right\} \\
&= \mathrm{Tr}\,\Sigma\, \mathrm{Tr} \left[Q_1^*Q_2 + Q_2'Q_1'^* \right] \ ,
\end{aligned} \tag{2.33}$$

taking account of the fact that quaternions have real traces, and using the relation:

$$\mathbb{M}\mathbb{M}^* = \begin{pmatrix} \mu_u & 0 \\ 0 & \mu_d \end{pmatrix} = \mathbb{1} \otimes \Sigma - iK \otimes \Delta \ , \quad \text{where} \tag{2.34}$$

$$\begin{cases} \Sigma = \frac{1}{2} (\mu_u + \mu_d) \\ \Delta = \frac{1}{2} (\mu_u - \mu_d) \end{cases} , \quad \begin{cases} \mu_u = M_u M_u^* \\ \mu_d = M_d M_d^* \end{cases} .$$

We showed that:

The $\mathbf{A}_{ew}$-bimodule $i\Omega_{D_q}A^1_{ew} = i\pi_q\left(\Omega(\mathbf{A}_{ew})^1\right)$ is isomorphic to $\mathbb{H}\oplus\mathbb{H}$ as a real vector space. With the notation

$$Q = \begin{pmatrix} H_2 & H^1 \\ -H_1 & H^2 \end{pmatrix}, \quad Q' = \begin{pmatrix} H'_2 & H'^1 \\ -H'_1 & H'^2 \end{pmatrix} \in \mathbb{H}$$

its elements [11]

$$\begin{pmatrix} & Q' \\ Q & \end{pmatrix} = \begin{pmatrix} 0 & \mathbb{M}^*\left(Q'\otimes \mathbb{I}_N\right) \\ (Q\otimes \mathbb{I}_N)\,\mathbb{M} & 0 \end{pmatrix}$$

$$= \begin{array}{c} \\ \\ \end{array}\begin{matrix} u_R & d_R & u_L & d_L \end{matrix} \tag{2.35}$$

$$= \begin{pmatrix} 0 & 0 & H'_2M^*_u & -H'_1M^*_u \\ 0 & 0 & H'_1M^*_d & H'^2M^*_d \\ H_2M_u & H^1M_d & 0 & 0 \\ -H_1M_u & H^2M_d & 0 & 0 \end{pmatrix}.$$

Concretely for $\rho = \sum_{i=1}^{k}(p_i,q_i)d(\pi_i,\chi_i) \in \Omega A^1_{ew},\ (p_i,q_i),(\pi_i,\chi_i)\in\mathbf{A}_{ew}$:

$$i\pi_q(\rho) = \begin{pmatrix} & Q' \\ Q & \end{pmatrix} \text{ with } \begin{cases} Q = \sum_{i=1}^{k} q_i(\pi_i-\chi_i) \\ Q' = \sum_{i=1}^{k} p_i(\chi_i-\pi_i) \end{cases}, \tag{2.36}$$

give rise to the following rules:

– *for the *-operation:*

$$\begin{pmatrix} & Q' \\ Q & \end{pmatrix}^* = \begin{pmatrix} & Q^* \\ Q'^* & \end{pmatrix}, \quad \begin{pmatrix} & Q' \\ Q & \end{pmatrix} \in \Omega_{D_q}A^1_{ew}\ ; \tag{2.37}$$

– *for the bimodule multiplication:*

$$(p_1,q_1)\begin{pmatrix} & Q' \\ Q & \end{pmatrix}(p_2q_2) = \begin{pmatrix} & p_1Q'q_2 \\ q_1Q\,p_2 & \end{pmatrix}, \tag{2.38}$$

$$(p_1,q_1),(p_2,q_2)\in\mathbf{A}_{ew}, \quad \begin{pmatrix} & Q' \\ Q & \end{pmatrix} \in \Omega_{D_q}A^1_{ew}\ ,$$

(this rule as well as the rule (2.67) is easily memorized by writing $\begin{pmatrix} p & \\ & q \end{pmatrix}$ for $(p,q)\in\mathbf{A}_{ew}$);

[11] The self-adjoint vector-potentials are obtained making $H'_1 = -H^1,\ H'_2 = H^2$.

– *for the differential in grade zero:*

$$id(p,q) = \begin{pmatrix} & q - p_{\text{diag}} \\ p_{\text{diag}} - q & \end{pmatrix}, \quad (p,q) \in \mathbf{A}_{ew} \; ; \tag{2.39}$$

– *for the scalar product in the real algebra case:*

$$\left(\begin{pmatrix} & Q_1' \\ Q_1 & \end{pmatrix} \begin{pmatrix} & Q_2' \\ Q_2 & \end{pmatrix} \right)_{D_q} = \operatorname{Tr} \Sigma \left[(Q_1, Q_2) + \left(Q_1', Q_2' \right) \right] \; ,$$

$$\begin{pmatrix} & Q_1' \\ Q_1 & \end{pmatrix}, \begin{pmatrix} & Q_2' \\ Q_2 & \end{pmatrix} \in \Omega_{D_q} A^1_{ew} \; ,$$

$$\text{with } \Sigma = \frac{1}{2} \left(M_u M_u^* + M_d M_d^* \right) \; . \tag{2.40}$$

(With our definition (**2.3**) for the complex structure of $\mathbb{H}$ defined by K, (q, q') becomes the real part of a hermitean scalar product with purely imaginary part $i(q, Kq')$).

2.4 The $\mathbf{A}_{ew}$-bimodule $\Omega_{D_q} A^2_{ew}$ of D_q-quantum two-forms

Recall (cf. [**1**]) that one has $\Omega_{D_q} A^2_{ew} = \pi_q \left(\Omega \left(\mathbf{A}_{ew} \right)^2 \right) / \pi_q \left(d\mathcal{K}^1 \right)$.
We first examine $\pi_q \left(d\mathcal{K}^1 \right)$: $\pi_q(d\rho)$ for $\rho = \sum_{i=1}^k (p_i, q_i) d(\pi_i, \chi_i)$ as in (2.25) follows by linearity from (2.23):

$$\pi_q(d\rho) = \tag{2.41}$$

$$\begin{pmatrix} \mathbb{M}^* \sum\limits_{i=1}^k (\mathbb{p}_i - \mathbb{q}_i) (\boldsymbol{\pi}_i - \boldsymbol{\chi}_i) \, \mathbb{M} & 0 \\ 0 & \sum\limits_{i=1}^k (\mathbb{p}_i - \mathbb{q}_i) \, \mathbb{M}\mathbb{M}^* v \, (\boldsymbol{\pi}_i - \boldsymbol{\chi}_i) \end{pmatrix},$$

$\pi_q \left(d\mathcal{K}^1 \right)$ thus consists of all elements of the above type for which:

$$\begin{aligned} \pi_q(\rho) = 0 \quad & \Leftrightarrow \quad \sum_{i=1}^k \mathbb{q}_i \left(\boldsymbol{\pi}_i - \boldsymbol{\chi}_i \right) = \sum_{i=1}^k \mathbb{p}_i \left(\boldsymbol{\chi}_i - \boldsymbol{\pi}_i \right) = 0 \\ & \Leftrightarrow \quad \sum_{i=1}^k q_i \left(\pi_i - \chi_i \right) = \sum_{i=1}^k p_i \left(\chi_i - \pi_i \right) = 0 \\ & \left(\Leftrightarrow \quad Q(\rho) = Q'(\rho) = 0 \right) \; . \end{aligned} \tag{2.42}$$

We observe that this implies vanishing of the upper left entry of the matrix r.h.s. of (2.41), as well as vanishing of the contribution of the first term r.h.s. of (2.34) to its lower right entry. Consequently one has:

$$\pi_q \left(d\mathcal{K}^1 \right) = \begin{pmatrix} 0 & 0 \\ 0 & \mathcal{R} \otimes \Delta \end{pmatrix} , \tag{2.43}$$

with

$$
\begin{aligned}
\mathcal{R} = \Big\{ -\frac{1}{2} i \sum_{j=1}^{k} (p_j - q_j) K (\pi_j - \chi_j); \quad p_j, \pi_j \in \mathbb{C}, \quad q_j, \chi_j \in \mathbb{H}, \\
\text{s. t.} \ \sum_{j=1}^{k} q_j(\pi_j - \chi_j) = \sum_{j=1}^{k} p_j(\chi_j - \pi_j) = 0 \Big\} \\
\supset i \Big\{ \sum_{j=1}^{k} q_j K \chi_i; \quad q_i, \chi_i \in \mathbb{H} \ \text{s. t.} \ \sum_{j=1}^{k} q_j \chi_j = 0 \Big\} = i\mathbb{H} \ ,
\end{aligned}
\tag{2.44}
$$

(indeed, the set{ } in the preceding line is manifestly an ideal of the field $\mathbb{H}$, hence coincides with $\mathbb{H}$ itself): we found that

$$
\pi_q \left(d\mathcal{K}^1 \right) = \begin{pmatrix} 0 & 0 \\ 0 & i\,\mathbb{H} \otimes \Delta \end{pmatrix} \ . \tag{2.45}
$$

On the other hand $\pi_q \left(\Omega \left(\mathbf{A}_{ew} \right)^2 \right)$ is obtained by linearity from (2.24): one has, for

$$
\theta = \sum_{i=1}^{k} (s_i, r_i) d(p_i, q_i) d(\pi_i, \chi_i), \quad (s_i, r_i), (p_i, q_i), (\pi_i, \chi_i) \in \mathbf{A}_{ew} \ , \tag{2.46}
$$

using (2.39):

$$
\begin{aligned}
&\pi_q(\theta) = \\
&\begin{pmatrix} \mathbb{M}^* \sum_{j=1}^{k} \mathbb{s}_j (\mathbb{p}_j - \mathbb{q}_j)(\boldsymbol{\pi}_j - \boldsymbol{\chi}_j) \, \mathbb{M} & 0 \\ 0 & \sum_{j=1}^{k} \mathbb{r}_j (\mathbb{p}_j - \mathbb{q}_j) \, \mathbb{M}\mathbb{M}^* (\boldsymbol{\pi}_j - \boldsymbol{\chi}_j) \end{pmatrix} \\
&= \begin{pmatrix} \mathbb{M}^*(Q \otimes \mathbb{I}_N) \, \mathbb{M} & 0 \\ 0 & Q' \otimes \Sigma + i \, Q'' \otimes \Delta \end{pmatrix} ,
\end{aligned}
\tag{2.47}
$$

expression which we shall denote $\begin{pmatrix} Q & \\ & Q', Q'' \end{pmatrix}$ as it is uniquely specified (cf. invertibility of $\mathbb{M}$) by the triple of quaternions:

$$
\begin{cases}
Q \ = \sum_{i=1}^{k} s_i (p_i - q_i)(\pi_i - \chi_i) \\
Q' \ = \sum_{i=1}^{k} r_i (p_i - q_i)(\pi_i - \chi_i) \\
Q'' = \sum_{i=1}^{k} r_i (p_i - q_i) K (\pi_i - \chi_i)
\end{cases} , \tag{2.48}
$$

which ranges through $\mathbb{H} \times \mathbb{H} \times \mathbb{H}$ as θ ranges through $\Omega(\mathbf{A}_{ew})^2$. Note that $\begin{pmatrix} Q & \\ & Q', Q'' \end{pmatrix}$ consists of the two terms:

$$\begin{pmatrix} Q & \\ & Q', Q'' \end{pmatrix} = \begin{pmatrix} \mathbb{M}^*(Q \otimes \mathbb{1}_N)\mathbb{M} & 0 \\ 0 & Q' \otimes \Sigma \end{pmatrix} + \begin{pmatrix} 0 & 0 \\ 0 & iQ'' \otimes \Delta \end{pmatrix} . \tag{2.49}$$

The second term belongs to $\pi_q(d\mathcal{K}^1)$ by (2.45). The first term, specified by a couple of quaternions, which we accordingly denote:

$$\begin{pmatrix} Q & \\ & Q' \end{pmatrix} = \begin{pmatrix} \mathbb{M}^*(Q \otimes \mathbb{1}_N)\mathbb{M} & 0 \\ 0 & Q' \otimes \Sigma \end{pmatrix}, \quad \left(= \begin{pmatrix} Q & \\ & Q', 0 \end{pmatrix} \right) \tag{2.50}$$

represents in fact univocally the class of $\begin{pmatrix} Q & \\ & Q', Q'' \end{pmatrix}$ modulo $\pi_q(d\mathcal{K}^1)$: we have indeed, with P_2 the projection for n=2 defined in section **[1]** with respect to the scalar product (1.13):

$$\begin{pmatrix} Q & \\ & Q' \end{pmatrix} = P_2 \begin{pmatrix} Q & \\ & Q', Q'' \end{pmatrix}; \quad Q, Q', Q'' \in \mathbb{H} , \tag{2.51}$$

as a consequence of the fact that the two terms in (2.49) have a vanishing scalar product since the expression

$$\mathrm{Tr}\left\{ \begin{pmatrix} \mathbb{M}^*(Q \otimes \mathbb{1}_N)\mathbb{M} & 0 \\ 0 & Q' \otimes \Sigma \end{pmatrix}^* \begin{pmatrix} 0 & 0 \\ 0 & iQ'' \otimes \Delta \end{pmatrix} \right\} = i\,\mathrm{Tr}\left(Q'Q''\right)\mathrm{Tr}\,(\Sigma\Delta) \tag{2.52}$$

has a vanishing real part. We shall henceforth consider the $\mathcal{A}$-bimodule $\Omega_{D_q} A^2_{ew}$ as consisting of pairs $\begin{pmatrix} Q & \\ & Q' \end{pmatrix} \in \mathbb{H} \oplus \mathbb{H}$.

From (2.51) it is obvious that $\begin{pmatrix} Q & \\ & Q' \end{pmatrix}^* = \begin{pmatrix} Q^* & \\ & Q'^* \end{pmatrix}$; and we have the bimodule multiplications:

$$\begin{aligned} (p,q) \begin{pmatrix} Q & \\ & Q' \end{pmatrix} &= \begin{pmatrix} \mathbb{p} & 0 \\ 0 & \mathbb{q} \end{pmatrix} \begin{pmatrix} \mathbb{M}^*(Q \otimes \mathbb{1}_N)\mathbb{M} & 0 \\ 0 & Q' \otimes \Sigma \end{pmatrix} \\ &= \begin{pmatrix} \mathbb{M}^*(pQ \otimes \mathbb{1}_N)\mathbb{M} & 0 \\ 0 & qQ' \otimes \Sigma \end{pmatrix} = \begin{pmatrix} pQ & \\ & qQ' \end{pmatrix} , \end{aligned} \tag{2.53}$$

$$\begin{aligned} \begin{pmatrix} Q & \\ & Q' \end{pmatrix} (p,q) &= \begin{pmatrix} \mathbb{M}^*(Q \otimes \mathbb{1}_N)\mathbb{M} & 0 \\ 0 & Q' \otimes \Sigma \end{pmatrix} \begin{pmatrix} \mathbb{p} & 0 \\ 0 & \mathbb{q} \end{pmatrix} \\ &= \begin{pmatrix} \mathbb{M}^*(Qp \otimes \mathbb{1}_N)\mathbb{M} & 0 \\ 0 & Q'q \otimes \Sigma \end{pmatrix} = \begin{pmatrix} Qp & \\ & Q'q \end{pmatrix} . \end{aligned} \tag{2.54}$$

The multiplication rule $\Omega_{D_q} A^1_{ew} \times \Omega_{D_q} A^1_{ew} \to \Omega_{D_q} A^2_{ew}$ is obtained as follows: one has:

$$\begin{pmatrix} & Q_1' \\ Q_1 & \end{pmatrix} \begin{pmatrix} & Q_2' \\ Q_2 & \end{pmatrix} = \begin{pmatrix} 0 & \mathbb{M}^* \mathbb{Q}_1' \\ \mathbb{Q}_1 \mathbb{M} & 0 \end{pmatrix} \begin{pmatrix} 0 & \mathbb{M}^* \mathbb{Q}_2' \\ \mathbb{Q}_2 \mathbb{M} & 0 \end{pmatrix}$$

$$= \begin{pmatrix} \mathbb{M}^* \mathbb{Q}_1' \mathbb{Q}_2 \mathbb{M} & 0 \\ 0 & \mathbb{Q}_1 \mathbb{M}\mathbb{M}^* \mathbb{Q}_2' \end{pmatrix} \tag{2.55}$$

$$= \begin{pmatrix} \mathbb{M}^* \left(Q_1' Q_2 \otimes \mathbb{1}_N \right) \mathbb{M} & 0 \\ 0 & (Q_1 \otimes \mathbb{1}_N)(\mathbb{1} \otimes \Sigma - iK \otimes \Delta)\left(Q_2' \otimes \mathbb{1}_N \right) \end{pmatrix} , \tag{2.56}$$

whence

$$P_2 \left\{ \begin{pmatrix} & Q_1' \\ Q_1 & \end{pmatrix} \begin{pmatrix} & Q_2' \\ Q_2 & \end{pmatrix} \right\} \tag{2.57}$$

$$= \begin{pmatrix} \mathbb{M}^* \left(Q_1' Q_2 \otimes \mathbb{1}_N \right) \mathbb{M} & 0 \\ 0 & Q_1 Q_2' \otimes \Sigma \end{pmatrix} = \begin{pmatrix} Q_1' Q_2 & \\ & Q_1 Q_2' \end{pmatrix} .$$

The rule for the first grade differential d: $\Omega_{D_q} A^1_{ew} \to \Omega_{D_q} A^2_{ew}$ is obtained by conferring (2.28), (2.29) rewritten as (2.36) with the relation obtained by specializing (2.51) to the case $s_i = r_i = \mathbb{1}$, $i = 1, \dots, k$:

$$P_2 \pi_q(d\rho) = d\mathcal{K}^1 \pi_q(\rho) = -i \begin{pmatrix} Q + Q' & \\ & Q + Q' \end{pmatrix} , \tag{2.58}$$

the comparison yielding the rule (2.70) below.

We now look at scalar products of elements of $\pi_q(\Omega A^2_{ew})$. Since elements $\begin{pmatrix} Q & \\ & 0,0 \end{pmatrix}$, $\begin{pmatrix} 0 & \\ & Q',0 \end{pmatrix}$, and $\begin{pmatrix} 0 & \\ & 0,Q'' \end{pmatrix}$ are obviously mutually orthogonal, it is enough to compute scalar products separately for those three types. We have:

$$\left(\begin{pmatrix} 0 & \\ & 0, Q_1'' \end{pmatrix} \begin{pmatrix} 0 & \\ & 0, Q_2'' \end{pmatrix} \right)_{D_q} = \mathrm{Re}\ \mathrm{Tr} \left\{ \begin{pmatrix} 0 & 0 \\ 0 & iQ_1'' \otimes \Delta \end{pmatrix}^* \begin{pmatrix} 0 & 0 \\ 0 & iQ_2'' \otimes \Delta \end{pmatrix} \right\}$$

$$= \mathrm{Tr}\ (\Delta^2)\ \mathrm{Tr} \left(Q_1''^* Q_2'' \right) = 2\mathrm{Tr}(\Delta^2)(Q_1'', Q_2'') , \tag{2.59}$$

further

$$\left(\begin{pmatrix} 0 & \\ & Q_1', 0 \end{pmatrix} \begin{pmatrix} 0 & \\ & Q_2', 0 \end{pmatrix} \right)_{D_q} = \mathrm{Re}\ \mathrm{Tr} \left\{ \begin{pmatrix} 0 & 0 \\ 0 & iQ_1' \otimes \Sigma \end{pmatrix}^* \begin{pmatrix} 0 & 0 \\ 0 & iQ_2' \otimes \Sigma \end{pmatrix} \right\}$$

$$= \mathrm{Tr}\ (\Sigma^2)\ \mathrm{Tr} \left(Q_1'^* Q_2' \right) = 2\mathrm{Tr}\ (\Sigma^2) \left(Q_1' Q_2' \right) . \tag{2.60}$$

Finally, for $Q_1 = \begin{pmatrix} \alpha_1 & \beta_1 \\ -\bar{\beta}_1 & \bar{\alpha}_1 \end{pmatrix}$, $Q_2 = \begin{pmatrix} \alpha_2 & \beta_2 \\ -\bar{\beta}_2 & \bar{\alpha}_2 \end{pmatrix}$:

$$\left(\begin{pmatrix} Q_1 & \\ & 0,0 \end{pmatrix} \begin{pmatrix} Q_2 & \\ & 0,0 \end{pmatrix} \right)_{D_q}$$

$$= \mathrm{Re\ Tr} \left\{ \begin{pmatrix} \mathbb{M}^* (Q_1 \otimes \mathbb{1}_N) \mathbb{M} & 0 \\ 0 & 0 \end{pmatrix}^* \begin{pmatrix} \mathbb{M}^* (Q_2 \otimes \mathbb{1}_N) \mathbb{M} & 0 \\ 0 & 0 \end{pmatrix} \right\}$$

$$= \mathrm{Re\ Tr} \left\{ \mathbb{M}\mathbb{M}^* (Q_1^* \otimes \mathbb{1}_N) \mathbb{M}\mathbb{M}^* (Q_2 \otimes \mathbb{1}_N) \right\}$$

$$= \mathrm{Re} \left\{ \mathrm{Tr} \left(\mu_u^2 \right) \mathrm{Tr} \left(E Q_1^* E Q_2 \right) + \mathrm{Tr} \left(\mu_d^2 \right) \mathrm{Tr} \left(F Q_1^* F Q_2 \right) \right.$$

$$\left. + \mathrm{Tr} \left(\mu_u \mu_d \right) \mathrm{Tr} \left(E Q_1^* F Q_2 + F Q_1^* E Q_2 \right) \right\}$$

$$= \frac{1}{2} \mathrm{Tr}(\mu_u^2 + \mu_d^2)(\bar{\alpha}_1 \alpha_2 + \alpha_1 \bar{\alpha}_2) + \mathrm{Tr}(\mu_u \mu_d)(\beta_1 \bar{\beta}_2 + \bar{\beta}_1 \beta_2) \ , \qquad (2.61)$$

where we used the fact that

$$\mathbb{M}\mathbb{M}^* = E \otimes M_u + F \otimes M_d, \left(E = \begin{pmatrix} 1 & 0 \\ 0 & 0 \end{pmatrix}, F = \begin{pmatrix} 0 & 0 \\ 0 & 1 \end{pmatrix} \right) \ . \qquad (2.62)$$

We showed that:
The $\mathbf{A}_{ew}$-bimodule $\Omega_{D_q} A_{ew}^2 = \pi_q \left(\Omega \left(\mathbf{A}_{ew} \right)^2 \right) / \pi_q \left(d\mathcal{K}^1 \right) \cong P_2 \pi_q \left(\Omega (\mathbf{A}_{ew})^2 \right)$ is isomorphic to $\mathbb{H} \oplus \mathbb{H}$ as a real vector space. Its elements

$$\begin{pmatrix} Q & \\ & Q' \end{pmatrix} = \begin{pmatrix} \mathbb{M}^* (Q \otimes \mathbb{1}_N) \mathbb{M} & 0 \\ 0 & Q' \otimes \Sigma \end{pmatrix}, \quad Q, Q' \in \mathbb{H} \qquad (2.63)$$

are obtained as the components $P_2 \begin{pmatrix} Q & \\ & Q', Q'' \end{pmatrix} = \begin{pmatrix} Q & \\ & Q', 0 \end{pmatrix}$ orthogonal to $\pi_q(d\mathcal{K}^1)$ of general elements of $\pi_q \left(\Omega(\mathbf{A}_{ew})^2 \right) \cong \mathbb{H} \oplus \mathbb{H} \oplus \mathbb{H}$:

$$\begin{pmatrix} Q & \\ & Q', Q'' \end{pmatrix} = \begin{pmatrix} \mathbb{M}^* (Q \otimes \mathbb{1}_N) \mathbb{M} & 0 \\ 0 & Q' \otimes \Sigma + i Q'' \otimes \Delta \end{pmatrix} \ . \qquad (2.64)$$

Concretely, for

$$\theta = \sum_{i=1}^{k} (s_i, r_i) d(p_i, q_i) d(\pi_i, \chi_i), \quad (s_i, r_i), (p_i, q_i), (\pi_i, \chi_i) \in \mathbf{A}_{ew} \ : \qquad (2.65)$$

$$\pi_q(\theta) = \begin{pmatrix} Q & \\ & Q', Q'' \end{pmatrix} \quad \text{with} \quad \begin{cases} Q &= \sum\limits_{i=1}^{k} s_i (p_i - q_i)(\pi_i - \chi_i) \\ Q' &= \sum\limits_{i=1}^{k} r_i (p_i - q_i)(\pi_i - \chi_i) \\ Q'' &= \sum\limits_{i=1}^{k} r_i (p_i - q_i) K (\pi_i - \chi_i) \end{cases} \ . \qquad (2.66)$$

One has the following rules:

– *for the* **-operation:*

$$\begin{pmatrix} Q & \\ & Q' \end{pmatrix}^* = \begin{pmatrix} Q^* & \\ & Q'^* \end{pmatrix}, \quad \begin{pmatrix} Q & \\ & Q' \end{pmatrix} \in \Omega_{D_q} A^2_{ew} \ ; \tag{2.67}$$

– *for the bimodule multiplication:*

$$(p_1, q_1) \begin{pmatrix} Q & \\ & Q' \end{pmatrix} (p_2, q_2) = \begin{pmatrix} p_1 Q p_2 & \\ & q_1 Q' q_2 \end{pmatrix} , \tag{2.68}$$
$$(p_1, q_1), \quad (p_2, q_2) \in \mathbf{A}_{ew}, \quad \begin{pmatrix} Q & \\ & Q' \end{pmatrix} \in \Omega_{D_q} A^2_{ew} \ ;$$

– *for the multiplication rule* $\Omega_{D_q} A^1_{ew} \times \Omega_{D_q} A^1_{ew} \to \Omega_{D_q} A^2_{ew}$*:*

$$\begin{pmatrix} & Q_1' \\ Q_1 & \end{pmatrix} \begin{pmatrix} & Q_2' \\ Q_2 & \end{pmatrix} = \begin{pmatrix} Q_1' Q_2 & \\ & Q_1 Q_2' \end{pmatrix}, \tag{2.69}$$
$$\begin{pmatrix} & Q_1' \\ Q_1 & \end{pmatrix}, \begin{pmatrix} & Q_2' \\ Q_2 & \end{pmatrix} \in \Omega_{D_q} A^1_{ew} \ ;$$

– *for the differential in grade one:*

$$id \begin{pmatrix} & Q' \\ Q & \end{pmatrix} = \begin{pmatrix} Q + Q' & \\ & Q + Q' \end{pmatrix}, \tag{2.70}$$
$$\begin{pmatrix} & Q' \\ Q & \end{pmatrix} \in \Omega_{D_q} A^1_{ew} \ ;$$

– *for the scalar product of the pair* $\begin{pmatrix} Q_1 & \\ & Q_1', Q_1'' \end{pmatrix} \begin{pmatrix} Q_2 & \\ & Q_2', Q_2'' \end{pmatrix} \in \Omega_{D_q} A^2_{ew}$ *with* $Q_1 = \begin{pmatrix} \alpha_1 & \beta_1 \\ -\bar{\beta}_1 & \bar{\alpha}_1 \end{pmatrix}$, $Q_2 = \begin{pmatrix} \alpha_2 & \beta_2 \\ -\bar{\beta}_2 & \bar{\alpha}_2 \end{pmatrix}$ (cf. (1.13) *in the real algebra case):*

$$\left(\begin{pmatrix} Q_1 & \\ & Q_1', Q_1'' \end{pmatrix} \begin{pmatrix} Q_2 & \\ & Q_2', Q_2'' \end{pmatrix} \right)_{D_q}$$
$$= 2\mathrm{Tr}\,(\sigma^2) \left(Q_1', Q_2' \right) + 2\mathrm{Tr}\,(\Delta^2) \left(Q_1'', Q_2'' \right)$$
$$+ \frac{1}{2} \mathrm{Tr}(\mu_u^2 + \mu_d^2)(\bar{\alpha}_1 \alpha_2 + \alpha_1 \bar{\alpha}_2) + \mathrm{Tr}(\mu_u \mu_d)(\beta_1 \bar{\beta}_2 + \bar{\beta}_1 \beta_2) \ . \tag{2.71}$$

(We recall the notation

$$\begin{cases} \Sigma = \frac{1}{2} (\mu_u + \mu_d) \\ \Delta = \frac{1}{2} (\mu_u - \mu_d) \end{cases} \quad \left\{ \begin{matrix} \mu_u = M_u M_u^* \\ \mu_d = M_d M_d^* \end{matrix} \right) \ . \tag{2.72}$$

2.5 Connections. Curvature. Yang-Mills action

The vector potentials are the antiself-adjoint elements of $\Omega_{D_q} A^1_{ew}$, of the form:

$$\begin{aligned} V &= -i\begin{pmatrix} & Q' \\ Q & \end{pmatrix} = -i\begin{pmatrix} & Q' \\ Q & \end{pmatrix}^{*} = -i\begin{pmatrix} & Q^* \\ Q'^* & \end{pmatrix} \\ &= -i\begin{pmatrix} & Q^* \\ Q & \end{pmatrix}, \quad Q \in \mathbb{H}\ , \end{aligned} \tag{2.73}$$

(cf. (2.35) and (2.37)). The corresponding curvature is:

$$\begin{aligned} \theta = dV + V^2 &= -id\begin{pmatrix} & Q^* \\ Q & \end{pmatrix} + (-i)^2\begin{pmatrix} & Q^* \\ Q & \end{pmatrix}\begin{pmatrix} & Q^* \\ Q & \end{pmatrix} \\ &= -\begin{pmatrix} Q + Q^* + Q^*Q & \\ & Q + Q^* + QQ^* \end{pmatrix} \\ &= -\left[|Q+1|^2 - 1\right]\begin{pmatrix} \mathbb{M}^*\mathbb{M} & 0 \\ 0 & \mathbb{1}\otimes\Sigma \end{pmatrix}\ . \end{aligned} \tag{2.74}$$

We showed that: *the vector potentials are of the form:*

$$V = -i\begin{pmatrix} & Q^* \\ Q & \end{pmatrix}, \quad Q \in \mathbb{H}\ , \tag{2.75}$$

with the corresponding curvature:

$$\theta = -\left[|Q+1|^2 - 1\right]\begin{pmatrix} \mathbb{M}^*\mathbb{M} & 0 \\ 0 & \mathbb{1}\otimes\Sigma \end{pmatrix}\ , \tag{2.76}$$

leading to the (reduced) Yang-Mills action:

$$\begin{aligned} YM(V) = \mathrm{Tr}\,(\theta^2) &= \mathrm{Tr}\left[\frac{3}{2}\left(\mu_u^2 + \mu_d^2\right) + \mu_u\mu_d\right]\left[|Q+1|^2 - 1\right]^2 \\ &= \mathrm{cst}\,\left(\Phi^2 - 1\right)^2\ , \end{aligned} \tag{2.77}$$

with the quaternion $\Phi = Q + 1$.

We see that, analogously to what happened in example [**1.9**], the action contains a Higgs-like object: but now it is not an irrealistic "embryonal Higgs", but actually the inner tensorial component of the postulated physical Higgs particle.

So much for the quarkonic spectral triple. We now conclude this section by having a look at the leptonic spectral triple: as mentioned above, the latter can be described as a modification (in fact a kind of an impoverishment) of the quarkonic spectral triple described above.

2.6 Remarks

(i): We recall that one passes from the quark operators (2.13)-(2.15) to lepton operators (2.9)-(2.11) through the process of **leptonic reduction** consisting of the following two steps applied to 4×4 matrices with entries in $M_N(\mathbb{C})$:

- 1) *effect the changes* $M_u \to 0$, $M_d \to M_e$;
- 2) *discard the first row and the first column of the* 4×4 *matrix.*

(ii): Embedding 3×3 matrices into 4×4 matrices both with entries in $M_N(\mathbb{C})$ as the latter's lower right corners bordered by zeros:

$$\begin{pmatrix} 0 & 0\,0\,0 \\ 0 & \\ 0 & S \\ 0 & \end{pmatrix}, \qquad S \in M_3(\mathbb{C}) \otimes M_N(\mathbb{C}) \ , \tag{2.78}$$

and considering the projection:

$$P = \begin{pmatrix} 0 & 0\ 0\ 0 \\ 0 & \\ 0 & \mathbb{1}_3 \\ 0 & \end{pmatrix} , \tag{2.79}$$

note that, for T a 4×4 matrix, PTP is obtained by replacing by zeros the entries in the first row and column of T. Note further that step 1) applied to D_q amounts to taking PD_qP and effecting there the change $M_d \to M_e$, this leading to $\begin{pmatrix} 0 & 0\ 0\ 0 \\ 0 & \\ 0 & D_l \\ 0 & \end{pmatrix}$. Note finally that $P\pi_q((p,q))P = \begin{pmatrix} 0 & 0 \quad 0 \quad 0 \\ 0 & \\ 0 & \pi_l((p,q)) \\ 0 & \end{pmatrix}$; $\pi_q(p,q)$ being unaffected by Step 1.

2.7 Lemma

One obtains $\pi_l(\omega)$, $\omega \in \Omega\mathbf{A}_{ew}{}^n$, *as the leptonic reduction of* $\pi_q(\omega)$. *For* $n \geq 1$, *step 1) applied to* $\pi_q(\omega)$ *yields in fact* $\begin{pmatrix} 0 & 0 \quad 0 \quad 0 \\ 0 & \\ 0 & \pi_1(\omega) \\ 0 & \end{pmatrix}$.

Proof
Step 1) and the procedure $T \to PTP$, both evidently commute with algebraic operations. Step 1) applied to

$$i^n \pi_q(a_0 da_1 \dots da_n) = \pi_q(a_0)[D_q, \pi_q(a_1)] \dots [D_q, \pi_q(a_n)]$$

amounts to making $D_q \to PD_qP$ followed by $M_d \to M_e$. Since $\pi_q(a_n)$ commutes with P, the first move applied to $[D_q, \pi_q(a_n)]$ yields:

$$[PD_qP, \pi_q(a_i)] = [PD_qP, P\pi_q(a_i)P] = P\,[D_q, \pi_q(a_i)]\,P \ , \tag{2.80}$$

which after the replacement equals $\begin{pmatrix} 0 & 0 & 0 & 0 \\ 0 & & & \\ 0 & & [D_l, \pi_l(a_n)] & \\ 0 & & & \end{pmatrix}$ by **[2.7]**(ii). Since multiplication from the left by $\pi_q(a_0)$ amounts to multiplying by $P\pi_q(a_0)P$, we see that step 1) applied to $i^n\pi_q(a_0 da_1 \dots da_n)$ yields

$$\begin{pmatrix} 0 & 0 & 0 & 0 \\ 0 & & & \\ 0 & & i^n\pi_l(a_0 da_1 \dots da_n) & \\ 0 & & & \end{pmatrix} . \tag{2.81}$$

The following result is relevant for the computation of the projection P_2 in the quark sector. Readers not interested in this are advised to skip it.

2.8 Proposition

(i): *Assume M_e invertible. We have the inclusion*

$$\mathcal{K}^1 = \Omega A^1_{ew} \cap \operatorname{Ker} \pi_q \subset \Omega A^1_{ew} \cap \operatorname{Ker} \pi_l \ , \tag{2.82}$$

hence $\mathcal{K}^1$ also equals:

$$\mathcal{K}^1 = \Omega A^1_{ew} \cap \operatorname{Ker} (\pi_q \oplus \pi_l) \ . \tag{2.83}$$

(ii): *We have that*

$$\pi_l(d\mathcal{K}^1) = \begin{pmatrix} 0 & 0 \\ 0 & i\mathbb{H} \otimes M_e M_e^* \end{pmatrix} , \tag{2.84}$$

whose elements thus result by leptonic reduction from those of $\pi_q(d\mathcal{K}^1)$.

Proof

(i) The changes $M_u \to 0$, $M_d \to M_e$ applied to $\mathbb{M}$ yield $F \otimes M_e$. Performing this change $\mathbb{M} \to F \otimes M_e$ on

$$i\pi_q(\rho) = \begin{pmatrix} 0 & \mathbb{M}^* \left[Q'(\rho) \otimes \mathbb{1}_N\right] \\ [Q(\rho) \otimes \mathbb{1}_N]\,\mathbb{M} & 0 \end{pmatrix} , \tag{2.85}$$

(cf. (2.25)) with

$$\begin{aligned} \mathbb{M}^* \left[Q'(\rho) \otimes \mathbb{1}_N\right] &= (F \otimes M_e^*) \left[Q'(\rho) \otimes \mathbb{1}_N\right] = FQ'(\rho) \otimes M_e^* \ , \\ \left[Q'(\rho) \otimes \mathbb{1}_N\right] \mathbb{M} &= [Q(\rho) \otimes \mathbb{1}_N] (F \otimes M_e) = Q(\rho)F \otimes M_e \ , \end{aligned} \tag{2.86}$$

we obtain the equivalence:

$$\pi_l(\rho) = 0 \quad \Leftrightarrow \quad FQ'(\rho) = Q(\rho)F = 0 \ , \tag{2.87}$$

which, conferred with (2.42), establishes (2.83).
(ii): We now apply leptonic reduction to (2.41) we have, in view of (2.29):

$$\begin{aligned}\sum_{j=1}^{k} \mathbb{M}^* \left[(\mathbb{p}_j - \mathbb{q}_j)(\pi_j - \chi_j)\right] \mathbb{M} &= -\mathbb{M}^*(\mathbb{Q}' + \mathbb{Q})\mathbb{M} \\ &= (F \otimes M_e^*) \left[(Q' + Q) \otimes \mathbb{1}_N\right] (F \otimes M_e) \\ &= F(Q' + Q)F \otimes M_e^* M_e \ ,\end{aligned} \tag{2.88}$$

vanishing on $\Omega A^1_{ew} \cap \operatorname{Ker} \pi_l$ by (2.87). On the other hand we have:

$$\begin{aligned}&\sum_{j=1}^{k} (\mathbb{p}_j - \mathbb{q}_j)\mathbb{M}\mathbb{M}^*(\pi_j - \chi_j) \\ &= \sum_{j=1}^{k} [(p_j - q_j) \otimes \mathbb{1}_N](F \otimes M_e M_e^*)[(\pi_j - \chi_j) \otimes \mathbb{1}_N] \\ &= \left[\sum_{j=1}^{k} (p_j - q_j)F(\pi_j - \chi_j)\right] \otimes M_e M_e^* \\ &= \frac{1}{2}\left[\sum_{j=1}^{k} (p_j - q_j)(\mathbb{1} + iK)(\pi_j - \chi_j\right] \otimes M_e M_e^* \\ &= \frac{1}{2}\left[-Q(\rho) - Q'(\rho) + i\sum_{j=1}^{k} q_j K(\pi_j - \chi_j)\right] \otimes M_e M_e^* \ ,\end{aligned} \tag{2.89}$$

by (2.82) in restriction to $\mathcal{K}^1$ equal to $\frac{1}{2} i \sum_{j=1}^{k} q_j K(\pi_j - \chi_j) \otimes M_e M_e^*$: however, as we saw in [**2.4**], the elements $\sum_{j=1}^{k} q_j \mathcal{K} \chi_j$ range through an ideal of $\mathbb{H}$, thus through the whole $\mathbb{H}$.

2.9 Conclusion

We conclude by mentioning the 2×2 matrix versions, with entries

$$\begin{pmatrix} M_1(\mathbb{C}) \otimes M_N(\mathbb{C}) & M(\mathbb{C}^2, \mathbb{C}) \otimes M_N(\mathbb{C}) \\ M(\mathbb{C}, \mathbb{C}^2) \otimes M_N(\mathbb{C}) & M_2(\mathbb{C}) \otimes M_N(\mathbb{C}) \end{pmatrix},$$

of the operators (2.9)-(2.11): we have:

$$\pi_l((p,q)) = \begin{matrix} \quad R \qquad\qquad L \\ \begin{pmatrix} p \otimes \mathbb{1}_N & 0 \\ 0 & q \otimes \mathbb{1}_N \end{pmatrix}\end{matrix} \quad \left(p \in \mathbb{C},\ q = \begin{pmatrix} a & b \\ -\bar{b} & \bar{a} \end{pmatrix} \in \mathbf{A}_{ew}\right) \ , \tag{2.90}$$

$$D_l = \begin{pmatrix} & R & L \\ & 0 & (0 M_e^*) \\ & \begin{pmatrix} 0 \\ M_e \end{pmatrix} & 0 \end{pmatrix} , \tag{2.91}$$

$$\chi_l = \begin{pmatrix} R & L \\ 1 \otimes \mathbb{1}_N & 0 \\ 0 & -\mathbb{1} \otimes \mathbb{1}_N \end{pmatrix} . \tag{2.92}$$

3 Sketch of the (Quantum Yang-Mills) Connes-Lott Model. Inadequacy of the Electroweak Sector in Isolation. Inevitability of Chromodynamics. Metric Dual Pairs

This section is a brief historical survey of the now abandoned Connes-Lott model (Yang-Mills access to the standard model). We offer it for providing the reader with a motivation for the successive definitions of the metric dual pair and S_0-real spectral triple which become natural by looking at their genesis. Knowledge of the Connes-Lott model is not logically necessary for reading the subsequent sections from [4] onward, to which the reader not caring for motivation can immediately pass, ignoring this section, which we advise him to consult casually, as a historical introduction.

3.1 Sketch of the Connes-Lott model

The idea of the Connes-Lott model is thoroughly natural: since, as we saw in section [**1**], the quantum Yang-Mills procedure reproduces the classical Yang-Mills in the electrodynamics case [**1.8**], and already exhibits a Higgs phenomenon in the two-point case [**1.9**] - furthermore lending itself canonically to system-tensorization [**1.10**] - one is naturally led to tenzorize the electrodynamics spectral triple by the (now hopefully realistic) finite "inner spectral triple" described in the preceding section [**2**] as embodying the flavour degrees of freedom, likewise exhibiting a Higgs phenomenon. This procedure incorporates the hitherto structureless "inner degrees of freedom" to an enriched (mildly) non-commutative basic space tied up with the elementary particle structure (well in the spirit of the Glashow-Salam-Weinberg synthesis of electrodynamics and weak interactions appearing as an enriched version of electrodynamics- we thus hope to get in this way the Glashow-Salam-Weinberg model of electroweak interactions. As using the tensor product in the sense of [**1.10**] of the spectral triples $\left(C^\infty(\mathbf{M}), L^2(\mathbb{S}_{\mathbf{M}}), \gamma^5, \widetilde{D}\right)$ (cf. [**1.1**] and $(\mathbf{A}_{ew} = \mathbb{C} \oplus \mathbb{H}, H, \chi, D)$ (cf. [**2**]) this first realistic attempt[12] works with the tensor-product algebra

[12] Not quite: Connes and Lott first attempted to use the two-point algebra (two-sheeted space) in combination with a more complicated projective-finite module describing the fermions – a project which they then abandoned for working with the algebra $\mathbf{A}_{ew}$ serving (as a right module over herself) as the "bundle" – more elegant and more like electrodynamics).

$\mathcal{A}_{ew} = C^\infty(\mathbf{M}) \otimes \mathbf{A}_{ew}$ acting as such on the tensor-product Hilbert space $L^2(\mathcal{S}_{\mathbf{M}}) \otimes H$ acted upon by compound chirality $\gamma^5 \otimes \chi$, and Dirac operator $\mathcal{D} = \widetilde{D} \otimes id + \gamma^5 \otimes D$. The computation of the corresponding quantum Yang-Mills action is a matter of (a somewhat weary) routine: one first works out the items of the quark sector, those of the lepton then resulting by leptonic reduction. The successive steps (explained for the quark sector) are the following:[13]
(i): find the general form of $\pi_{\mathcal{D}}(\Omega\mathcal{A}_{ew})$ in grade one and two: in this one is helped by the following tensorial decompositions: with the shorthand $C^\infty(\mathbf{M}) = \mathbb{A}$ we have [8]:

$$\pi_{\mathcal{D}_q}(\Omega\mathcal{A}_{ew}{}^1) = \pi_{\widetilde{D}}(\Omega\mathbb{A}^1) \otimes \pi_{D_q}(\mathbf{A}_{ew}) + \pi_{\widetilde{D}}(\mathbb{A})\gamma^5 \otimes \pi_{D_q}(\Omega A^1_{ew}) \ , \tag{3.1}$$

i.e.[14]

$$\begin{aligned}\Omega_{\mathcal{D}}\mathcal{A}_{ew}{}^1 &= \Omega_{\widetilde{D}}(\mathbb{A}^1) \otimes A^1_{ew} + \mathbb{A}\gamma^5 \otimes \Omega_{D_q}A^1_{ew} \\ &= \gamma(\Omega(\mathbf{M})^1) \otimes \mathbf{A}_{ew} + C^\infty(\mathbf{M}) \otimes \Omega_{D_q}A^1_{ew} \ \ \mathit{and}\end{aligned} \tag{3.2}$$

$$\begin{aligned}\pi_{\mathcal{D}_q}(\Omega_{\mathcal{D}}\mathcal{A}_{ew}{}^2) &= \pi_{\widetilde{D}}(\Omega\mathbb{A}^2) \otimes \pi_{D_q}(\mathbf{A}_{ew}) \\ &+ \pi_{\widetilde{D}}(\Omega\mathbb{A}^1) \otimes \gamma^5\pi_{D_q}(\Omega A^1_{ew}) + \pi_{\widetilde{D}}(\mathbb{A}) \otimes \pi_{D_q}(\Omega A^2_{ew}) \\ &= \gamma(\Omega(\mathbf{M})^2) \otimes \pi_{D_q}(\mathbf{A}_{ew}) + \pi\gamma(\Omega(\mathbf{M})^1) \otimes \gamma^5\pi_{D_q}(\Omega A^1_{ew}) \\ &+ C^\infty(\mathbf{M}) \otimes \pi_{D_q}(\Omega A^2_{ew}) \ ;\end{aligned} \tag{3.3}$$

(ii): work out the projection P_2 (in fact quite involved - a somewhat nightmarish complexity which Thomas Schücker avenges by calling this step "removal of the junk");
(iii): compute the curvature (somewhat involved);
(iv): stick the square of the curvature under the integral [**1.5**].

This procedure at first sight appears amazingly successful in that it yields *exactly all the terms of the complicated Glashow-Salam-Weinberg Lagrangian* (hardly fortuitous!). But a second look at the action of the gauge group reveals that the *hypercharges are wrong* - an embarrassment then providentially turned into an advantage in that *compels us to append chromodynamics for correcting the wrong hypercharges*! Incorporating chromodynamics is anyway necessary to describe nature: we need to append the gluons, which, since the gluonic fields commute with the electroweak fields, are naturally introduced by tensoring the electroweak inner algebra $\mathbf{A}_{ew} = \mathbb{C} \oplus \mathbb{H}$ by a "gluonic algebra" $\mathbf{B}_{\text{chrom}} = \mathbb{C} \oplus M_3(\mathbb{C})$ – here we are not in the favourable situation which we had with $\mathbb{H}$ yielding exactly $SU(2)$ as its group of unitaries, $M_3(\mathbb{C})$ yields $U(3)$ instead, compelling us to pass to the required $SU(3)$ by a trick[15]:

[13] The reader interested in the details can consult e.g. [16].
[14] These formulae in fact hold in all grades for any tensor product of spectral triple [8].
[15] Merely mentioned in this sketchy survey – the details require the computation of the action of the gauge group cf. [16].

a mathematically correct, but physically heuristic subsequent "modular correction" (in fact the $\mathbb{C}$-summand of $\mathbf{B}_{\text{chrom}}$ was taken for providing maneuverability for this). As for the Hilbert space, we shall naturally replace the quark inner space H_q by $H_q \otimes \mathbb{C}^3$, this threefold multiplicity giving the quarks their color degree of freedom connected with gluons, whilst the lepton inner space H_l remains unchanged (i.e. replaced by $H_l \otimes \mathbb{C}$ - the leptons are colourless). With $(p', q) \in \mathbf{B}_{\text{chrom}}$, $p' \in \mathbb{C}$, $q \in M_3(\mathbb{C})$ we shall then naturally let q and p' act on the respective right tensorial factors $\mathbb{C}^3$ and $\mathbb{C}$ above, whilst the quark and lepton Dirac operators D_q and D_l are now acting only on the tensorial factors H_q and H_l (in the same way as before in [**2**]):[16] this ensures the vectorial nature of the gluonic fields (no Higgs phenomenon in chromodynamics) by the simple fact that, since Dirac operator "entirely on the electroweak side" commutes with representative of $\mathbf{B}_{\text{chrom}}$, $\Omega(\mathbf{B}_{\text{chrom}})$ vanishes but in zero grade, causing the gluonic quantum forms to lie in $\gamma(\Omega(\mathbf{M})) \otimes \mathbf{B}_{\text{chrom}}$. Now, performing the steps (i) trough (iv) above on the tensor product of the space-time spectral triple by this modified "inner spectral triple of the full standard model" combining the fermion and the gluon inner degrees of freedom, the miracle is that we can, at the same time, perform our "modular correction" and correct for the wrong hypercharges! (a priori not all warranted). *Noncommutative geometry not only allows, but demands the construction of the full standard model* (a canonical fact from the mathematical point of view as a manifestation of *noncommutative Poincaré duality*). We are now at a point where noncommutative geometry gives an elegant formulation (without changing a iota to its structure, except interesting constraints) to the (Lagrangian aspect of the) full standard model of elementary particles - in the first place removing the hitherto unpleasantly heuristic nature of the Higgs particle. The situation could appear entirely satisfactory but for some conceptual flaws which we discuss in [**3.3**] below. Before doing this we give in the next paragraph a mathematical exegesis of our combined inner spectral triple. If we set $\mathbf{A}_{ew} = \mathbf{B}'$, $\mathbf{B}_{\text{chrom}} = \mathbf{B}''$, and, $\pi'(b') = \pi_{\mathcal{D}}(b' \otimes \mathbb{1}_{\text{chrom}})$, $\pi''(b'') = \pi_{\mathcal{D}}(\mathbb{1}_{ew} \otimes b'')$, $b' \in \mathbf{B}'$, $b'' \in \mathbf{B}''$, we see that the latter is an example of the generic notion of "real metric dual pair " as given by:

3.2 Definition (metric dual pairs)

$(\mathbf{B}' \otimes \mathbf{B}'', (\mathbf{H}, D, \chi))$ is a **real metric dual pair** whenever $\mathbf{B}'$ and $\mathbf{B}''$ are *-algebras over $\mathbb{R}$ with π' and π'' respective real *-representations of $\mathbf{B}'$ and $\mathbf{B}''$ by even operators on $\mathbf{H}$ such that

$$[\pi'(b'), \pi''(b'')] = 0, \qquad b' \in \mathbf{B}', b'' \in \mathbf{B}'' , \tag{3.4}$$

and D is an odd self-adjoint operator of $\mathbf{H}$ such that all commutators $[D, \pi'(b')\pi''(b'')]$, $b' \in \mathbf{B}'$, $b'' \in \mathbf{B}''$, are bounded, D^{-1} is a compact operator,

[16] The reader who would prefer to see formulae is referred to [**5.2**] below.

and one has (**"first-order condition"**):

$$[[D, \pi'(b')], \pi''(b'')] = 0, \qquad b' \in \mathbf{B}', \, b'' \in \mathbf{B}''^{17} \tag{3.5}$$

The real metric dual pair is called **finite** whenever $\mathbf{H}$ is finite-dimensional. (Note that owing to (3.4) the Jacobi identity implies that the requirement (3.5) is symmetrical in $\mathbf{B}'$ and $\mathbf{B}''$).[18]

Our next paragraph discusses some drawbacks of the above Connes-Lott theory- a discussion which preludes to the notion of "real spectral triple" to be discussed in the next section.

3.3 Conceptual flaws

Interesting and successful as it is, the Connes-Lott theory has however imperfections:

- the notion of metric dual pair pertains only to inner space (since one does not wish to tensorially double space-time);
- the relevant connections are not all the connections, but the "biconnections" of the type $\nabla_{ew} \otimes id + id \otimes \nabla_{\text{chrom}}$ ("remembering" the tensorial splitting $(\mathbb{C} \otimes \mathbb{H}) \otimes (\mathbb{C} \oplus M_3(\mathbb{C}))$. [19] This subsidiary assumption ruptures our initial doctrine of having the geometry encoded by the spectral triple (stricto sensu), with no restriction on connections;
- the theory understresses particle-antiparticle (change-conjugation) symmetry;
- the "modular correction" is heuristic, not conceptual, thus insufficiently understood despite the link with anomaly-freeness [11];
- the elimination of the "junk" may appear anesthetically complicated.

These conceptual flaws will be remedied (only partially what concerns the "modular correction") by replacing our "metric dual pair" by the corresponding "S_0-real triple" as explained in the next section.

3.4 The Higgs mass

One of the very interesting features of the Connes-Lott model is the fact that it predicts the Higgs mass at tree-level (in contradistinction to the usual theory in which the Higgs mass is structureless (a sad "floating anchor").This was

[17] This situation is subsumed by requiring: (i): that $(\mathbf{B}' \otimes \mathbf{B}'', \mathbf{H}, D)$ be an even spectral triple as in [**1.2**] with $\mathbf{B}' \otimes \mathbf{B}''$ and $\pi' \otimes \pi''$ real; (ii): the first-order condition (3.5).

[18] This and the fact that $\mathcal{D}$ commutes with $\pi_{\mathcal{D}}(\mathbf{B}_{\text{chrom}})$ establishes the first-order condition (3.5) for our combined inner spectral triple (condition (3.5) being obvious).

[19] Connes motivated the use of biconnections by a gauge-group argument [9]. We attempted a more basic justification [20]. The problem evaporates whilst passing to the real spectral triples.

recognized at an early stage of the theory [4][5].Now, in this respect, one has the choice between two opposite attitudes: -either, as advocated in [9], be as little committed possible in the choice of the coupling constant, requiring only consistence with first principles (gauge invariance, positivity): this leads to a multi-parameter, matrix coupling constant. This doctrine has been intensively studied [10][19][22][24][25], with th striking conclusion that the tree-level prediction of the Higgs mass (or rather the ratio of the Higgs to the top mass) is largely independent of the multidimensional choice of coupling constant; -or adopt the more appealing philosophy that the most symmetric (most esthetic) choice of coupling constant.should deliver a fundamental theory describing the primal matter.This was the attitude of [7] (interpreting the advocated "grand unification of a new kind" as relevance to "primal matter") and is then the doctrine to which one is *forced* (as I think happily) by the new spectral action theory, as we shall see in section [**7**].

4 Real Spectral Triples. The Special Case of S_0-Real Spectral Triples versus Metric Dual Pairs

4.1 Definition

With $\mathcal{A}$ a *-algebra over $\mathbb{R}$, a **d-dimensional real spectral triple** $(\mathcal{A},(\mathbf{H},D,\chi),J)$ is the data of:

- a $\mathbb{Z}/2$-graded complex Hilbert space $\mathbf{H} = \mathbf{H}^0 \oplus \mathbf{H}^1$ (with grading involution χ);
- a faithful ($\mathbb{R}$-linear) *-representation $\mathcal{A} \ni a \to \underline{a} \in B(\mathbf{H})$ of $\mathcal{A}$ by even bounded operators;
- an odd self-adjoint operator D of $\mathbf{H}$ such that all $[D,\underline{a}]$, $\underline{a} \in \mathcal{A}$, are bounded, and D^{-1} is compact such that $D^{\text{-d}}$ has its discrete eigenvalues μ_n such that $\sum_{n=1}^{N} \mu_n = O(\ln N)$; [20]
- moreover possessing **a real structure modulo 8** in the following sense: there is an antilinear operator J of $\mathbf{H}$ such that

(i): $JD = DJ$
(ii): $J^*J = \mathbb{1} = JJ^*$ (i.e. J is antiunitary=antilinear unitary)
(iii): $J^2 = \varepsilon\mathbb{1}$ (i.e. $J^{-1} = \varepsilon J$)
(iv): $J\chi = \varepsilon' \chi J$
(v): $J\underline{b}J^{-1}$, $b \in \mathcal{A}$, commutes with $\underline{a}$ and $[D,\underline{a}]$, $a \in \mathcal{A}$ (hence with $\pi_D(\Omega\mathcal{A})$) .
Here ε and ε' are given as follows in terms of d:

$$\begin{array}{ccccc} \mathrm{d} & 0 & 2 & 4 & 6 \\ \varepsilon & 1 & -1 & -1 & 1 \\ \varepsilon' & 1 & -1 & 1 & -1 \end{array} \tag{4.1}$$

[20] These three first requirements are a variant of the definition of even spectral triples in [**1.1**] where it is now assumed that the algebra is real and the representation real and faithful.

Note that the second equality [**4.1**](ii) is implied by the first in conjunction with [**4.1**](iii) and [**4.1**](ii) meaning that

$$(J\xi, J\eta) = (J^*\xi, J^*\eta) = (\eta, \xi), \quad \eta,\, \xi \in H \ . \tag{4.2}$$

4.2 Remark

With ($\mathcal{A}$,(**H**,D,χ),J) a real spectral triple, **H** becomes a $\mathcal{A}$-bimodule by setting:

$$\begin{cases} a\psi = \underline{a}\psi \\ \psi b = J\underline{b}^* J^{-1}\psi, \qquad a,\, b \in \mathcal{A},\ \psi \in H \ . \end{cases} \tag{4.3}$$

Proof

Owing to [**4.1**],(v) one has $a(\psi b) = (a\psi)b$ (written $a\psi b$), and:

$$a'(a\psi b)b'\psi = \underline{a}'Jb'J^{-1}\underline{a}J\underline{b}^*J^{-1}\psi = \underline{a}'\underline{a}J\underline{b}'^*J^{-1}J\underline{b}^*J^{-1}\psi$$

$$= \underline{a}'\underline{a}J\underline{b}'^*\underline{b}^*J^{-1}\psi = \underline{a}'\underline{a}J(\underline{bb}')^*J^{-1}\psi = (a'a)\psi(b'b) \ . \tag{4.4}$$

4.3 Proposition (the classical case)

Let **M** *be a* d*-dimensional compact spin manifold, where* d=2m=$2\begin{cases}2\mu \\ 2\mu+1\end{cases}$ *with spin bundle* $\mathbb{S}_{\mathbf{M}}$, *Dirac operator* $\widetilde{D} = i\gamma^\mu\widetilde{\nabla}_\mu$, *and chirality* $\chi = \gamma^{2\mathrm{d}+1} = i^m\gamma_1\cdots\gamma_{2m}$. Let $L^2(\mathbb{S}_{\mathbf{M}})$ *be the Hilbert space of square-integrable spinors equipped with the scalar product*

$$(\psi, \psi') = \int B_+(\psi(x), \psi'(x))dv, \quad \psi, \psi' \in L^2(\mathbb{S}_{\mathbf{M}})$$

, B_+ *the positive-definite scalar product of the Euclidean spinors, dv the volume form of* **M**, and *let* $J = C$ *be the Euclidean charge-conjugation. Then*

$$(\mathcal{A} = C^\infty(\mathbf{M}, \mathbb{R}),\ (L^2(\mathbb{S}_{\mathbf{M}}), \widetilde{D}, \gamma^{2d+1}), C) \tag{4.5}$$

is a d*-dimensional real spectral triple.* [21]

Proof

Based on the following information: one has, with * denoting the adjoint with respect to B_+:

$$\begin{cases} (a) & \gamma(u)^* = \gamma(u), & u \in T_{\mathbf{M}} \\ (b) & C^{-1}\gamma(u)C = -\gamma(u), & u \in T_{\mathbf{M}} \\ (c) & C^2 = \varepsilon = \begin{cases} (-1)^\mu, & m = 2\mu \\ -(-1)^\mu, & m = 2\mu+1 \end{cases} \\ (d) & C^* = \varepsilon C \end{cases} , \tag{4.6}$$

[21] We denote by $T_{\mathbf{M}}$ the tangent bundle of **M**. The Clifford bundle $\mathbb{C}1(\mathbf{M})$ is defined with $\gamma(u)^2 = (u,u)$, $u \in T_{\mathbf{M}}$. $\widetilde{\nabla}_\mu$ denotes the spin covariant derivative. The *Dirac-Lichnerowicz-Atiyah-Singer operator* $\widetilde{D}$ is known to be self-adjoint, and one has $C\widetilde{D} = \widetilde{D}C$.

with the ensuing table:

$$\begin{array}{lrrrr} \mathrm{d} & 0 & 2 & 4 & 6 \\ m & 0 & 1 & 2 & 3 \\ \mu & 0 & 0 & 1 & 1 \\ \varepsilon & 1 & -1 & -1 & 1 \\ \varepsilon' = (-1)^m & 1 & -1 & 1 & -1 \end{array} \tag{4.7}$$

4.4 Definitions

Given a real spectral triple $(\mathbb{A}, (\mathbf{H}, D, \chi), J)$:
(i): We define as usual the **gauge group** as the group

$$G = \{u \in \mathbb{A};\ u^* u = uu^* = \mathbb{I}\} \tag{4.8}$$

of unitaries of $\mathbb{A}$.
(ii): The **adjoint action** $G \ni u \to adu \in \operatorname{End} H$ $(H \ni \psi \to^u \psi \in H)$ of G on $\mathbf{H}$ is defined as : [22]

$$adu = \underline{u} J \underline{u} J^* \ (= J \underline{u} J^* \underline{u} = \underline{u} J^* \underline{u} J = J^* \underline{u} J \underline{u}), \quad u \in G\ . \tag{4.9}$$

(iii): The subspace of **Majorana spinors** is by definition that of fixpoint under J:

$$H^J = \{\psi \in H,\ J\psi = \psi\}\ . \tag{4.10}$$

(iv): We specify as follows the action $A \to {}^u A$ of the gauge group on connection-forms A:

$${}^u A = \underline{u} A \underline{u}^* + \underline{u}\,[D, \underline{u}^*], \quad u \in G\ . \tag{4.11}$$

4.5 Lemma

(i): *Definition* (4.9) *yields a unitary action of* G: *one has:*

$$ad(uv) = adu \cdot adv, \quad (\text{i.e.}^u\,({}^v\psi) = {}^{uv}\psi,\ \psi \in H)\,,\ u, v \in G\ , \tag{4.12}$$

$$ad\,(u^*) = (adu)^{-1}\,, \quad \left(\text{i.e.} \left({}^u\psi, {}^u\psi'\right) = \left(\psi, \psi'\right), \quad \psi, \psi' \in H\right), u \in G\ . \tag{4.13}$$

(ii): *This action commutes with* J:

$$adu \cdot J = J \cdot adu, \quad (\text{i.e.}^u\,(J\psi) = J\,({}^u\psi)\,,\ \psi \in H)\,, \quad u \in G\ , \tag{4.14}$$

and thus leaves stable the subset of Majorana spinors.
Proof
(i): One has by [**4.1**](ii) and [**4.1**](v):

$$adu \cdot adv = \underline{u} J \underline{u} J^* J \underline{v} J^* \underline{v} = \underline{u} J \underline{uv} J^* \underline{v} = J \underline{uv} J^* \underline{uv} = ad(uv)\ , \tag{4.15}$$

and adu is unitary since u is unitary and J and J^* are antiunitary.
(ii): One has by [**4.1**](ii):

$$adu \cdot J = J \underline{u} J^* \underline{u} J = J \cdot adu, \quad u \in G\ . \tag{4.16}$$

[22] Terminology justified by remark [**4.2**] - note that the fermions now transform according to the adjoint rather tan to the fundamental representation of the gauge group:

4.6 Remark

With $(\mathcal{A}, (\mathbf{H}, D, \chi), J)$ a d-dimensional real spectral triple as in [**4.1**], and $A \in \Omega_D(\mathcal{A})$, $(\mathcal{A}, (\mathbf{H}, D_A, \chi), J)$, where $D_A = D + A + J^* A J$, is real spectral triple of the same kind.

4.7 Proposition

Defining the **fermionic action of real spectral triples** *as the following functional of* $\psi \in H$ *and the connection one-form* A:

$$I_F(\psi, A) = (\psi, D_A \psi) = (\psi, (D + A + J^* A J)\psi) \ , \tag{4.17}$$

we have gauge invariance of I_F:

$$I_F(\psi, A) = I_F({}^u\psi, {}^uA), \quad \psi \in H,\ u \in G \ , \tag{4.18}$$

in other terms:

$$\begin{cases} ({}^u\psi, (D + {}^uA + J^*\, {}^uAJ)\, {}^u\psi) = (\psi, (D + A + J^* A J)\psi) \\ \qquad \text{or} \ \ ({}^u\psi, D_{{}^uA}\, {}^u\psi) = (\psi, D_A \psi), \quad \psi \in H,\ u \in G \end{cases} . \tag{4.19}$$

The proof results from:

4.8 Lemma

We have for $u \in G$ with $adu = \underline{u} J \underline{u} J^*$:

$$(adu)^* D (adu) = D + [\,\underline{u}^*, D\,]\,\underline{u} + J^*[\,\underline{u}^*, D\,]\,\underline{u} J \ , \tag{4.20}$$

$$(adu)^*\, {}^uA (adu) = A - [\,\underline{u}^*, D\,]\,\underline{u} \ , \tag{4.21}$$

$$(adu)^* (J^*\, {}^uAJ)(adu) = J^* A J - J^*[\,\underline{u}^*, D\,]\,\underline{u} J \ . \tag{4.22}$$

Proof
(i): Using $J^*J = JJ^* = \underline{u}^*\underline{u} = \underline{u}\underline{u}^* = \mathbb{1}$, whence $(J\underline{u}J^*)^*(J\underline{u}J^*) = \mathbb{1}$, and $J^*DJ = D$, and noting that $\underline{u}^* D \underline{u} = [\,\underline{u}^*, D\,]\,\underline{u} + D$, one has:

$$\begin{aligned}(\underline{u}J\underline{u}J^*)^* D (\underline{u}J\underline{u}J^*) &= J\underline{u}^* J^* \underline{u}^* D \underline{u} J \underline{u} J^* \\ &= J\underline{u}^* J^* ([\,\underline{u}^*, D\,]\underline{u} + D) J \underline{u} J^* \\ &= [\,\underline{u}^*, D\,]\,\underline{u} + J\underline{u}^* J^* D J \underline{u} J^* \\ &= [\,\underline{u}^*, D\,]\,\underline{u} + J\underline{u}^* D \underline{u} J^* \\ &= [\,\underline{u}^*, D\,]\,\underline{u} + J([\,\underline{u}^*, D\,]\,\underline{u} + D) J^* \\ &= [\,\underline{u}^*, D\,]\,\underline{u} + J[\,\underline{u}^*, D\,]\,\underline{u} J^* + D \ , \end{aligned} \tag{4.23}$$

further, since $J\mathcal{A}J^*$ commutes with $\pi_D(\Omega\mathcal{A})$:

$$\begin{aligned}(\underline{u}J\underline{u}J^*)^*\, {}^uA (\underline{u}J\underline{u}J^*) &= J\underline{u}^* J^* \underline{u}^*\, {}^uA \underline{u} J \underline{u} J^* \\ &= \underline{u}^*\, {}^uA \underline{u} = \underline{u}^* (\underline{u} A \underline{u}^* + \underline{u}[D, \underline{u}^*])\, \underline{u} \\ &= A + [D, \underline{u}^*]\, \underline{u} \ . \end{aligned} \tag{4.24}$$

Moreover:

$$\begin{aligned}(\underline{u}J\underline{u}J^*)^* J^{*\,u}AJ(\underline{u}J\underline{u}J^*) &= J\underline{u}^* J^* \underline{u}^* J^{*u}AJ\underline{u}J\underline{u}J^* \\ &= J\underline{u}^{*\,u}AJ^*\underline{u}^* J^* J\underline{u}J\underline{u}J^* = J\underline{u}^{*\,u}A\underline{u}J^* \\ &= J\underline{u}^*(\underline{u}A\underline{u}^* + \underline{u}\,[D,\underline{u}^*])\,\underline{u}\,J^* \\ &= JAJ^* + J[D,\underline{u}^*]\,\underline{u}\,J^* \;. \end{aligned} \tag{4.25}$$

4.9 Proposition (tensor product of real spectral triple , the first of which of even half-dimension, the second of zero-dimension)

With $(\mathcal{A}_1, (\mathbf{H}_1, D_1, \chi_1), J_1)$ *and* $(\mathcal{A}_2, (\mathbf{H}_2, D_2, \chi_2), J_2)$ *real spectral triples of respective dimensions mod 8* d_1 *and* d_2 *assuming* $\mathrm{d}_1 = 0$ *or* 4, *and* $\mathrm{d}_2 = 0$ (*thus* $\varepsilon_1^{'} = \varepsilon_2 = \varepsilon_2^{'} = 1$), *one gets a real spectral triple* $(\mathcal{A}, \mathbb{H}, J)$, $\mathbb{H} = (H, D, \chi,)$ *of dimension* d_1 *by setting:*

$$\left\{\begin{aligned} \mathcal{A} &= \mathcal{A}_1 \otimes_{\mathbb{R}} \mathcal{A}_2 \\ H &= H_1 \otimes_{\mathbb{C}} H_2 \\ \chi &= \chi_1 \otimes \chi_2 \\ \underline{a_1 a_2} &= \underline{a}_1 \otimes \underline{a}_2, \; a_1 \in \mathcal{A}_1, \; a_2 \in \mathcal{A}_2 \\ D &= D_1 \otimes id_2 + \chi_1 \otimes D_2 \\ J &= J_1 \otimes J_2 \end{aligned}\right. \tag{4.26}$$

Proof

We know (cf. [**1.10**]) that $(\mathcal{A}, (\mathbf{H}, D, \chi), J)$ is a $(\mathrm{d}_1 + \mathrm{d}_2)$- dimensional even spectral triple. It thus suffices to check axioms (i) through (v) in [**4.1**]. One has, using $\varepsilon_1^{'} = 1$:

$$\begin{aligned} JD &= (J_1 \otimes J_2)(D_1 \otimes id_2 + \chi_1 \otimes D_2) = J_1 D_1 \otimes J_2 + J_1\chi_1 \otimes J_2 D_2 \\ &= D_1 J_1 \otimes J_2 + \varepsilon_1^{'}\chi_1 J_1 \otimes D_2 J_2 = D_1 J_1 \otimes J_2 + \chi_1 J_1 \otimes D_2 J_2 \\ &= (D_1 \otimes id_2 + \chi_1 \otimes D_2)(J_1 \otimes J_2) = DJ \;. \end{aligned} \tag{4.27}$$

Check of (ii): (we noted that the first equation suffices):

$$\begin{aligned} J^* J &= (J_1 \otimes J_2)^*(J_1 \otimes J_2) = (J_1^* \otimes J_2^*)(J_1 \otimes J_2) \\ &= J_1^* J_1 \otimes J_2^* J_2 = id_1 \otimes id_2 \;. \end{aligned} \tag{4.28}$$

Check of (iii):

$$J^2 = (J_1 \otimes J_2)^2 = J_1^2 \otimes J_2^2 = \varepsilon_1 id_1 \otimes \varepsilon_2 id_2 = \varepsilon_2 id_1 \otimes id_2 = \varepsilon_1\varepsilon_2 id = \varepsilon_1 id \;. \tag{4.29}$$

Check of (iv):

$$\begin{aligned} J\chi &= (J_1 \otimes J_2)(\chi_1 \otimes \chi_2) = J_1\chi_1 \otimes J_2\chi_2 = \varepsilon_1^{'}\chi_1 J_1 \otimes \varepsilon_2^{'}\chi_2 J_2 \\ &= \varepsilon_1^{'}\chi_1 J_1 \otimes \chi_2 J_2 = \varepsilon_1^{'}(\chi_1 \otimes \chi_2)(J_1 \otimes J_2) = \varepsilon_1^{'}\chi J \;. \end{aligned} \tag{4.30}$$

Check of (v): one has, for a_1, $b_1 \in \mathbb{A}_1$, a_2, $b_2 \in \mathbb{A}_2$, on the one hand:

$$\begin{aligned}
\underline{a_1a_2}J\underline{b_1b_2}J^{-1} &= (\underline{a}_1 \otimes \underline{a}_2)(J_1 \otimes J_2)(\underline{b}_1 \otimes \underline{b}_2)(J_1 \otimes J_2)^{-1} \\
&= \underline{a}_1 J_1 \underline{b}_1 J_1^{-1} \otimes \underline{a}_2 J_2 \underline{b}_2 J_2^{-1} \\
&= J_1 \underline{b}_1 J_1^{-1} \underline{a}_1 \otimes J_2 \underline{b}_2 J_2^{-1} \underline{a}_2 \\
&= (J_1 \otimes J_2)(\underline{b}_1 \otimes \underline{b}_2)(J_1 \otimes J_2)^{-1}(\underline{a}_1 \otimes \underline{a}_2) \\
&= J\underline{b_1b_2}J^{-1}\underline{a_1a_2} \ ,
\end{aligned} \tag{4.31}$$

and, on the other:

$$\begin{aligned}
&[D, \underline{a_1a_2}]J\underline{b_1b_2}J^{-1} \\
&= [D_1 \otimes id_2 + \chi_1 \otimes D_2, \underline{a}_1 \otimes \underline{a}_2][J_1\underline{b}_1 J_1^{-1} \otimes J_2 \underline{b}_2 J_2^{-1}] \\
&= \{[D_1, \underline{a}_1] \otimes \underline{a}_2 + \chi_1 \underline{a}_1 \otimes [D_2, \underline{a}_2]\} [J_1\underline{b}_1 J_1^{-1} \otimes J_2 \underline{b}_2 J_2^{-1}] \\
&= [D_1, \underline{a}_1] J_1\underline{b}_1 J_1^{-1} \otimes \underline{a}_2 J_2 \underline{b}_2 J_2^{-1} + \chi_1 \underline{a}_1 J_1 \underline{b}_1 J_1^{-1} \otimes [D_2, \underline{a}_2] J_2 \underline{b}_2 J_2^{-1} \\
&= J_1\underline{b}_1 J_1^{-1}[D_1, \underline{a}_1] \otimes J_2 \underline{b}_2 J_2^{-1} \underline{a}_2 + J_1 \underline{b}_1 J_1^{-1} \chi_1 \underline{a}_1 \otimes J_2 \underline{b}_2 J_2^{-1}[D_2, \underline{a}_2] \\
&= [J_1\underline{b}_1 J_1^{-1} \otimes J_2 \underline{b}_2 J_2^{-1}] \{[D_1, \underline{a}_1] \otimes \underline{a}_2 + \chi_1 \underline{a}_1 \otimes [D_2, \underline{a}_2]\} \\
&= [J_1\underline{b}_1 J_1^{-1} \otimes J_2 \underline{b}_2 J_2^{-1}][D_1 \otimes id_2 + \chi_1 \otimes D_2, \underline{a}_1 \otimes \underline{a}_2] \\
&= J\underline{b_1b_2}J^{-1}[D, \underline{a_1a_2}] \ .
\end{aligned} \tag{4.32}$$

The next result allows to turn metric dual pairs of real algebras into so called S_0-real spectral triples: a procedure which Alain Connes uses to pass from the Connes-Lott model to the superior " spectral standard model" (cf. [13]) and sections below of these lectures) - This procedure corresponds to a generic situation which we now describe.

4.10 Definitions

(i): A real spectral triple $(\mathbb{A}, (\mathbb{H}, \mathbb{D}, \boldsymbol{\chi}), \boldsymbol{J})$, with $\boldsymbol{\pi}$ the corresponding *-representation of $\mathbb{A}$ on $\mathbb{H}$, is called **S_0-real** whenever it comes with a projection $P = P^2 = P^* = \mathbb{1} - \overline{P}$, with corresponding direct-sum splitting $\mathbb{H} = \mathbf{H} \oplus \overline{\mathbf{H}}$, $\mathbf{H} = P\mathbb{H}$, $\overline{\mathbf{H}} = \overline{P}\mathbb{H}$, in such a way that:

$$\begin{cases}
P\boldsymbol{J} + \boldsymbol{J}P = \boldsymbol{J} \ (\Leftrightarrow P\boldsymbol{J} = \boldsymbol{J}\overline{P} \Leftrightarrow \overline{P}\boldsymbol{J} = \boldsymbol{J}P) & \Leftrightarrow \ \boldsymbol{J} \text{ exchanges } \mathbf{H} \text{ and } \overline{\mathbf{H}} \\
P\boldsymbol{\chi} = \boldsymbol{\chi}P \quad (= P\boldsymbol{\chi}P) & \Leftrightarrow \ \boldsymbol{\chi} \text{ leaves } \mathbf{H} \text{ and } \overline{\mathbf{H}} \text{ stable} \\
P\mathbb{D} = \mathbb{D}P \quad (= P\mathbb{D}P) & \Leftrightarrow \ \mathbb{D} \text{ leaves } \mathbf{H} \text{ and } \overline{\mathbf{H}} \text{ stable} \\
P\boldsymbol{\pi}(a) = \boldsymbol{\pi}(a)P \quad (= P\boldsymbol{\pi}(a)P) & \Leftrightarrow \ \boldsymbol{\pi}(a) \text{ leaves } \mathbf{H} \text{ and } \overline{\mathbf{H}} \text{ stable}
\end{cases} \tag{4.33}$$

(ii): $(\mathbf{B}^{'} \otimes \mathbf{B}^{''}, (\mathbf{H}, D, \chi))$ is a **real metric dual pair** whenever $\mathbf{B}^{'}$ and $\mathbf{B}^{''}$ are *-algebras over $\mathbb{R}$ with $\pi^{'}$ and $\pi^{''}$ respective real *-representations of $\mathbf{B}^{'}$ and $\mathbf{B}^{''}$ by even operators on $\mathbf{H}$ such that

$$[\pi^{'}(b^{'}), \pi^{''}(b^{''})] = 0, \quad b^{'} \in \mathbf{B}^{'}, \ b^{''} \in \mathbf{B}^{''} \ , \tag{4.34}$$

and D is an odd self-adjoint operator of $\mathbf{H}$ such that all commutators $[D, \pi'(b')\pi''(b'')]$, $b' \in \mathbf{B}'$, $b'' \in \mathbf{B}''$, are bounded, D^{-1} is a compact operator, and one has ("**first-order condition**"):

$$[[D, \pi'(b')], \pi''(b'')] = 0, \quad b' \in \mathbf{B}', \, b'' \in \mathbf{B}'' \, , \tag{4.35}$$

(note that owing to (4.34) the Jacobi identity implies that the requirement (4.35) is symmetrical in $\mathbf{B}'$ and $\mathbf{B}''$). [23] The real metric dual pair is called **finite** whenever $\mathbf{H}$ is finite-dimensional.[24]

4.11 Proposition

(i): *Given a real metric dual pair* $(\mathbf{B}' \otimes \mathbf{B}'', (\mathbf{H}, D, \chi))$, *we get as follows a* S_0- *real spectral triple* $(\mathbb{A}, (\mathbb{H}, \mathbb{D}, \boldsymbol{\chi}), \mathbb{J}, P)$ *with* $\varepsilon = \varepsilon' = 1$: *set, with* $b' \in \mathbf{B}'$, $b'' \in \mathbf{B}''$:

(a): $\mathbb{A} = \mathbf{B}' \oplus \mathbf{B}''$ (direct sum of algebras);
(b): $\mathbb{H} = \mathbf{H} \oplus \overline{\mathbf{H}}$, $\overline{\mathbf{H}}$ the conjugate Hilbert space of $\mathbf{H}$; [25]
(c): $\mathbb{J}(\xi \oplus \overline{\eta}) = \eta \oplus \overline{\xi}, \quad \xi, \, \eta \in \mathbf{H}$;
(d): $\boldsymbol{\chi} = \chi \oplus \overline{\chi}$;
(f): $\mathbb{D} = D \oplus \overline{D}$;
(e): $\boldsymbol{\pi}(b' \oplus b'') = \pi'(b') \oplus \overline{\pi''(b'')}$;

(4.36)

Here we use the notations $\overline{T} = JTJ$, $T \in \operatorname{End}\mathbf{H}$ with $J : \; J\xi = \overline{\xi}, \; J\overline{\xi} = \xi$, $\xi \in \mathbf{H}$, so that $(S \oplus \overline{T})(\xi \oplus \overline{\eta}) = (S\xi \oplus \overline{T\eta})$, $\xi, \eta \in \mathbf{H}$. Note that we have the straightforward properties:

$$\begin{cases} \left(\alpha S + \alpha' S'\right) \oplus \overline{(\alpha T + \alpha' T')} = \alpha \left(S \oplus \overline{T}\right) + \alpha' \left(S' \oplus \overline{T}'\right) \\ \left(SS' \oplus \overline{TT'}\right) = \left(S \oplus \overline{T}\right)\left(S' \oplus \overline{T}'\right) \\ \left(S^* \oplus \overline{T}^*\right) = \left(S \oplus T\right)^* \\ \mathbb{J}\left(S \oplus \overline{T}\right) = \left(T \oplus \overline{S}\right)\mathbb{J} \qquad \begin{cases} S, S', T, T' \in \operatorname{End}\mathbf{H} \\ \alpha, \alpha' \in \mathbb{R} \end{cases} . \end{cases} \tag{4.37}$$

[23] This situation is subsumed by requiring:
(i): that $(\mathbf{B}' \otimes \mathbf{B}'', H, D)$, $\mathbf{B}''$ be an even spectral triple as in [**1.1**], with $\mathbf{B}' \otimes \mathbf{B}''$ and $\pi' \otimes \pi''$ are real;
(ii): the first order condition (4.35).

[24] Our subsequent utilization of the concept of real metric dual pair in physics is confined to the finite case, possibly the only case of interest. The argument in the proof of [**4.11**],(i) below does not use finiteness but, since it yields $\varepsilon = \varepsilon' = 0$, for obtaining a bona-fide (forcibly zero-dimensional) real spectral triple.

[25] $\overline{\mathbf{H}}$ **is the conjugate Hilbert space** of $\mathbf{H}$, i.e. $\mathbf{H}$ and $\overline{\mathbf{H}}$ are the same sets: $\mathbf{H} \ni \xi \overset{J}{\Leftrightarrow} \overline{\xi} \in \overline{\mathbf{H}}$, with the linear structure $\overline{\alpha\xi + \beta\eta} = \overline{\alpha\xi} + \overline{\beta\eta}$, $\alpha, \beta \in \mathbb{C}$, and the scalar product $(\overline{\xi}, \overline{\eta}) = (\eta, \xi)$, $\xi, \, \eta \in \mathbf{H}$.

(ii): *Conversely, given a zero-dimensional S_0-real spectral triple* $(\mathbb{A}, (\mathbb{H}, \mathbb{D}, \boldsymbol{\chi}), \boldsymbol{J}, P)$, *with* $\mathbf{H} = P\mathbb{H}$, $\overline{\mathbf{H}} = \overline{P}\mathbb{H}$, *and* $\mathbf{B}' = \operatorname{Ker} \boldsymbol{\pi}|_{\overline{\mathbf{H}}}$ *and* $\mathbf{B}'' = \operatorname{Ker} \boldsymbol{\pi}|_{\mathbf{H}}$, *setting, with* $b' \in \mathbf{B}'$, $b'' \in \mathbf{B}''$:

$$\begin{cases} (a) : \chi = \text{ restriction of } \boldsymbol{\chi} \text{ to } \mathbf{H} \\ (b) : D = \text{ restriction of } \boldsymbol{\chi} \text{ to } \mathbf{H} \\ (c) : \pi'(b') = \text{ restriction of } \boldsymbol{\pi}(b' \oplus 0) \quad \text{ to } \mathbf{H}, b' \oplus 0 \in \mathbf{B}' \oplus \mathbf{B}'' \\ (d) : \pi''(b'') = \text{ restriction of } \boldsymbol{\pi}(0 \oplus b'') \quad \text{ to } \mathbf{H}, 0 \oplus b'' \in \mathbf{B}' \oplus \mathbf{B}'' \end{cases}, \tag{4.38}$$

yield a finite real metric dual pair $(\mathbf{B}' \otimes \mathbf{B}'', (\mathbf{H}, D, \chi))$.

(iii): *The correspondences* (i) *and* (ii) *are inverse of each other, hence yielding a bijection* $(\mathbf{B}' \otimes \mathbf{B}'', (\mathbf{H}, D, \chi)) \leftrightarrow (\mathbb{A}, (\mathbb{H}, \mathbb{D}, \boldsymbol{\chi}), \boldsymbol{J}, P)$ *between finite real metric dual pair and zero-dimensional S_0-real spectral triples.*

Proof

(i): It follows immediately from (4.36) that $\boldsymbol{J}$ is an antilinear antiunitary involution. Consequences of (4.37):

- the first line entails $\mathbb{R}$-linearity of $\boldsymbol{\chi}$, $\mathbb{D}$, $\boldsymbol{\pi}(b' \oplus b'')$, $b' \in \mathbf{B}'$ $b'' \in \mathbf{B}''$, and of the map $\boldsymbol{\pi}$;
- the second line entails that $\boldsymbol{\chi}$ is a $\mathbb{Z}/2$-grading commuting with $\boldsymbol{\pi}(\mathbb{A})$ and anticommuting with $\mathbb{D}$, further that $\boldsymbol{\pi}$ is multiplicative and that one has:

$$\begin{aligned} \left[\mathbb{D}, \boldsymbol{\pi}\left(b' \oplus b''\right)\right] &= \left[D, \pi'\left(b'\right)\right] + \left[\overline{D}, \overline{\pi''(b'')}\right] \\ &= \left[\mathbb{D}, \boldsymbol{\pi}\left(b'\right)\right] + \left[\overline{\mathbb{D}, \boldsymbol{\pi}\,(b'')}\right] \quad ; \end{aligned} \tag{4.39}$$

- the third line entails that $\boldsymbol{\pi}$ is *-preserving and that $\boldsymbol{\chi}$ and $\mathbb{D}$ are self-adjoint;
- the fourth line entails that $\boldsymbol{J}$ commutes with $\mathbb{D}$ and with $\boldsymbol{\chi}$, further that one has

$$\boldsymbol{J}\boldsymbol{\pi}(b' \oplus b'')\boldsymbol{J} = \pi''(b'') \oplus \overline{\pi'(b')}, \quad b' \in \mathbf{B}', \, b'' \in \mathbf{B}'' \ . \tag{4.40}$$

This then entails commutation of $\boldsymbol{J}\boldsymbol{\pi}(\mathbb{A})\boldsymbol{J}$ and $[\overline{\mathbb{D}, \boldsymbol{\pi}(\mathbb{A})}]$.

We checked thus far that $(\mathbb{A}, (\mathbb{H}, \mathbb{D}, \boldsymbol{\chi}), \boldsymbol{J})$ is a real spectral triple. The fact that it is S_0-real with P the projection on $\mathbf{H}$ is obvious from (4.36).

(ii): Since $\boldsymbol{\pi}$ is faithful and $\mathbb{H} = \mathbf{H} \oplus \overline{\mathbf{H}}$, it is clear that $\mathbb{A} = \mathbf{B}' \oplus \mathbf{B}''$. The respective restriction χ, D, and $\pi'(b')$ to $\mathbf{H}$ of $\boldsymbol{\chi}, \mathbb{D}, \boldsymbol{\pi}(b' \oplus 0)$ are clearly a self-adjoint involution, a self-adjoint operator and the value for $b' \in \mathbf{B}'$ of a $\mathbb{R}$-linear bounded *-representation of $\mathbf{B}'$, χ commuting with $\pi'(b')$ and anti-commuting with D. Since $\boldsymbol{J}$ is involutory and commutes with $\boldsymbol{\chi}$, the restriction $\pi''(b'')$ of $\boldsymbol{J}\boldsymbol{\pi}(0 \oplus b')\boldsymbol{J}$ to $\mathbf{H}$ is the value for $b'' \in \mathbf{B}''$ of a $\mathbb{R}$-linear bounded *-representation of $\mathbf{B}''$ which commutes with χ since $\boldsymbol{\chi}$ commutes with $\boldsymbol{J}$ and $\boldsymbol{\pi}(0 \oplus b'')$. Since $\boldsymbol{J}\boldsymbol{\pi}(\mathbb{A})\boldsymbol{J}$ commutes with $\boldsymbol{\pi}(\mathbb{A})$, $\pi'(b')$ and $\pi''(b'')$ commute. The boundedness of the elements of $[\mathbb{D}, \boldsymbol{\pi}(\mathbb{A})]$ implies that of those of $[D, \pi'(\mathbf{B}')]$ and $[D, \pi''(\mathbf{B}'')]$

(for the latter observe that $[D,\pi''(b'')]$=restriction to $\mathbf{H}$ of $\mathbb{J}[\mathbb{D}, \boldsymbol{\pi}(0\oplus b'')]\mathbb{J}$. Remains to check the first-order condition (4.35): now $\left[[D,\pi'(b')],\pi''(b'')\right]$ vanishes as the restriction to $\mathbf{H}$ of $\left[\left[\mathbb{D}, \boldsymbol{\pi}\left(b'\oplus 0\right)\right], \mathbb{J}\boldsymbol{\pi}\left(0\oplus b''\right)\mathbb{J}\right]$.

The following fact will serve us for the "modular correction of the gauge group", cf. the "coalescence subgroup" in [**5.4**] below.

4.12 Remark-definition

Given a zero-dimensional S_0-real spectral triple

$$(\mathbb{A} = \mathbf{B}'\oplus\mathbf{B}'', (\mathbb{H},\mathbb{D},\boldsymbol{\chi}), \mathbb{J}, P) \tag{4.41}$$

with $\mathbf{B}' = \mathbf{A}'\oplus\mathbf{A}_0'$ and $\mathbf{B}'' = \mathbf{A}''\oplus\mathbf{A}_0''$ with $\mathbf{A}_0'\cong\mathbf{A}_0''\cong\mathbf{A}_0$ its **compression** $(\mathbb{A} = \mathbf{A}'\oplus\mathbf{A}''\oplus\mathbf{A}_0, (\mathbb{H},\mathbb{D},\boldsymbol{\chi}), \mathbb{J}, P)$ by $\mathbf{A}_0$ is the zero-dimensional S_0-real spectral triple obtained by identifying $\mathbf{A}_0'\cong\mathbf{A}_0''\cong\mathbf{A}_0$.

Proof: This identification has no bearing on the argumentation of the construction [**4.11**](i), but for the check of the $\mathbb{R}$-linearity of the assignment

$$a'\oplus a''\oplus a_0 \to \boldsymbol{\pi}(a'\oplus a''\oplus a_0) = \pi'(a'+a_0)\oplus\overline{\pi''(a''+a_0)}:$$

the latter however clearly proceeds from the fact that the r.h.s. of the last equation is a direct summand (this identification could not be effected without destroying linearity in the dual-metric-pair version of the S_0-real spectral triple).

4.13 Remark

Given a finite real metric dual pair $(\mathbf{B}'\otimes\mathbf{B}'', (\mathbf{H}, D,\chi))$ and a zero-dimensional S_0-real spectral triple $(\mathbb{A}, (\mathbb{H},\mathbb{D},\boldsymbol{\chi}), \mathbb{J}, P)$ corresponding to each other as in [**4.11**](iii) above, and with $G = \{u\in\mathbb{A};\ u^*u = uu^* = \mathbb{1}\} = G'\times G''$ the corresponding **gauge group**,[26]

$$G' = \left\{v'\in\mathbf{B}';\ v'^*v' = v'v'^* = \mathbb{1}\right\},\ G'' = \left\{v''\in\mathbf{B}'';\ v''^*v'' = v''v''^* = \mathbb{1}\right\}$$

the restriction of $ad(v',v'')$ to $\mathbf{H}$ coïncides with $\pi'(v')\otimes\pi''(v'')$. In the situation of the compression [**4.12**] the latter then becomes $\pi'(u_0u')\otimes\pi''(u_0u'')$, $u_0\in\mathbb{A}_0$, $u'\in\mathbb{A}'$, $u''\in\mathbb{A}''$.

5 The Inner S_0-Real Spectral Triple of the Standard Model

In what follows we describe the inner space real triple $(\mathbf{A},\mathbf{H},J)$ with algebra $\mathbf{A} = \mathbb{C}\oplus\mathbb{H}\oplus M_3(\mathbb{C})$ obtained by the device [**4.10**] from the inner space real metric dual pair [**3.2**] with pair of algebras $\mathbf{A}_{ew} = \mathbb{C}\oplus\mathbf{H}$ and $\mathbf{B}_{\text{chrom}} = \mathbb{C}\oplus M_3(\mathbb{C})$. We first recall the latter for the convenience of the reader.

[26] G is the gauge group of the S_0-real spectral triple $(\mathbb{A}, (\mathbb{H},\mathbb{D},\boldsymbol{\chi}), \mathbb{J}, P)$ in the sense [**4.11**].

5.1 Reminder. The inner real metric dual pair

Recalling that we identify the real algebra $\mathbb{C}$ of complex numbers with the real algebra $\mathbb{H}_{\text{diag}}$ of diagonal quaternions: $\mathbb{C} \ni p \leftrightarrow \begin{pmatrix} \bar{p} & 0 \\ 0 & p \end{pmatrix} \in \mathbb{H}_{\text{diag}}$, our inner metric dual pair $(\mathbf{A}_{ew} \otimes \mathbf{B}_{\text{chrom}}, H, D)$ is the direct sum of the following quarkonic, resp. leptonic dual pair:

	Quarkonic	*Leptonic*
Hilbert space:	36-dimensional	9-dimensional

$$H_q = [(\underset{u_R\, d_R}{\mathbb{C}_R^2} \oplus \underset{u_L\, d_L}{\mathbb{C}_L^2}) \otimes \mathbb{C}^N] \otimes \underset{\text{color}}{\mathbb{C}^3} \qquad H_l = [(\underset{e_R}{\mathbb{C}_R^1} \oplus \underset{\nu_L\, e_L}{\mathbb{C}_L^2}) \otimes \mathbb{C}^N] \otimes \underset{\text{no color}}{\mathbb{C}}$$

(type of particle indicated under direct summands, tensorial factor $\mathbb{C}^N$ for the N families of fermions). Endomorphisms written as:

4×4 matrices with entries in $M_N(\mathbb{C})$ $\qquad$ 3×3 matrices with entries in $M_N(\mathbb{C})$

– grading (= parity):

$$\chi_q = \begin{pmatrix} \mathbb{I}_N & 0 & 0 & 0 \\ 0 & \mathbb{I}_N & 0 & 0 \\ 0 & 0 & -\mathbb{I}_N & 0 \\ 0 & 0 & 0 & -\mathbb{I}_N \end{pmatrix} \otimes id, \quad \chi_l = \begin{pmatrix} \mathbb{I}_N & 0 & 0 \\ 0 & -\mathbb{I}_N & 0 \\ 0 & 0 & -\mathbb{I}_N \end{pmatrix} \otimes id \; ; \tag{5.1}$$

(columns labelled $u_R\ d_R\ u_L\ d_L$, resp. $e_R\ \nu_L\ e_L$)

– representative of $\left(p = \begin{pmatrix} \bar{p} & 0 \\ 0 & p \end{pmatrix}, \; q = \begin{pmatrix} a & b \\ -\bar{b} & \bar{a} \end{pmatrix}\right) \in \mathbf{A}_{ew} = \mathbb{C} \oplus \mathbb{H}$:

$$\pi_q((p,q)) = \begin{pmatrix} \bar{p}\mathbb{I}_N & 0 & 0 & 0 \\ 0 & p\mathbb{I}_N & 0 & 0 \\ 0 & 0 & a\mathbb{I}_N & b\mathbb{I}_N \\ 0 & 0 & -\bar{b}\mathbb{I}_N & \bar{a}\mathbb{I}_N \end{pmatrix} \otimes id, \quad \pi_l((p,q)) = \begin{pmatrix} p\mathbb{I}_N & 0 & 0 \\ 0 & a\mathbb{I}_N & b\mathbb{I}_N \\ 0 & -\bar{b}\mathbb{I}_N & \bar{a}\mathbb{I}_N \end{pmatrix} \otimes id \; ; \tag{5.2}$$

(columns labelled $u_R\ d_R\ u_L\ d_L$, resp. $e_R\ \nu_L\ e_L$)

– representative of $(p', m) \in \mathbf{B}_{\text{chrom}} = \mathbb{C} \otimes M_3(\mathbb{C})$:

$$\pi_q((p', m)) = [id \text{ of } (\mathbb{C}_R^2 \oplus \mathbb{C}_L^2) \otimes \mathbb{C}^N] \otimes m \; , \qquad \pi_l((p', m)) = [id \text{ of } (\mathbb{C}_R^1 \oplus \mathbb{C}_L^2) \otimes \mathbb{C}^N] \otimes p' = p' \; ; \tag{5.3}$$

– generalized Dirac operator:

$$D_q = \begin{pmatrix} 0 & 0 & M_u^* & 0 \\ 0 & 0 & 0 & M_d^* \\ M_u & 0 & 0 & 0 \\ 0 & M_d & 0 & 0 \end{pmatrix} \otimes id, \quad D_l = \begin{pmatrix} 0 & 0 & M_e \\ 0 & 0 & 0 \\ M_e & 0 & 0 \end{pmatrix} \otimes id \; ; \tag{5.4}$$

(columns labelled $u_R\ d_R\ u_L\ d_L$, resp. $e_R\ \nu_L\ e_L$)

where: M_e is a diagonal strictly positive matrix with eigenvalues the masses of the electron, muon, and tau; and M_u is a diagonal strictly positive matrix with eigenvalues the masses of the upper, charmed, and top quark; and $M_d = C|M_d|$, $|M_d|$ a diagonal strictly positive matrix with eigenvalues the masses of the lower, strange, and bottom quark, C the unitary **Kobayashi-Maskawa matrix** (we comply to the common usage of choosing our fermion mass-matrices such that M_e and M_u *are diagonal positive matrices, whilst* $M_d = C|M_d|$,*with* C (the **Kobayashi-Maskawa matrix**) *unitary and* $|M_d|$ *strictly positive.* Furthermore we assume that *all fermion masses are different* (the eigenvalues of M_e, M_u and $|M_d|$ consists of positive numbers (the masses of leptons and quarks) all different from one another - experiment!). We further assume that *no eigenstate of* $|M_d|$ *is an eigenstate of* C (experiment!).

Remark. One passes from the quarkonic matrices to the corresponding leptonic matrices through the changes $M_u \to 0$, $M_d \to M_e$ followed by restriction to the right-lower corner 3×3 matrix. This procedure applied to a 4×4 matrix depending upon M_n and M_d is called **leptonic reduction.**

Two-by-two matrix versions

Quarkonic
2×2 matrices with entries in
$$M_2(\mathbb{C}) \otimes M_N(\mathbb{C})$$

Leptonic
2×2 matrices with entries
$$\begin{pmatrix} M_1(\mathbb{C}) \otimes M_N(\mathbb{C}) & M(\mathbb{C}^2, \mathbb{C}) \otimes M_N(\mathbb{C}) \\ M(\mathbb{C}, \mathbb{C}^2) \otimes M_N\mathbb{C} & M_2(\mathbb{C}) \otimes M_N(\mathbb{C}) \end{pmatrix}$$

– grading:

$$\chi_q = \overset{\begin{matrix} R \qquad\qquad & L \end{matrix}}{\begin{pmatrix} \mathbb{I}\otimes\mathbb{I}_N & 0 \\ 0 & -\mathbb{I}\otimes\mathbb{I}_N \end{pmatrix}} \qquad \chi_l = \overset{\begin{matrix} R \qquad\qquad & L \end{matrix}}{\begin{pmatrix} 1\otimes\mathbb{I}_N & 0 \\ 0 & -\mathbb{I}\otimes\mathbb{I}_N \end{pmatrix}} \quad ; \tag{5.5}$$

– representative of $(p, q) \in \mathbb{C} \otimes \mathbb{H}$:

$$\pi_q((p,q)) = \overset{\begin{matrix} R \qquad\qquad & L \end{matrix}}{\begin{pmatrix} \begin{pmatrix} \bar{p} & 0 \\ 0 & p \end{pmatrix} \otimes \mathbb{I}_N & 0 \\ 0 & q \otimes \mathbb{I}_N \end{pmatrix}} \qquad \pi_l((p,q)) = \overset{\begin{matrix} R \qquad\qquad & L \end{matrix}}{\begin{pmatrix} p\otimes\mathbb{I}_N & 0 \\ 0 & q\otimes\mathbb{I}_N \end{pmatrix}} \quad ; \tag{5.6}$$

– generalized Dirac operator:

$$D_q = \overset{\begin{matrix} R & L \end{matrix}}{\begin{pmatrix} 0 & \mathbb{M}^* \\ \mathbb{M} & 0 \end{pmatrix}} \qquad D_l = \overset{\begin{matrix} R \qquad\qquad & L \end{matrix}}{\begin{pmatrix} 0 & (0 M_e^*) \\ \begin{pmatrix} 0 \\ M_e \end{pmatrix} & 0 \end{pmatrix}} \quad , \tag{5.7}$$

where

$$\begin{cases} \mathbb{M} = \begin{pmatrix} M_u & 0 \\ 0 & M_d \end{pmatrix} = E \otimes M_u + F \otimes M_d \\ \text{with } E = \begin{pmatrix} 1 & 0 \\ 0 & 0 \end{pmatrix}, \text{ and } F = \begin{pmatrix} 0 & 0 \\ 0 & 1 \end{pmatrix} \end{cases} \quad . \tag{5.8}$$

5.2 Definition (the inner-space S_0-real spectral triple)

The **algebra** is now the direct sum:

$$\mathbf{A} = \mathbb{C} \oplus \mathbb{H} \oplus M_3(\mathbb{C}) = \{(p,q,m);\ p \in \mathbb{C},\ q \in \mathbb{H},\ m \in M_3(\mathbb{C})\}\ ; \tag{5.9}$$

(thus with gauge group

$$\begin{aligned} G &= \{u = (u,v,w) \in \mathbf{A};\ u^*u = uu^* = \mathbb{I},\ v^*v = vv^* = \mathbb{I},\ w^*w = ww^* = \mathbb{I}\} \\ &= U(1) \times SU(2) \times U(3)\ )\ . \end{aligned} \tag{5.10}$$

The **inner-space real spectral triple** is then $(\mathbf{A}, (\mathcal{H}, \chi, \mathcal{D}), J)$, where the Hilbert space $\mathcal{H}$ is the direct sum of H and its conjugate Hilbert space $\overline{H}$:

$$\mathcal{H} = H \oplus \overline{H} \tag{5.11}$$

with the K-cycle $(\mathcal{H}, \chi, \mathcal{D})$ specified as follows: one has acting on $\psi = (\xi, \bar{\eta}) \in \mathcal{H}$ (we use throughout the same notation for the element (p,q,m) of $\mathbf{A}$ and its action on $\mathcal{H}$):

$$J(\xi \oplus \bar{\eta}) = \eta \oplus \bar{\xi}\ , \tag{5.12}$$

$$\chi(\xi \oplus \bar{\eta}) = \chi\xi \oplus \overline{\chi\eta}\ , \tag{5.13}$$

$$\begin{aligned} (p,q,m)(\xi \oplus \bar{\eta}) &= (p,q)\xi \oplus \overline{(p,m)\eta} \\ J(p,q,m)J(\xi, \bar{\eta}) &= (p,m)\xi \oplus \overline{(p,q)\eta}, \quad (p,q,m) \in \mathbf{A}\ , \end{aligned} \tag{5.14}$$

$$\mathcal{D}(\xi \oplus \bar{\eta}) = \mathcal{D}\xi \oplus \overline{\mathcal{D}\eta}\ , \tag{5.15}$$

with the real structure: (note that, in the notation of [**4.10**], (5.12) - (5.14) read:

$$\chi = \chi \oplus \bar{\chi}\ , \tag{5.16}$$

$$\begin{aligned} (p,q,m) &= (p,q) \oplus \overline{(p,m)} \\ J(p,q,m)J &= (p,m) \oplus \overline{(p,q)}, \quad (p,q,m) \in \mathbf{A}\ , \end{aligned} \tag{5.17}$$

$$\mathcal{D} = D \oplus \overline{D}\ \ )\ . \tag{5.18}$$

5.3 Proposition

With the definition [**5.2**] $(\mathbf{A}, (\mathcal{H}, \chi, \mathcal{D}), \mathcal{J})$ *is a 0-dimensional spectral triple in the sense that* $(\mathcal{H}, \chi, \mathcal{D})$ *is an even 0-dimensional K-cycle of* $\mathbf{A}$, *and one has:* $J^2 = \mathbb{I}$, $J\mathcal{D} = \mathcal{D}J$, $J\chi = \chi J$, $[a, Ja'J] = 0$ *and* $[[\mathcal{D}, a], Ja'J] = 0$ *for all* $a, a' \in \mathbf{A}$.

Proof

First apply [**4.11**] with $\mathbf{B}' = \mathbb{C} \oplus \mathbb{H}$, $\mathbf{B}'' = \mathbb{C} \oplus M_3(\mathbb{C})$, taking account of (5.16) - (5.18), then effect the compression by $\mathbb{C}$ as described in [**4.12**] (the required properties also result from inspection of the matrix expression in [**5.5**] below).

5.4 Proposition (action of the gauge group, weak isotopic spin and hypercharge)

(i): *The adjoint action of the gauge group on $\mathcal{H}$ is as follows: one has for* $(u, v, w) \in G = U(1) \times SU(2) \times U(3)$:

$$ad(u,v,w)(\xi,\bar{\eta}) = ((u,v)(u,w)\xi, \overline{(u,v)(u,w)\eta}), \quad (\xi,\bar{\eta}) \in \mathcal{H} \ , \tag{5.19}$$

(corresponding to the matrix:

$$\begin{array}{cc} \begin{array}{ccc} u_R\, d_R & u_L\, d_L & \end{array} & \begin{array}{cc} e_R & \nu_L\, e_L \end{array} \\ \begin{pmatrix} \begin{pmatrix} \bar{u} & 0 \\ 0 & u \end{pmatrix} \otimes \mathbb{I}_N & 0 \\ 0 & v \otimes \mathbb{I}_N \end{pmatrix} \otimes w & \\ & u \begin{pmatrix} u \otimes \mathbb{I}_N & 0 \\ 0 & v \otimes_N \end{pmatrix} \end{array} \ , \tag{5.20}$$

acting on H - the complex-conjugate matrix acts on $\overline{H}$).

(ii): *The one-parameter subgroups* $t \to (\mathbb{I}, e^{iVt}, \mathbb{I})$, $t \to (e^{iUt}, \mathbb{I}, \mathbb{I})$, $t \to (\mathbb{I}, \mathbb{I}, e^{iWt})$ *are accordingly infinitesimally represented as follows*:

	u_R	d_R	u_L	d_L	e_R	ν_L	e_L
adV	0	0	1	−1	0	1	−1
adU	−1	1	0	0	2	1	1
adW	1	1	1	1	0	0	0

(5.21)

Comparison with the following values of T_3 (weak isotopic spin) *and* Y (hypercharge - we also plotted the charge $Q = T_3 + \frac{1}{2}Y$):

	u_R	d_R	u_L	d_L	e_R	ν_L	e_L
T_3	0	0	$\frac{1}{2}$	$-\frac{1}{2}$	0	$\frac{1}{2}$	$-\frac{1}{2}$
Y	$\frac{4}{3}$	$-\frac{2}{3}$	$\frac{1}{3}$	$\frac{1}{3}$	−2	−1	−1
Q	$\frac{2}{3}$	$-\frac{1}{3}$	$\frac{2}{3}$	$-\frac{1}{2}$	−1	0	−1

(5.22)

yields $T_3 = \frac{1}{2}\,\mathrm{adV}$, $Y = \mathrm{adU} - \frac{1}{3}\mathrm{adW}$. *Thus the "modular correction subgroup"* $\{(u, \mathbb{I}, u^{1/3}),\ u \in \mathbb{C}\}$ of G *yields the wanted gauge-group* $U(1) \times SU(2) \times U(3)$.

Proof

(i): Apply (4.40) (or use (5.14) to see that for $U = (u, v, w)$ one has:

$$U(\xi,\bar\eta) = ((u,v)\xi, \overline{(u,w)\eta}) \text{ and } JUJ(\xi,\bar\eta) = ((u,w)\xi, \overline{(u,v)\eta}) \ ,$$

whence (5.19) since (u, v) and (v, w) commute.

(ii): Obvious from (5.20) and (5.22)

In the remainder of this section we exhibit the matrix from of the quoted items. Since the respective basis of H and $\overline{H}$ are J-related, the matrix of $\overline{T} = JTJ \in \operatorname{End} \overline{H}$ is the complex-conjugate of the matrix of $T \in \operatorname{End} H$.

5.5 Matrix-form of $\pi_{\mathcal{D}}(\mathrm{A})$, $J\pi_{\mathcal{D}}(\mathrm{A})J$, $ad(G)$ and $\mathcal{D}$.

Matrix-form of (p, q, m), $p = \begin{pmatrix}\bar p & 0\\ 0 & p\end{pmatrix} \in \mathbb{H}_{\mathrm{diag}} \cong \mathbb{C}$, $q \in \mathbb{H}$, $m \in M_3(\mathbb{C})$:

$$
\begin{array}{lcc}
 & u_R\, d_R \qquad\qquad u_L\, d_L & e_R \qquad \nu_L\, e_L \\
(p,q,m)|_H = & \begin{pmatrix} \begin{pmatrix}\bar p & 0\\ 0 & p\end{pmatrix}\otimes \mathbb{1}_N & 0 \\ 0 & q\otimes\mathbb{1}_N \end{pmatrix}\otimes \mathbb{1}_3 & \\
 & & \begin{pmatrix} p\otimes \mathbb{1}_N & 0 \\ 0 & q\otimes \mathbb{1}_N\end{pmatrix}
\end{array} , \tag{5.23}
$$

$$
\begin{array}{lcc}
 & \bar u_R\, \bar d_R \qquad \bar u_L\, \bar d_L & \bar e_R \qquad \bar\nu_L\, \bar e_L \\
(p,q,m)|_{\overline{H}} = & \begin{pmatrix} \mathbb{1}_2\otimes\mathbb{1}_N & 0 \\ 0 & \mathbb{1}_2\otimes\mathbb{1}_N \end{pmatrix}\otimes \bar m & \\
 & & \bar p \begin{pmatrix} \mathbb{1}_2\otimes \mathbb{1}_N & 0 \\ 0 & \mathbb{1}_2\otimes \mathbb{1}_N\end{pmatrix}
\end{array} .
$$

Matrix-form of $J(p, q, m)J$, $p \in \mathbb{H}_{\mathrm{diag}} \cong \mathbb{C}$, $q \in \mathbb{H}$, $m \in M_3(\mathbb{C})$:

$$
\begin{array}{lcc}
 & u_R\, d_R \qquad u_L\, d_L & e_R \qquad \nu_L\, e_L \\
J(p,q,m)J|_H = & \begin{pmatrix} \mathbb{1}_2\otimes\mathbb{1}_N & 0 \\ 0 & \mathbb{1}_2\otimes\mathbb{1}_N \end{pmatrix}\otimes m & \\
 & & p \begin{pmatrix} \mathbb{1}_1\otimes \mathbb{1}_N & 0 \\ 0 & \mathbb{1}_2\otimes \mathbb{1}_N\end{pmatrix}
\end{array} , \tag{5.24}
$$

$$
\begin{array}{lcc}
 & \bar u_R\, \bar d_R \qquad\qquad \bar u_L\, \bar d_L & \bar e_R \qquad \bar\nu_L\, \bar e_L \\
J(p,q,m)J|_{\overline{H}} = & \begin{pmatrix} \begin{pmatrix}\bar p & 0\\ 0 & p\end{pmatrix}\otimes \mathbb{1}_N & 0 \\ 0 & q\otimes\mathbb{1}_N \end{pmatrix}^{-}\otimes \mathbb{1}_3 & \\
 & & \begin{pmatrix} p\otimes \mathbb{1}_N & 0 \\ 0 & q\otimes \mathbb{1}_N\end{pmatrix}^{-}
\end{array} .
$$

Matrix-form of representatives $ad(u, v, w)$ of G:

$$
\begin{array}{l}
ad(u,v,w) = J(u,v,w)J(u,v,w)|_H \\
\qquad \begin{array}{cccc} u_R\; d_R & u_L\; d_L & e_R & \nu_L\; e_L \end{array} \\
= \begin{pmatrix} \begin{pmatrix} \bar u & 0 \\ 0 & u \end{pmatrix} \otimes \mathbb{1}_N & 0 \\ 0 & v \otimes \mathbb{1}_N \end{pmatrix} \otimes w \\
\qquad\qquad\qquad\qquad u \begin{pmatrix} u \otimes \mathbb{1}_N & 0 \\ 0 & v \otimes \mathbb{1}_N \end{pmatrix} ,
\end{array}
\tag{5.25}
$$

$$
\begin{array}{l}
ad(\bar u,\bar v,\bar w) = J(u,v,w)\bar J(\bar u,v,w)|_{\overline{H}} \\
\qquad \begin{array}{cccc} \bar u_R\; \bar d_R & \bar u_L\; \bar d_L & \bar e_R & \bar\nu_L\; \bar e_L \end{array} \\
= \overline{\begin{pmatrix} \begin{pmatrix} \bar u & 0 \\ 0 & u \end{pmatrix} \otimes \mathbb{1}_N & 0 \\ 0 & v \otimes \mathbb{1}_N \end{pmatrix}} \otimes \bar w \\
\qquad\qquad\qquad\qquad \bar u\, \overline{\begin{pmatrix} u \otimes \mathbb{1}_N & 0 \\ 0 & v \otimes \mathbb{1}_N \end{pmatrix}} .
\end{array}
$$

Matrix-form of the generalized Dirac operator:

$$
\begin{array}{l}
\qquad\qquad \begin{array}{cccc} u_R\; d_R & u_L\; d_L & e_R & \nu_L\; e_L \end{array} \\
D_q|_H = \begin{pmatrix} 0 & \begin{pmatrix} M_u^* & 0 \\ 0 & M_d^* \end{pmatrix} \\ \begin{pmatrix} M_u & 0 \\ 0 & M_d \end{pmatrix} & 0 \end{pmatrix} \otimes \mathbb{1}_3 \\
\qquad\qquad\qquad\qquad D_l|_H = \begin{pmatrix} 0 & (0 M_e^*) \\ \begin{pmatrix} 0 \\ M_e \end{pmatrix} & 0 \end{pmatrix} ,
\end{array}
\tag{5.26}
$$

$$
\begin{array}{l}
\qquad\qquad \begin{array}{cccc} \bar u_R\; \bar d_R & \bar u_L\; \bar d_L & \bar e_R & \bar\nu_L\; \bar e_L \end{array} \\
\overline{D}_q|_{\overline{H}} = \overline{\begin{pmatrix} 0 & \begin{pmatrix} M_u^* & 0 \\ 0 & M_d^* \end{pmatrix} \\ \begin{pmatrix} M_u & 0 \\ 0 & M_d \end{pmatrix} & 0 \end{pmatrix}} \otimes \mathbb{1}_3 \\
\qquad\qquad\qquad\qquad \overline{D}_l|_{\overline{H}} = \overline{\begin{pmatrix} 0 & (0 M_e^*) \\ \begin{pmatrix} 0 \\ M_e \end{pmatrix} & 0 \end{pmatrix}}
\end{array}
$$

5.6 Matrix-form of quantum one-forms

$A = \begin{pmatrix} & Q' \\ Q & \end{pmatrix} \in \pi_{\mathcal{D}}(i\Omega(\mathbf{A})^1)$, $Q = \begin{pmatrix} H_2 & H^1 \\ -H_1 & H^2 \end{pmatrix}$, $Q' = \begin{pmatrix} H_2' & H'^1 \\ -H_1' & H'^2 \end{pmatrix} \in \mathbb{H}$, has in restriction to H the matrix: [27]

[27] The self-adjoint vector-potential are obtained making $H_1' = -H^1$, $H_2' = H^2$.

$$A = \begin{pmatrix} 0 & 0 & H_2'M_u^* & -H_1'M_u^* \\ 0 & 0 & H_1'M_d^* & H'^2M_d^* \\ H_2M_u & H^1M_d & 0 & 0 \\ -H_1M_u & H^2M_d & 0 & 0 \end{pmatrix} \quad (\text{columns } u_R, d_R, u_L, d_L)$$

$$\begin{pmatrix} 0 & H_1'M_d^* & H'^2M_d^* \\ H_1M_d & 0 & 0 \\ H_2M_d & 0 & 0 \end{pmatrix} \quad (\text{columns } e_R, \nu_L, e_L) \tag{5.27}$$

and vanishes on $\overline{H}$.

Proof

The first statement is, for the quark part, a rewriting of (2.35), the lepton part resulting by leptonic reduction (cf. [**2.7**]). Vanishing of the $\overline{H}$-components of one-forms (for that matter, of all n-forms with $n \geq 1$) stems from commutation of $\mathcal{D}$ with the elements $p, 0, m$ of the chromodynamics algebra.

6 The S_0-Real Spectral Triple of the Full Standard Model. The Covariant Dirac Operator

We now tensor the zero-dimensional real spectral triple $(\mathbf{A}, (\mathcal{H}, \chi, \mathcal{D}), \mathcal{J})$ described in the preceding section [**5**] by the space-time (classical) real spectral triple $(\mathbb{A} = C^\infty(\mathbf{M}, \mathbb{R}), (L^2(\mathbb{S}_\mathbf{M}), \gamma^5, \widetilde{D}, C)$ in [**4.3**] where space-time is a compact Riemannian spin 4-manifold $\mathbf{M}$.[28]

6.1 Definitions

(i): **The full standard-model algebra $\mathbb{A}$** is the tensor product of the space-time algebra $\mathbb{A} = C^\infty(\mathbf{M})$ by the inner-space algebra $\mathbf{A} = \mathbb{C} \oplus \mathbb{H} \oplus M_3(\mathbb{C})$:[29]

$$\mathbb{A} = C^\infty(\mathbf{M}, \mathbb{R}) \otimes \mathbf{A} \ ; \tag{6.1}$$

(in other terms:

$$\mathbb{A} = C^\infty(\mathbf{M}, \mathbf{A}) = \{(p, q, m);\ p \in C^\infty(\mathbf{M}, \mathbb{C}),\ q \in C^\infty(\mathbf{M}, \mathbb{H}),\ m \in C^\infty(\mathbf{M}, M_3(\mathbb{C}))\} \ , \tag{6.2}$$

with gauge group:

$$\mathcal{G} = C^\infty(\mathbf{M}, U(\mathbf{A})) = \{(u, v, w);\ u \in C^\infty(\mathbf{M}, U(1)),\ v \in C^\infty(\mathbf{M}, SU(2)),\ w \in C^\infty(\mathbf{M}, U(3))\} ; \tag{6.3}$$

[28] See [**4.3**] for details concerning Riemannian spin space.

[29] Here $\mathbb{C}$ and $M_3(\mathbb{C})$ are taken as algebras over $\mathbb{R}$, as is (necessarily) the quaternion algebra $\mathbb{H}$.

(ii):**The full standard-model S_0-real spectral triple**

$$(\mathcal{A}, (\mathcal{H} = \underline{\mathbb{H}} \oplus \overline{\mathbb{H}}, \boldsymbol{\chi}, \boldsymbol{D}), \mathcal{J})$$

is the tensor product of the zero-dimensional inner S_0-real spectral triple

$$(\mathbf{A}, (\mathcal{H}, \chi, \mathcal{D}), \mathcal{J})$$

described in **[5]** by the classical real spectral triple described in **[4.3]** where d=4 (cf. **[4.9]**). As such it is a 4-dimensional real spectral triple (cf. **[4.9]** with $d_1 = 4$, $d_2 = 0$). Specifically:

- the $\mathbb{Z}/2$-graded complex Hilbert space $\mathcal{H}$ is the tensor product (over $\mathbb{C}$):

$$\mathcal{H} = L^2(\mathcal{S}_{\mathbf{M}}) \otimes \mathcal{H} \quad \text{with grading involution} \quad \boldsymbol{\chi} = \gamma^5 \otimes \chi \ . \tag{6.4}$$

  This will allow us below to write the endomorphisms of $\mathcal{H}$ as matrices analogous to those in section **[5.1]** but now with entries endomorphisms of $L^2(\mathcal{S}_{\mathbf{M}})$. Corresponding to the direct-sum splitting $\mathcal{H} = H \oplus \overline{H}$ of the inner Hilbert space, the full Hilbert space splits as $\mathcal{H} = \underline{\mathbb{H}} \oplus \overline{\mathbb{H}}$ into **a particle Hilbert space** $\underline{\mathbb{H}}$ and **a antiparticle Hilbert space** $\underline{\mathbb{H}}$ conjugate of each other;
- the conjugation is the tensor product of the charge-conjugation C of Euclidean electrodynamics (cf. **[4.3]**) by the conjugation J of the standard model real inner spectral triple:

$$\boldsymbol{J} = C \otimes J \quad (\boldsymbol{J} = -\boldsymbol{J}^{-1} = -\boldsymbol{J}^* \text{ exchanges } \underline{\mathbb{H}} \text{ and } \overline{\mathbb{H}}) \ ; \tag{6.5}$$

- the representation $\pi_{\boldsymbol{D}}$ of $\mathcal{A}$ is the tensor product of the representation of $\mathbb{A}$ on the square-integrable spinors and the representation of $\mathbf{A}$ on $\mathcal{H}$:

$$\pi_{\boldsymbol{D}} = \pi_{\widetilde{D}} \otimes \pi_{\mathcal{D}} \ ; \tag{6.6}$$

- the generalized Dirac operator is:

$$\boldsymbol{D} = \widetilde{D} \otimes id_{\mathcal{H}} + \gamma^5 \otimes \mathcal{D} = \mathbb{D} \oplus \boldsymbol{J}\, \mathbb{D}\, \boldsymbol{J}^* \ , \tag{6.7}$$

  where $\mathbb{D}$ is the restriction of $\boldsymbol{D}$ to $\underline{\mathbb{H}}$.

6.2 Reminder (tensor product structure of quantum forms)

We recall that we have the following tensor-decomposition of quantum one-forms:

$$\pi_{\boldsymbol{D}}(\Omega\mathcal{A}^1) = \pi_{\widetilde{D}}(\Omega\mathbb{A}^1) \otimes \pi_{\mathcal{D}}(\mathbf{A}) \oplus \pi_{\widetilde{D}}(\mathbb{A})\gamma^5 \otimes \pi_{\mathcal{D}}(\Omega\mathbf{A}^1) \ , \tag{6.8}$$

(in other terms

$$\Omega_{\boldsymbol{D}}\left(\mathcal{A}^1\right) = \gamma\left(\Omega(\mathbf{M})^1\right) \otimes \mathbf{A} \oplus C^\infty(\mathbf{M}, \mathbb{R})\gamma^5 \otimes \Omega_{\mathcal{D}}\mathbf{A}^1 \Big) \ , \tag{6.9}$$

(for these results we refer to [8]). We now give the matrix form of the operators of the theory (with entries tensor products in the obvious sense), resulting via (6.6)-(6.9) from the inner space matrix forms of the previous section [5]. Notice that, apart from $\boldsymbol{J}$, which exchanges $\underline{\mathbb{H}}$ and $\overline{\mathbb{H}}$, all operators of the theory preserve $\underline{\mathbb{H}}$ and $\overline{\mathbb{H}}$. Our choice of mutually $\boldsymbol{J}$-related basis in $\underline{\mathbb{H}}$ and $\overline{\mathbb{H}}$ will make the matrix of $\overline{T} = \boldsymbol{J}T\boldsymbol{J}^* \in \mathrm{End}\,\underline{\mathbb{H}}$ result from that of $T \in \mathrm{End}\,\overline{\mathbb{H}}$ by applying $ad\boldsymbol{J} = adC \otimes adJ$, with C acting as indicated in [**4.3**] and adJ causing the complex conjugation of matrices. The hurried reader is invited to browse quickly for a general picture of through the paragraphs [**6.3**] through [**6.5**], and concentrate on the subsequent sections [**6.6**], [**6.7**] describing ab initio the covariant Dirac operator $\boldsymbol{D}_A$ solely needed for the computation of the standard-model action (as well bosonic, cf. [**7**], as fermionic, cf. [**8**]) – and $ad\boldsymbol{J}$-invariant, hence of the form $T \oplus \overline{T}$, therefore specified by its restriction $\mathbb{D}_A$ to the particle Hilbert space $\underline{\mathbb{H}}$).

6.3 Matrix form of $\pi_{\mathbb{D}}(\boldsymbol{\mathcal{A}})$, $\boldsymbol{\mathcal{J}}\pi_{\boldsymbol{D}}(\boldsymbol{\mathcal{A}})\boldsymbol{\mathcal{J}}^*$ and $ad(G)$

For $(p,q,m) \in \boldsymbol{\mathcal{A}}$, with $p \in \mathbb{C}$, $q \in C^\infty(\mathbf{M},\mathbb{H})$, $m \in C^\infty(\mathbf{M}, M_3(\mathbb{C}))$:

$$
\pi_{\boldsymbol{D}}(p,q,m)|_H = \begin{array}{c}
\qquad u_R\, d_R \qquad\qquad u_L\, d_L \qquad\qquad\qquad e_R \qquad \nu_L\, e_L \\
\begin{pmatrix}
\begin{pmatrix} \begin{pmatrix} \bar{p} & 0 \\ 0 & p \end{pmatrix} \otimes \mathbb{1}_N & 0 \\ 0 & q \otimes \mathbb{1}_N \end{pmatrix} \otimes \mathbb{1}_3 & \\
& \begin{pmatrix} p \otimes \mathbb{1}_N & 0 \\ 0 & q \otimes \mathbb{1}_N \end{pmatrix}
\end{pmatrix}
\end{array} , \tag{6.10}
$$

$$
\pi_{\boldsymbol{D}}(p,q,m)|_{\overline{H}} = \begin{array}{c}
\qquad \bar{u}_R\, \bar{d}_R \qquad\qquad \bar{u}_L\, \bar{d}_L \qquad\qquad\qquad \bar{e}_R \qquad \bar{\nu}_L\, \bar{e}_L \\
\begin{pmatrix}
\begin{pmatrix} \mathbb{1}_2 \otimes \mathbb{1}_N & 0 \\ 0 & \mathbb{1}_2 \otimes \mathbb{1}_N \end{pmatrix} \otimes \bar{m} & \\
& \bar{p} \begin{pmatrix} \mathbb{1}_2 \otimes \mathbb{1}_N & 0 \\ 0 & \mathbb{1}_2 \otimes \mathbb{1}_N \end{pmatrix}
\end{pmatrix}
\end{array} .
$$

$$
\pi_{\boldsymbol{D}}(p,q,m)\boldsymbol{J}^*|_H = \begin{array}{c}
\qquad u_R\, d_R \qquad\qquad u_L\, d_L \qquad\qquad\qquad e_R \qquad \nu_L\, e_L \\
\begin{pmatrix}
\begin{pmatrix} \mathbb{1}_2 \otimes \mathbb{1}_N & 0 \\ 0 & \mathbb{1}_2 \otimes \mathbb{1}_N \end{pmatrix} \otimes m & \\
& p \begin{pmatrix} \mathbb{1}_1 \otimes \mathbb{1}_N & 0 \\ 0 & \mathbb{1}_2 \otimes \mathbb{1}_N \end{pmatrix}
\end{pmatrix}
\end{array} , \tag{6.11}
$$

$$
\boldsymbol{J}\pi_{\boldsymbol{D}}(p,q,m)\boldsymbol{J}^*|_{\overline{H}} = \begin{array}{c}
\qquad \bar{u}_R\, \bar{d}_R \qquad\qquad \bar{u}_L\, \bar{d}_L \qquad\qquad\qquad \bar{e}_R \qquad \bar{\nu}_L\, \bar{e}_L \\
\begin{pmatrix}
\begin{pmatrix} \begin{pmatrix} \bar{p} & 0 \\ 0 & p \end{pmatrix} \otimes \mathbb{1}_N & 0 \\ 0 & q \otimes \mathbb{1}_N \end{pmatrix}^{-} \otimes \mathbb{1}_3 & \\
& \begin{pmatrix} p \otimes \mathbb{1}_N & 0 \\ 0 & q \otimes \mathbb{1}_N \end{pmatrix}^{-}
\end{pmatrix}
\end{array} .
$$

For $(u,v,w)\in\mathcal{G}$ with $u\in C^\infty(\mathbf{M},U(1))$, $v\in C^\infty(\mathbf{M},SU(2))$, $w\in C^\infty(\mathbf{M},U(3))$ we have:

$$ad(u,v,w) = \pi_{\mathcal{D}}(u,v,w)\mathcal{J}\pi_{\mathcal{D}}(u,v,w)\mathcal{J}^*|_H$$

$$\begin{array}{ll} \quad u_R\ d_R \qquad\quad u_L\ d_L & \qquad e_R \qquad \nu_L\ e_L \\ = \left(\begin{pmatrix} \bar u & 0 \\ 0 & u \end{pmatrix}\otimes \mathbb{I}_N \quad 0 \atop 0 \qquad\quad v\otimes\mathbb{I}_N \right)\otimes w & \\ & \bar u \begin{pmatrix} u\otimes\mathbb{I}_N & 0 \\ 0 & v\otimes\mathbb{I}_N \end{pmatrix}, \end{array} \tag{6.12}$$

$$ad(u,v,\bar w) = \pi_{\mathcal{D}}(u,v,\bar w)\mathcal{J}\pi_{\mathcal{D}}(u,v,w)\mathcal{J}^*|_{\overline{H}}$$

$$\begin{array}{ll} \quad \bar u_R\ \bar d_R \qquad\quad \bar u_L\ \bar d_L & \qquad \bar e_R \qquad \bar\nu_L\ \bar e_L \\ = \left(\begin{pmatrix} \bar u & 0 \\ 0 & u \end{pmatrix}\otimes \mathbb{I}_N \quad 0 \atop 0 \qquad\quad v\otimes\mathbb{I}_N \right)^-\otimes \bar w & \\ & u \begin{pmatrix} \bar u\otimes\mathbb{I}_N & 0 \\ 0 & v\otimes\mathbb{I}_N \end{pmatrix}^- . \end{array}$$

Proof: (6.10), resp. (6.11) follow from (5.23), resp. (5.24) via (6.2); (6.12) follows from (5.25) via (6.3).

6.4 Matrix-form of $\mathcal{D}$

$$\mathcal{D}|_H =$$

$$\begin{array}{ll} \qquad u_R \qquad d_R \qquad\qquad\qquad u_L\ d_L & \qquad e_R \qquad\qquad \nu_L\ e_L \\ \begin{pmatrix} \begin{pmatrix} \widetilde D\otimes\mathbb{I}_N & 0 \\ 0 & \widetilde D\otimes\mathbb{I}_N \end{pmatrix} & \gamma^5\otimes\begin{pmatrix} M_u^* & 0 \\ 0 & M_d^* \end{pmatrix} \\ \gamma^5\otimes\begin{pmatrix} M_u & 0 \\ 0 & M_d \end{pmatrix} & \begin{pmatrix} \widetilde D\otimes\mathbb{I}_N & 0 \\ 0 & \widetilde D\otimes\mathbb{I}_N \end{pmatrix} \end{pmatrix}\otimes\mathbb{I}_3 & \\ & \begin{pmatrix} \widetilde D & (0 \quad M_e^*) \\ \begin{pmatrix} 0 \\ M_e \end{pmatrix} & \begin{pmatrix} \widetilde D\otimes\mathbb{I}_N & 0 \\ 0 & \widetilde D\otimes\mathbb{I}_N \end{pmatrix} \end{pmatrix} \end{array} \tag{6.13}$$

$$
\mathcal{D}|_{\overline{H}} =
\begin{array}{cc}
\begin{array}{cccc} \bar{u}_R & \bar{d}_R & \bar{u}_L & \bar{d}_L \end{array} & \begin{array}{ccc} \bar{e}_R & \bar{\nu}_L & \bar{e}_L \end{array} \\
\begin{pmatrix} \begin{pmatrix} \widetilde{D}\otimes \mathbb{I}_N & 0 \\ 0 & \widetilde{D}\otimes \mathbb{I}_N \end{pmatrix} & \gamma^5 \otimes \begin{pmatrix} M_u^* & 0 \\ 0 & M_d^* \end{pmatrix}^- \\ \gamma^5 \otimes \begin{pmatrix} M_u & 0 \\ 0 & M_d \end{pmatrix}^- & \begin{pmatrix} \widetilde{D}\otimes \mathbb{I}_N & 0 \\ 0 & \widetilde{D}\otimes \mathbb{I}_N \end{pmatrix} \end{pmatrix} \otimes \mathbb{I}_3 & \\
 & \begin{pmatrix} \widetilde{D} & (0 \quad M_e^*)^- \\ \begin{pmatrix} 0 \\ M_e \end{pmatrix}^- & \begin{pmatrix} \widetilde{D}\otimes \mathbb{I}_N & 0 \\ 0 & \widetilde{D}\otimes \mathbb{I}_N \end{pmatrix} \end{pmatrix}
\end{array}
$$

Proof: follows from (6.7).

6.5 Matrices of $\pi_{\mathcal{D}}(i\Omega\mathcal{A}^1)$

$A \in \Omega_{\mathcal{D}}\mathcal{A}^1 = \pi_{\mathcal{D}}(i\Omega\mathcal{A}^1)$ has the matrix (stenography in term of 2×2 matrices):

$$
A|_H = (\mathbf{a},\, \mathbf{b}^{\cdot}.,\, \mathbf{H},\, \mathbf{H}',\, \mathbf{c}_0,\, \mathbf{c}^{\cdot}) =
$$

$$
\begin{array}{cc}
\begin{array}{cccc} u_R & d_R & u_L & d_L \end{array} & \begin{array}{ccc} e_R & \nu_L & e_L \end{array} \\
\begin{pmatrix} \gamma(\mathbf{a}^{\cdot}.)\otimes 1_N & \mathbb{M}^*(\gamma^5\mathbf{H}'\otimes \mathbb{I}_N) \\ (\gamma^5\mathbf{H}\otimes \mathbb{I}_N)\mathbb{M} & \gamma(\mathbf{b}^{\cdot}.)\otimes \mathbb{I}_N \end{pmatrix} & \text{Leptonic reduction of latter}
\end{array}
\qquad (6.14)
$$

$$
A|_{\overline{H}}(\mathbf{a},\, \mathbf{b}^{\cdot}.,\, \mathbf{H},\, \mathbf{H}',\, \mathbf{c}_0,\, \mathbf{c}^{\cdot}) =
$$

$$
\begin{array}{ccccccc} \bar{u}_R & \bar{d}_R & \bar{u}_L & \bar{d}_L & \bar{e}_R & \bar{\nu}_L & \bar{e}_L \end{array}
$$

$$
\frac{i}{2}\gamma(\mathbf{c}_0)\mathbb{I}_2\otimes \mathbb{I}_N\otimes \mathbb{I}_3 + \frac{i}{2}\gamma(\mathbf{c}^a)\mathbb{I}_2\otimes \mathbb{I}_N\otimes \frac{\lambda}{2}\ .
$$

Here:
- $\mathbf{a}$ and $\mathbf{c}_0$ are classical U(1)-connections-one-form: $\mathbf{a} \in \Omega(\mathbf{M},\mathbb{C})^1$;
- $\mathbf{b}^{\cdot}.$ is a classical U(2)-connections-one-form:

$$
\mathbf{b}^{\cdot}. = \begin{pmatrix} \mathbf{b}_1^1 & \mathbf{b}_2^1 = \bar{\mathbf{b}}_1^2 \\ \mathbf{b}_1^2 & \mathbf{b}_2^2 = -\bar{\mathbf{b}}_1^1 \end{pmatrix} \in \Omega(\mathbf{M}, i\,\mathbb{H})^1\ ;
$$

- $\mathbf{c}^{\cdot} = (\mathbf{c}^a)_{a=1,\dots,8}$ is SU(3))-connections-one-form (the λ_a are the eight Gell-Man matrices). We used the shorthands:

$$\gamma(\mathbf{a}^{\cdot}.) = \begin{pmatrix} -\gamma(\bar{\mathbf{a}}) & 0 \\ 0 & \gamma(\mathbf{a}) \end{pmatrix} \qquad \left(\mathbf{a}^{\cdot}. = \begin{pmatrix} -\bar{\mathbf{a}} & 0 \\ 0 & \mathbf{a} \end{pmatrix} \in \Omega\left(\mathbf{M}, i\,\mathbb{H}_{\text{diag}}\right)^1\right) , \tag{6.15}$$

$$\gamma(\mathbf{b}^{\cdot}.) = \begin{pmatrix} \gamma(\mathbf{b}_1^1) & \gamma(\mathbf{b}_2^1) \\ \gamma(\mathbf{b}_1^2) & \gamma\mathbf{b}_2^2 \end{pmatrix} . \tag{6.16}$$

The vector potentials antihermitean elements of $\Omega_{\mathcal{D}_q}\mathcal{A}^1$, singled out as the hermitean $A = i\rho$ above (s.t. $\underline{\mathbf{a}} = \mathbf{a}$, $\mathbf{b}_2^2 = -\mathbf{b}_1^1$, $\mathbf{H}' = \mathbf{H}^*$), are accordingly specified by quintuples

$$A = A(\mathbf{a}, \mathbf{b}, \mathbf{c}_0, \mathbf{c}^{\cdot}, \mathbf{H}) , \tag{6.17}$$

of a hermitean U(1)-connections-one-form $\mathbf{a}$, a hermitean SU(2)-connections-one-form $\mathbf{b}^{\cdot}.$, a second hermitean U(1)-connections-one-form $\mathbf{c}_0$ and a hermitean SU(3)-connections-one-form $\mathbf{c}^{\cdot}$, and a doublet field identified with an $\mathbf{H} \in C^\infty(\mathbf{M}, \mathbb{H})$:

$$\mathbf{H} = \begin{pmatrix} \mathbf{H}_2 & \mathbf{H}^1 \\ -\mathbf{H}_1 & \mathbf{H}^2 \end{pmatrix} \leftrightarrow \mathbf{H}^{\cdot} = (\mathbf{H}^1, \mathbf{H}^2) \leftrightarrow \quad \mathbf{H}. = (\mathbf{H}_1, \mathbf{H}_2) = (\overline{\mathbf{H}}^1, \overline{\mathbf{H}}^2) . \tag{6.18}$$

Proof follows from (2.29) via (6.9).

The remainder of this section is devoted to a description of the covariant Dirac operator $\mathcal{D}_A = \mathcal{D} + A + \mathcal{J}A\mathcal{J}^*$, the item needed for the computation of both the bosonic and the fermionic actions of the standard model, cf. [**7**], resp. [**8**]. As mentioned above it is enough to specify the matrix of the restriction of $\mathcal{D}_A$ to the particle subspace $\underline{\mathbb{H}}$, the restriction to $\overline{\mathbb{H}}$ resulting from the latter by application of $ad\mathcal{J}$, tensor product of adC and adJ, with C acting as indicated in [**4.3**] and adJ causing the complex conjugation of matrices.

6.6 Matrix form of the covariant Dirac operator $\mathcal{D}_A = \mathcal{D} + \mathcal{A} + \mathcal{J}\mathcal{A}\mathcal{J}^*$

We give separately the quark-, respectively lepton-matrices of the three terms of the restriction $\mathbb{D}_A$ of $\mathcal{D}_A$ to $\underline{\mathbb{H}}$: a "zoomed" rehearsal (using (4 × 4, resp. 3 × 3 matrices) of the compact formulae above written in terms 2 × 2 matrices. We have the quark matrices:

$$D_q = \begin{pmatrix} \tilde{D} \otimes \mathbb{I}_N & 0 & \gamma^5 \otimes M_u^* & 0 \\ 0 & \tilde{D} \otimes \mathbb{I}_N & 0 & \gamma^5 \otimes M_d^* \\ \gamma^5 \otimes M_u & 0 & \tilde{D} \otimes \mathbb{I}_N & 0 \\ 0 & \gamma^5 \otimes M_d & 0 & \tilde{D} \otimes \mathbb{I}_N \end{pmatrix} \otimes \mathbb{I}_3 \quad \text{(columns: } u_R, d_R, u_L, d_L\text{)} . \tag{6.19}$$

Restr. to $\underline{\mathbb{H}}$ of

$$A_q = \begin{pmatrix} -\gamma(\mathbf{a}) \otimes \mathbb{I}_N & 0 & \mathbf{H}^2\gamma^5 \otimes M_u^* & -\mathbf{H}^1\gamma^5 \otimes M_u^* \\ 0 & \gamma(\mathbf{a}) \otimes \mathbb{I}_N & \mathbf{H}_1\gamma^5 \otimes M_d^* & \mathbf{H}_2\gamma^5 \otimes M_d^* \\ \mathbf{H}_2\gamma^5 \otimes M_u & \mathbf{H}^1\gamma^5 \otimes M_d & \gamma(\mathbf{b}_1^1) \otimes \mathbb{I}_N & \gamma(\mathbf{b}_2^1) \otimes \mathbb{I}_N \\ -\mathbf{H}_1\gamma^5 \otimes M_u & \mathbf{H}^2\gamma^5 \otimes M_d & \gamma(\mathbf{b}_1^2) \otimes \mathbb{I}_N & \gamma(\mathbf{b}_2^2) \otimes \mathbb{I}_N \end{pmatrix} \otimes \mathbb{I}_3 . \tag{6.20}$$

(columns: u_R, d_R, u_L, d_L)

Restr. to $\underline{\mathbb{H}}$ of

$$(\mathcal{J}A\mathcal{J}^*)_q = \begin{pmatrix} \gamma(\mathbf{c}_0)\otimes\mathbb{I}_N & 0 & 0 & 0 \\ 0 & \gamma(\mathbf{c}_0)\otimes\mathbb{I}_N & 0 & 0 \\ 0 & 0 & \gamma(\mathbf{c}_0)\otimes\mathbb{I}_N & 0 \\ 0 & 0 & 0 & \gamma(\mathbf{c}_0)\otimes\mathbb{I}_N \end{pmatrix} \otimes \mathbb{I}_3$$
$$+ \begin{pmatrix} \gamma(\mathbf{c}^a)\otimes\mathbb{I}_N & 0 & 0 & 0 \\ 0 & \gamma(\mathbf{c}^a)\otimes\mathbb{I}_N & 0 & 0 \\ 0 & 0 & \gamma(\mathbf{c}^a)\otimes\mathbb{I}_N & 0 \\ 0 & 0 & 0 & \gamma(\mathbf{c}^a)\otimes\mathbb{I}_N \end{pmatrix} \otimes \frac{\lambda_a}{2} . \tag{6.21}$$

(columns: u_R, d_R, u_L, d_L)

where

- $\mathbf{a}$ is a hermitean U(1)-connection-one-form;
- $\mathbf{b}^{\cdot}_{\cdot}$ a hermitean SU(2)-connection-one-form;
- $\mathbf{H}$ a doublet field identified with $\mathbf{H}^{\cdot}_{\cdot} = \begin{pmatrix} \mathbf{H}_2 & \mathbf{H}^1 \\ -\mathbf{H}_1 & \mathbf{H}^2 \end{pmatrix} \in C^\infty(\mathbf{M}, \mathbb{H})$;
- $\mathbf{c}_0$ is a second hermitean U(1)-connection-one-form;
- $\mathbf{c}^{\cdot}_{\cdot}$ a hermitean SU(3)-connection-one-form.

With this notation we have the lepton matrices (obtainable through lepton reduction):

$$D_l = \begin{pmatrix} \widetilde{D} \otimes \mathbb{I}_N & 0 & \gamma^5 \otimes M_e^* \\ 0 & \widetilde{D} \otimes \mathbb{I}_N & 0 \\ \gamma^5 \otimes M_e & 0 & \widetilde{D} \otimes \mathbb{I}_N \end{pmatrix} . \tag{6.22}$$

(columns: e_R, ν_L, e_L)

Restr. to $\underline{\mathbb{H}}$ of

$$A_l = \begin{pmatrix} \gamma(\mathbf{a}) \otimes \mathbb{I}_N & \mathbf{H}_1\gamma^5 \otimes M_e^* & \mathbf{H}_2\gamma^5 \otimes M_e^* \\ \mathbf{H}^1\gamma^5 \otimes M_e & \gamma(\mathbf{b}_1^1) \otimes \mathbb{I}_N & \gamma(\mathbf{b}_2^1) \otimes \mathbb{I}_N \\ \mathbf{H}^2\gamma^5 \otimes M_e & \gamma(\mathbf{b}_1^2) \otimes \mathbb{I}_N & \gamma(\mathbf{b}_2^2) \otimes \mathbb{I}_N \end{pmatrix} . \tag{6.23}$$

(columns: e_R, ν_L, e_L)

Restr. to $\underline{\mathbb{H}}$ of

$$(\mathcal{J}A\mathcal{J}^*)_l = \begin{pmatrix} \gamma(\mathbf{a})\otimes \mathbb{1}_N & 0 & 0 \\ 0 & \gamma(\mathbf{a})\otimes \mathbb{1}_N & 0 \\ 0 & 0 & \gamma(\mathbf{a})\otimes \mathbb{1}_N \end{pmatrix} . \tag{6.24}$$

(columns labelled e_R, ν_L, e_L)

6.7 Conversion into classical objects: the covariant Dirac operator as a differential operator

Tensoring the classical S_0-real spectral triple by the finite-dimensional inner S_0-real spectral triple amounts to decorating the fiber of the spin bundle, which becomes 90-dimensional, and turns the usual Dirac operator into a generalized Dirac operator. We now study this decorated bundle and covariant generalized Dirac operator $\mathcal{D}_A$ as classical differential- theoretic objects. In fact it suffices to look at the particle-parts $\underline{\mathbb{H}}$ and $\mathbb{D}_A$ of these objects, the anti-particle parts resulting by applying the particle-antiparticle conjugation $\mathcal{J}$.

¿From now on we shall use instead of the particle Hilbert space[30] $\underline{\mathbb{H}}$ its smooth dense sub-$C^\infty(\mathbf{M})$-module $\underline{\mathbb{E}} = \mathcal{S}(\mathbf{M}) \otimes H$, left invariant by all the operators under consideration. We have the following situation:

(i): $\underline{\mathbb{E}}$ is a finite-projective $C^\infty(\mathbf{M})$-module, expressible as the tensor product

$$\underline{\mathbb{E}} = \mathcal{S}(\mathbf{M})\otimes_{C^\infty(\mathbf{M})}\mathbf{E} \quad \text{with} \quad \mathbf{E} = C^\infty(\mathbf{M})\otimes H \ , \tag{6.25}$$

in fact a twisted Clifford module $(\underline{\mathbb{E}}, c)$ under the $\mathbb{Z}/2$-grading χ and the Clifford action[31]

$$c(\gamma(\lambda)) = \gamma(\lambda)\otimes id_{\mathbf{E}}, \quad \gamma(\lambda)\in \mathbb{C}1(\mathbf{M}),\ \lambda\in\Omega(\mathbf{M}), \tag{6.26}$$

and split in a direct sum of a quarkonic and the leptonic $C^\infty(\mathbf{M})$-module according to the decomposition $\mathbf{E} = \mathbf{E}_q \oplus \mathbf{E}_l$, where $\mathbf{E}_q = C^\infty(\mathbf{M})\otimes H_q$ and $\mathbf{E}_l = C^\infty(\mathbf{M})\otimes H_l$.

(ii): We have $\mathbb{D}_A = D^{\nabla} + \Phi$, direct sum $(\mathbb{D}_A)_q \oplus (\mathbb{D}_A)_l$ of the quark and the lepton parts:

$$\begin{cases} (\mathbb{D}_A)_q = (D^{\nabla})_q + \Phi_q \\ (\mathbb{D}_A)_l = (D^{\nabla})_l + \Phi_l \end{cases} \quad \text{with} \quad \begin{cases} (D^{\nabla})_q = ic^\mu \nabla_{q\mu} \\ (D^{\nabla})_l = ic^\mu \nabla_{l\mu} \end{cases} , \tag{6.27}$$

where:

– the endomorphisms Φ_q, Φ_l of $\mathbb{E}$ respectively act on the quark and lepton subspaces as the matrices:[32]

$$\Phi_q = \begin{pmatrix} 0 & 0 & \Phi^2\gamma^5\otimes M_u^* & -\Phi^1\gamma^5\otimes M_u^* \\ 0 & 0 & \Phi_1\gamma^5\otimes M_d^* & \Phi_2\gamma^5\otimes M_d^* \\ \Phi_2\gamma^5\otimes M_u & \Phi^1\gamma^5\otimes M_d & 0 & 0 \\ -\Phi_1\gamma^5\otimes M_u & \Phi^2\gamma^5\otimes M_d & 0 & 0 \end{pmatrix}\otimes \mathbb{1}_3 , \tag{6.28}$$

(columns labelled u_R, d_R, u_L, d_L)

[30] which it suffices to consider by charge-conjugation symmetry.

[31] See [B]. We recall that for us γ is a map from contravariant tensors (in particular $\Omega(\mathbf{M})$) to End $\Omega(\mathbf{M})^1$.

[32] Note that Φ anticommutes with the c^μ.

$$\boldsymbol{\Phi}_l = \begin{pmatrix} 0 & \Phi_1\gamma^5 \otimes M_e^* & \Phi_2\gamma^5 \otimes M_e^* \\ \Phi^1\gamma^5 \otimes M_e & 0 & 0 \\ \Phi^2\gamma^5 \otimes M_e & 0 & 0 \end{pmatrix} \quad (\text{columns: } e_R,\ \nu_L,\ e_L), \tag{6.29}$$

where $\Phi_i = \overline{\Phi}^i \in C^\infty(\mathbf{M}, \mathbb{C})$; $\boldsymbol{\Phi}^{\cdot}. = \mathbf{H}^{\cdot}. + \mathbf{I}$;[33]
- $D^{\mathbb{V}}$ with local expression $ic^\mu \mathbb{V}_\mu$, $c^\mu = \gamma(\mathbf{d}x^\mu)$, $\mathbb{V}_\mu = \mathbb{V}_{\partial\mu}$, is the Dirac operator of the connexion $\mathbb{V}$ of $\mathbb{E}$, tensor-product:

$$\mathbb{V} = \widetilde{\nabla} \otimes id_{\mathbf{E}} + id_{\mathbb{S}(\mathbf{M})} \otimes \nabla^{\mathbf{E}} : \quad \mathbb{V}_\xi = \widetilde{\nabla}_\xi \otimes id_{\mathbf{E}} + id_{\mathbb{S}(\mathbf{M})} \otimes \nabla^{\mathbf{E}}_\xi \tag{6.30}$$

for each vector field ξ of the spin connexion $\widetilde{\nabla}$ of $\mathbb{S}(\mathbf{M})$ by the connexion $\nabla^{\mathbf{E}}$ of $\mathbf{E}$ specified as follows: $\nabla^{\mathbf{E}}$ is the direct sum $\nabla^{\mathbf{E}_q} \oplus \nabla^{\mathbf{E}_l}$ of a quark and a lepton connexion acting respectively on the quark and lepton frames as the sum of the exterior derivative and the matrices:

$$\begin{aligned} id_{\mathbb{S}(\mathbf{M})} \otimes (\nabla^{\mathbf{E}_q} - \partial)_\mu = \\ -i \begin{pmatrix} (-\mathbf{a}_\mu + \mathbf{c}^0_\mu) \otimes \mathbb{I}_N & 0 & 0 & 0 \\ 0 & (\mathbf{a}_\mu + \mathbf{c}^0_\mu) \otimes \mathbb{I}_N & 0 & 0 \\ 0 & 0 & (\mathbf{b}^1_{1\mu} + \mathbf{a}_\mu) \otimes \mathbb{I}_N & \mathbf{b}^1_{2\mu} \otimes \mathbb{I}_N \\ 0 & 0 & \mathbf{b}^2_{1\mu} \otimes \mathbb{I}_N & (\mathbf{b}^2_{2\mu} + \mathbf{a}_\mu) \otimes \mathbb{I}_N \end{pmatrix} \otimes \mathbb{I}_3 \\ -i \begin{pmatrix} \mathbf{c}^a_\mu \otimes \mathbb{I}_N & 0 & 0 & 0 \\ 0 & \mathbf{c}^a_\mu \otimes \mathbb{I}_N & 0 & 0 \\ 0 & 0 & \mathbf{c}^a_\mu \otimes \mathbb{I}_N & 0 \\ 0 & 0 & 0 & \mathbf{c}^a_\mu \otimes \mathbb{I}_N \end{pmatrix} \otimes \frac{\lambda_a}{2}, \end{aligned} \quad (\text{columns: } u_R,\ d_R,\ u_L,\ d_L) \tag{6.31}$$

$$id_{\mathbb{S}(\mathbf{M})} \otimes (\nabla^{\mathbf{E}_l} - \partial)_\mu = -i \begin{pmatrix} 2\mathbf{a}_\mu \otimes \mathbb{I}_N & 0 & 0 \\ 0 & (\mathbf{b}^1_{1\mu} + \mathbf{a}_\mu) \otimes \mathbb{I}_N & \mathbf{b}^1_{2\mu} \otimes \mathbb{I}_N \\ 0 & \mathbf{b}^2_{1\mu} \otimes \mathbb{I}_N & (\mathbf{b}^2_{2\mu} + \mathbf{a}_\mu) \otimes \mathbb{I}_N \end{pmatrix} . \quad (\text{columns: } e_R,\ \nu_L,\ e_L) \tag{6.32}$$

Here:[34]
-$\mathbf{a}$ and $\mathbf{c}_0$ are classical U(1)-vector-potential: $\bar{\mathbf{a}} = \mathbf{a} \in \Omega(\mathbf{M}, \mathbb{C})^1$;
-$\mathbf{b}^{\cdot}.$ is a classical U(2)-vector-potential:

$$\mathbf{b}^{\cdot}. = \begin{pmatrix} \mathbf{b}^1_1 = \bar{\mathbf{b}}^1_1 & \mathbf{b}^1_2 = \bar{\mathbf{b}}^2_1 \\ \mathbf{b}^2_1 & \mathbf{b}^2_2 = -\mathbf{b}^1_1 \end{pmatrix} \in \Omega(\mathbf{M}, i\mathbb{H}_{\text{traceless}})^1 \ ;$$

---

[33] Note that $\boldsymbol{\Phi}^{\cdot}. = \begin{pmatrix} \Phi_2 & \Phi^1 \\ -\Phi_1 & \Phi^2 \end{pmatrix} \in C^\infty(\mathbf{M}, \mathbb{H})$.

[34] Following the physicists' usage we multiply by i our connexion one-form to make them self-adjoint ("vector potential"). Note that a quaternion is antihermitean iff it is traceless.

-$\mathbf{c}^{\cdot} = (\mathbf{c}^a)_{a=1,\dots,8}$ is SU(3))-vector-potential (the λ_a are the eight Gell-Man matrices).

Modular correction: we shall, in accordance to [**5.4**](ii) heuristically set $\mathbf{c}_0 = -\frac{1}{3}\mathbf{a}$.

Note that $\mathbb{V}$ is a **Clifford connection** in the sense:

$$[\mathbb{V}_\mu, c(\gamma(\lambda))] = c\left(\gamma\left(\nabla_\mu^{\mathbf{M}}\lambda\right)\right), \quad \lambda \in \Omega(\mathbf{M})^1 \tag{6.33}$$

owing to (6.26), (6.30), and the known Clifford connection property of the spin connexion $\tilde{\nabla}$, namely: $[\tilde{\nabla}_\mu, \gamma(\lambda)] = \gamma\left(\nabla_\mu^{\mathbf{M}}\lambda\right), \quad \lambda \in \Omega(\mathbf{M})^1$.

Proof

$\mathbb{E} = \mathcal{S}(\mathbf{M}) \otimes H$ is the finite-projective $C^\infty(\mathbf{M})$-module pull-back of the $\mathbb{C}$-module H by the $C^\infty(\mathbf{M}) - \mathbb{C}$-bimodule $\mathcal{S}(\mathbf{M})$, thus obviously expressible as the tensor product of $C^\infty(\mathbf{M})$-modules $\mathbb{E} = \mathcal{S}(\mathbf{M}) \otimes_{C^\infty(\mathbf{M})} \mathbf{E}$. The action (6.26) of $\mathbb{C}1(\mathbf{M})$ on $\mathbb{E}$ then makes it a Clifford module $(\mathbb{E}, c)$: indeed $\mathbb{E}$ is a $\mathbb{Z}/2$-graded $\mathbb{C}1(\mathbf{M})$-module owing to the Clifford relations $c^\mu c^\nu + c^\nu c^\mu = g^{\mu\nu}$. The remaining claims follow from the matrix from of the Dirac operators $\mathcal{D}_q$, $\mathcal{D}_l$ and of the vector-potentials A_q, A_l, $(\mathcal{J}A\mathcal{J}^*)_q$, $(\mathcal{J}A\mathcal{J}^*)_l$, cf. (6.19)-(6.24)

The covariant Dirac operator $\mathcal{D}_A$ (for that matter $\mathbb{D}_A$) enters the spectral action through its square which, as the square of a generalized Dirac operator, is a generalized Laplacian. We now describe the canonical splitting $\Delta^{\mathbb{V}} + E$ of the latter which is needed for the heat-kernel expansion computation of section [7] below (cf. **Appendices** [A] and [B]); and list matrix expression of terms appearing in of E which we shall need in this computation (the reader can omit to read this until then).

6.8 Canonical decomposition of $\mathbb{D}_A^2$

(i): We have the canonical splitting: $\mathbb{D}_A^2 = \Delta^{\mathbb{V}} + E$ of the generalized Laplacian $\mathbb{D}_A^2$ as the sum of the connection-Laplacian $\Delta^{\mathbb{V}} = -g^{\mu\nu}(\mathbb{V}_\mu\mathbb{V}_\nu - \Gamma_{\mu\nu}^\alpha \mathbb{V}^\alpha)$ plus the endomorphism:[35]

$$\begin{aligned} E &= \frac{1}{4}\mathbf{s}\mathbb{1} - \frac{1}{2}c\left(R^{\mathbf{E}}\right) + ic^\mu[\mathbb{V}_\mu, \Phi] + \Phi^2 \\ \text{with} \quad & c\left(R^{\mathbf{E}}\right) = -\gamma^\mu\gamma^\nu \otimes R^{\mathbf{E}}(e_\mu, e_\nu) \ , \end{aligned} \tag{6.34}$$

[35] Cf. [B]. Note that $[\mathbb{V}_\mu, \Phi]$ lies in $\mathrm{End}_{\mathbb{A}}(\mathbb{E})$ as the commutator of a ∂_μ-derivation and a 0-derivation.

where $\mathbf{s}$ is the scalar curvature of $\mathbf{M}$, and $R^{\mathbf{E}}$ is the curvature of $\nabla^{\mathbf{E}}$.
(ii): We have the following matrix expressions:

$$c^\mu[\mathbb{V}_{q\mu}, \boldsymbol{\Phi}_q] = \begin{array}{cccc} u_R & d_R & u_L & d_L \end{array}$$
$$\begin{pmatrix} 0 & 0 & \gamma(\mathbf{D}\Phi^2)\gamma^5 \otimes M_u^* & -\gamma(\mathbf{D}\Phi^1)\gamma^5 \otimes M_u^* \\ 0 & 0 & \gamma(\mathbf{D}\Phi_1)\gamma^5 \otimes M_d^* & \gamma(\mathbf{D}\Phi_2)\gamma^5 \otimes M_d^* \\ \gamma(\mathbf{D}\Phi_2)\gamma^5 \otimes M_u & \gamma(\mathbf{D}\Phi^1)\gamma^5 \otimes M_d & 0 & 0 \\ -\gamma(\mathbf{D}\Phi_1)\gamma^5 \otimes M_u & \gamma(\mathbf{D}\Phi^2)\gamma^5 \otimes M_d & 0 & 0 \end{pmatrix} \otimes \mathbb{I}_3 \tag{6.35}$$

$$c^\mu[\mathbb{V}_{l\mu}, \boldsymbol{\Phi}_l] = \begin{array}{ccc} e_R & \nu_L & e_L \end{array}$$
$$\begin{pmatrix} 0 & \gamma(\mathbf{D}\Phi_1)\gamma^5 \otimes M_e^* & \gamma(\mathbf{D}\Phi_2)\gamma^5 \otimes M_e^* \\ \gamma(\mathbf{D}\Phi^1)\gamma^5 \otimes M_e & 0 & 0 \\ \gamma(\mathbf{D}\Phi^2)\gamma^5 \otimes M_e^* & 0 & 0 \end{pmatrix}, \tag{6.36}$$

where

$$\begin{cases} \mathbf{D}\Phi^j = \mathbf{d}\Phi^j + i(\mathbf{a}\Phi^j - \mathbf{b}^j_k\Phi^k) \\ \mathbf{D}\Phi_j = \mathbf{d}\Phi_j - i(\mathbf{a}\Phi_j - \mathbf{b}^k_j\Phi_k) \end{cases}, \quad j = 1, 2;$$
$$\left(\text{i.e.} \quad \begin{cases} \mathbf{D}\Phi^{\cdot} = \mathbf{d}\Phi^{\cdot} + i\left(\mathbf{a} - \mathbf{b}^a \frac{\tau_a}{2}\right)\Phi^{\cdot} \\ \mathbf{D}\Phi_{\cdot} = \mathbf{d}\Phi_{\cdot} - i\Phi_{\cdot}\left(\mathbf{a} - \mathbf{b}^a \frac{\tau_a}{2}\right) \end{cases} \right). \tag{6.37}$$

(iii): *We have the following quark matrix expressions:*

$$\boldsymbol{\Phi}_q = \begin{pmatrix} 0 & \Phi^{\cdot}{}_{\cdot}{}^{*} \\ \Phi^{\cdot}{}_{\cdot} & 0 \end{pmatrix} \otimes \mathbb{I}_3 , \tag{6.38}$$

with:

$$\Phi^{\cdot}{}_{\cdot}{}^{*} = \begin{pmatrix} \Phi^2\gamma^5 \otimes M_u^* & -\Phi^1\gamma^5 \otimes M_u^* \\ \Phi_1\gamma^5 \otimes M_d^* & \Phi_2\gamma^5 \otimes M_d^* \end{pmatrix},$$
$$\Phi^{\cdot}{}_{\cdot} = \begin{pmatrix} \Phi_2\gamma^5 \otimes M_u & \Phi^1\gamma^5 \otimes M_d \\ -\Phi_1\gamma^5 \otimes M_u & \Phi^2\gamma^5 \otimes M_d \end{pmatrix}, \tag{6.39}$$

thus

$$\boldsymbol{\Phi}_q^2 = \begin{pmatrix} \Phi^{\cdot}{}_{\cdot}{}^{*}\Phi^{\cdot}{}_{\cdot} & 0 \\ 0 & \Phi^{\cdot}{}_{\cdot}\Phi^{\cdot}{}_{\cdot}{}^{*} \end{pmatrix} \otimes \mathbb{I}_3 , \tag{6.40}$$

with:

$$\Phi^{\cdot}{}_{\cdot}{}^{*}\Phi^{\cdot}{}_{\cdot} = |\Phi|^2 \begin{pmatrix} \mathbb{I} \otimes M_u M_u^* & 0 \\ 0 & \mathbb{I} \otimes M_d M_d^* \end{pmatrix}, \quad (|\Phi|^2 = \Phi_1\Phi^1 + \Phi_2\Phi^2) , \tag{6.41}$$

and the lepton matrix expressions:

$$\boldsymbol{\Phi}_l = \begin{array}{c} \begin{array}{ccc} e_R & \nu_L & e_L \end{array} \\ \begin{pmatrix} 0 & \Phi_1\gamma^5 \otimes M_e^* & \Phi_2\gamma^5 \otimes M_e^* \\ \Phi^1\gamma^5 \otimes M_e & 0 & 0 \\ \Phi^2\gamma^5 \otimes M_e & 0 & 0 \end{pmatrix} \end{array}, \tag{6.42}$$

$$\boldsymbol{\Phi}_l^2 = \begin{pmatrix} |\Phi|^2 \mathbb{I} \otimes M_e^* M_e & 0 & 0 \\ 0 & \Phi_1\Phi^1 \mathbb{I} \otimes M_e M_e^* & \Phi_2\Phi^1 \mathbb{I} \otimes M_e M_e^* \\ 0 & \Phi_1\Phi^2 \mathbb{I} \otimes M_e M_e^* & \Phi_2\Phi^2 \mathbb{I} \otimes M_e M_e^* \end{pmatrix} \quad (6.43)$$

(columns labelled e_R, ν_L, e_L)

Proof:
(i): We have, owing to anticommutativity of $\boldsymbol{\Phi}$ and the c^μ:

$$\mathbb{D}^2_{\mathbb{A}} = (ic^\mu \mathbb{V}_\mu + \boldsymbol{\Phi})(ic^\nu \mathbb{V}_\nu + \boldsymbol{\Phi}) = -c^\mu \mathbb{V}_\mu c^\nu \mathbb{V}_\nu + ic^\mu \mathbb{V}_\mu \boldsymbol{\Phi} + i\boldsymbol{\Phi} c^\mu \mathbb{V}_\mu + \boldsymbol{\Phi}^2$$
$$= \left(D^{\mathbb{V}}\right)^2 + ic^\mu\left[\mathbb{V}_\mu, \boldsymbol{\Phi}\right] + \boldsymbol{\Phi}^2 = \triangle^{\mathbb{V}} + \frac{1}{4}\mathbf{s}\mathbb{I} - \frac{1}{2}c\left(R^{\mathbf{E}}\right) + ic^\mu\left[\mathbb{V}_\mu, \boldsymbol{\Phi}\right] + \boldsymbol{\Phi}^2 \quad (6.44)$$

where we plugged in the Lichnerowicz formula for the square of $\mathbb{D}$, cf [**A**]:

$$\left(D^{\mathbb{V}}\right)^2 = \triangle^{\mathbb{V}} + \frac{1}{4}\mathbf{s}\mathbb{I} - \frac{1}{2}c\left(R^{\mathbf{E}}\right) . \quad (6.45)$$

(ii): The expressions (6.35)-(6.36) are computed using (6.28)-(6.29) and (6.31)-(6.32). Observe that $\boldsymbol{\Phi}$ commutes with the spin-connection one-form since the latter commutes with γ^5. It also commutes with the gluon-connection one-forms whose matrices are diagonal with entries Clifford scalars. Thus it suffices to compute $\left[id_{\mathbb{S}(\mathbf{M})} \otimes \left(\nabla_q^{'\mathbf{E}} - \partial\right)_\mu, \boldsymbol{\Phi}\right]$,with $\nabla^{'\mathbf{E}}$ obtained from $\nabla^{\mathbf{E}}$ by deleting the gluon-connection.
Check of (6.36):

$$\left[id_{\mathbb{S}(\mathbf{M})} \otimes \left(\nabla_l^{'\mathbf{E}} - \partial\right)_\mu\right]\boldsymbol{\Phi}_l = -iX_l \times Y_l$$
$$= -i \begin{pmatrix} 0 & \mathbf{a}_\mu \Phi_1 \gamma^5 \otimes M_e^* & \mathbf{a}_\mu \Phi_2 \gamma^5 \otimes M_e^* \\ (\mathbf{b}^1_{1\mu}\Phi^1 + \mathbf{b}^1_{2\mu}\Phi^2)\gamma^5 \otimes M_e & 0 & 0 \\ (\mathbf{b}^2_{1\mu}\Phi^1 + \mathbf{b}^2_{2\mu}\Phi^2)\gamma^5 \otimes M_e & 0 & 0 \end{pmatrix} ,(6.46)$$

where:

$$X_l = \begin{pmatrix} \mathbf{a}_\mu \otimes \mathbb{I}_N & 0 & 0 \\ 0 & \mathbf{b}^1_{1\mu} \otimes \mathbb{I}_N & \mathbf{b}^1_{2\mu} \otimes \mathbb{I}_N \\ 0 & \mathbf{b}^2_{1\mu} \otimes \mathbb{I}_N & \mathbf{b}^2_{2\mu} \otimes \mathbb{I}_N \end{pmatrix} ,$$

and

$$Y_l = \begin{pmatrix} 0 & \Phi_1\gamma^5 \otimes M_e^* & \Phi_2\gamma^5 \otimes M_e^* \\ \Phi^1\gamma^5 \otimes M_e & 0 & 0 \\ \Phi^2\gamma^5 \otimes M_e & 0 & 0 \end{pmatrix} .$$

$$\boldsymbol{\Phi}_l\left[id_{\mathbb{S}(\mathbf{M})} \otimes \left(\nabla_l^{'\mathbf{E}} - \partial\right)_\mu\right] = -iY_l \times X_l = -i\begin{pmatrix} 0 & S_l \\ T_l & 0 \end{pmatrix} , \quad (6.47)$$

with

$$S_l = \left((\Phi_1\mathbf{b}^1_{1\mu} + \Phi_2\mathbf{b}^2_{1\mu})\gamma^5 \otimes M_e^* \ (\Phi_1\mathbf{b}^1_{2\mu} + \Phi_2\mathbf{b}^2_{2\mu})\gamma^5 \otimes M_e^* \right) ,$$

$$T_l = \begin{pmatrix} \Phi^1\mathbf{a}_\mu\gamma^5 \otimes M_e \\ \Phi^2\mathbf{a}_\mu\gamma^5 \otimes M_e \end{pmatrix} .$$

$$[id_{\mathbb{S}(\mathbf{M})} \otimes \left(\nabla_l'^{\mathbf{E}} - \partial\right)_\mu, \boldsymbol{\Phi}_l] = \begin{pmatrix} 0 & \widetilde{S}_l \\ \widetilde{T}_l & 0 \end{pmatrix} , \tag{6.48}$$

with

$$\widetilde{S}_l = \left((\mathbf{a}_\mu \Phi_1 - \Phi_1 \mathbf{b}^1_{1\mu} - \Phi_2 \mathbf{b}^2_{1\mu})\gamma^5 \otimes M_e^* \; (\mathbf{a}_\mu \Phi_2 - \Phi_1 \mathbf{b}^1_{2\mu} - \Phi_2 \mathbf{b}^2_{2\mu})\gamma^5 \otimes M_e^* \right),$$

$$\widetilde{T}_l = \begin{pmatrix} (\mathbf{b}^1_{1\mu}\Phi^1 + \mathbf{b}^1_{2\mu}\Phi^2 - \Phi^1 \mathbf{a}_\mu)\gamma^5 \otimes M_e \\ (\mathbf{b}^2_{1\mu}\Phi^1 + \mathbf{b}^2_{2\mu}\Phi^2 - \Phi^1 \mathbf{a}_\mu)\gamma^5 \otimes M_e \end{pmatrix} .$$

$$[\mathbb{V}_{l\mu}, \boldsymbol{\Phi}_l] = \begin{pmatrix} 0 & (\mathbf{D}\Phi_1)_\mu \gamma^5 \otimes M_e^* & (\mathbf{D}\Phi_2)_\mu \gamma^5 \otimes M_e^* \\ \gamma^5 (\mathbf{D}\Phi^1)_\mu \otimes M_e & 0 & 0 \\ \gamma^5 (\mathbf{D}\Phi^2)_\mu \otimes M_e & 0 & 0 \end{pmatrix} , \tag{6.49}$$

whence (6.36). Check of (6.35):

$$[id_{\mathbb{S}(\mathbf{M})} \otimes \left(\nabla_l'^{\mathbf{E}} - \partial\right)_\mu]\boldsymbol{\Phi}_q = -iX_q \times Y_q \tag{6.50}$$

$$= -i \begin{pmatrix} 0 & 0 & -\mathbf{a}_\mu \Phi^2 \gamma^5 \otimes M_u^* & \mathbf{a}_\mu \Phi^1 \gamma^5 \otimes M_u^* \\ 0 & 0 & \mathbf{a}_\mu \Phi_1 \gamma^5 \otimes M_d^* & \mathbf{a}_\mu \Phi_2 \gamma^5 \otimes M_d^* \\ -(\mathbf{b}^k_{2\mu}\Phi_k)\gamma^5 \otimes M_u & (\mathbf{b}^1_{k\mu}\Phi^k)\gamma^5 \otimes M_d & 0 & 0 \\ (\mathbf{b}^k_{1\mu}\Phi_k)\gamma^5 \otimes M_u & (\mathbf{b}^2_{k\mu}\Phi^k)\gamma^5 \otimes M_d & 0 & 0 \end{pmatrix} \otimes \mathbb{I}_3$$

where

$$X_q = \begin{pmatrix} -\mathbf{a}_\mu \otimes \mathbb{I}_N & 0 & 0 & 0 \\ 0 & \mathbf{a}_\mu \otimes \mathbb{I}_N & 0 & 0 \\ 0 & 0 & \mathbf{b}^1_{1\mu} \otimes \mathbb{I}_N & \mathbf{b}^1_{2\mu} \otimes \mathbb{I}_N \\ 0 & 0 & \mathbf{b}^2_{1\mu} \otimes \mathbb{I}_N & \mathbf{b}^2_{2\mu} \otimes \mathbb{I}_N \end{pmatrix} ,$$

$$Y_q = \begin{pmatrix} 0 & 0 & \Phi^2 \gamma^5 \otimes M_u^* & -\Phi^1 \gamma^5 \otimes M_u^* \\ 0 & 0 & \Phi_1 \gamma^5 \otimes M_d^* & \Phi_2 \gamma^5 \otimes M_d^* \\ \Phi_2 \gamma^5 \otimes M_u & \Phi^1 \gamma^5 \otimes M_d & 0 & 0 \\ -\Phi_1 \gamma^5 \otimes M_u & \Phi^2 \gamma^5 \otimes M_d & 0 & 0 \end{pmatrix} \otimes \mathbb{I}_3 .$$

$$\boldsymbol{\Phi}_q [id_{\mathbb{S}(\mathbf{M})} \otimes \left(\nabla_q'^{\mathbf{E}} - \partial\right)_\mu] = -i\,(Y_q \times X_q) \otimes \mathbb{I}_3 = -i \begin{pmatrix} 0 & S_q \\ T_q & 0 \end{pmatrix} \otimes \mathbb{I}_3 , \tag{6.51}$$

with

$$S_q = \begin{pmatrix} -(\mathbf{b}^2_{2\mu}\Phi^2 + \mathbf{b}^2_{1\mu}\Phi^1)\gamma^5 \otimes M_u^* & (\mathbf{b}^1_{2\mu}\Phi^2 + \mathbf{b}^1_{1\mu}\Phi_1)\gamma^5 \otimes M_u^* \\ (\mathbf{b}^1_{1\mu}\Phi_1 + \mathbf{b}^2_{1\mu}\Phi_2)\gamma^5 \otimes M_d^* & (\mathbf{b}^1_{2\mu}\Phi_1 + \mathbf{b}^2_{2\mu}\Phi_2)\gamma^5 \otimes M_d^* \end{pmatrix} ,$$

$$T_q = \begin{pmatrix} -\mathbf{a}_\mu \Phi_2 \gamma^5 \otimes M_u & \mathbf{a}_\mu \Phi^1 \gamma^5 \otimes M_d \\ \mathbf{a}_\mu \Phi_1 \gamma^5 \otimes M_u & \mathbf{a}_\mu \Phi^2 \gamma^5 \otimes M_d \end{pmatrix} .$$

$$i[id_{\mathbb{S}(\mathbf{M})} \otimes \left(\nabla_q'^{\mathbf{E}} - \partial\right)_\mu, \boldsymbol{\Phi}_q] = \begin{pmatrix} 0 & \widetilde{S}_q \\ \widetilde{T}_q & 0 \end{pmatrix} \otimes \mathbb{I}_3 , \tag{6.52}$$

with

$$\widetilde{S}_q = \begin{pmatrix} -(\mathbf{a}_\mu \Phi^2 - \mathbf{b}^2_{k\mu} \Phi^k)\gamma^5 \otimes M_u^* & (\mathbf{a}_\mu \Phi^1 - \mathbf{b}^1_{k\mu} \Phi^k)\gamma^5 \otimes M_u^* \\ (\mathbf{a}_\mu \Phi_1 - \mathbf{b}_{1k\mu} \Phi^k)\gamma^5 \otimes M_d^* & (\mathbf{a}_\mu \Phi_2 - \mathbf{b}_{2k\mu} \Phi^k)\gamma^5 \otimes M_d^* \end{pmatrix},$$

$$\widetilde{T}_q = \begin{pmatrix} (\mathbf{a}_\mu \Phi_2 - \mathbf{b}^k_{2\mu} \Phi_k)\gamma^5 \otimes M_u & -(\mathbf{a}_\mu \Phi^1 - \mathbf{b}^1_{k\mu} \Phi^k)\gamma^5 \otimes M_d \\ -(\mathbf{a}_\mu \Phi_1 - \mathbf{b}_{1k\mu} \Phi^k)\gamma^5 \otimes M_u & -(\mathbf{a}_\mu \Phi^2 - \mathbf{b}^2_{k\mu} \Phi^k)\gamma^5 \otimes M_d \end{pmatrix},$$

whence (6.35).
(iii): Immediate verifications.

7 The Spectral Action and its Heat-Kernel Asymptotic Expansion. Classical Unification of Standard Model and Gravitation

Alain Connes realized in [32] that "the gauge bosons are fluctuations of the gravitational metric" in the following sense: a gravitational action written as the (ordinary) trace of $F\left(\frac{1}{\Lambda}\mathcal{D}_A^2\right)$, $\mathcal{D}_A$ the generalized covariant Dirac operator, F an appropriate approximation of the characteristic function of interval [0,1], Λ a cut-off of the order of the inverse square of the Planck length, turns out to yield asymptotically in decreasing powers of Λ (heat-kernel expansion, cf. (7.2) below): a constant Λ^2-term (gravitational constant), then a Λ-term sum of the Einstein-Hilbert action plus a multiple of the square of the Higgs-field, finally a Λ-independent term sum of a multiple of the squared Weyl operator plus a standard model action of a grand-unification flavour which Connes views as describing the primal matter - hence to be subjected to a renormalization-group treatment for an attempt to compute the Higgs mass (assuming the existence of a great desert!). The fact that the above-mentioned 8 terms have the right relative signs is a strong indication that they appear within a significant context.

In this section we expound the details of the calculation of the three first terms of the heat-expansion of the spectral action, a calculation only sketched in the Chemseddine-Connes paper.

I may evoke my personal history with the subject: at the incitation of Alain Connes, I had previously checked that the second term of the heat expansion embodies the Einstein-Hilbert action (by direct computation of the (synonymous) Wodcziki residue of the inverse square of the Dirac operator ([29] - see also [30]). I then naturally attempted to get at the same time electrodynamics by using instead of $\widetilde{D}$ the covariant Dirac operator $\widetilde{D} + A$: alas the photon connexion A drops out of this calculation (also noticed in [30]: six extra terms cancel each other!) - a fact that I sadly recorded writing that "the two theories seem to repel each other at this level" (indeed, conversely, the Yang-Mills procedure ignores the Levy-Civita connection). Alain Connes and Ali Chamseddine revealed me in [32],[33] that I should have dived a level deeper in heat-kernel expansion! I had all tools in my hands to compute the third heat expansion term. My failure to do this (wrong choice of priorities!) leaves me with a bitter regret.[36]

Our notation relative to the 4-dimensional smooth compact oriented spin manifold $\mathbf{M}$ and its spin bundle is that of [**1.1**] and [**4.3**] above, in particular $\tilde{\nabla}$ denotes the spin connexion. We first define the Chamseddine-Connes spectral action.

7.1 Definition

The **spectral action** is defined in [32], [33] as

$$I_B(g,A) = (4\pi)^{-2}\mathrm{Tr}\, F\left(\frac{1}{\Lambda}\mathcal{D}_A^2\right) = \int_{\mathbf{M}} I_B(x,g,A)dv \ , \tag{7.1}$$

where $\mathcal{D}_A = \mathcal{D} + A + \mathcal{J} A \mathcal{J}^*$ is the "covariant Dirac operator" of the S_0-real spectral triple of the standard model as described in [**6**] above. F is a function: $\mathbb{R}_+^* \to \mathbb{R}$ such that $F\left(\frac{1}{\Lambda}\mathcal{D}_A^2\right)$ is trace-class, Λ the inverse square of the Plank length acting as a cut-off: $\Lambda = l_p^{-2}$ if, as we shall assume, the function F has its support in the unit interval [0,1]: we shall in fact choose for F a smooth approximation of the characteristic function of [0,1].

We note that $\mathcal{D}_A$ splits as a direct sum $\mathbb{D}_A \oplus \mathcal{J}\mathbb{D}_A\mathcal{J}^*$ corresponding to the splitting $\mathcal{H} = \underline{\mathbb{H}} \oplus \overline{\mathbb{H}}$ into the particle Hilbert space $\underline{\mathbb{H}}$ and the antiparticle Hilbert space $\underline{\mathbb{H}}$ giving the same contribution to the spectral action which, up to a factor 2, can be computed replacing $\mathcal{D}_A$ by $\mathbb{D}_A$ and $\mathcal{H}$ by $\underline{\mathbb{H}}$: this is what we do in what follows.

As an elliptic operator acting on the smooth dense subbundle $\underline{\mathbb{E}} = \mathcal{S}(\mathrm{M}) \otimes H$ of $\underline{\mathbb{H}}$, $\mathbb{D}_A$ has a pure point spectrum bounded below, and extends uniquely to a self-adjoint operator, $\mathbb{D}_A$ of $\underline{\mathbb{H}}$. Moreover its eigenstates lie in $\underline{\mathbb{E}}$ on which we concentrate in the sequel, forgetting $\underline{\mathbb{H}}$. The same facts prevail for $\mathbb{D}_A^2$. We recall (cf. [**6.7**] that the module $\underline{\mathbb{E}}$ can be written as the tensor product $\underline{\mathbb{E}} = \mathcal{S}(\mathrm{M}) \otimes_{C^\infty(\mathbf{M})} \mathbf{E}$ ($\mathbf{E} = C^\infty(\mathbf{M}) \otimes H$) of modules over $\mathbf{M}$, and that $\mathbb{D}_A$ is a generalized Dirac operator for the Clifford-bundle structure of $\underline{\mathbb{E}}$ with Clifford action $c = \gamma \otimes id_{\mathbf{E}}$: $\mathbb{D}_A^2$ is thus is a generalized Laplacian.

Note also that, once the cut-off Λ is fixed, the expression (7.1) merely depends upon the (discrete) eigenvalues of the operator $\mathcal{D}_A$, which appear in this

[36] Another point of personal history: I was asked by Nuclear Physics to referee the Chamseddine-Connes paper. Competent to the extent of having shortly redone and corroborated their undisclosed calculations, I sent an enthusiastic recommendation which was willingly ignored: the paper was rejected "for lack of experimental corroboration" (in contradistinction to the experimentally highly corroborated strings abundant in Nuclear Physics! and despite the impressing relative-sign agreement of 8 terms!). I hope to live to the day when this story will owe me the aura of a modern prophet gaged by postmedieval churchmen!

context as fundamental dynamical variables (a perspective examined in [39]).[37] The spectral action is thus invariant under all "isospectral" transformation (= leaving those eigenvalues unchanged) - an enlargement of the Einstein invariance requirement including both the diffeomorphism and the gauge groups.

In this section we achieve contact with the traditional dynamical variables of the theory of elementary particles and of gravitation whose Lagrangians will appear to be jointly encoded in (7.1).

We state the result, the bulk of this section being then devoted to its proof:

7.2 Asymptotic expansion of the spectral action

The three first term of the asymptotic development in Λ^{-2} of the spectral action-density read:

$$\begin{aligned}
I_B(x,g,A) &= 90(2f_0)\Lambda^2 - f_2\Lambda\left\{15\mathbf{s} - 8A_f|\Phi|^2\right\} \\
&+ f_4\left\{\frac{80}{9}N\mathbf{f}_{\mu\nu}\mathbf{f}^{\mu\nu} + \frac{4}{3}N\mathbf{h}^{s}_{\mu\nu}\mathbf{h}^{\mu\nu}_{s} + \frac{4}{3}N\mathbf{g}^{a}_{\mu\nu}\mathbf{g}^{\mu\nu}_{a}\right. \\
&\left. + 4A_f|\mathbf{D}\Phi|^2 + \frac{2}{3}A_f\mathbf{s}|\Phi|^2 + 4B_f|\Phi|^4 - \frac{9}{4}\mathbf{C}^2\right\} \\
&+ \text{the surface term } f_4\left\{11\pi^2\chi_4 + \frac{8}{3}\Delta\mathbf{s} + \frac{4}{3}A_f\Delta(|\Phi|^2)\right\}\ ,
\end{aligned} \tag{7.2}$$

where $f_0 = \int F(u)udu$, $f_2 = \int F(u)du$, $F_4 = F(0)$ (note that for $F = \chi_{[0,1]}$ - non licit, but indicative choice - one has $2f_0 = f_2 = f_4 = 1$), and:

$$\begin{cases} A_f = \mathrm{Tr}_N[3(M_u^*M_u + M_d^*M_d) + M_e^*M_e] \\ B_f = \mathrm{Tr}_N[3(M_u^*M_uM_u^*M_u + M_d^*M_dM_d^*M_d) + M_e^*M_eM_e^*M_e]\ . \end{cases} \tag{7.3}$$

Here:

- A is the quantum potential;
- **f** is a classical U(1)-curvature: $\bar{\mathbf{f}} = \mathbf{f} \in \Omega(\mathbf{M},\mathbb{C})^2$;
- $\mathbf{h}^{\cdot}.$ is a classical U(2)-curvatures:

$$\mathbf{h}^{\cdot}. = \begin{pmatrix} \mathbf{h}_1^1 = \bar{\mathbf{h}}_1^1 & \mathbf{h}_2^1 = \bar{\mathbf{h}}_1^2 \\ \mathbf{h}_1^2 & \mathbf{h}_2^2 = -\mathbf{h}_1^1 \end{pmatrix} \in \Omega(\mathbf{M}, i\,\mathbb{H}_{\text{traceless}})^2\ ;$$

- $\mathbf{g}^{\cdot} = (\mathbf{g}^a)\frac{\lambda_a}{2}$ is a SU(3)-curvatures (the λ_a, $a = 1,\ldots,8$ are the Gell-Man matrices);
- $\Phi^{\cdot}. = \begin{pmatrix} \Phi_2 & \Phi_1 \\ -\Phi_1 & \Phi^2 \end{pmatrix} \in C^\infty(\mathbf{M},\mathbb{H})$ is a Higgs field;
- **C** is the Weyl tensor;

[37] Note that our choice of a function F with support [0,1] cuts off the eigenvalues of $\mathcal{D}_A$ (or, for that matter, of $\mathbb{D}_A$) at the inverse Planck length l_p^{-1}.

– $\chi_4 dv$ is the Euler form.

This result agrees with the Chamseddine-Connes paper [33][38].
Correspondence of notation:

[Chamseddine-Connes]	The present lectures	
H	$\boldsymbol{\Phi}$	
$g_{01}B_{\mu},\ g_{01}B_{\mu\nu}$	$-2\mathbf{a}_{\mu},\ -2\mathbf{f}_{\mu}$	
$g_{02}A^{\cdot}_{\mu},\ g_{02}F^{\cdot}_{\mu\nu}$	$\mathbf{b}^{\cdot}_{\mu},\ \mathbf{h}^{\cdot}_{\mu}$	
$g_{03}G^{\cdot}_{\mu},\ g_{03}C^{\cdot}_{\mu\nu}$	$\mathbf{c}^{\cdot}_{\mu},\ \mathbf{g}^{\cdot}_{\mu}$	(7.4)
m_0^2	Λ	
R	$-\mathbf{s}$	
y^2	$\frac{1}{3}A_f$	
z^2	$\frac{1}{3}B_f$	

The fact that $\mathcal{D}^2{}_A$ is a generalized Laplacian makes available the heat-kernel asymptotic expansion of $\exp\left(-\frac{1}{\Lambda}\mathcal{D}^2_A\right)$, with corresponding expansion of the spectral action (7.1) as shown by the following Laplace-transform argument.

7.3 Remark

Let G be the Laplace transform of the function F above:

$$F(u) = \int_0^{+\infty} e^{-tu} G(t)dt, \quad u \in \mathbb{R}^+ \ , \tag{7.5}$$

extended by functional calculus to the generalized Laplacian P:

$$F(P) = \int_0^{+\infty} e^{-tP} G(t)dt \ . \tag{7.6}$$

We have the following asymptotic expansion in t:

$$\operatorname{Tr}[F(P)] = \int_0^{+\infty} \operatorname{Tr}(e^{-tP})G(t)dt = \sum_{i\in\mathbb{N}} \int_0^{+\infty} t^{i-2}G(t)dt \int_{\mathbf{M}} a_{2i}(x,P)dv \ , \tag{7.7}$$

obtained by plugging in the heat-kernel expansion

$$\operatorname{Tr}(e^{-tP}) = \sum_{i\in\mathbb{N}} t^{i-2} \int_{\mathbf{M}} a_{2i}(x,P)dv \tag{7.8}$$

[38] Up to the change $z^2 \to \operatorname{Tr}[|k_0^d|^4 + |k_0^u|^4 + \frac{1}{3}|k_0^e|^4]$ in [33] which we suggest.

(cf. [**A.1**] with d=4 and $\delta = 2$). We list the following values of the numerical integrals in (7.7):

$$\begin{cases} \int_0^{+\infty} t^{-2}G(t)dt = \int_0^{+\infty} F(u)u\,du = f_0 \\ \int_0^{+\infty} t^{-1}G(t)dt = \int_0^{+\infty} F(u)du = f_2 \\ \int_0^{+\infty} G(t)dt = F(0) = f_4 \\ \int_0^{+\infty} t^k G(t)dt = (-i)^k F^k(0) \end{cases} \tag{7.9}$$

Applying this to $P = \mathbb{D}_A^2$ with $t = \frac{1}{\Lambda}$ and plugging in the values of the heat-expansion coefficients as listed by Guilkey [37], cf. [**A.4**] we get the following

7.4 Spectral action computation program

With f_0, f_2, f_4 as above, we have to compute:

$$\operatorname{Tr} F\left(\frac{1}{\Lambda}\mathbb{D}_A^2\right) = \int_{\mathbf{M}} \{\Lambda^2 f_0 a_0(x, \mathbb{D}_A^2) + \Lambda f_2 a_2(x, \mathbb{D}_A^2) + f_4 a_4(x, \mathbb{D}_A^2)\}\, dv + O(\Lambda^{-1}) , \tag{7.10}$$

with

$$a_0(x, \mathbb{D}_A^2) = (4\pi)^{-2}\operatorname{Tr}_x(\mathbb{I}) , \tag{7.11}$$

$$a_2(x, \mathbb{D}_A^2) = (4\pi)^{-2}\operatorname{Tr}_x\left(\frac{1}{6}\mathbf{s}\mathbb{I} - E\right) , \tag{7.12}$$

$$a_4(x, \mathbb{D}_A^2) = (4\pi)^{-2}\frac{1}{360}\operatorname{Tr}_x\{5\mathbf{s}^2\mathbb{I} - 2\mathbf{r}^2\mathbb{I} + 2\mathbf{R}^2\mathbb{I} - 60\mathbf{s}E + 180E^2 + 30\mathbb{R}_{\mu\nu}\mathbb{R}^{\mu\nu}\} \tag{7.13}$$

where we omitted in $\{\}$ the surface-term $12\mathbf{s};^{\alpha}{}_{\alpha}\mathbb{I}$, $-60E;^{\alpha}{}_{\alpha}$.
Here:
- $\mathbf{R}$, $\mathbf{r}$, resp. $\mathbf{s}$ are the respective Levi-Civita Riemann tensor, Ricci tensor, and scalar curvature of **M**, with $\mathbf{r}^2 = \mathbf{r}_{\mu\nu}\mathbf{r}^{\mu\nu}$, $\mathbf{R}^2 = \mathbf{R}_{\mu\nu\alpha\beta}\mathbf{R}^{\mu\nu\alpha\beta}$;
- the endomorphism $E \in \operatorname{End}\mathbb{E}$ and the endomorphism-valued curvature 2-tensor $\mathbb{R}$ of the connexion $\mathbb{V}$ of $\mathbb{E}$ proceed from the canonical splitting $\mathbb{D}_A^2 = \triangle^{\mathbb{V}} + E$ in a connection-Laplacian and an endomorphism (cf.[**B.3**]) which we computed in [**6.8**] with the following results: $\mathbb{V}$ is the Clifford connection of $\mathbb{E}$ (cf. (6.30)):

$$\mathbb{V} = \tilde{\nabla} \otimes id_{\mathbf{E}} + id_{\mathbb{S}(\mathbf{M})} \otimes \nabla^{\mathbf{E}} , \tag{7.14}$$

tensor product of the spin connexion $\tilde{\nabla}$ of $\mathbb{S}(\mathbf{M})$ by connexion $\nabla^{\mathbf{E}}$ of $\mathbf{E}$ (with curvature 2-tensor $R^{\mathbf{E}}$ specified as the direct sum $\nabla_q^{\mathbf{E}} \oplus \nabla_l^{\mathbf{E}}$ of a quark and a lepton connexion acting respectively on the quark and lepton subspaces as the sum of the exterior derivative and the matrices: (6.31) and (6.32). Whilst E is the endomorphism (cf. (6.34)):

$$E = \frac{1}{4}\mathbf{s}\mathbb{I} - \frac{1}{2}c\left(R^{\mathbf{E}}\right) + ic^{\mu}\left[\nabla_{\mu}, \boldsymbol{\Phi}\right] + \boldsymbol{\Phi}^2$$
$$\text{with} \quad c\left(R^{\mathbf{E}}\right) = -\gamma^{\mu}\gamma^{\nu} \otimes R^{\mathbf{E}}\left(e_{\mu}, e_{\nu}\right) \ , \tag{7.15}$$

with the quark, resp. lepton matrices of the two last term given by (6.35) and, resp. (6.36).

The rest of our section is devoted to the computation. We first compute the traces in (7.11)-(7.13). The result is as follows:[39]

7.5 Computation of fiber-traces

We have the following traces on the fiber $\mathbb{E}_x$ of $x \in \mathbf{M}$:

$$(a) \qquad \begin{cases} \mathrm{Tr}_x \mathbb{I}_q = 144 \\ \mathrm{Tr}_x \mathbb{I}_l = 36 \\ \mathrm{Tr}_x \mathbb{I} = 180 \\ \mathrm{Tr}_x^{\mathbf{E}} \mathbb{I} = 45 \end{cases} ,$$

$$(b) \qquad \begin{cases} \mathrm{Tr}_x(\boldsymbol{\Phi}_q^2) = 8A_q|\Phi|^2 & \text{with } A_q = 3\mathrm{Tr}(\mu_u + \mu_d), & \mu_u = M_u M_u^* \\ \mathrm{Tr}_x(\boldsymbol{\Phi}_l^2) = 8A_l|\Phi|^2 & \text{with } A_l = \mathrm{Tr}\mu_e, & \mu_d = M_d M_d^* \\ \mathrm{Tr}_x(\boldsymbol{\Phi}^2) = 8A_f|\Phi|^2 & \text{with } A_f = \mathrm{Tr}(3(\mu_u + \mu_d) + \mu_e) & \mu_e = M_e M_e^* , \end{cases} , \tag{7.16}$$

$$(c) \qquad \begin{cases} \mathrm{Tr}_x(\boldsymbol{\Phi}_q^4) = 8B_q|\Phi|^4 & \text{with } B_q = 3\mathrm{Tr}(\mu_u^2 + \mu_d^2), \\ \mathrm{Tr}_x(\boldsymbol{\Phi}_l^4) = 8B_l|\Phi|^4 & \text{with } B_l = \mathrm{Tr}\mu_e^2, \quad |\Phi|^2 = \Phi_1\Phi^1 + \Phi_2\Phi^2 \\ \mathrm{Tr}_x(\boldsymbol{\Phi}^4) = 8B_f|\Phi|^4 & \text{with } B_f = \mathrm{Tr}(3(\mu_u^2 + \mu_d^2) + \mu_e^2), \end{cases} ,$$

$$(d) \qquad \begin{array}{l} \mathrm{Tr}_x(c^{\mu}[\nabla_{q\mu}, \boldsymbol{\Phi}_q]) = \mathrm{Tr}_x(c^{\mu}[\nabla_{l\mu}, \boldsymbol{\Phi}_l]) = \\ \mathrm{Tr}_x(c^{\mu}[\nabla_{\mu}, \boldsymbol{\Phi}]) = \mathrm{Tr}_x(\boldsymbol{\Phi}^2 c^{\mu}[\nabla_{\mu}, \boldsymbol{\Phi}]) = 0 \end{array} ,$$

$$(e) \qquad \begin{cases} \mathrm{Tr}_x\left\{(ic^{\mu}[\nabla_{q\mu}, \boldsymbol{\Phi}_q])^2\right\} = 8A_q|\mathbf{D}\Phi|^2 \\ \mathrm{Tr}_x\left\{i(c^{\mu}[\nabla_{l\mu}, \boldsymbol{\Phi}_l])^2\right\} = 8A_l|\mathbf{D}\Phi|^2 \\ \mathrm{Tr}_x\left\{i(c^{\mu}[\nabla_{\mu}, \boldsymbol{\Phi}])^2\right\} = 8A_f|\mathbf{D}\Phi|^2 \end{cases} ,$$

$$(f) \qquad \mathrm{Tr}_x c\left(R^{\mathbf{E}}\right) = \mathrm{Tr}_x\left\{c\left(R^{\mathbf{E}}\right) c^{\mu}\left[\nabla_{\mu}, \boldsymbol{\Phi}\right]\right\} = \mathrm{Tr}_x\left\{c\left(R^{\mathbf{E}}\right)\boldsymbol{\Phi}^2\right\} = 0 \ ,$$

[39] We use the shorthand Tr_x for the trace on the fiber $\mathbb{E}_x$, and denote the trace on the fiber $\mathbf{E}_x$ by $\mathrm{Tr}_x^{\mathbf{E}}$.

$$(g) \qquad \mathrm{Tr}_x\left[c\left(R^{\mathbf{E}}\right)^2\right] = -8\mathrm{Tr}_x^{\mathbf{E}}\left(R^{\mathbf{E}}_{\mu\nu}R^{\mathbf{E}\,\mu\nu}\right)$$

$$(h) \qquad \mathrm{Tr}_x\mathbb{R}_{\mu\nu}\mathbb{R}^{\mu\nu} = -\frac{45}{2}\mathbf{R}^2 + 4\mathrm{Tr}_x^{\mathbf{E}}\left(R^{\mathbf{E}}_{\mu\nu}R^{\mathbf{E}\,\mu\nu}\right)$$

$$(k) \qquad \begin{cases} \mathrm{Tr}_x^{\mathbf{E}}\left(R^{\mathbf{E}}_{q\mu\nu}R^{\mathbf{E}\,q\mu\nu}\right) = \frac{22}{3}N\mathbf{f}_{\mu\nu}\mathbf{f}^{\mu\nu} + \frac{3}{2}N\mathbf{h}^s_{\mu\nu}\mathbf{h}_s^{\mu\nu} + 2N\mathbf{g}^a_{\mu\nu}\mathbf{g}_a^{\mu\nu} \\ \mathrm{Tr}_x^{\mathbf{E}}\left(R^{\mathbf{E}}_{l\mu\nu}R^{\mathbf{E}\,l\mu\nu}\right) = \frac{18}{3}N\mathbf{f}_{\mu\nu}\mathbf{f}^{\mu\nu} + \frac{N}{2}\mathbf{h}^s_{\mu\nu}\mathbf{h}_s^{\mu\nu} \\ \mathrm{Tr}_x^{\mathbf{E}}\left(R^{\mathbf{E}}_{\mu\nu}R^{\mathbf{E}\,\mu\nu}\right) = 2\left(\frac{20}{3}N\mathbf{f}_{\mu\nu}\mathbf{f}^{\mu\nu} + N\mathbf{h}^s_{\mu\nu}\mathbf{h}_s^{\mu\nu} + N\mathbf{g}^a_{\mu\nu}\mathbf{g}_a^{\mu\nu}\right) \end{cases} .$$

After modular adjustment (for the definition of $\mathbf{f}$, $\mathbf{h}^{\cdot}_{\cdot}$, $\mathbf{g}^{\cdot}$ see ([**7.2**] above)

Proof

Check of (a):

– one has rank $\mathbb{S}_{\mathbf{M}} = 4$, $\dim H_q = 36$, $\dim H_l = 9$;

– thus:

$$\mathrm{Tr}_x\mathbb{I}_q = \mathrm{rank}\,\mathbb{E}_q = 4\times 36 = 144,\ \mathrm{Tr}_x\mathbb{I}_l = \mathrm{rank}\,\mathbb{E}_l = 4\times 9 = 36,\ \mathrm{Tr}_x\mathbb{I} = 180,$$
$$\text{and } \mathrm{Tr}_x^{\mathbf{E}}\mathbb{I} = \mathrm{rank}\,\mathbf{E} = 45\ .$$

Check of (b):

– from (6.41) we have

$$\mathrm{Tr}\,(\Phi^{\cdot}{}_{\cdot}{}^{*}\Phi^{\cdot}{}_{\cdot}) = |\Phi|^2\mathrm{Tr}\,(\mu_u + \mu_d)\ ;$$

– from (6.40):

$$\mathrm{Tr}_x(\boldsymbol{\Phi}_q^2) = 4\times 3\,[\mathrm{Tr}\,(\Phi^{\cdot}{}_{\cdot}{}^{*}\Phi^{\cdot}{}_{\cdot}) + \mathrm{Tr}\,(\Phi^{\cdot}{}_{\cdot}{}^{*}\Phi^{\cdot}{}_{\cdot})] = 8\times 3\,\mathrm{Tr}\,(\Phi^{\cdot}{}_{\cdot}{}^{*}\Phi^{\cdot}{}_{\cdot}) = 8\times 3|\Phi|^2\,\mathrm{Tr}\,(\mu_u+\mu_d) = 8A_q|\Phi|^2\ ;$$

– from (6.43) we have

$$\mathrm{Tr}_x(\boldsymbol{\Phi}_l^2) = 4\times(|\Phi|^2 + \Phi_1\Phi^1 + \Phi_2\Phi^2)\mathrm{Tr}\,(M_eM_e^*) = 8|\Phi|^2\mathrm{Tr}(\mu_e) = 8A_l|\Phi|^2 .$$

Check of (c):

from (6.43) we have:

$$\Phi^{\cdot}{}_{\cdot}{}^{*}\Phi^{\cdot}{}_{\cdot}\quad \Phi^{\cdot}{}_{\cdot}{}^{*}\Phi^{\cdot}{}_{\cdot} = |\Phi|^4\begin{pmatrix}\mathbb{I}\otimes\mu_u^2 & 0\\ 0 & \mathbb{I}\otimes\mu_d^2\end{pmatrix}\ , \tag{7.17}$$

whence $\mathrm{Tr}_x\,(\Phi^{\cdot}.{}^*\Phi^{\cdot}.\quad \Phi^{\cdot}.{}^*\Phi^{\cdot}.) = |\Phi|^4(\mu_u^2+\mu_d^2)$. We have then from (6.40):

$$\boldsymbol{\Phi}_q^4 = \begin{pmatrix} \Phi^{\cdot}.{}^*\Phi^{\cdot}. \quad \Phi^{\cdot}.{}^*\Phi^{\cdot}. & 0 \\ 0 & \Phi^{\cdot}.\Phi^{\cdot}.{}^* \quad \Phi^{\cdot}.\Phi^{\cdot}.{}^* \end{pmatrix} \otimes \mathbb{1}_3 \tag{7.18}$$

whence

$$\mathrm{Tr}_x(\boldsymbol{\Phi}_q^4) = 4\times 3\times 2\mathrm{Tr}(\Phi^{\cdot}.{}^*\Phi^{\cdot}.\quad \Phi^{\cdot}.{}^*\Phi^{\cdot}.) = 8\times 3|\Phi|^4(\mu_u^2+\mu_d^2) = 8B_q|\Phi|^4.$$

¿From (6.43) we have:

$$\boldsymbol{\Phi}_l^4 = |\Phi|^2 \begin{pmatrix} |\Phi|^2\mathbb{1}\otimes\widetilde{M} & 0 & 0 \\ 0 & \Phi_1\Phi^1\mathbb{1}\otimes\widetilde{\widetilde{M}} & \Phi_1\Phi^2\mathbb{1}\otimes\widetilde{\widetilde{M}} \\ 0 & \Phi_2\Phi^1\mathbb{1}\otimes\widetilde{\widetilde{M}} & \Phi_2\Phi^2\otimes\widetilde{\widetilde{M}} \end{pmatrix} \tag{7.19}$$

where $\widetilde{M} = M_e^*M_eM_e^*M_e$ and $\widetilde{\widetilde{M}} = M_eM_e^*M_eM_e^*$,

whence

$$\mathrm{Tr}_x\left(\boldsymbol{\Phi}_l^4\right) = 4|\Phi|^2\left[|\Phi|^2+\Phi_1\Phi^1+\Phi_2\Phi^2\right]\mathrm{Tr}\,(M_e^*M_eM_e^*M_e) = 8|\Phi|^4\mathrm{Tr}(\mu_e^2) = 8B_l|\Phi|^4$$

Check of (d): due to the fact, cf. formulae (6.35), (6.36) (quoted in (7.22), (7.23) below, that $c^\mu[\mathbb{V}_\mu,\boldsymbol{\Phi}]$ is antidiagonal (whilst $\boldsymbol{\Phi}^2$ is diagonal, cf. (6.40), (6.43)).
Check of (e):

we found, cf. (6.35), that $c^\mu[\mathbb{V}_{q\mu},\boldsymbol{\Phi}_q] = \begin{pmatrix} 0 & X \\ Y & 0 \end{pmatrix}\otimes\mathbb{1}_3$ with:

$$X = \begin{pmatrix} \gamma(\mathbf{D}\Phi^2)\gamma^5\otimes M_u^* & -\gamma(\mathbf{D}\Phi^1)\gamma^5\otimes M_u^* \\ \gamma(\mathbf{D}\Phi_1)\gamma^5\otimes M_d^* & \gamma(\mathbf{D}\Phi_2)\gamma^5\otimes M_d^* \end{pmatrix},$$
$$Y = \begin{pmatrix} \gamma(\mathbf{D}\Phi_2)\gamma^5\otimes M_u & \gamma(\mathbf{D}\Phi^1)\gamma^5\otimes M_d \\ -\gamma(\mathbf{D}\Phi_1)\gamma^5\otimes M_u & \gamma(\mathbf{D}\Phi^2)\gamma^5\otimes M_d \end{pmatrix}, \tag{7.20}$$

thus $\{c^\mu[\mathbb{V}_{q\mu},\boldsymbol{\Phi}_q]\}^2 = \begin{pmatrix} XY & 0 \\ 0 & YX \end{pmatrix}\otimes\mathbb{1}_3$ with trace $2\times 3\,\mathrm{Tr}\,(XY)$, where:

$$-XY = \begin{pmatrix} Z_{11} & Z_{12} \\ Z_{21} & Z_{22} \end{pmatrix},$$

$$\begin{aligned} Z_{11} &= \left[\gamma(\mathbf{D}\Phi^2)\gamma(\mathbf{D}\Phi_2)+\gamma(\mathbf{D}\Phi^1)\gamma(\mathbf{D}\Phi_1)\right]\otimes M_u^*M_u \\ Z_{12} &= \left[\gamma(\mathbf{D}\Phi^2)\gamma(\mathbf{D}\Phi^1)-\gamma(\mathbf{D}\Phi^1)\gamma(\mathbf{D}\Phi^2)\right]\otimes M_u^*M_d \\ Z_{21} &= \left[\gamma(\mathbf{D}\Phi_1)\gamma(\mathbf{D}\Phi_2)-\gamma(\mathbf{D}\Phi_2)\gamma(\mathbf{D}\Phi_1)\right]\otimes M_d^*M_u \\ Z_{22} &= \left[\gamma(\mathbf{D}\Phi_1)\gamma(\mathbf{D}\Phi^1)+\gamma(\mathbf{D}\Phi_2)\gamma(\mathbf{D}\Phi^2)\right]\otimes M_d^*M_d \end{aligned}, \tag{7.21}$$

whence $\mathrm{Tr}(XY) = -2|\mathbf{D}\Phi|^2\mathrm{Tr}(M_u^*M_u+M_d^*M_d) = -|\mathbf{D}\Phi|^2\mathrm{Tr}(\mu_u+\mu_d)$. We conclude that:

$$\mathrm{Tr}_x\left\{\left(ic^\mu\left[\mathbb{V}_q\mu,\boldsymbol{\Phi}_q\right]\right)^2\right\} = 4\times 2\times 3\mathrm{Tr}(XY) = 4\times 3\times 3|\mathbf{D}\Phi|^2\mathrm{Tr}(\mu_u+\mu_d) = 8A_q|\mathbf{D}\Phi|^2.$$

We found on the other hand, cf. (6.20), that

$$c^\mu[\mathbb{V}_{l\mu},\boldsymbol{\Phi}_l] = \begin{pmatrix} 0 & \gamma(\mathbf{D}\Phi_1)\gamma^5\otimes M_e^* & \gamma(\mathbf{D}\Phi_2)\gamma^5\otimes M_e^* \\ \gamma(\mathbf{D}\Phi^1)\gamma^5\otimes M_e & 0 & 0 \\ \gamma(\mathbf{D}\Phi^2)\gamma^5\otimes M_e^* & 0 & 0 \end{pmatrix} \tag{7.22}$$

implying

$$-c^\mu[\mathbb{V}_{l\mu},\boldsymbol{\Phi}_l]^2 = \begin{pmatrix} X_{11}\otimes M_e^*M_e & 0 & 0 \\ 0 & X_{22}\otimes M_e^*M_e & X_{23}\otimes M_e^*M_e \\ 0 & X_{32}\otimes M_e^*M_e & X_{33}\otimes M_e^*M_e \end{pmatrix} , \tag{7.23}$$

with $X_{11} = X_{22} = X_{33} = (\mathbf{D}\Phi^2)\gamma(\mathbf{D}\Phi_2) + \gamma(\mathbf{D}\Phi^1)\gamma(\mathbf{D}\Phi_1) = 2|\mathbf{D}\Phi|^2$. We conclude that

$$\mathrm{Tr}_x\left\{(ic^\mu[\mathbb{V}_{l\mu},\boldsymbol{\Phi}_l])^2\right\} = 4\times 2|\mathbf{D}\Phi|^2\mathrm{Tr}(M_e^*M_e) = 8|\mathbf{D}\Phi|^2\mathrm{Tr}\,\mu_e = 8A_l|\mathbf{D}\Phi|^2 .$$

Check of (f): due to the fact that the Clifford traces vanish.
Check of (g):

$$\begin{aligned}
\mathrm{Tr}_x\left[c\left(R^{\mathbf{E}}\right)^2\right] &= \mathrm{Tr}_x\left[\left(\gamma^\mu\gamma^\nu\otimes R^{\mathbf{E}}\,(e_\mu,e_\nu)\right)^2\right] \\
&= \mathrm{Tr}_x\left[\gamma^\mu\gamma^\nu\gamma^\alpha\gamma^\beta\otimes R^{\mathbf{E}}\,(e_\mu,e_\nu)\,R^{\mathbf{E}}\,(e_\alpha,e_\beta)\right] \\
&= 4\left(g^{\mu\nu}g^{\alpha\beta} + g^{\mu\beta}g^{\nu\alpha} - g^{\mu\alpha}g^{\nu\beta}\right)\mathrm{Tr}_x^{\mathbf{E}}\left(R^{\mathbf{E}}_{\mu\nu}R^{\mathbf{E}}_{\alpha\beta}\right) \\
&= 4(g^{\mu\nu}g^{\alpha\beta} - g^{\mu\alpha}g^{\nu\beta})\mathrm{Tr}_x^{\mathbf{E}}\left(R^{\mathbf{E}}_{\mu\nu}R^{\mathbf{E}}_{\alpha\beta}\right) \\
&= 4\mathrm{Tr}_x^{\mathbf{E}}\left(R^{\mathbf{E}\beta}_{\nu}R^{\mathbf{E}\nu}_{\beta} - R^{\mathbf{E}\alpha}_{\nu}R^{\mathbf{E}\nu}_{\alpha}\right) \\
&= 4\mathrm{Tr}_x^{\mathbf{E}}\left(R^{\mathbf{E}}_{\beta\nu}R^{\mathbf{E}\nu\beta} - R^{\mathbf{E}}_{\alpha\nu}R^{\mathbf{E}\alpha\nu}\right) = -8\mathrm{Tr}_x^{\mathbf{E}}\left(R^{\mathbf{E}}_{\mu\nu}R^{\mathbf{E}\mu\nu}\right) .
\end{aligned} \tag{7.24}$$

Check of (h): using the fact that $\mathbb{R}_{\mu\nu} = \widetilde{R}_{\mu\nu}\otimes id_{\mathbf{E}} + id_{\mathbb{S}_{\mathbf{M}}}\otimes R^{\mathbf{E}}_{\mu\nu}$, cf.[**C.1**], we have:

$$\begin{aligned}
&\mathrm{Tr}_x\left[\mathbb{R}_{\mu\nu}\mathbb{R}^{\mu\nu}\right] \\
&= \mathrm{Tr}_x\left[\left(\widetilde{R}_{\mu\nu}\otimes id_{\mathbf{E}} + id_{\mathbb{S}_{\mathbf{M}}}\otimes R^{\mathbf{E}}_{\mu\nu}\right)\left(\widetilde{R}^{\mu\nu}\otimes id_{\mathbf{E}} + id_{\mathbb{S}_{\mathbf{M}}}\otimes R^{\mathbf{E}\mu\nu}\right)\right] \\
&= \mathrm{Tr}_x\left(\widetilde{R}_{\mu\nu}\widetilde{R}^{\mu\nu}\otimes id_{\mathbf{E}}\right) + \mathrm{Tr}_x\left(id_{\mathbb{S}_{\mathbf{M}}}\otimes R^{\mathbf{E}}_{\mu\nu}R^{\mathbf{E}\mu\nu}\right) + 2\mathrm{Tr}_x\left(\widetilde{R}_{\mu\nu}\otimes R^{\mathbf{E}\mu\nu}\right) \\
&= 45\mathrm{Tr}_{\mathbb{S}_{\mathbf{M}}}\left(\widetilde{R}_{\mu\nu}\widetilde{R}^{\mu\nu}\right) + 4\mathrm{Tr}_x^{\mathbf{E}}\left(R^{\mathbf{E}}_{\mu\nu}R^{\mathbf{E}\mu\nu}\right) \\
&= -\frac{45}{2}\mathbf{R}^2 + 4\mathrm{Tr}_x{}^{\mathbf{E}}\left(R^{\mathbf{E}}_{\mu\nu}R^{\mathbf{E}\mu\nu}\right) ,
\end{aligned} \tag{7.25}$$

indeed the curvature $\widetilde{R}_{\mu\nu} = \frac{1}{4}\mathbf{R}_{\mu\nu\alpha\beta}\gamma^\alpha\gamma^\beta$ of the spin connection is traceless and fulfills:

$$\begin{aligned}
&\mathrm{Tr}_{\mathbf{\$_M}}\left(\widetilde{R}_{\mu\nu}\widetilde{R}^{\mu\nu}\right)\\
&= \frac{1}{16}\mathbf{R}^{\mu\nu}_{\sigma\tau}\mathbf{R}_{\mu\nu\alpha\beta}\left[\mathrm{Tr}_x\left(\gamma^\alpha\gamma^\beta\gamma^\sigma\gamma^\tau\right) = 4\left(g^{\alpha\beta}g^{\sigma\tau} + g^{\alpha\tau}g^{\beta\sigma} - g^{\alpha\sigma}g^{\beta\tau}\right)\right]\\
&= \frac{1}{4}\left(\mathbf{R}^{\tau\sigma}_{\mu\nu}\mathbf{R}^{\mu\nu}_{\sigma\tau} - \mathbf{R}^{\sigma\tau}_{\mu\nu}\mathbf{R}^{\mu\nu}_{\sigma\tau}\right) = -\frac{1}{2}\mathbf{R}^{\sigma\tau}_{\mu\nu}\mathbf{R}^{\mu\nu}_{\sigma\tau} = -\frac{1}{2}\mathbf{R}^2 \ . \qquad (7.26)
\end{aligned}$$

Check of (k): We have from (6.31):

$$iR^{\mathbf{E}}_{q\mu\nu} = \begin{pmatrix} -\mathbf{f}_{\mu\nu}\otimes\mathbb{I}_N & 0 & 0 & 0\\ 0 & \mathbf{f}_{\mu\nu}\otimes\mathbb{I}_N & 0 & 0\\ 0 & 0 & \mathbf{h}^1_{1\mu\nu}\otimes\mathbb{I}_N & \mathbf{h}^1_{2\mu\nu}\otimes\mathbb{I}_N\\ 0 & 0 & \mathbf{h}^2_{1\mu\nu}\otimes\mathbb{I}_N & \mathbf{h}^2_{2\mu\nu}\otimes\mathbb{I}_N \end{pmatrix}\otimes\mathbb{I}_3 \qquad (7.27)$$

(columns labelled u_R, d_R, u_L, d_L)

$$+\mathbf{g}^0_{\mu\nu}\mathbb{I}_4\otimes\mathbb{I}_N\otimes\mathbb{I}_3 + \mathbf{g}^a_{\mu\nu}\mathbb{I}_4\otimes\mathbb{I}_N\otimes\frac{\lambda_a}{2} \ ,$$

transformed by the modular adjustment $\mathbf{g}^0_{\mu\nu} = -\frac{1}{3}\mathbf{f}_{\mu\nu}$ into:

$$iR^{\mathbf{E}}_{q\mu\nu} = \begin{pmatrix} -\frac{4}{3}\mathbf{f}_{\mu\nu}\otimes\mathbb{I}_N & 0 & 0 & 0\\ 0 & \frac{2}{3}\mathbf{f}_{\mu\nu}\otimes\mathbb{I}_N & 0 & 0\\ 0 & 0 & \underline{\mathbf{h}}^1_{1\mu\nu}\otimes\mathbb{I}_N & \underline{\mathbf{h}}^1_{2\mu\nu}\otimes\mathbb{I}_N\\ 0 & 0 & \underline{\mathbf{h}}^2_{1\mu\nu}\otimes\mathbb{I}_N & \underline{\mathbf{h}}^2_{2\mu\nu}\otimes\mathbb{I}_N \end{pmatrix}\otimes\mathbb{I}_3 \qquad (7.28)$$

(columns labelled u_R, d_R, u_L, d_L)

$$+\mathbf{g}^a_{\mu\nu}\mathbb{I}_4\otimes\mathbb{I}_N\otimes\frac{\lambda_a}{2} \ ,$$

and

$$iR^{\mathbf{E}}_{l\mu\nu} = \begin{pmatrix} 2\mathbf{f}_{\mu\nu}\otimes\mathbb{I}_N & 0 & 0\\ 0 & \overline{\mathbf{h}}^1_{1\mu\nu}\otimes\mathbb{I}_N & \overline{\mathbf{h}}^1_{2\mu\nu}\otimes\mathbb{I}_N\\ 0 & \overline{\mathbf{h}}^2_{1\mu\nu}\otimes\mathbb{I}_N & \overline{\mathbf{h}}^2_{2\mu\nu}\otimes\mathbb{I}_N \end{pmatrix} \ , \qquad (7.29)$$

(columns labelled e_R, ν_l, e_L)

where $\underline{\mathbf{h}}^i_k = \mathbf{h}^i_k - \frac{1}{3}\mathbf{f}\delta^i_k$, and $\overline{\mathbf{h}} = \mathbf{h} + \mathbf{f}\delta^i_k$. We thus have after some algebra:

$$-R^{\mathbf{E}}_{q\mu\nu}R^{\mathbf{E}\mu\nu}_q =$$

$$\begin{pmatrix} \frac{16}{9}\mathbf{f}_{\mu\nu}\mathbf{f}^{\mu\nu}\otimes\mathbb{I}_N & 0 & 0 & 0\\ 0 & \frac{4}{9}\mathbf{f}_{\mu\nu}\mathbf{f}^{\mu\nu}\otimes\mathbb{I}_N & 0 & 0\\ 0 & 0 & \underline{\mathbf{h}}^1_{i\mu\nu}\underline{\mathbf{h}}^{i\mu\nu}_1\otimes\mathbb{I}_N & \underline{\mathbf{h}}^1_{i\mu\nu}\underline{\mathbf{h}}^{i\mu\nu}_2\otimes\mathbb{I}_N\\ 0 & 0 & \underline{\mathbf{h}}^2_{i\mu\nu}\underline{\mathbf{h}}^{i\mu\nu}_1\otimes\mathbb{I}_N & \underline{\mathbf{h}}^2_{i\mu\nu}\underline{\mathbf{h}}^{i\mu\nu}_2\otimes\mathbb{I}_N \end{pmatrix}\otimes\mathbb{I}_3 \qquad (7.30)$$

(columns labelled u_R, d_R, u_L, d_L)

$$+\mathbf{g}^a_{\mu\nu}\mathbf{g}^{\mu\nu}_a\mathbb{I}_4\otimes\mathbb{I}_N\otimes\mathbb{I}_3 \ ,$$

and from (6.32):

$$-R^{\mathbf{E}}_{l\mu\nu}R^{\mathbf{E}\mu\nu}_{l} = \begin{pmatrix} 4\mathbf{f}_{\mu\nu}\mathbf{f}^{\mu\nu}\otimes \mathbb{1}_N & 0 & 0 \\ 0 & \overline{\mathbf{h}}^{1}_{i\mu\nu}\overline{\mathbf{h}}^{i\mu\nu}_{1}\otimes \mathbb{1}_N & \overline{\mathbf{h}}^{1}_{i\mu\nu}\overline{\mathbf{h}}^{i\mu\nu}_{2}\otimes \mathbb{1}_N \\ 0 & \overline{\mathbf{h}}^{2}_{i\mu\nu}\overline{\mathbf{h}}^{i\mu\nu}_{1}\otimes \mathbb{1}_N & \overline{\mathbf{h}}^{2}_{i\mu\nu}\overline{\mathbf{h}}^{i\mu\nu}_{2}\otimes \mathbb{1}_N \end{pmatrix} . \tag{7.31}$$

(columns labelled e_R, ν_l, e_L)

Quark contribution to $\mathrm{Tr}_x{}^{\mathbf{E}}(R'^{\mathbf{E}}_{\mu\nu}R'^{\mathbf{E}\mu\nu})$: for the computation of the tensor-product traces $\mathrm{Tr}_p\otimes\mathrm{Tr}_3$, we shall use the following elementary fact: we have, due to the tracelessness of the λ_a and the relation $\mathrm{Tr}(\lambda_a\lambda_b)= 2\delta_{ab}$:

$$\left\{\mathrm{Tr}_p\otimes\mathrm{Tr}_3\right\}\left[M\otimes\mathbb{1}_3 + N^a\otimes\frac{\lambda_a}{2}\right]^2 = 3\mathrm{Tr}_p M^2 + \frac{1}{2}\mathrm{Tr}_p N^a N_a \quad M,\, N^a \in M_p(\mathbb{C}) . \tag{7.32}$$

¿From that follows that we have no mixed electroweak-chromodynamics term, and that the gluonic contribution is:

$$2N\mathbf{g}^a_{\mu\nu}\mathbf{g}^{\mu\nu}_a . \tag{7.33}$$

Noting that we have, by the fact that the $\mathbf{h}^i_k$ are traceless:

$$\underline{\mathbf{h}}^i_{k\mu\nu}\underline{\mathbf{h}}^{k\mu\nu}_i = \mathbf{h}^i_{k\mu\nu}\mathbf{h}^{k\mu\nu}_i + \frac{2}{9}\mathbf{f}_{\mu\nu}\mathbf{f}^{\mu\nu} , \tag{7.34}$$

we get the electroweak contribution:

$$3N\left(\frac{16}{9}\mathbf{f}_{\mu\nu}\mathbf{f}^{\mu\nu} + \frac{4}{9}\mathbf{h}^i_{k\mu\nu}\mathbf{h}^{k\mu\nu}_i + \frac{2}{9}\mathbf{f}_{\mu\nu}\mathbf{f}^{\mu\nu}\right) = \frac{22}{3}N\mathbf{f}_{\mu\nu}\mathbf{f}^{\mu\nu} + \frac{3}{2}N\mathbf{h}^s_{\mu\nu}\mathbf{h}^{\mu\nu}_s , \tag{7.35}$$

(we used the fact that

$$\mathbf{h}^s_{\mu\nu}\mathbf{h}^{\mu\nu}_s = 2\mathbf{h}^i_{k\mu\nu}\mathbf{h}^{k\mu\nu}_i \;) \tag{7.36}$$

Lepton contribution to $\mathrm{Tr}_x{}^{\mathbf{E}}\left(R^{\mathbf{E}}_{\mu\nu}R^{\mathbf{E}\mu\nu}\right)$: using

$$\overline{\mathbf{h}}^i_{k\mu\nu}\overline{\mathbf{h}}^{k\mu\nu}_i = \mathbf{h}^i_{k\mu\nu}\mathbf{h}^{k\mu\nu}_i + 2\mathbf{f}_\mu\nu\mathbf{f}^{\mu\nu} , \tag{7.37}$$

we get, using again (7.36):

$$N\left(4\mathbf{f}^{\mu\nu}_{\mu\nu} + \mathbf{h}^i_{k\mu\nu}\mathbf{h}^{k\mu\nu}_i + 2\mathbf{f}_{\mu\nu}\mathbf{f}^{\mu\nu}\right) = 6N\mathbf{f}_{\mu\nu}\mathbf{f}^{\mu\nu} + \frac{1}{2}N\mathbf{h}^s_{\mu\nu}\mathbf{h}^{\mu\nu}_s . \tag{7.38}$$

Summing up this checks (k).

7.6 Gathering the pieces

Noting that linear combination of the above results yields:

$$\mathrm{Tr}_x(E) = 45\mathbf{s} + 8A|\Phi|^2 \ , \tag{7.39}$$

$$\mathrm{Tr}_x\left(E^2\right) = \frac{45}{4}\mathbf{s}^2 + 8A|\mathbf{D}\Phi|^2 + 8B|\Phi^4| + 4\mathbf{s}A|\Phi|^2 \ . \tag{7.40}$$

We now have the ingredients of the computation of our heat-kernel expansion.
Computation of $a_2(x, \mathbb{D}_A^2)$: we have $(4\pi)^2 a_0(x, \mathbb{D}_A^2) = \mathrm{Tr}_x(\mathbb{I}) = 180$.
Computation of $a_2(x, \mathbb{D}_A^2)$: we have:

$$(4\pi)^2 a_2(x, \mathbb{D}_A^2) = \mathrm{Tr}_x\left(\frac{1}{6}\mathbf{s}(\mathbb{I}) - E\right) - 15\mathbf{s} - 8A_f|\Phi|^2 \ .$$

Computation of $a_4(x, \mathbb{D}_A^2)$: we have:

$$\begin{aligned}
&(4\pi)^2 a_4(x, \mathbb{D}_A^2)\\
&= \frac{1}{360}\mathrm{Tr}_x\{5\mathbf{s}^2\mathbb{I} - 2\mathbf{r}^2\mathbb{I} + 2\mathbf{R}^2\mathbb{I} - 60\mathbf{s}E + 180E^2 + 30\mathbb{R}_{\mu\nu}\mathbb{R}^{\mu\nu}\}\\
&= \frac{1}{2}(5\mathbf{s}^2 - 8\mathbf{r}^2 - 7\mathbf{R}^2) - \frac{1}{6}\mathbf{s}\left(45\mathbf{s} + 8A_f|\Phi|^2\right)\\
&+\frac{1}{2}\left[\frac{45}{4}\mathbf{s}^2 + 8A_f|\mathbf{D}\Phi|^2 + 8B_f|\Phi|^4 + 4A_f\mathbf{s}|\Phi|^2\right]\\
&+\frac{1}{12}\left[-\frac{45}{2}\mathbf{R}^2 + 4(R_{\mu\nu}^{\mathbf{E}} R^{\mathbf{E}\mu\nu})\right]\\
&= \frac{1}{8}\left(5\mathbf{s}^2 - 8\mathbf{r}^2 - 7\mathbf{R}^2\right) + \frac{2}{3}A_f\mathbf{s}|\Phi|^2 + 4A_f|\mathbf{D}\Phi|^2 + 4B_f|\Phi|^4\\
&-\frac{2}{3}\mathrm{Tr}_x{}^{\mathbf{E}}\left(R_{\mu\nu}^{\mathbf{E}} R^{\mathbf{E}\mu\nu}\right) = 10\pi^2\chi_4 - \frac{9}{4}\mathbf{C}^2 + \frac{2}{3}A_f\mathbf{s}|\Phi|^2\\
&+4B_f|\Phi|^4 - \frac{4}{3}\left(20N\mathbf{f}_{\mu\nu}\mathbf{f}^{\mu\nu} + N\mathbf{h}_{\mu\nu}^s\mathbf{h}_s^{\mu\nu} + N\mathbf{g}_{\mu\nu}^a\mathbf{g}_a^{\mu\nu}\right) \ .
\end{aligned} \tag{7.41}$$

Here we effected the replacement:

$$5\mathbf{s}^2 - 8\mathbf{r}^2 - 7\mathbf{R}^2 = 88\pi^2\chi_4 - 18\mathbf{C}^2 \ , \tag{7.42}$$

where $\mathbf{C}^2 = \mathbf{R}^2 - 2\mathbf{r}^2 + \frac{1}{3}\mathbf{s}^2$ is the square of the Weyl tensor, and we subsequently suppress $\chi_4 = 2(4\pi)^{-2}\left(\mathbf{R}^2 - 4\mathbf{r}^2 + \mathbf{s}^2\right)$, since $\chi_4 dv$ is the Euler form, which, by the Gauss-Bonnet theorem, does not contribute to the action. We proved:

$$(4\pi)^2 a_2(x, \mathbb{D}_A^2) = 180 \tag{7.43}$$

$$(4\pi)^2 a_2(x, \mathbb{D}_A^2) = -\frac{45}{3}\mathbf{s} - 8A_f|\Phi|^2 \tag{7.44}$$

$$(4\pi)^2 a_4(x, \mathbb{D}_A^2) = 10\pi^2\chi_4 - \frac{9}{4}\mathbf{C}^2 + \frac{2}{3}A_f \mathbf{s}|\mathbf{D}\Phi|^2 + 4B_f|\Phi|^4 - \frac{4}{3}\left(\frac{20}{3}N\mathbf{f}_{\mu\nu}\mathbf{f}^{\mu\nu} + N\mathbf{h}^s_{\mu\nu}\mathbf{h}^{\mu\nu}_s + N\mathbf{g}^a_{\mu\nu}\mathbf{g}^{\mu\nu}_a\right) . \quad (7.45)$$

Plugging this into (7.10) we get the tree first significant lines of the action (7.2). The fourth line results as follows:

7.7 Remark

The sum of the surface terms which we discarded in the expression of the action is the following:

$$11\pi^2\chi_4 + \frac{8}{3}\triangle\,\mathbf{s} + \frac{4}{3}A_f\,\triangle\,(|\Phi|^2) , \quad (7.46)$$

where χ_4 is the Euler form.

Proof:

All the surface-term come from $a_4(x, \mathbb{D}_A^2)$. The χ_4-term is associated to the square of the Weyl tensor in the expression (7.43). The surface-term which we discarded from $a_4(x, \mathbb{D}_A^2)$ inside the bracket $\{\}$ (7.13) sum up to:

$$\begin{aligned}
&\frac{1}{360}\mathrm{Tr}_x\left\{12\mathbf{s};^\alpha{}_\alpha\,\mathbb{I} - 60E;^\alpha{}_\alpha\right\} \\
&= \frac{1}{360}\triangle\,\mathrm{Tr}_x\left\{-12\mathbf{s}\mathbb{I} + 60\left(\frac{1}{4}\mathbf{s}\mathbb{I} - \frac{1}{2}c\left(R^{\mathbf{E}}\right) + ic^\mu\left[\mathbb{V}_\mu, \boldsymbol{\Phi}\right] + \boldsymbol{\Phi}^2\right)\right\} \\
&= \frac{1}{360}\triangle\,\mathrm{Tr}_x\left\{3\mathbf{s}\mathbb{I} + 60\boldsymbol{\Phi}^2\right\} = \frac{1}{360}\triangle\left\{3\mathbf{s}\ \mathrm{rank}\ \mathbf{V} + 60\times 8A_f|\Phi|^2\right\} \\
&= \frac{8}{3}\triangle\,\mathbf{s} + \frac{4}{3}A_f\,\triangle\,(|\Phi|^2) .
\end{aligned} \quad (7.47)$$

8 Tree-Approximation Results

We identify the bosonic part of the "spectral action" with N=3:

$$I_B(x, g, A) = 4\mathbf{g}^a_{\mu\nu}\mathbf{g}^{\mu\nu}_a + 4\mathbf{h}^s_{\mu\nu}\mathbf{h}^{\mu\nu}_s + \frac{80}{3}\mathbf{f}_{\mu\nu}\mathbf{f}^{\mu\nu} + 90\Lambda^4 - 15\Lambda^2\mathbf{s} + \frac{2}{3}A_f\mathbf{s}|\Phi|^2 - \frac{9}{4}\mathbf{C}^2 + 4A_f|\mathbf{D}\Phi|^2 + 4B_f|\Phi|^4 - 8\Lambda^2A_f|\Phi|^2 , \quad (8.1)$$

with the Lagrangian density of the full Euclidean standard model:

$$\mathcal{L}_{\text{stand}} = \frac{1}{4}G^a_{\mu\nu}G^{\mu\nu}_a + \frac{1}{4}W^s_{\mu\nu}W^{\mu\nu}_s + \frac{1}{4}B_{\mu\nu}B^{\mu\nu} + (D_\mu\phi)^*(D^\mu\phi) - \frac{\mu^2}{\nu^2}(\phi^*\phi)^2 + \mu^2\phi^*\phi , \quad (8.2)$$

where

$$\begin{cases} G^a_{\mu\nu} &= \partial_\mu G^a_\nu - \partial_\nu G^a_\mu + g_3 f_{abc}G^b_\mu G^c_\nu \\ W^{\mu\nu}_s &= \partial_\mu W^s_\nu - \partial_\nu W^s_\mu + g_2\varepsilon_{srt}W^r_\mu W^t_\nu \\ B_{\mu\nu} &= \partial_\mu B_\nu - \partial_\nu B_\mu \end{cases} . \quad (8.3)$$

We recall the relationships of the parameters in (8.2) with boson masses and Weinberg angle:

$$\begin{cases} M_H &= \sqrt{2\mu} \\ M_Z &= \frac{1}{2}vg \\ M_W &= M_Z \cos\theta_W = \frac{1}{2}vg_2 \\ \operatorname{tg}\theta_W &= \frac{g_1}{g_2} \end{cases} . \tag{8.4}$$

We also recall that the Lagrangian (8.1), resp. (8.2) are assorted with the covariant derivatives:

$$\begin{cases} \mathbf{D}_\mu &= \tilde{\nabla}_\mu + i(\mathbf{a}_\mu - \mathbf{b}^s_\mu \frac{\tau_s}{2}) \\ \mathbf{D}^{Rl}_\mu &= \tilde{\nabla}_\mu - 2i\mathbf{a}_\mu \\ \mathbf{D}^{Ll}_\mu &= \tilde{\nabla}_\mu - i\mathbf{a}_\mu - i\mathbf{b}^s_\mu \frac{\tau_s}{2} \\ \mathbf{D}^{Ru}_\mu &= \tilde{\nabla}_\mu + \frac{4}{3}i\mathbf{a}_\mu - i\mathbf{c}^a_\mu \frac{\lambda_a}{2} \\ \mathbf{D}^{Rd}_\mu &= \tilde{\nabla}_\mu - \frac{2}{3}i\mathbf{a}_\mu - i\mathbf{c}^a_\mu \frac{\lambda_a}{2} \\ \mathbf{D}^{Lq} &= \tilde{\nabla}_\mu + \frac{1}{3}i\mathbf{a}_\mu - i\mathbf{b}^s_\mu \frac{\tau_s}{2} - i\mathbf{c}^a_\mu \frac{\lambda_a}{2} \end{cases} , \tag{8.5}$$

respectively

$$\begin{cases} D_\mu &= \partial_\mu - i\frac{g_1}{2}B_\mu - ig_2 W^s_\mu \frac{\tau_s}{2} \\ D^{Rl}_\mu &= \partial_\mu + ig_1 B_\mu \\ D^{Ll}_\mu &= \partial_\mu + i\frac{g_1}{2}B_\mu - ig_2 W^s_\mu \frac{\tau_s}{2} \\ D^{Ru}_\mu &= \partial_\mu - 2i\frac{g_1}{3}B_\mu - ig_3 G^a_\mu \frac{\lambda_a}{2} \\ D^{Rd}_\mu &= \partial_\mu + i\frac{g_1}{3}B_\mu - ig_3 G^a_\mu \frac{\lambda_a}{2} \\ D^{Lq}_\mu &= \partial_\mu - i\frac{g_1}{6}B_\mu - ig_2 W^s_\mu \frac{\tau_s}{2} - ig_3 G^a_\mu \frac{\lambda_a}{2} \end{cases} . \tag{8.6}$$

Identifying the latter is synonymous with the identifications:

$$\begin{cases} \mathbf{c}^a_\mu = g_3 G^a_\mu & a = 1, \dots, 8 \\ \mathbf{b}^s_\mu = g_2 W^s_\mu & s = 1, 2, 3 \\ \mathbf{a}_\mu = -\frac{1}{2}g_1 B_\mu & \end{cases} , \tag{8.7}$$

implying:

$$\begin{cases} \mathbf{g}^a_{\mu\nu} = g_3 G^a_{\mu\nu} & a = 1, \dots, 8 \\ \mathbf{h}^s_{\mu\nu} = g_2 W^s_{\mu\nu} & s = 1, 2, 3 \\ \mathbf{f}_{\mu\nu} = -\frac{1}{2}g_1 B_{\mu\nu} & \end{cases} , \tag{8.8}$$

whereby the first line (7.1) becomes

$$4\left(g_3^2 G^a_{\mu\nu} G^{\mu\nu}_a + g_2^2 W^s_{\mu\nu} W^{\mu\nu}_s + \frac{5}{3} g_1^2 B_{\mu\nu} B^{\mu\nu}\right) . \tag{8.9}$$

Conferring this with the first line (8.2) implies

$$g_2 = g_3 , \tag{8.10}$$

and

$$\frac{g_1^2}{g_2^2} = \mathrm{tg}^2\theta_W = \frac{3}{5}, \quad \text{i.e.} \quad \sin^2\theta_W = \frac{3}{8} , \tag{8.11}$$

and further shows that we should identify $I_B(x, g, A)$ with $16 g_2^2 \mathcal{L}_{\text{stand}}$. One has then the following relations:

– identifying the kinetic Higgs terms:

$$\Phi = 2 A_f^{-1/2} g_2 \phi ; \tag{8.12}$$

– identifying the quadratic Higgs terms:

$$\mu^2 = 2\Lambda^2 \quad \text{i.e.} \quad M_H = 2\Lambda ; \tag{8.13}$$

– identifying the quartic Higgs terms:

$$v = \frac{A_f}{2 B_f^{1/2}} \mu \quad \text{i.e.} \quad M_W = \frac{1}{2} v g_2 = \frac{A_f}{2\sqrt{2 B_f}} \Lambda . \tag{8.14}$$

Thus, assuming dominance of the top mass implying $A_f / \sqrt{B_f} = \sqrt{3}$:

$$M_W = \frac{\sqrt{3}}{2\sqrt{2}} \Lambda . \tag{8.15}$$

The above constraints have to be subjected to a renormalization group treatment for extracting the information on the Higgs mass. This requires the conversion of the above $U(1) \times SU(2) \times SU(3)$-invariant Lagrangian into the stable symmetry-broken Lagrangian. For lack of place we refrain from giving the details of this computation for which we refer the reader to the paper [36]. Then the Higgs mass is predicted as: $M_H = 182 \mp 17$ Gev.

9 Fermionic Action

In what follows we compute the fermionic action density [**4.7**] of the S_0-real spectral triple $(\mathcal{A}, (\mathcal{H} = \underline{\mathbb{H}} \oplus \overline{\mathbb{H}}, \chi, \mathcal{D}), \mathcal{J})$ of the full standard model (cf.[**6**]): namely, with $(\cdot,\cdot)$ is the Hilbert scalar product of Euclidean spinor fields:

$$\mathcal{L}_F(\Psi_{\rm tot}, A) = (\Psi_{\rm tot}, \mathcal{D}_A \Psi_{\rm tot}), \qquad \Psi_{\rm tot} \in \mathcal{H}, \tag{9.1}$$

by charge-conjugation symmetry twice its particle-space part:

$$\mathbb{L}_F(\underline{\Psi}, A) = (\underline{\Psi}, \mathbb{D}\underline{\Psi}), \qquad \underline{\Psi} \in \underline{\mathbb{H}}, \tag{9.2}$$

(here $\underline{\Psi}$ is the component of $\Psi_{\rm tot}$ in the particle-space $\underline{\mathbb{H}}$, to which $\mathcal{D}_A$ restricts as $\mathbb{D}_A$: one has thus $\Psi_{\rm tot} = \underline{\Psi} + J\underline{\Psi}$ and $\mathcal{D}_A = \mathbb{D}_A + J\mathbb{D}_A J^*$ (cf,[**6**]). $\mathbb{L}_F(\underline{\Psi}, A)$ is the sum of its respective lepton and quark parts: with Ψ, $\mathcal{Q}$ the respective lepton and quark components of $\underline{\Psi}$:where

$$\begin{aligned} \mathbb{L}_{F\ell}(\Psi, A) &= (\Psi, \mathbb{D}_A \Psi) \\ \mathbb{L}_{Fq}(\mathcal{Q}, A) &= (\mathcal{Q}, \mathbb{D}_A \mathcal{Q}) \end{aligned} \tag{9.3}$$

9.1 Labeling the basis of $\underline{\mathbb{H}}$

The fermionic field is the sum $\Psi + \mathcal{Q}$ of

– the leptonic field $\Psi \in L^2(\mathbb{S}_{\mathbf{M}}) \otimes \underline{H_l}$ spanned by the:

$$\begin{aligned} (\Psi_f{}^R)_{f=1,\dots,N} &\in L^2(\mathbb{S}_{\mathbf{M}}) \otimes \mathbb{C}^1{}_{\rm hyp} \otimes \mathbb{C}^N \\ (\Psi_f{}^{Li})_{f=1,\dots,N,\ i=1,2} &\in L^2(\mathbb{S}_{\mathbf{M}}) \otimes \mathbb{C}^2{}_{\rm iso} \otimes \mathbb{C}^N, \end{aligned} \tag{9.4}$$

– the quark field $\mathcal{Q} \in L^2(\mathbb{S}\mathrm{M}) \otimes \underline{H_q}$ spanned by the:

$$\begin{aligned} (\mathcal{Q}_f{}^{Rum})_{f=1,\dots,N,\ m=1,2,3} &\in L^2(\mathbb{S}_{\mathbf{M}}) \otimes \mathbb{C}^1{}_{\rm hyp} \otimes \mathbb{C}^N \otimes \mathbb{C}^3{}_{\rm colour} \\ (\mathcal{Q}_f{}^{Rdm})_{f=1,\dots,N,\ m=1,2,3} &\in L^2(\mathbb{S}_{\mathbf{M}}) \otimes \mathbb{C}^1{}_{\rm hyp} \otimes \mathbb{C}^N \otimes \mathbb{C}^3{}_{\rm colour} \\ (\mathcal{Q}_f{}^{Lim})_{f=1,\dots,N,\ i=1,2,\ m=1,2,3} &\in L^2(\mathbb{S}_{\mathbf{M}}) \otimes \mathbb{C}^2{}_{\rm iso} \otimes \mathbb{C}^N \otimes \mathbb{C}^3{}_{\rm colour} \end{aligned} \tag{9.5}$$

Here $L^2(\mathbb{S}_{\mathbf{M}})$ is the Hilbert space of Euclidean square integrable spinors, $\mathbb{C}^1{}_{\rm hyp}$ indicates a $U(1)$-(hypercharge) singlet, $\mathbb{C}^2{}_{\rm iso}$ a $SU(2)$-(isospin) doublet, R and L stand respectively for right and left, $\mathbb{C}^3{}_{\rm colour}$ corresponds to the quark colour degrees of freedom, $\mathbb{C}^N$ to N fermion families with f the fermion-family index (present experimental evidence yields $N = 3$, corresponding to the respective electron, muon, and tau families). The spinor fields correspond to the following particle-types (indicated in the case of the electron family):

$$\begin{array}{lllll} \text{Leptons} & \Psi^R : e_R & & \Psi_f{}^{L1} : \nu_L & \Psi_f{}^{L2} : e_R \\ \text{Quarks} & \mathcal{Q}_f{}^{Ru} : u_R & \mathcal{Q}_f{}^{Rd} : d_R & \mathcal{Q}_f{}^{L1} : u_L & \mathcal{Q}_f{}^{L2} : d_L \end{array} . \tag{9.6}$$

9.2 Euclidean fermionic action: the Fermi and Yukawa terms

The result of the computation of equations (9.3) is as follows: with $(\cdot,\cdot)$ the Euclidean scalar product of $L^2(\$_{\mathrm{M}})$:

- the leptonic action density is the sum $\mathbb{L}_{F\ell} = \mathbb{L}_{F\ell}$ Fermi $+\mathbb{L}_{F\ell}$ Yukawa of the **Fermi term**

$$\mathbb{L}_{\ell} \text{ Yukawa} = \sum_{f=1,\dots,N} \{(\Psi_f{}^R, i\gamma^\mu \mathbf{D}^{R\ell}{}_\mu \Psi_f{}^R) + (\Psi_f{}^L, i\gamma^\mu \mathbf{D}^{L\ell}{}_\mu \Psi_f{}^L)\} \quad (9.7)$$

with the covariant derivatives:

$$\mathbf{D}^{Rl}_\mu = \tilde{\nabla}_\mu - 2i\mathbf{a}_\mu \quad (9.8)$$

$$\mathbf{D}^{Ll}_\mu = \tilde{\nabla}_\mu - i\mathbf{a}_\mu - i\mathbf{b}_\mu = \tilde{\nabla}_\mu - i\mathbf{a}_\mu - i\mathbf{b}^s_\mu \frac{\tau_s}{2}, \quad (9.9)$$

and the **Yukawa term**

$$\mathbb{L}_{F\ell} \text{ Yukawa} = \textstyle\sum_{f_1,f_2=1,\dots,N,\ i=1,2} \{M_e{}^*{}_{f_1 f_2}(\Psi_{f_1}{}^R, \overline{\Phi}{}^i \gamma^5 \Psi_{f_2}{}^{Li}) + M_{e f_1 f_2}(\Psi_{f_1}{}^{Li}\overline{\Phi}{}^i, \gamma^5 \Psi_{f_2}{}^R)\}. \quad (9.10)$$

- the quark action density is the sum $\mathbb{L}_{Fq} = \mathbb{L}_{Fq}$ Fermi $+\mathbb{L}_{Fq}$ Yukawa of the **Fermi term**

$$\mathbb{L}_{Fq} \text{ Fermi} = \textstyle\sum_{f=1,\dots,N,\ m=1,2,3} \{(\mathbf{Q}_f{}^{Ru}{}_m, i\gamma_\mu \mathbf{D}^{Ru}{}_\mu \mathbf{Q}_{Rum}) + (\mathbf{Q}_f{}^{Rd}{}_m, i\gamma_\mu \mathbf{D}^{Rd}{}_\mu \mathbf{Q}_{Rdm}) + (\mathbf{Q}_f{}^{L}{}_m, i\gamma_\mu \mathbf{D}^{Lq}{}_\mu \mathbf{Q}_{Lm})\}, \quad (9.11)$$

with the covariant derivatives:

$$\mathbf{D}^{Ru}_\mu = \tilde{\nabla}_\mu + \frac{4}{3}i\mathbf{a}_\mu - i\mathbf{c}_\mu = \tilde{\nabla}_\mu + \frac{4}{3}i\mathbf{a}_\mu - i\mathbf{c}^a_\mu \frac{\lambda_a}{2} \quad (9.12)$$

$$\mathbf{D}^{Rd}_\mu = \tilde{\nabla}_\mu - \frac{2}{3}i\mathbf{a}_\mu - i\mathbf{c}_\mu = \tilde{\nabla}_\mu - \frac{2}{3}i\mathbf{a}_\mu - i\mathbf{c}^a_\mu \frac{\lambda_a}{2} \quad (9.13)$$

$$\mathbf{D}^{Lq} = \tilde{\nabla}_\mu + \frac{1}{3}i\mathbf{a}_\mu - i\mathbf{b}_\mu - i\mathbf{c}_\mu = \tilde{\nabla}_\mu + \frac{1}{3}i\mathbf{a}_\mu - i\mathbf{b}^s_\mu \frac{\tau_s}{2} - i\mathbf{c}^a_\mu \frac{\lambda_a}{2}, \quad (9.14)$$

and the **Yukawa term** (where $\tilde{\Phi}^1 = \Phi^2$, $\tilde{\Phi}^2 = -\Phi^1$):

$$\mathbb{L}_{Fq} \text{ Yukawa} = \textstyle\sum_{f_1,f_2=1,\dots,N,\ i=1,2,\ m=1,2,3} \{M_{d f_1 f_2}(\mathbf{Q}_{f_1}{}^{Lim}\overline{\Phi}{}^i, \gamma^5 \mathbf{Q}_{f_2}{}^{Rdm}) + M_d{}^*{}_{f_1 f_2}(\mathbf{Q}_{f_1}{}^{Rdm}, \overline{\Phi}{}^i \gamma^5 \mathbf{Q}_{f_2}{}^{Lim}) + M_{u f_1 f_2}(\mathbf{Q}_{f_1}{}^{Lim}\tilde{\Phi}{}^i, \gamma^5 \mathbf{Q}_{f_2}{}^{Rum}) + M_u{}^*{}_{f_1 f_2}(\mathbf{Q}_{f_1}{}^{Rum}, \tilde{\Phi}{}^i \gamma^5 \mathbf{Q}_{f_2}{}^{Lim})\}. \quad (9.15)$$

Proof: We showed in [**6.7**](ii) that the covariant Dirac operator $\mathbb{D}_A$ acts on the tensor-product bundle $\underline{\mathbb{E}} = \mathcal{S}(\mathbf{M}) \otimes_{C^\infty(\mathbf{M})} \mathbf{E}$ as the sum $\mathbb{D}_A = D^{\mathbb{V}} + \boldsymbol{\Phi}$ of the endomorphism $\boldsymbol{\Phi}$ and the Dirac operator $D^{\mathbb{V}} = (\gamma^\mu \otimes id_{\mathbf{E}})\mathbb{V}_\mu$ of the connection $\mathbb{V} = \tilde{\nabla} \otimes id_{\mathbf{E}} + id_{\mathcal{S}(\mathbf{M})} \otimes \nabla^{\mathbf{E}}$, tensor-product of the spin connection $\tilde{\nabla}$ and the inner-space connection $\nabla^{\mathbf{E}}$. One has separately: $(\mathbb{D}_A)_\ell = (D^{\mathbb{V}})_\ell + \boldsymbol{\Phi}_\ell$, $(\mathbb{D}_A)_q = (D^{\mathbb{V}})_q + \boldsymbol{\Phi}_q$ with the Yukawa terms stemming from $\boldsymbol{\Phi}_\ell$ resp. $\boldsymbol{\Phi}_q$ whilst the Fermi terms stem from $(D^{\mathbb{V}})_\ell$ resp. $(D^{\mathbb{V}})_q$. The expressions above then follow by inspection from the results at the end of [**6,7**], which we reproduce here in the modular-corrected form using the precise labeling [**9.1**] of particle space:

$$
id_{\mathcal{S}(\mathbf{M})} \otimes (\nabla^{\mathbf{E}_q} - \partial)_\mu =
$$

$$
\begin{matrix} \boldsymbol{Q}_f{}^{Rum} & \boldsymbol{Q}_f{}^{Rdm} & \boldsymbol{Q}_f{}^{L1m} & \boldsymbol{Q}_f{}^{L2m} \end{matrix}
$$

$$
-i \begin{pmatrix} -\frac{4}{3}\mathbf{a}_\mu \otimes \mathbb{1}_N & 0 & 0 & 0 \\ 0 & \frac{2}{3}\mathbf{a}_\mu \otimes \mathbb{1}_N & 0 & 0 \\ 0 & 0 & (\mathbf{b}^1_{1\mu} - \frac{1}{3}\mathbf{a}_\mu) \otimes \mathbb{1}_N & \mathbf{b}^1_{2\mu} \otimes \mathbb{1}_N \\ 0 & 0 & \mathbf{b}^2_{1\mu} \otimes \mathbb{1}_N & (\mathbf{b}^2_{2\mu} - \frac{1}{3}\mathbf{a}_\mu) \otimes \mathbb{1}_N \end{pmatrix} \otimes (\delta^n{}_m)
$$

$$
\begin{matrix} \boldsymbol{Q}_f{}^{Rum} & \boldsymbol{Q}_f{}^{Rdm} & \boldsymbol{Q}_f{}^{L1m} & \boldsymbol{Q}_f{}^{L2m} \end{matrix}
$$

$$
-i \begin{pmatrix} \mathbf{c}^a_\mu \otimes \mathbb{1}_N & 0 & 0 & 0 \\ 0 & \mathbf{c}^a_\mu \otimes \mathbb{1}_N & 0 & 0 \\ 0 & 0 & \mathbf{c}^a_\mu \otimes \mathbb{1}_N & 0 \\ 0 & 0 & 0 & \mathbf{c}^a_\mu \otimes \mathbb{1}_N \end{pmatrix} \otimes \frac{1}{2}\lambda_a{}^n{}_m \,, \tag{9.16}
$$

$$
\begin{matrix} \boldsymbol{Q}_f{}^{Rum} & \boldsymbol{Q}_f{}^{Rdm} & \boldsymbol{Q}_f{}^{L1m} & \boldsymbol{Q}_f{}^{L2m} \end{matrix}
$$

$$
\boldsymbol{\Phi}_q = \begin{pmatrix} 0 & 0 & \Phi^2\gamma^5 \otimes M_u^* & -\Phi^1\gamma^5 \otimes M_u^* \\ 0 & 0 & \Phi_1\gamma^5 \otimes M_d^* & \Phi_2\gamma^5 \otimes M_d^* \\ \Phi_2\gamma^5 \otimes M_u & \Phi^1\gamma^5 \otimes M_d & 0 & 0 \\ -\Phi_1\gamma^5 \otimes M_u & \Phi^2\gamma^5 \otimes M_d & 0 & 0 \end{pmatrix} \otimes (\delta^n{}_m) \,, \tag{9.17}
$$

$$
\begin{matrix} \boldsymbol{\Psi}_f{}^{R} & \boldsymbol{\Psi}_f{}^{L1} & \boldsymbol{\Psi}_f{}^{L2} \end{matrix}
$$

$$
id_{\mathcal{S}(\mathbf{M})} \otimes (\nabla^{\mathbf{E}_l} - \partial)_\mu = -i \begin{pmatrix} 2\mathbf{a}_\mu \otimes \mathbb{1}_N & 0 & 0 \\ 0 & (\mathbf{b}^1_{1\mu} + \mathbf{a}_\mu) \otimes \mathbb{1}_N & \mathbf{b}^1_{2\mu} \otimes \mathbb{1}_N \\ 0 & \mathbf{b}^2_{1\mu} \otimes \mathbb{1}_N & (\mathbf{b}^2_{2\mu} + \mathbf{a}_\mu) \otimes \mathbb{1}_N \end{pmatrix} , \tag{9.18}
$$

$$
\begin{matrix} \boldsymbol{\Psi}_f{}^{R} & \boldsymbol{\Psi}_f{}^{L1} & \boldsymbol{\Psi}_f{}^{L2} \end{matrix}
$$

$$
\boldsymbol{\Phi}_l = \begin{pmatrix} 0 & \Phi_1\gamma^5 \otimes M_e^* & \Phi_2\gamma^5 \otimes M_e^* \\ \Phi^1\gamma^5 \otimes M_e & 0 & 0 \\ \Phi^2\gamma^5 \otimes M_e & 0 & 0 \end{pmatrix} . \tag{9.19}
$$

9.3 The ensuing Minkowskian fermionic action

The conversion into a minkowskian action (undoing the fermion "Wick rotation") arises in two steps:

- step 1: trade the Euclidean scalar product $(\cdot,\cdot)$ for the minkowskian one $(\cdot\gamma^0,\cdot)$.
- step 2: affect the right-handed particle-fields with the chiral factor $\frac{1}{2}(\gamma^5 + id)$.

The results [**9.2**] then yield the fermionic action of the standard model.

10 Does the Inner Spectral Triple of the Full Standard Model Proceed from a Quantum Group? The Finite $U_q(s\ell 2)$ at Third Root of Unity, its Regular Representation and "Hopf Bar-Operation"

The noncommutative theory version of the classical (=tree-) level of the standard model of elementary particles synthesized with the Lagrangian of gravitation as we presented it in the previous section [7] is physically satisfactory (up to future agreement with yet unperformed experiments of the tree-approximation results [7] subjected to the renormalization- group). However the object we have at present is not mathematically conceptually autonomous (as e.g. the Maxwell equations) since the inner spectral triple was tailored so as to fit (dictated by) phenomenology. Hence it would be conceptually rewarding to derive this object mathematically from first principles. In [13] Alain Connes advanced the fascinating idea that the finite quantum group $U_q(s\ell 2)$ for $q = e^{2\pi/3}$, whose semi-simple quotient $\mathbf{M}_1(\mathbb{C}) \oplus \mathbf{M}_1(\mathbb{C}) \oplus \mathbf{M}_1\mathbb{C}$ strangely resembles the algebra $\mathbf{M}_1(\mathbb{C}) \oplus \mathbb{H} \oplus \mathbf{M}_1\mathbb{C}$ of our finite spectral triple, might in fact dictate the latter. If this, or something of the same nature, was true, then, in addition to providing the aforementioned conceptual autonomy of the classical level of the standard model, this would open the following fascinating perspectives:

- The requirement that the Dirac operator be equivariant w.r.t. the relevant quantum group should produce constraints on the fermion mass-matrices whose renormalization-group treatment could then be conferred with the long-known masses of the fermions, a subject on which the usual theory is sadly mute!
- One would be tempted to stick into the Lagrangian-creating paradigm the whole quantum group rather than its semi-simple quotient, expecting then the emergence of an extension of the standard model (of a generalized super-symmetric type) improving upon the present mathematically not appealing supersymmetric standard model.

In what follows we present the "Hopf bar-operation" of $U_q(s\ell 2)$ for q of unit modulus, passing to the finite-dimensional quotients $\mathbf{H}_N$ for q a primitive root of unity. We hope that this element of structure will serve us in the pursuit of the above programme, since the bar-operation of matrices is essentially used in the construction of the S_q-real spectral triple of the standard model. For more details the reader is referred to [48].

10.1 The Hopf Algebra H_1 and its regular representation

Definition.

(i): Let $q \in \mathbb{C}$, $q \neq 0$, $q \neq 1$. Defining the unital algebra $U_q(s\ell 2)$ over $\mathbb{C}$ by symbols K, K^{-1}, E, F, and the relations:

$$\begin{cases} KK^{-1} = K^{-1}K = \mathbb{1} \\ KE = q^2 EK \\ KF = q^{-2}FK \\ EF - FE = \dfrac{K - K^{-1}}{q - q^{-1}} \end{cases}, \tag{10.1}$$

we recall that $U_q(s\ell 2)$ is a Hopf algebra with coproduct Δ, counit ε, and antipode S specified as follows:

$$(a)\begin{cases} \Delta\mathbb{1} = \mathbb{1}\otimes\mathbb{1} \\ \Delta K = K\otimes K \\ \Delta K^{-1} = K^{-1}\otimes K^{-1} \\ \Delta E = E\otimes\mathbb{1} + K\otimes E \\ \Delta F = F\otimes K^{-1} + \mathbb{1}\otimes F \end{cases} \quad (b)\begin{cases} \varepsilon\mathbb{1} = 1 \\ \varepsilon K = 1 \\ \varepsilon K^{-1} = 1 \\ \varepsilon E = 0 \\ \varepsilon F = 0 \end{cases} \quad (c)\begin{cases} S\mathbb{1} = \mathbb{1} \\ SK = K^{-1} \\ SK^{-1} = K \\ SE = -K^{-1}E \\ SF = -FK \end{cases}, \tag{10.2}$$

moreover equipped with a *-operation * specified by:

$$\begin{cases} E^* = F \\ F^* = E \\ K^* = K^{-1} \end{cases} \tag{10.3}$$

The following *Casimir operator* then belongs to the center of $U_q(s\ell 2)$:

$$C = FE = \frac{qK + q^{-1}K^{-1}}{(q-q^{-1})^2} = EF + \frac{q^{-1}K + qK^{-1}}{(q-q^{-1})^2} \tag{10.4}$$

(ii): Assume that q is a primitive p^{th} root of the identity, $p > 2$ odd. Then the ideal $\mathbf{I}$ of $U_q(s\ell 2)$ generated by

$$E^{p'},\ F^{p'},\ K^{Np'} - \mathbb{1},\ p' = \begin{cases} p & \text{if } p \text{ is odd} \\ \frac{1}{2}p & \text{if } p \text{ is even} \end{cases}$$

is a Hopf ideal, thus yielding a quotient Hopf algebra $\mathbf{H}_N = U_q(s\ell 2)/\mathbf{I}_N$ linearly spanned by the monomials $F^pE^qK^r$, p, q ranging over $0, 1, \ldots, p' - 1$, and r over $0, 1, \ldots, p'N - 1$: $\mathbf{H}_N$ is thus of dimension Np'^3.

Proposition (the Hopf bar-operation Γ).
Let q be of unit modulus. Then there is a unique antilinear and multiplicative map $\Gamma : a \to \bar{a}$ from $U_q(s\ell 2)$ to $U_q(s\ell 2)$ such that

$$\begin{cases} \overline{E} = -qKF & (= -q^{-1}FK) \\ \overline{F} = -qK^{-1}E & (= -q^{-1}EK^{-1} \\ \overline{K} = K \ . \end{cases} \tag{10.5}$$

This map is involutive, commutes with the antipode:

$$\overline{Sa} = S\bar{a}, \quad a \in U_q(s\ell 2) \ , \tag{10.6}$$

commutes with the coproduct in the sense:

$$\Delta\Gamma(a) = (\Gamma \otimes \Gamma)\Delta a \tag{10.7}$$

is intertwined by ε with the complex conjugation of $\mathbb{C}$:

$$\varepsilon(\bar{a}) = \overline{\varepsilon(a)}, \quad a \in \mathbf{H}_N \ , \tag{10.8}$$

commutes with the involution *:

$$\Gamma(a^*) = \Gamma(a)^* \tag{10.9}$$

and preserves the Casimir element:

$$\overline{C} = C \tag{10.10}$$

(in fact we have:

$$\begin{cases} \Gamma(EF) = FE \\ \Gamma(FE) = EF \ . \end{cases} \quad) \tag{10.11}$$

(ii): Assume now that q is a primitive p^{th} root of the identity, $p > 2$. Then the Hopf ideal $\mathbf{I}_N$ of $U_q(s\ell 2)$ is invariant under Γ which thus induces an involution on the quotient $\mathbf{H}_N$. The above results (10.1) through (10.10) then hold with the change $U_q(s\ell 2) \to \mathbf{H}_N$.
Proof: Straightforward verifications, of these properties, after straightforward check of the coherence of the definition (10.5).

Corollary (the real Hopf subalgebra $\mathbf{H}_N^{\text{real}}$ such that $\mathbf{H}_N = \mathbf{H}_N^{\text{real}} \oplus i\mathbf{H}_N^{\text{real}}$).

(i): The set $\mathbf{H}_N^{\text{real}} = \{a \in \mathbf{H}_N;\ \Gamma a = a\}$ of fixpoints of Γ in $\mathbf{H}_N$ is a real subspace of $\mathbf{H}_N$ closed under products, stable under the antipode S, mapped into $\mathbb{R}$ by ε, and mapped into $\mathbf{H}_N^{\text{real}} \otimes_{\mathbb{R}} \mathbf{H}_N^{\text{real}}$ by Δ: thus $(\mathbf{H}_N^{\text{real}}, \cdot, \Delta, \varepsilon, S)$ is a real Hopf algebra.

(ii): Consequently the complex Hopf algebra $\mathbf{H}_N$ is the complexification of the real Hopf algebra $\mathbf{H}_N^{\text{real}}$ in the sense that $\mathbf{H}_N = \mathbf{H}_N^{\text{real}} \oplus i\mathbf{H}_N^{\text{real}}$ as a vector space, with the

- algebra product $(a^{'} \oplus ia^{''}(b^{'} \oplus ib^{''} = (a^{'}b^{'} - a^{''}b^{''}) \oplus i(a^{'}b^{''} + a^{''}b^{'})$;
- counit $\varepsilon(a^{'} \oplus ia^{''}) = \varepsilon(a^{'}) + i\varepsilon(a^{''})$;
- antipode $S(a^{'} \oplus ia^{''}) = S(a^{'}) + iS(a^{''})$;
- coproduct $\Delta(a^{'} \oplus ia^{''}) = \Delta(a^{'}) + i\,\Delta\,(a^{''})$.

Proof: in fact proceeds from:
Remarks.
(i):The passage from $\mathbf{H}_N$ to $\mathbf{H}_N^{\mathrm{real}}$ corresponds to the following general fact: for a complex finite-dimensional Hopf algebra $(\mathbf{H}, \Delta, \varepsilon, S)$ the following are the same:

- a multiplicative antilinear involution Γs.t. $S\Gamma = \Delta\Gamma(a) = (\Gamma \otimes \Gamma)\,\Delta\, a$, and $\varepsilon(\Gamma a) = \overline{\varepsilon(a)},\ a \in \mathbf{H}$;
- a real Hopf subalgebra $\mathbf{H}^{\mathrm{real}}$ of $\mathbf{H}$ such that $\mathbf{H} = \mathbf{H}^{\mathrm{real}} \oplus i\mathbf{H}^{\mathrm{real}}$, whereby $\mathbf{H}^{\mathrm{real}}$ is the fixpoint of $\mathbf{H}$ for Γ, and $\Gamma(a^{'} + ia^{''}) = a^{'} - ia^{''},\ a^{'}, a^{''} \in \mathbf{H}^{\mathrm{real}}$.

Calling such a structure a **real spine if H** we then have that:
(ii): with **H** and **K** two complex Hopf algebras in strict duality a real spine of **H** yields a real spine of **K** and vice versa, the corresponding involutions being transposed of each other.

10.2 The case of $\mathbf{H}_1$: complexification versus regular representation

The reader will find in [42] (proofs in [48]) description of $\mathbf{H}_1$ and if its regular representation which splits as follows: $\mathbf{H}_1 = \mathbf{F}_1 \oplus \mathbf{F}_2 \oplus \mathbf{M} \oplus \mathbf{N}$, $\mathbf{F}_1, \mathbf{F}_2$ real subalgebras respectively isomorphic to $\mathbf{M}_1(\mathbb{C})$ and $\mathbf{M}_2(\mathbb{C})$, $\mathbf{M}$ a principal ideal isomorphic to $\mathbf{M}_3(\mathbb{C})$, $\mathbf{N}$ the 13-dimensional nilradical. **Proposition.**
One has $\mathbf{H}_1^{\mathrm{real}} = \mathbf{F}_1^{\mathrm{real}} \oplus \mathbf{F}_2^{\mathrm{real}} \oplus \mathbf{M}^{\mathrm{real}} \oplus \mathbf{N}^{\mathrm{real}}$, where:

- $\mathbf{F}_1^{\mathrm{real}} = \mathbb{R} f_0$ is a real sub-algebra of $\mathbf{H}_1$ isomorphic as such to $\mathbb{R}$;
- $\mathbf{F}_2^{\mathrm{real}}$ of real dimension 4 is a real sub-algebra of $\mathbf{H}_1$ isomorphic as such to $\mathbb{H}$, the algebra of quaternions;
- $\mathbf{M}^{\mathrm{real}}$ of real dimension 9 is a principal ideal of $\mathbf{H}_1^{\mathrm{real}}$ isomorphic to $\mathbf{M}_3(\mathbb{R})$ as an algebra;
- $\mathbf{N}^{\mathrm{real}}$ with real dimension 13 is the radical of $\mathbf{H}_1^{\mathrm{real}}$.

A Heat-Kernel Expansion

M is in what follows a d-dimensional smooth oriented compact manifold without boundary for which we use the notation of [**7**], in particular $\mathcal{A} = C^\infty(\mathbf{M})$.

- **V** is a smooth vector-bundle over **M**, with $\mathcal{A}$-module of smooth section **E**. The fiber-trace of **V** at $x \in \mathbf{M}$ is denoted by Tr_x;
- $\mathrm{Ell}_+^{>0}(\mathbf{E})$ denotes the set of elliptic pseudo-differential operators: $\mathbf{E} \to \mathbf{E}$ of positive order having a positive-definite principal symbol.

A.1 General expansion result

Let $P \in \mathrm{Ell}_+^{\geq 0}(\mathbf{E})$ be order $\delta > 0$. We recall that e^{-tP}, $t > 0$, is then a smoothing operator whose (thus smooth) kernel we denote by $K(t,x,y)$. The restriction of K to the diagonal has the following expansion for $t = 0$:

$$K(t,x,x) \cong \sum_{j=0}^{\infty} t^{(j-d)/\delta} e_j(x,P) \quad \text{with } e_j(x,P) \in \mathrm{End}(\mathbf{E}),\ x \in \mathbf{M} \tag{1.1}$$

in the following sense: given $k \in \mathbb{N}$, there is $n(k) \in \mathbb{N}$ and $C_k > 0$ with

$$\left|K(t,x,x) - \sum_{j=0}^{\infty} t^{(j-d)/\delta} e_j(x,P)\right|_{\infty,k} < C_k t^k \ , \tag{1.2}$$

(where $|f|_{\infty,k} = \sup_x \sum_{|\alpha|} \leq k\, |D^\alpha f|$) implying the expansion:

$$\begin{aligned} \mathrm{Tr}(e^{-tP}) = \textstyle\int_{\mathbf{M}} \mathrm{Tr}_x(K(t,x,x))dv &\cong \sum_{j=0}^{\infty} t^{(j-d)/\delta} \textstyle\int_{\mathbf{M}} a_j(x,P)dv \\ &\cong \sum_{i=0}^{\infty} t^{(2i-d)/\delta} a_{2i}(P) \ , \end{aligned} \tag{1.3}$$

where

$$a_j(x,P) = \mathrm{Tr}_x(e_j(x,P)) \ , \tag{1.4}$$

and the $a_j(x,P)$ vanish for j odd.

A.2 Properties of the $e_j(x,P)$

(i): With $\mathbf{E}$, $\mathbf{E}'$ over $\mathbf{M}$ acted upon by $P \in \mathrm{Ell}_+^{\geq 0}(\mathbf{E})$ resp. $P' \in \mathrm{Ell}_+^{\geq 0}(\mathbf{E}')$ we have $P \oplus P' \in \mathrm{Ell}_+^{\geq 0}(\mathbf{E} \oplus \mathbf{E}')$, and

$$e_j(x, P \oplus P') = e_j(x,P) \oplus e_j(x,P') \ . \tag{1.5}$$

A.3 Weyl's theorem

Proof for $P = -\left(\frac{d}{d\theta}\right)^2$ acting on S^1: the eigenvalues of P are the n^2 (with eigenvectors $\exp(in\theta)$), $n \in \mathbb{Z}$, such that:

$$\mathrm{Tr}(e^{-tP}) = \sum_{j=0}^{\infty} e^{-\mathrm{tn}^2} \cong \int_{-\infty}^{+\infty} e^{-tx^2} dx = t^{-1/2}\, \pi^{1/2} \ , \tag{1.6}$$

which, equated to

$$t^{(0-1)/2} \int a_0(\theta,P)d\theta = 2\pi t^{-1/2} a_0(x,P) \tag{1.7}$$

yields $a_0(\theta,P) = (4\pi t^-)^{1/2}$. In the general case we get by **[B]**(i),(ii) $a_0(\theta,P) = (4\pi t^-)^{d/2}$ rank $\mathbf{V}$.

A.4 Generalized Laplacians

For $\triangle \in \mathrm{Lapl}\mathbf{E}$ with horizontal connexion $\nabla^{\triangle}$ and canonical decomposition $\triangle = \triangle^{\nabla^{\triangle}} + E$ (cf. [B]) the heat expansion up to order 5 is given as follows:

$$a_0(x, \triangle) = (4\pi)^{-d/2}\mathrm{Tr}_x(\mathbb{1}) \qquad \text{(Weyl's Theorem)} . \tag{1.8}$$

$$a_2(x, \triangle) = (4\pi)^{-d/2}\mathrm{Tr}_x\left(-E + \frac{1}{6}\mathbf{R}^{\mu\nu}_{\mu\nu}\right) = (4\pi)^{-d/2}\mathrm{Tr}_x\left(\frac{1}{6}\mathbf{s} - E\right) , \tag{1.9}$$

$$\begin{aligned} a_4(x, \triangle) &= (4\pi)^{-d/2}\frac{1}{360}\mathrm{Tr}_x\Big\{12\mathbf{R}^{\mu\nu}_{\mu\nu;}{}^{\alpha}{}_{\alpha} + 5\mathbf{R}^{\mu\nu}_{\mu\nu}\mathbf{R}^{\alpha\beta}_{\alpha\beta} - 2\mathbf{R}^{\mu\nu}_{\mu\alpha}\mathbf{R}^{\beta\alpha}_{\beta\nu} \\ &\quad +2\mathbf{R}^{\mu\nu\alpha\beta}\mathbf{R}_{\mu\nu\alpha\beta} - 60\mathbf{R}^{\alpha\beta}_{\alpha\beta}E + 180E^2 - 60E_{;}{}^{\alpha}{}_{\alpha} + 30\Omega^{\mu\nu}\Omega_{\mu\nu}\Big\} \\ &= (4\pi)^{-d/2}\frac{1}{360}\mathrm{Tr}_x\left\{12\mathbf{s}_{;}{}^{\alpha}{}_{\alpha} + 5\mathbf{s}^2 - 2\mathbf{r}^{\nu}_{\alpha}\mathbf{r}^{\alpha}_{\nu}\right\} \\ &\quad +2\mathbf{R}^{\mu\nu\alpha\beta}\mathbf{R}_{\mu\nu\alpha\beta} - 60\mathbf{s}E + 180E^2 - 60E_{;}{}^{\alpha}{}_{\alpha} + 30\Omega_{\mu\nu}\Omega^{\mu\nu}\Big\} , \end{aligned} \tag{1.10}$$

where:

- $\mathbf{R}$ is the (Levi-Civita) Riemann-Christoffel tensor of $\mathbf{M}, \mathbf{r}$ the Ricci-tensor, $\mathbf{s}$ the scalar curvature;
- E is as stated above;
- Ω is the curvature-tensor (with values in End($\mathbf{E}$)) of the horizontal connexion $\nabla^{\triangle}$ of $\mathbf{V}$.

For the proof of (1.9) and (1.10) we refer to Gilkey's book [33], Theorem 4.1.6., p.336. Our E is Gilkey's $-E$, our $\mathbf{s}$ is Gilkey's τ.

B Generalized Laplacians

In what follows $\mathbf{M}$ is a Riemannian manifold with the notation in [7], in particular we write $C^{\infty}(\mathbf{M}, \mathbb{R}) = \mathcal{A}$. We denote by $\nabla^{\mathbf{M}}$ the Levi-Civita connection of $\mathbf{M}$. We first define generalized Laplacians and the more restricted connection-Laplacians.

B.1 Definitions

Let $\mathcal{E}$ be a smooth vector-bundle over $\mathbf{M}$, with $\mathcal{A}$-module of smooth sections $\mathbf{E}$. $\mathrm{Diff}^n\mathbf{E}$ denotes the set of differential operators of degree n on $\mathbf{E}$.
(i): A **generalized Laplacian if E** is a second-order differential operator H of $\mathbf{E}$ with principal symbol:

$$\sigma_2(H)(\mathbf{d}f) = -\frac{1}{2}[f, [f, H]] = -\frac{1}{2}[[H, f], f] = |\mathbf{d}f|^2, \quad f \in \mathcal{A} , \tag{2.1}$$

(implying by polarization via Jacobi identity:[40]

$$[[H,f],g]=[[H,g],f]=-2(\mathbf{d}f,\mathbf{d}g),\quad f,\,g\in \mathcal{A} . \tag{2.2}$$

Local formulation: $\triangle \in \mathrm{Diff}^2\mathbf{E}$ is a generalized Laplacian iff one has for the coordinate patch (x_μ) of $\mathbf{M}$ and the trivializing frame (e_i) of $\mathbf{E}$:

$$\triangle = -g^{\mu\nu}\partial_\mu\partial_\nu + \triangle_1 \qquad \text{with } \triangle_1 \in \mathrm{Diff}^1\mathbf{E} . \tag{2.3}$$

The set of generalized Laplacian of $\mathbf{E}$ is denoted Lapl$\mathbf{E}$.
(ii): With ∇ a connection of $\mathbf{E}$, the following composition of $\mathbb{R}$-linear maps:

$$\mathbf{E} \xrightarrow{\nabla} \Omega(\mathbf{M})^1 \otimes \mathbf{E} \xrightarrow{id\otimes\nabla + \nabla^M\otimes id} \Omega(\mathbf{M})^1\otimes\Omega(\mathbf{M})^1\otimes\mathbf{E} \xrightarrow{-g(.\,,.)\otimes id} \mathbf{E} , \tag{2.4}$$

defines the **connection-Laplacian** $\triangle^\nabla$ **of** ∇, locally given as follows in the coordinate patch $\{x_\mu\}$ of $\mathbf{M}$ and the trivializing frame $\{e_i\}$ of $\mathbf{E}$: with $\nabla_\alpha = \nabla_{\partial_\alpha}$, and $\Gamma^\alpha_{\mu\nu}$ with the Christoffel symbols:[41]

$$\triangle^\nabla = -g^{\mu\nu}(\nabla_\mu\nabla_\nu - \Gamma^\alpha_{\mu\nu}\nabla_\alpha) , \tag{2.5}$$

(implying $\triangle^\nabla \in$ Lapl$\mathbf{E}$). Note that (2.5) is the contraction

$$\triangle^\nabla = -g^{\mu\nu}\nabla_{\partial_\mu,\partial_\nu} , \tag{2.6}$$

where

$$\nabla_{X,Y} = \nabla_X\nabla_Y - \nabla_{\nabla^M_X Y},\quad X,\,Y\in\chi(\mathbf{M}) . \tag{2.7}$$

The set of connection-Laplacians of $\mathbf{E}$ is denoted Connlapl$\mathbf{E}$.
(iii): The **scalar Laplacian** $\triangle$ of $\mathbf{M}$ corresponds to the case $\mathbf{E} = \mathcal{A}$, with ∇ the flat connexion:

$$\begin{aligned}\triangle f = -g\left(\nabla^M_{(\cdot)}\mathbf{d}f(\cdot)\right) &= -g^{ij}\left(\nabla^M_{e_i}\mathbf{d}f(e_j)\right) = -g^{ij}\left(\left[\nabla^M_{e_i}\mathbf{d}f\right](e_j)\right)\\ &= -g^{ij}\left(\partial_i\partial_j - \Gamma^k_{ij}\partial_k\right)f,\quad f\in\mathcal{A} .\end{aligned} \tag{2.8}$$

(Note that one has

$$\begin{aligned}\left[\nabla^M_{\partial_i}\mathbf{d}f\right](\partial_j) = \left[\nabla^M_{\partial_i}\partial_k f\,dx^k\right](\partial_j) &= \partial_i\partial_k f\,dx^k(\partial_j) - \partial_k f\,\Gamma^k_{ij}dx^j\\ &= \partial_i\partial_j f - \partial_k f\,\Gamma^k_{ij}dx^j ,\end{aligned} \tag{2.9}$$

and, as an immediate consequence of (2.8):

$$\triangle(fg) = f\triangle(g) - 2(\mathbf{d}f,\mathbf{d}g) + g\triangle(f),\quad f,\,g\in\mathcal{A} . \tag{2.10}$$

[40] In what follows we consistently identify elements of $\mathcal{A}$ with their multiplicative action on $\mathbf{E}$.

[41] We check this below in the proof of [**B.2**].

B.2 Remarks

(i) The coherence of definition (2.4) is a consequence of the $\mathcal{A}$-linearity of the tensor product of connexions. Alternatively it follows from the "tensor property":

$$\nabla_{fX,gY} = fg\nabla_{X,Y}, \quad X,\, Y \in \chi(\mathbf{M}), \quad f,\, g \in \mathcal{A} \ , \tag{2.11}$$

resulting from:

$$\begin{cases} \nabla_{fX}\nabla_{gY} - fg\nabla_X\nabla_Y = f(Xg)\nabla_Y \\ \nabla^M_{fX}gY - fg\nabla^M_X Y = f(Xg)Y \\ \nabla_{\nabla^M_{fX}gY} - fg\nabla_{\nabla^M_X Y} = f(Xg)\nabla_Y \end{cases} \quad X,\, Y \in \chi(\mathbf{M}), \quad f,\, g \in \mathcal{A} \ . \tag{2.12}$$

(ii): With $\nabla^+_{X,Y} = \frac{1}{2}(\nabla_X\nabla_Y + \nabla_Y\nabla_X)$ and $\nabla^-_{X,Y} = \frac{1}{2}(\nabla_X\nabla_Y - \nabla_Y\nabla_X)$ one has:

$$\Delta^\nabla = -g^{\mu\nu}\nabla^+_{\partial_\mu,\partial_\nu} \ , \tag{2.13}$$

whilst

$$\nabla^-_{X,Y} = \frac{1}{2}R_{X,Y} = [\nabla_X, \nabla_Y] - \nabla_{[X,Y]} \ . \tag{2.14}$$

(iii): One has:

$$-i\sigma_1(\mathbf{d}v) = [v, \Delta^\nabla] = 2\nabla_{\mathrm{gradv}} - \Delta v, \quad v \in \mathcal{A} \ , \tag{2.15}$$

i.e. Δ^∇ determines in turn ∇ as follows:

$$\nabla_{\mathrm{ugradv}} = \frac{1}{2}u\left\{[v, \Delta^\nabla] + \Delta v\right\} \ ; \tag{2.16}$$

thus *the connection and the connection-Laplacian are one-to-one*: *we have a bijection*: $\mathrm{Conn}\mathbf{E} \ni \nabla \leftrightarrow \Delta^\nabla \in \mathrm{Connlapl}\mathbf{E}$.

Proof: Check of (2.5): with $\eta \in \mathbf{E}$ one has $\nabla\eta = \mathbf{d}x^\mu \otimes \nabla_\mu\eta$, thus, using the fact that $\nabla^M \mathbf{d}x^\alpha = -\Gamma^\alpha_{\mu\nu}\mathbf{d}x^\mu \otimes \mathbf{d}x^\nu$:

$$\left(\nabla^M \otimes id + id \otimes \nabla\right)\nabla\eta = \mathbf{d}x^\mu \otimes \mathbf{d}x^\mu \otimes \nabla_\mu\nabla_\nu\eta - \Gamma^\alpha_{\mu\nu}\mathbf{d}x^\mu \otimes \mathbf{d}x^\nu \otimes \nabla_\alpha\eta \ , \tag{2.17}$$

whence (2.5), applying $-g(.,.) \otimes id$.

Check of (i): (2.11) follows from (2.12), checked as follows: we have, using the derivation rules $\nabla_X g = g\nabla_X + Xg$, $\nabla^M_X g = g\nabla^M_Y + Xg$,:

$$\begin{cases} \nabla_{fX}\nabla_{gY} = f\nabla_X g\nabla_Y = fg\nabla_X\nabla_Y + f(Xg)\nabla_Y \\ \nabla^M_{fX}gY = f\nabla^M_X gY = fg\nabla^M_X Y + f(Xg)Y \\ \nabla_{\nabla^M_{fX}gY} = \nabla_{\nabla^M_X Y + f(Xg)Y} = fg\nabla_{\nabla^M_X Y} + f(Xg)\nabla_Y \end{cases} \ . \tag{2.18}$$

Check of (ii): (2.13) holds owing to $g^{\mu\nu} = g^{\nu\mu}$. Check of (2.14): we have, since the torsion $\nabla^M_X Y - \nabla^M_Y X - [X,Y]$ vanishes:

$$\begin{aligned} \nabla^-_{X,Y} &= \frac{1}{2}(\nabla_X\nabla_Y - \nabla_Y\nabla_X) + \frac{1}{2}\nabla_{\nabla^M_X Y} - \nabla^M_X Y \\ &= \frac{1}{2}\nabla_{[X,Y]} + \frac{1}{2}R_{X,Y} - \frac{1}{2}\nabla_{[X,Y]} = \frac{1}{2}R_{X,Y} \end{aligned} \ . \tag{2.19}$$

Check of (iii): we have:

$$\begin{aligned}[v, \triangle^{\nabla}] &= \left[g^{\mu\nu}\left(\nabla_\mu\nabla_\nu - \Gamma^\alpha_{\mu\nu}\nabla_\alpha\right), v\right] = g^{\mu\nu}\left\{[\nabla_\mu\nabla_\nu, v] - \Gamma^\alpha_{\mu\nu}[\nabla_\alpha, v]\right\} \\ &= g^{\mu\nu}\left\{\nabla_\mu[\nabla_\nu, v] + [\nabla_\mu, v]\nabla_\nu - \Gamma^\alpha_{\mu\nu}[\nabla_\alpha, v]\right\} \\ &= g^{\mu\nu}\left\{\nabla_\mu\partial_\nu v + \partial_\mu v\nabla_\nu - \Gamma^\alpha_{\mu\nu}\partial_\alpha v\right\} \\ &= g^{\mu\nu}\left\{\partial_\nu v\nabla_\mu + \partial_\mu v\nabla_\nu + \partial_\mu\partial_\nu v - \Gamma^\alpha_{\mu\nu}\partial_\alpha v\right\} \\ &= 2g^{\mu\nu}\partial_\mu v\nabla_\nu - \triangle v = (\mathrm{gradv})^\nu\nabla_\nu - \triangle v = 2\nabla_{\mathrm{gradv}} - \triangle v\end{aligned} \quad .$$

(2.20)

We now show that besides the above bijection between connections and connection-Laplacians (cf. [**B.2**] we have a bijection between connections and generalized Laplacians modulo endomorphisms.

B.3 Proposition-Definition

Let $\mathcal{E}$ be a smooth vector-bundle over $\mathbf{M}$,*with* $\mathcal{A}$*-module of smooth section* $\mathbf{E}$*. And let* $H \in \mathrm{Lapl}\mathbf{E}$.*Then, identifying elements of* $\mathcal{A}$ *with their multiplicative action on* $\mathbf{E}$:

(i): *H determines both:*

- *a connection ∇^H of* $\mathbf{E}$ *called* **the horizontal connection of** *H, specified by:*

$$\nabla^H_{\mathrm{ugradv}} = \frac{1}{2}u\{[v, H] + \triangle v\} \left(= \frac{1}{2}u\{-i\sigma_2(H)(\mathbf{d}v) + \triangle v\}\right), \quad u,\, v \in \mathcal{A} \ , \tag{B.20}$$

where $\triangle$ denotes the scalar Laplacian;

- *an element $\Phi^H \in End\mathcal{A}$ called* **the endomorphism of** *H, given by the difference:*

$$\Phi^H = H - \triangle^{\nabla^H} \quad . \tag{2.21}$$

(ii): *In fact the splitting $H = \triangle^{\nabla^H} + \Phi^H$ is unique: one has the implication:*

$$H = \triangle + \Phi \text{ with } \triangle \in \mathrm{Connlapl}\mathbf{E},\ \Phi \in \mathrm{End}\mathcal{A} \longrightarrow \quad \triangle = \triangle^{\nabla^H},\ \Phi = \Phi^H \ . \tag{2.22}$$

(iii): *Consequently* $\mathrm{Lap}\mathbf{E}$ *is a fiber with basis the bijectively related* $\mathrm{Conn}\mathbf{E} \approx \mathrm{Connlapl}\mathbf{E} \approx \sigma_2(\mathrm{Lap}\mathbf{E})$, *and fiber* $\mathrm{End}\mathcal{A}$ *acting on* $\mathrm{Lap}\mathbf{E}$ *by translation*: $\mathrm{Lap}\mathbf{E} \ni \triangle \to \triangle + \Phi$.

For the proof we need the following result which we proved in (cf. [**2**]).

B.4 Lemma

Let $\mathbf{E}$ *be a projective-finite* $\mathcal{A}$*-module, and let ϕ be a* $\mathbf{C}$*-linear map:* $\mathcal{A} \to \mathrm{End}_{\mathbf{C}}\mathbf{E}$ *fulfilling:*

$$\phi(fg) = f\phi(g) + g\phi(f), \quad f,\, g \in \mathcal{A} \ . \tag{2.23}$$

Then with $\chi(\mathbf{M})$ *the Lee algebra of vectors fields on* $\mathbf{M}$: (i): *there is a:* $\mathcal{A}$*-linear map* $\Phi : \chi(\mathbf{M}) \ni X \to \Phi_X \in \mathrm{End}_{\mathbf{C}}\, \mathbf{E}$:

$$\phi(f) = \Phi_{\mathrm{gradf}}, \quad f \in \mathcal{A}, \quad (\text{hence} \quad g\phi(f) = \Phi_{\mathrm{gradf}}, \quad f \in \mathcal{A}) \; ; \tag{2.24}$$

(ii): *moreover*:

- Φ_X *is a* X*-derivation of* $\mathbf{E}$ *for all* $X \in \chi(\mathbf{M})$ (i.e. Φ is a connection of $\mathbf{E}$) *iff* ϕ *fulfills*:
$$[\phi(f), a] = (\mathbf{d}f, \mathbf{d}a), \quad f,\ a \in \mathcal{A} \; ; \tag{2.25}$$
- Φ_X *is* $\mathcal{A}$*-linear for all* $X \in \chi(\mathbf{M})$ (i.e. $\Phi_X \in \mathrm{End}_{\mathcal{A}}\mathbf{E}$) (*iff* ϕ *is* $\mathcal{A}$*-linear*):
$$[\phi(f), a] = 0 \; . \tag{2.26}$$

Proof of [B.3]:
(i): With $\phi(u) = \frac{1}{2}\{\Delta u - [H, u]\}$ $u \in \mathcal{A}$, the preceding Lemma reduces the proof to checking the properties (2.24) and (2.25). Now we have, using (2.2):

$$\begin{aligned} [H, uv] &= u[H, v] + [H, u]v = u[H, v] + v[H, u] + [[H, u], v] && (2.27)\\ &= u[H, v] + v[H, u] - 2(\mathbf{d}u, \mathbf{d}f),\ u,\ v \in \mathcal{A}\,, && (2.28) \end{aligned}$$

whilst (2.29) said that:

$$\Delta(uv) = u \,\Delta\, (v) + v \,\Delta\, (u) - 2(\mathbf{d}u, \mathbf{d}v) \; , \tag{2.29}$$

property (2.19) then follows by difference. As for property (2.25), since multiplication by Δu is $\mathcal{A}$-linear, it boils down to (2.2). Now H and Δ^{∇^H} have the same horizontal connection ∇^H, thus $\sigma_2(H) = \sigma_2(\Delta^{\nabla^H})$, and the difference $H - \Delta^{\nabla^H}$ commutes with every $f \in \mathcal{A}$ and thus lies in $\mathrm{End}\mathcal{A}$.
(ii): $H = \Delta + \Phi = \Delta^{\nabla^H} + \Phi^H$ implies that the connexion-Laplacians Δ and Δ^{∇^H} have the same horizontal connection: they must thus coincide by [**B.2**](iii).

B.5 Proposition (local description of [B.3])

In the coordinate patch $\{x_\mu\}$ where

$$H = -g^{\mu\nu}\partial_\mu\partial_\nu + B^\mu\partial_\mu + C \; , \tag{2.30}$$

we have

$$\begin{aligned} \frac{1}{2}\{\Delta v \; - \; [H, v]\} &= (g^{\mu\nu}\partial_\mu v)\,\partial_\nu\psi + \frac{1}{2}\left[(g^{\mu\nu}\Gamma^\alpha_{\mu\nu} - B^\mu)\,\partial_\mu v\right]\psi \\ &= (\mathrm{dradv})^\mu\partial_\mu\psi + \frac{1}{2}\left[(g^{\alpha\beta}\Gamma_{\alpha\beta,\mu} - B_\mu)\,(\mathrm{gradv})^\mu\right]\psi,\ \phi \in \mathbf{E} \; , \end{aligned} \tag{2.31}$$

thus the canonical decomposition $H = \Delta^{\nabla^H} + \Phi^H$ with:

$$\nabla^H_\mu = \partial_\mu + \Phi_\mu \ \text{ with } \Phi_\mu = \frac{1}{2}\left(g^{\alpha\beta}\Gamma_{\alpha\beta,\mu} - B_\mu\right), \quad \Phi^\mu = -\frac{1}{2}\left(\Gamma^\mu_\alpha\beta - B^\mu\right) \; , \tag{2.32}$$

and:

$$\Phi^H = C + g^{\mu\nu}\left[\partial_\mu \Phi_\nu + \Phi_\mu \Phi_\nu - g^{\mu\nu}\Gamma^\alpha_{\mu\nu}\Phi_\alpha\right] \ . \tag{2.33}$$

Proof

With $H = -g^{\mu\nu}\partial_\mu\partial_\nu + B^\mu\partial_\mu + C$ we have, for $\psi \in \mathbf{E}$:

$$\begin{aligned}
[H, v]\,\psi &= [-g^{\mu\nu}\partial_\mu\partial_\nu + B^\mu\partial_\mu v]\,\psi \\
&= v g^{\mu\nu}\partial_\mu\partial_\nu\psi - g^{\mu\nu}\partial_\mu\partial_\nu(v\psi) + B^\mu\partial_\mu(v\psi) - vB^\mu\partial_\mu\psi \\
&= g^{\mu\nu}[v\partial_\mu\partial_\nu\psi - \partial_\mu((\partial_\nu v)\psi + v\partial_\nu\psi)] + (B^\mu\partial_\mu v)\psi \\
&= g^{\mu\nu}[v\partial_\mu\partial_\nu\psi - (\partial_\mu\partial_\nu v)\psi - (\partial_\nu v)\partial_\mu\psi - (\partial_\mu v)\partial_\nu\psi - v\partial_\mu\partial_\nu\psi] + (B^\mu\partial_\mu v)\psi \\
&= -g^{\mu\nu}[(\partial_\mu\partial_\nu v)\psi + 2(\partial_\mu v)\partial_\nu\psi] + (B^\mu\partial_\mu v)\psi \ ,
\end{aligned} \tag{2.34}$$

whilst, by (2.8): $(\Delta v)\psi = -g^{\mu\nu}\left(\partial_\mu\partial_\nu v - \Gamma^\alpha_{\mu\nu}\partial_\alpha v\right)\psi$, whence (2.32) by difference. Since $\nabla^H_\mu = \nabla^H_{\partial_\mu}$, we then obtain by choosing v such that

$$\text{gradv} = \partial_\mu \quad \Leftrightarrow \quad (\text{gradv})^\nu_\mu = \delta^\nu_\mu \quad \Leftrightarrow \quad dv = g\partial_\mu = g_{\sigma\mu}dx^\sigma \ . \tag{2.35}$$

Check of (2.18): we now have reminding that $\Phi^\mu = -\dfrac{1}{2}\left(\Gamma^\mu_{\alpha\beta} - B^\mu\right)$:

$$\begin{aligned}
\Delta^{\nabla^H} &= -g^{\mu\nu}\left(\nabla^H_\mu\nabla^H_\nu - \Gamma^\alpha_{\mu\nu}\nabla^H_\alpha\right) = -g^{\mu\nu}[(\partial_\mu + \Phi_\mu)(\partial_\nu + \Phi_\nu) - \Gamma^\alpha_{\mu\nu}(\partial_\alpha + \Phi_\alpha)] \\
&= -g^{\mu\nu}\left[\partial_\mu\partial_\nu + \partial_\mu\Phi_\nu + \Phi_\nu\partial_\mu + \Phi_\mu\partial_\nu + \Phi_\mu\Phi_\nu - \Gamma^\alpha_{\mu\nu}\partial_\alpha - \Gamma^\alpha_{\mu\nu}\Phi_\alpha\right] \\
&= -g^{\mu\nu}\partial_\mu\partial_\nu + \left(g^{\alpha\beta}\Gamma^\mu_{\alpha\beta} - 2\Phi^\mu\right)\partial_\mu - g^{\mu\nu}[\partial_\mu\Phi_\nu + \Phi_\mu\Phi_\nu - g^{\mu\nu}\Gamma^\alpha_{\mu\nu}\Phi_\alpha] \\
&= -g^{\mu\nu}\partial_\mu\partial_\nu + B^\mu\partial_\mu - g^{\mu\nu}[\partial_\mu\Phi_\nu + \Phi_\mu\Phi_\nu - g^{\mu\nu}\Gamma^\alpha_{\mu\nu}\Phi_\alpha] \\
&= H - C - g^{\mu\nu}[\partial_\mu\Phi_\nu + \Phi_\mu\Phi_\nu - g^{\mu\nu}\Gamma^\alpha_{\mu\nu}\Phi_\alpha] \ .
\end{aligned} \tag{2.36}$$

C Clifford Modules. Clifford Connections. Bochner and Lichnerowicz Formulae

C.1

A **Clifford module** $(\mathbb{E}, c)$ over a Riemannian spin manifold $\mathbf{M}$ is a finite-projective $\mathbb{Z}/2$-graded $C^\infty(\mathbf{M})$-module which is a graded $\mathbb{C}1(\mathbf{M})$-module with $\mathbb{C}1(\mathbf{M})$-action c ($C^\infty(\mathbf{M})$ is considered as the zero-grade part of $\mathbb{C}l(\mathbf{M})$).[42] A **Clifford connection** $\mathbb{V}$ of the Clifford module $(\mathbb{E},c)$ is a connection $\mathbb{V}$ of $\mathbb{E}$ which: (a): is even and (b): fulfills:

$$[\mathbb{V}_\mu, c(\gamma(\lambda))] = c(\gamma\left(\nabla^{\mathbf{M}}_\mu\lambda\right)), \qquad \lambda \in \Omega(\mathbf{M})^1 \ . \tag{3.1}$$

The corresponding Dirac operator $D^{\mathbb{V}}$ is then defined by the local expression

$$D^{\mathbb{V}} = ic^\mu\mathbb{V}\mu, \ c^\mu = \gamma\left(\mathbf{d}x^\mu\right), \ \mathbb{V}_\mu = \mathbb{V}_{\partial_\mu} \ .$$

[42] Clifford modules are the modules of sections of **Clifford bundles**, the two notions being equivalent.

Examples:
- the module of section $\mathbb{S}(\mathbf{M})$ of the spin bundle of $\mathbf{M}$ acted upon by $\mathbb{C}l(\mathbf{M})$. The spin connexion is then a Clifford connection:

$$[\tilde{\nabla}_\mu, \gamma(\lambda)] = \gamma\left(\nabla^{\mathbf{M}}_\mu \lambda\right), \qquad \lambda \in \Omega(\mathbf{M})^1 \ ; \tag{3.2}$$

- given a $\mathbb{Z}/2$-graded $C^\infty(\mathbf{M})$-module $\mathbf{E}$, the **twisted module** $\mathbb{S}(\mathbf{M}) \otimes_{C^\infty(\mathbf{M})} \mathbf{E}$ (by the twisting module $\mathbf{E}$), equipped with the tensor-product $\mathbb{Z}/2$-grading and the Clifford action:

$$c(\gamma(\lambda)) = \gamma(\lambda) \otimes id_{\mathbf{E}},\ \gamma(\lambda) \in \mathbb{C}l(\mathbf{M}),\ \lambda \in \Omega(\mathbf{M}) \ . \tag{3.3}$$

With $\nabla^{\mathbf{E}}$ an even connexion of $\mathbf{E}$, one the gets the Clifford **compound connexion**:

$$\mathbb{V} = \tilde{\nabla} \otimes id_{\mathbf{E}} + id_{\mathbb{S}(\mathbf{M})} \otimes \nabla^{\mathbf{E}} : \quad \mathbb{V}_\xi = \tilde{\nabla}_\xi \otimes id_{\mathbf{E}} + id_{\mathbb{S}(\mathbf{M})} \otimes \nabla^{\mathbf{E}}\xi \tag{3.4}$$

for each vector field ξ.

C.2

Let $D^{\mathbb{V}} = ic^\mu \mathbb{V}\mu$ be the Dirac operator on the Clifford bundle $(\mathbb{E}, c)$ associated with a Clifford connection $\mathbb{V}$ with curvature $\mathbb{R}$: one has the **Bochner formula**:

$$\mathbb{D}^2 = \Delta^{\mathbb{V}} - \frac{1}{2}c(\mathbb{R}), \quad \text{where} \quad c(\mathbb{R}) = c^\mu c^\nu \mathbb{R}_{\mu\nu} \ . \tag{3.5}$$

Proof: one has

$$\begin{aligned}
\mathbb{D}^2 &= (ic^\mu \mathbb{V}_\mu)(ic^\nu \mathbb{V}_\nu) = -c^\mu c^\nu \mathbb{V}_\mu \mathbb{V}_\nu - c^\mu [\mathbb{V}_\mu, c^\nu] \mathbb{V}_\nu \\
&= -c^\mu c^\nu \mathbb{V}_\mu \mathbb{V}_\nu + \Gamma^\nu_{\mu\alpha} c^\mu c^\alpha \mathbb{V}_\mu \\
&= -\frac{1}{2}\left(c^\mu c^\nu + c^\mu c^\nu\right) \mathbb{V}_\mu \mathbb{V}_\nu - \frac{1}{2}\left(c^\mu c^\nu - c^\mu c^\nu\right) \mathbb{V}_\mu \mathbb{V}_\nu \\
&+ \frac{1}{2}\left(c^\mu c^\nu + c^\mu c^\nu\right) \Gamma^\alpha_{\mu\nu} \mathbb{V}_\mu \\
&= -g^{\mu\nu}\left(\mathbb{V}_\mu \mathbb{V}_\nu - \Gamma^\nu_{\mu\alpha} \mathbb{V}_\alpha\right) - \frac{1}{2} c^\mu c^\nu \left(\mathbb{V}_\mu \mathbb{V}_\nu - \mathbb{V}_\nu \mathbb{V}_\mu\right) \\
&= -g^{\mu\nu}\left(\mathbb{V}_\mu \mathbb{V}_\nu - \Gamma^\alpha_{\mu\nu} \mathbb{V}_\alpha\right) - \frac{1}{2} c^\mu c^\nu \mathbb{R}\left(\partial_\mu, \partial_\nu\right) \ ,
\end{aligned} \tag{3.6}$$

where we took account of the facts that $[\partial_\mu, \partial_\nu] = 0$, and

$$[\mathbb{V}_\mu, c^\nu] = (c \circ \gamma)(\nabla^{\mathbf{M}}_\mu dx^\nu) = -(c \circ \gamma)(\Gamma^\nu_{\mu\alpha} dx^\alpha) = -\Gamma^\nu_{\mu\alpha} c^\alpha \ . \tag{3.7}$$

C.3

For the Dirac operator $\mathbb{D}^{\mathbb{V}}$, acting on the twisted bundle $\mathcal{S}(\mathbf{M})\otimes\mathbf{E}$ associated to the compound connection $\mathbb{V}=\tilde{\nabla}\otimes id_{\mathbf{E}}+id_{\mathcal{S}(\mathbf{M})}\otimes\nabla^{\mathbf{E}}$, $\nabla^{\mathbf{E}}$ a connection of $\mathbf{E}$ with curvature $\mathbf{R}$ we have the **Lichnerowicz formula:**

$$\mathbb{D}^2=\Delta^{\mathbb{V}}+\frac{1}{4}\mathbf{s}\mathbb{I}-\frac{1}{2}c\left(R^{\mathbf{E}}\right)\qquad\text{with } c(R^{\mathbf{E}})=\gamma^i\gamma^j\ ,\otimes R^{\mathbf{E}}(e_i,\varepsilon_j)\tag{3.8}$$

where $\mathbf{s}$ is the scalar curvature, and $\{e_i,e^j\}_{i=1,\dots,\mathrm{d}}$ a local orthonormal frame.
Proof: Plugging $\mathbb{R}=R^{\mathbf{E}}+\tilde{R}$ in (3.5), we get:

$$\begin{aligned}\mathbb{D}^2&=\Delta-\frac{1}{2}c\left(\mathbb{R}\right)=\Delta-\frac{1}{2}c\left(R^{\mathbf{E}}\right)-\frac{1}{2}c\left(\tilde{R}\right)\\&=\Delta+\frac{1}{2}c\left(F^{\mathbf{E}}\right)-\frac{1}{8}R_{ijmn}c^mc^nc^ic^j\ ,\end{aligned}\tag{3.9}$$

taking account of the fact that:

$$R_{mn}=\frac{1}{2}R_{ijmn}\varepsilon^i\wedge\varepsilon^j=R_{ijmn}(\varepsilon^i\otimes\varepsilon^j-\varepsilon^j\otimes\varepsilon^i)=R_{ijmn}\varepsilon^i\otimes\varepsilon^j\ ,\tag{3.10}$$

whence $c\left(R_{mn}\right)=R_{ijmn}c^ic^j$, whence $c\left(\tilde{R}\right)=\frac{1}{4}R_{ijmn}c^mc^nc^ic^j$. Now, owing to the orthonormality of the ε^i, and to the Clifford relation we have that:[43]

$$c^mc^nc^i=\frac{1}{6}\sum_{\sigma\in S_3}\chi(\sigma)c^{\sigma m}c^{\sigma n}c^{\sigma i}-\delta^{ni}c^m+\delta^{mi}c^n-\delta^{mn}c^i\ ,\tag{3.11}$$

implying

$$\begin{aligned}&R_{ijmn}c^mc^nc^ic^j=R_{mnij}c^mc^nc^ic^j=R_{mnij}[-\delta^{ni}c^m+\delta^{mi}c^n]c^j\\&=2R_{mnij}\delta^{mi}c^nc^J=R_{mnij}\delta^{mi}(c^nc^j+c^jc^n)\\&=-2R_{mnij}g^{mi}g^{nj}=2R^{ij}_{ij}=-2\mathbf{s}\ ,\end{aligned}\tag{3.12}$$

which turns (3.9) into (3.8) (we took account of the relations $R_{mnij}+R_{nimj}+Rimnj=0$ and $R_{mnij}+R_{nmij}=0$ making the first, resp. the last term of (3.12) ineffective; and also of the fact that R_{inij} is symmetric in n and j).

D Weyl Tensor

$\mathbf{M}$ is in what follows a d-dimensional smooth manifold for which we use the notation of [**4.3**].

[43] One verifies that this holds (i): for m, n, i all different (r.h.s. reducing to its first term); (ii): for $m\neq n=i$ (r.h.s. reducing to its second term); (iii): for $n\neq m=i$ (r.h.s. reducing to its third term); (iv): for $i\neq m=n$ (r.h.s. reducing to its fourth term).

D.1 Definition

Denoting respectively by $\mathbf{R}$, $\mathbf{r}$ and $\mathbf{s}$ the respective Riemann-Christoffel tensor, Ricci-tensor, and scalar curvature; and with

$$\varepsilon^{\mu\nu}_{\alpha\beta} = \delta^{\mu}_{\alpha}\delta^{\nu}_{\beta} - \delta^{\nu}_{\alpha}\delta^{\mu}_{\beta} \quad \text{i.e.} \quad \varepsilon_{\mu\nu\alpha\beta} = g_{\mu\alpha}g_{\nu\beta} - g_{\nu\alpha}g_{\mu\beta} \ , \tag{4.1}$$

we define the **Weyl tensor C** as follows:

$$\begin{aligned} &\mathbf{C}^{\mu\nu}_{\alpha\beta} = \mathbf{R}^{\mu\nu}_{\alpha\beta} + A\eta^{\mu\nu}_{\alpha\beta} + B\mathbf{s}\varepsilon^{\mu\nu}_{\alpha\beta} \\ &\text{with } A = -\frac{1}{d-2},\ B = \frac{1}{(d-1)(d-2)} \ , \end{aligned} \tag{4.2}$$

i.e.

$$\mathbf{C}_{\mu\nu\alpha\beta} = \mathbf{R}_{\mu\nu\alpha\beta} + A\eta_{\mu\nu\alpha\beta} + B\mathbf{s}\varepsilon_{\mu\nu\alpha\beta} \ , \tag{4.3}$$

where

$$\eta^{\mu\nu}_{\alpha\beta} = \mathbf{r}^{\mu}_{\alpha}\delta^{\nu}_{\beta} - \mathbf{r}^{\nu}_{\alpha}\delta^{\mu}_{\beta} + \mathbf{r}^{\nu}_{\beta}\delta^{\mu}_{\alpha} - \mathbf{r}^{\mu}_{\beta}\delta^{\nu}_{\alpha} \ , \tag{4.4}$$

i.e.

$$\eta_{\mu\nu\alpha\beta} = \mathbf{r}_{\mu\alpha}g_{\nu\beta} - \mathbf{r}_{\nu\alpha}g_{\mu\beta} + \mathbf{r}_{\nu\beta}g_{\mu\alpha} - \mathbf{r}_{\mu\beta}g_{\nu\alpha} \tag{4.5}$$

D.2 Lemma

The Weyl tensor $\mathbf{C}$ has the same symmetry properties as $\mathbf{R}$:

$$\begin{cases} \mathbf{C}_{\mu\nu\alpha\beta} = -\mathbf{C}_{\nu\mu\alpha\beta} = \mathbf{C}_{\mu\nu\beta\alpha} = \mathbf{C}_{\alpha\beta\mu\nu} \\ \mathbf{C}_{\mu\nu\alpha\beta} + \mathbf{C}_{\mu\alpha\beta\nu} + \mathbf{C}_{\mu\beta\nu\alpha} = 0 \end{cases} \tag{4.6}$$

Proof: The symmetry properties are obvious for the first and last term r.h.s. of (4.3). And we have, on the one hand:

$$\begin{aligned} \eta^{\mu\nu\alpha\beta} &= \mathbf{r}_{\mu\alpha}g_{\nu\beta} - \mathbf{r}_{\nu\alpha}g_{\mu\beta} + \mathbf{r}_{\nu\beta}g_{\mu\alpha} - \mathbf{r}_{\mu\beta}g_{\nu\alpha} \\ &= \mathbf{r}_{\mu\alpha}g_{\nu\beta} - \mathbf{r}_{\nu\alpha}g_{\mu\beta} \quad - (\beta \leftrightarrow \nu) && (4.7) \\ &= \mathbf{r}_{\mu\alpha}g_{\nu\beta} - \mathbf{r}_{\nu\alpha}g_{\mu\beta} \quad - (\alpha \leftrightarrow \beta) && (4.8) \\ &= \mathbf{r}_{\mu\alpha}g_{\nu\beta} - \mathbf{r}_{\nu\alpha}g_{\mu\beta} \quad (\mu,\alpha \leftrightarrow \nu,\beta) \ ; && (4.9) \end{aligned}$$

and on the other:

$$\begin{aligned} \eta_{\mu\nu\alpha\beta} + \eta_{\mu\alpha\beta\nu} + \eta_{\mu\beta\nu\alpha} &= \mathbf{r}_{\mu\alpha}g_{\nu\beta} - \mathbf{r}_{\nu\alpha}g_{\mu\beta} + \mathbf{r}_{\nu\beta}g_{\mu\alpha} - \mathbf{r}_{\mu\beta}g_{\nu\alpha} \\ &= \mathbf{r}_{\mu\beta}g_{\alpha\nu} - \mathbf{r}_{\alpha\beta}g_{\mu\nu} + \mathbf{r}_{\alpha\nu}g_{\mu\beta} - \mathbf{r}_{\mu\nu}g_{\alpha\beta} \\ &= \mathbf{r}_{\mu\nu}g_{\beta\alpha} - \mathbf{r}_{\beta\nu}g_{\mu\alpha} + \mathbf{r}_{\beta\alpha}g_{\mu\nu} - \mathbf{r}_{\mu\alpha}g_{\beta\nu} = 0 \end{aligned} \tag{4.10}$$

D.3 Lemma (square of the Weyl tensor).

We have, with $\mathbf{R}^2 = \mathbf{R}_{\mu\nu\alpha\beta}\mathbf{R}^{\mu\nu\alpha\beta}$, $\mathbf{r}^2 = \mathbf{r}^\mu_\nu \mathbf{r}^\nu_\mu$:

$$\mathbf{C}^2 = \mathbf{C}_{\mu\nu\alpha\beta}\mathbf{C}^{\mu\nu\alpha\beta} = \mathbf{R}^2 + 8A(A+1)\mathbf{r}^2 + 4(A^2 + B + 12AB + 12B^2)\mathbf{s}^2 \ . \tag{4.11}$$

In particular, for d=4, we have:

$$\mathbf{C}^2 = \mathbf{R}^2 - 2\mathbf{r}^2 + \frac{1}{3}\mathbf{s}^2 \ . \tag{4.12}$$

Proof:
We have for any 4-tensor T:

$$\varepsilon_{\mu\nu\alpha\beta}T^{\mu\nu\alpha\beta} = (g_{\mu\alpha}g_{\nu\beta} - g_{\nu\alpha}g_{\mu\beta})T^{\mu\nu\alpha\beta} = T^{\mu\nu}_{\mu\nu} - T^{\mu\nu}_{\nu\mu} \ ; \tag{4.13}$$

hence

$$\varepsilon_{\mu\nu\alpha\beta}\varepsilon^{\mu\nu\alpha\beta} = \varepsilon^{\mu\nu}_{\mu\nu} - \varepsilon^{\mu\nu}_{\nu\mu} = 2\varepsilon^{\mu\nu}_{\mu\nu} = 2(\delta^\mu_\mu\delta^\nu_\nu - \delta^\nu_\mu\delta^\mu_\nu) = 2(\mathrm{d}\cdot\mathrm{d} - \mathrm{d}) = 2\mathrm{d}(\mathrm{d}-1) \ , \tag{4.14}$$

$$\varepsilon_{\mu\nu\alpha\beta}\mathbf{R}^{\mu\nu\alpha\beta} = \mathbf{R}^{\mu\nu}_{\mu\nu} = 2\mathbf{s} \ , \tag{4.15}$$

and

$$\begin{aligned}\varepsilon_{\mu\nu\alpha\beta}\mathbf{R}^{\mu\nu\alpha\beta} &= 2\eta^{\mu\mu}_{\mu\nu} = 2(\mathbf{r}^\mu_\nu\delta^\nu_\nu - \mathbf{r}^\nu_\mu\delta^\mu_\nu + \mathbf{r}^\nu_\nu\delta^\mu_\mu - \mathbf{r}^\mu_\nu\delta^\nu_\nu) \\ &= 2(4\mathbf{s} - \mathbf{s} + 4\mathbf{s} - \mathbf{s}) = 12\mathbf{s} \ .\end{aligned} \tag{4.16}$$

On the other hand we have:

$$\begin{aligned}&\boldsymbol{\eta}_{\mu\nu\alpha\beta}\boldsymbol{\eta}^{\mu\nu\alpha\beta} = \eta^{\mu\nu}_{\alpha\beta}\eta^{\alpha\beta}_{\mu\nu} \\ &= [\mathbf{r}^\mu_\alpha\delta^\nu_\beta - \mathbf{r}^\nu_\alpha\delta^\mu_\beta + \mathbf{r}^\nu_\beta\delta^\mu_\alpha - \mathbf{r}^\mu_\beta\delta^\nu_\alpha][\mathbf{r}^\alpha_\mu\delta^\beta_\nu - \mathbf{r}^\alpha_\nu\delta^\beta_\mu + \mathbf{r}^\beta_\nu\delta^\alpha_\mu - \mathbf{r}^\beta_\mu\delta^\alpha_\nu] \\ &= \mathbf{r}^\mu_\alpha\delta^\nu_\beta[\mathbf{r}^\alpha_\mu\delta^\beta_\nu - \mathbf{r}^\alpha_\nu\delta^\beta_\mu + \mathbf{r}^\beta_\mu\delta^\alpha_\mu - \mathbf{r}^\beta_\mu\delta^\alpha_\nu] - \mathbf{r}^\mu_\alpha\delta^\mu_\beta[\mathbf{r}^\alpha_\mu\delta^\beta_\nu - \mathbf{r}^\alpha_\nu\delta^\beta_\mu + \mathbf{r}^\beta_\nu\delta^\alpha_\mu - \mathbf{r}^\beta_\mu\delta^\alpha_\nu] \\ &+\mathbf{r}^\nu_\beta\delta^\mu_\alpha[\mathbf{r}^\alpha_\mu\delta^\beta_\nu - \mathbf{r}^\alpha_\nu\delta^\beta_\mu + \mathbf{r}^\beta_\nu\delta^\alpha_\mu - \mathbf{r}^\beta_\mu\delta^\alpha_\nu] - \mathbf{r}^\mu_\beta\delta^\nu_\alpha[\mathbf{r}^\alpha_\mu\delta^\beta_\nu - \mathbf{r}^\alpha_\nu\delta^\beta_\mu + \mathbf{r}^\beta_\nu\delta^\alpha_\mu - \mathbf{r}^\beta_\mu\delta^\alpha_\nu] \\ &= \mathbf{r}^\mu_\alpha\mathbf{r}^\alpha_\mu\delta^\nu_\nu - \mathbf{r}^\mu_\alpha\delta^\nu_\mu\mathbf{r}^\alpha_\mu + \mathbf{r}^\mu_\mu\mathbf{r}^\nu_\nu - \mathbf{r}^\mu_\nu\mathbf{r}^\nu_\mu - \mathbf{r}^\beta_\alpha\mathbf{r}^\alpha_\beta + \mathbf{r}^\nu_\alpha\delta^\mu_\mu\mathbf{r}^\alpha_\nu - \mathbf{r}^\nu_\mu\mathbf{r}^\nu_\mu + \mathbf{r}^\nu_\nu\mathbf{r}^\mu_\mu \\ &+\mathbf{r}^\nu_\nu\mathbf{r}^\mu_\mu - \mathbf{r}^\nu_\mu\mathbf{r}^\mu_\nu + \mathbf{r}^\nu_\beta\delta^\mu_\mu\mathbf{r}^\beta_\nu - \mathbf{r}^\alpha_\beta\mathbf{r}^\beta_\alpha + \mathbf{r}^\mu_\nu\mathbf{r}^\nu_\mu + \mathbf{r}^\mu_\mu\mathbf{r}^\nu_\nu - \mathbf{r}^\alpha_\beta\mathbf{r}^\beta_\alpha + \mathbf{r}^\mu_\beta\delta^\nu_\nu\mathbf{r}^\beta_\mu \\ &= 4(2\mathbf{r}^2 + \mathbf{s}^2) = 8\mathbf{r}^2 + 4\mathbf{s}^2 \ .\end{aligned} \tag{4.17}$$

$$\begin{aligned}\mathbf{R}_{\mu\nu\alpha\beta}\boldsymbol{\eta}^{\mu\nu\alpha\beta} &= \mathbf{R}_{\mu\nu\alpha\beta}[\mathbf{r}^{\mu\alpha}g^{\nu\beta} - \mathbf{r}^{\nu\alpha}g^{\mu\beta} + \mathbf{r}^{\nu\beta}g^{\mu\alpha} - \mathbf{r}^{\mu\beta}g^{\nu\alpha}] \\ &= \mathbf{R}^{\beta}_{\mu\alpha\beta}\mathbf{r}^{\mu\alpha} - \mathbf{R}^{\beta}_{\nu\alpha\beta}\mathbf{r}^{\nu\alpha} + \mathbf{R}^{\mu}_{\mu\nu\beta}\mathbf{r}^{\nu\beta}\mathbf{R}^{\nu}_{\mu\nu\beta}\mathbf{r}^{\mu\beta} \\ &= \mathbf{R}^{\beta}_{\mu\beta\alpha}\mathbf{r}^{\mu\alpha} - \mathbf{R}^{\beta}_{\nu\beta\alpha}\mathbf{r}^{\nu\alpha} + \mathbf{R}^{\mu}_{\nu\mu\beta}\mathbf{r}^{\nu\beta}\mathbf{R}^{\nu}_{\mu\nu\beta}\mathbf{r}^{\mu\beta} \\ &= \mathbf{r}_{\mu\alpha}\mathbf{r}^{\mu\alpha} + \mathbf{r}_{\nu\alpha}\mathbf{r}^{\nu\alpha} + \mathbf{r}_{\nu\beta}\mathbf{r}^{\nu\beta}\mathbf{r}_{\mu\beta}\mathbf{r}^{\mu\beta} = 4\mathbf{r}^2 \ .\end{aligned} \tag{4.18}$$

We have then, since $\mathbf{C}^{\mu\nu}_{\alpha\beta} = \mathbf{R}^{\mu\nu}_{\alpha\beta} - A\eta^{\mu\nu}_{\alpha\beta} + B\mathbf{s}\varepsilon^{\mu\nu}_{\alpha\beta}$:

$$\begin{aligned}
\mathbf{C}^2 &= \mathbf{C}_{\mu\nu\alpha\beta}\mathbf{C}^{\mu\nu\alpha\beta} \\
&= [\mathbf{R}_{\mu\nu\alpha\beta} + A\eta_{\mu\nu\alpha\beta} + B\mathbf{s}\varepsilon_{\mu\nu\alpha\beta}][\mathbf{R}^{\mu\nu\alpha\beta} + A\eta^{\mu\nu\alpha\beta} + B\mathbf{s}\varepsilon^{\mu\nu\alpha\beta}] \\
&= \mathbf{R}_{\mu\nu\alpha\beta}\mathbf{R}^{\mu\nu\alpha\beta} + 2A\mathbf{R}_{\mu\nu\alpha\beta}\eta^{\mu\nu\alpha\beta} + 2B\mathbf{R}_{\mu\nu\alpha\beta}sg\varepsilon^{\mu\nu\alpha\beta} \\
&\quad + A^2\eta_{\mu\nu\alpha\beta}\eta^{\mu\nu\alpha\beta} + 2AB\eta_{\mu\nu\alpha\beta}\mathbf{s}\varepsilon^{\mu\nu\alpha\beta} + B^2\mathbf{s}^2\varepsilon_{\mu\nu\alpha\beta}\varepsilon^{\mu\nu\alpha\beta} \\
&= \mathbf{R}^2 + 2A\cdot 4\mathbf{r}^2 + 2B\mathbf{s}\cdot 2\mathbf{s} + A^2(8\mathbf{r}^2 + 4\mathbf{s}^2) + 2AB\mathbf{s}\cdot 12\mathbf{s} + 2\mathrm{d}(\mathrm{d}-1)B^2\mathbf{s}^2 \\
&= \mathbf{R}^2 + (8A + 8A^2)\mathbf{r}^2 + (4B + 4A^2 + 24AB + 2\mathrm{d}(\mathrm{d}-1)B^2)\mathbf{s}^2 \\
&= \mathbf{R}^2 + 8A(A+1)\mathbf{r}^2 + 4(B + A^2 + 6AB + 6B^2)\mathbf{s}^2 \ . \qquad (4.19)
\end{aligned}$$

yielding (4.12) for d=4.

Bibliography

Expository papers are signaled with a *.

[0] A. Connes, Noncommutative Geometry.
Academic Press, (1994)

[1] A.Connes, Spectral sequence and homology of currents for operator algebras.
Math. Forschungsinst.. Oberwolfach, Tagungsbericht 42181; Funktionalanalysis C-Algebren, 27/-3/10 (1981)*

[2] A.Connes, Essay on physics and non-commutative geometry. The Interface of Mathematics and Particle Physics.
Clarendon Press, Oxford,1990

[3] A. Connes and J. Lott, Particle models and noncommutative geometry.
Nucl. Phys. B 18(1990), suppl,29-47(1991)

[4]* A. Connes and J. Lott, The metric aspect of noncommutative geometry.
Proceedings of the July 1991 Cargèse school of Theoretical Physics.

[5] D. Kastler and T. Schücker, Remarks on Alain Connes' approach to the standard model in noncommutative geometry.
Theoret. Math. Phys. 92 (1992) 523

[6] J.C. Varilly and J.M. Gracia-Bonda, Connes' noncommutative differential geometry and the Standard Model.
J. Geom. Phys 12 (1992) 223

[7] D. Kastler, Towards extracting physical predictions from Alain Connes' version of the standard model (the new grand unification?).
Istanbul workshop Operator algebras, Mathematical physics and Low dimensional topology, A.K. Peters (1993).

[8] D. Kastler and D.Testard, Quantum forms of tensor products.
Comm.Math.Phys.155 (1993)135

[9]* A. Connes, Noncommutative geometry and physics.
Proceedings of 1992 Les Houches school of Theoretical Physics, North Holland (1995)

[10] B. Iochum, D. Kastler and T. Schücker, Fuzzy mass relations for the Higgs.
J. Math. Phys. 36 (1995) 232

[11] E. Alvarez, J. Garcia-Bondia and C.P. Martin, Anomaly cancellation and the gauge group of the standard model in noncommutative geometry. *Phys.Lett. B364 (1995) 33*

[12] J.M. Gracia-Bondia, Connes' interpretation of the Standard Model and massive neutrinos.
Phys. Lett. B 351 (1995) 510

[13] A. Connes, Noncommutative geometry and reality.
J. Math. Phys. 36 (1995) 6194

[14] M. Takesaki, Tomita's theory of modular Hilbert algebras and its applications. *Lecture Notes in Math. 128 (Springer 1970)*

[14'] D. Kastler, KMS in nutshell.
**Enlarged Proceedings of the ISI GUCCIA Workshop on quantum groups, noncommutative geometry and fundamental physical interactions, December 1997. (Nova Science Publishers, Inc., Commack, New York)*

[15] M.F. Atiyah, K-theory and reality.
Quart. J. Math. Oxford (2) 17 (1966) 367

[16] D. Kastler, A detailed account of Alain Connes' version of the standard model in noncommutative differential geometry.
III Rev. Math. Phys. 8 (1996) 103

[17] B. Iochum and T. Schücker, Yang-Mills-Higgs versus Connes-Lott.
Comm. Math. Phys. 178 (1996) 1

[18] R. Asquith, Noncommutative geometry and the strong force.
Phys. Lett. B 366 (1996) 220

[19] D. Kastler and T. Schücker, A detailed account of Alain Connes' version of the standard model in noncommutative differential geometry.
IV Rev. Math. Phys. 8 (1996) 205

[20] D. Kastler, Dual pairs of quantum spaces.
Lett. Math. Phys. 37 (1996) 329

[21] A. Connes, Brisure de symmétrie spontanée et géométrie du point de vue spectral.
Exposé Bourbaki 816 (1995-96)

[22] D. Kastler and T. Schücker, The standard model a la Connes-Lott.
J. Geom. Phys. 388 (1996) 1

[23] J.M. Gracia-Bondia, J.C. Varilly and C.P. Martin, The Standard Model as a noncommutative geometry: the low mass regime.
Phys. Rep.

[24] B. Iochum, D. Kastler and T. Schücker, Standard model and fuzzy mass relations.
"Geometry and nature", Contemp. Math. 203 (1997) 175 (H.Nencka and J.P.Bourguignon eds)

[25] L. Carminati, B. Iochum, D. Kastler and T. Schücker, The noncommutative constraints on the standard model à la Connes.
J. Math. Phys. 38 (1997) 1269

[26] L. Carminati, B. Iochum, D. Kastler and T. Schücker, Noncommutative Yang-Mills and noncommutative relativity. A bridge over troubled water.

Preprint hep-th/9706105

[27] T. Krajewski, Classification of finite spectral triples.
(to appear in J. Geom. and Phys.)

[28] M. Paschke and A. Sitarz, Discrete spectral triples and their symmetries.
Preprint q-alg/9612029

[29] D. Kastler, The Dirac operator and gravitation.
Comm.Math.Phys. 166 (1995) 633

[30] W. Kalau and M. Walze, Gravity, noncommutative geometry and the Wodzicki reside.
J.Geom.Phys. 16 (1995) 327

[31] J. Madore, Noncommutative geometry and its physical applications.
London. Math. Soc. Lecture Note Series 206. (Cambridge University press 1995)

[32] A. Connes, Gravity coupled with matter and the foundation of noncommutative geometry.
Comm.Math.Phys. 117 (1996) 673

[33] A. Chamseddine and A. Connes, The spectral action principle.
Comm. Math. Phys. 186 (1996) 731

[34]* R. Brout, Notes on Connes construction of the standard model.
Preprint hep-th/9706200 (Bruxelles 1996)

[35] B. Iochum, D. Kastler and T. Schücker, On the universal Chamseddine-Connes action.
J. Math. Phys. 38 (1997) 10

[36] L. Carminati, B. Iochum, D. Kastler and T. Schücker, On Connes new principle of general relativity: can spinors hear the forces of space-time? Operator algebras and quantum field theory.
S. Doplicher at al. eds. (International Press 1997)

[37] P. Gilkey, Invariance theory, that heat equation and the Atiyah-Singer index theorem.
(CCr Press 1995)

[38] T. Schücker, Geometry and forces.
Lecture notes, School on noncommutative geometry. Monsaraz, Portugal, Sept., 1997

[39] D. Kastler, Noncommutative geometry and fundamental physical interactions.
School on noncommutative geometry.. Monsaraz, Portugal, Sept., 1997

[40] G. Landi and C. Rovelli, General relativity in terms of Dirac eigenvalues.
Phys. Rev. Lett. 3051 (1997)

[41] G. Landi, An introduction to noncommutative spaces and their geometries
Springer-Verlag Berlin, Heidelberg 1997

[42] A.Yu. Alekseev, D.V. Glushenkov and A.V. Lyakovskaya, Regular representation of the quantum group $\mathrm{sl_q}(2)$ (q is a root of unit).
St. Petersburg Math. J. 6 5 (1994) 969

[43] D. Kastler, Notes on $\mathrm{Sl_q}(2)$.
Enlarged Proceeding of the ISI GUCCIA Workshop on quantum groups,

noncommutative geometry and fundamental physical interactions, December 1997. (Nova Science Publishers, Inc., Commack, New York)

[44] L. Dabrowski, P.M. Hajac and P. Siniscalco, Explicit Hopf-Galois description of $SL_{\exp(2i\pi/3)}(2)$-induced Frobenius homomorphisms.
Enlarged Proceeding of the ISI GUCCIA Workshop on quantum groups, noncommutative geometry and fundamental physical interactions, December 1997 (Nova Science Publishers, Inc., Commack, New York)

[45] L. Dabrowski, F. Nesti and P. Siniscalco, A finite quantum symmetry of $M_3(3, \mathbb{C})$.
SISSA 63/97/FM (1997)

[46] A. Connes, D. Kreimer, Hopf algebras, renormalization and noncommutative geometry.
Quantum field theory/Perspective and Prospective. C.DeWitt-Morette, J.B. Zuber eds. Kluwer (1999)

[47] T. Krajewski and R. Wulkenhaar, On Kreimer's Hopf algebra structure of Feynman graphs.
*CPT-98/P.3639. hep-th/9805098, Eur.Phys.J.*C *7 (1999) 697*

An Introduction to Noncommutative Geometry

J. Madore

LPT, Batiment 211, Université de Paris-Sud
F-91405 Orsay (France)

Abstract. A review is made of some recent results in noncommutative geometry, including its use as a regularization procedure. Efforts to add a gravitational field to noncommutative models of space-time are also reviewed. Special emphasis is placed on the case which could be considered as the noncommutative analogue of a parallelizable space-time.

1 Introduction and Motivation

Simply stated, a noncommutative space-time is a space-time in which the 'coordinates' do not commute. One typically replaces the four Minkowski coordinates x^μ by four generators q^μ of a noncommutative algebra which satisfy commutation relations of the form

$$[q^\mu, q^\nu] = i\bar{k} q^{\mu\nu}. \tag{1.1}$$

The parameter $\bar{k}$ is a fundamental area scale which we shall suppose to be of the order of the Planck area:

$$\bar{k} \simeq \mu_P^{-2} = G\hbar.$$

There is however no need for this assumption; the experimental bounds would be much larger. Equation (1.1) contains little information about the algebra. If the right-hand side does not vanish it states that at least some of the q^μ do not commute. It states also that it is possible to identify the original coordinates with the generators q^μ in the limit $\bar{k} \to 0$:

$$\lim_{\bar{k}\to 0} q^\mu = x^\mu. \tag{1.2}$$

For mathematical simplicity we shall suppose this to be the case although one could include a singular 'renormalization constant' Z and replace (1.2) by an equation of the form

$$\lim_{\bar{k}\to 0} q^\mu = Z\, x^\mu. \tag{1.3}$$

If, as we shall argue, gravity acts as a universal regulator for ultraviolet divergences then one could reasonably expect the limit $\bar{k} \to 0$ to be a singular limit.

Perhaps not the simplest but certainly the most familiar example of a noncommutative 'space' is the quantized version of a 2-dimensional phase space, described by the 'coordinates' p and q. This example has the advantage of illustrating what is for us the essential interest of the relation of the form (1.1) as expressed in the Heisenberg uncertainty relations. Since one cannot measure simultaneously p and q to arbitrary precision quantum phase space has no longer a notion of a point. It can however be thought of as divided into cells of volume $2\pi\hbar$. If the classical phase space is of finite total volume there will be a finite number of cells and the quantum system will have a finite number of possible states. A 'function' then on quantum phase space will be defined by a finite number of values and can be represented by a matrix. Since points have been replaced by cells we shall refer to a noncommutative 'space' as a 'fuzzy space'.

By analogy with quantum mechanics we shall suppose that the generators q^μ can be represented as hermitian operators on some complex Hilbert space. The presence of the factor i in (1.1) implies that the $q^{\mu\nu}$ are also hermitian operators. The q^μ have real eigenvalues but because of the relations (1.1) they cannot be simultaneously diagonalized; points are ill-defined and space-time consists of elementary cells of volume $(2\pi k\!\!\!{}^-)^2$. Now when a physicist calculates a Feynman diagram he is forced to place a cut-off Λ on the momentum variables in the integrand. This means that he renounces any interest in regions of space-time of volume less than Λ^{-4}. As Λ becomes larger and larger the forbidden region becomes smaller and smaller but it can never be made to vanish. There is a fundamental length scale, much larger than the Planck length, below which the notion of a point is of no practical importance. The simplest and most elegant, if certainly not the only, way of introducing such a scale in a Lorentz-invariant way is through the introduction of the 'coordinates' q^μ. The analogues of the Heisenberg uncertainty relations imply then that

$$\Lambda^2 k\!\!\!{}^- \lesssim 1.$$

The existence of a forbidden region around each point in space-time means that the standard description of Minkowski space as a 4-dimensional continuum is redundant; there are too many points. Heisenberg already in the early days of quantum field theory proposed to replace the continuum by a lattice structure. A lattice however breaks Poincaré invariance and can hardly be considered as fundamental. It was Snyder [119] who first had the idea of using non-commuting coordinates to mimic a discrete structure in a covariant way.

As a simple illustration of how a 'space' can be 'discrete' in some sense and still covariant under the action of a continuous symmetry group one can consider the ordinary round 2-sphere, which has acting on it the rotational group SO_3. As a simple example of a lattice structure one can consider two points on the sphere, for example the north and south poles. One immediately notices of course that by choosing the two points one has broken the rotational invariance. It can be restored at the expense of commutativity. The set of functions on the two points can be identified with the algebra of diagonal 2×2 matrices, each of the two entries on the diagonal corresponding to a possible value of a function at one of the two points. Now an action of a group on the lattice is equivalent to an action of the group on the matrices and there can obviously be no non-trivial action of the group SO_3 on the algebra of diagonal 2×2 matrices. However if one extends the algebra to the noncommutative algebra of all 2×2 matrices one recovers the invariance. The two points, so to speak, have been smeared out over the surface of a sphere; they are replaced by two cells. An 'observable' is an hermitian 2×2 matrix and has therefore two real eigenvalues, which are its values on the two cells. Although what we have just done has nothing to do with Planck's constant it is similar to the procedure of replacing a classical spin which can take two values by a quantum spin of total spin $1/2$. Only the latter is invariant under the rotation group. By replacing the spin $1/2$ by arbitrary spin s one can describe a 'lattice structure' of $n = 2s + 1$ points in an SO_3-invariant manner. The algebra becomes then the algebra M_n of $n \times n$ complex matrices.

It is to be stressed that we shall here modify the structure of Minkowski space-time but maintain covariance under the action of the Poincaré group. A fuzzy space-time looks then like a solid which has a homogeneous distribution of dislocations but no disclinations. We can pursue this solid-state analogy and think of the ordinary Minkowski coordinates as macroscopic order parameters obtained by coarse-graining over scales less than the fundamental scale. They break down and must be replaced by elements of some noncommutative algebra when one considers phenomena on these scales. It might be argued that since we have made space-time 'noncommutative' we ought to do the same with the Poincaré group. This logic leads naturally to the notion of a q-deformed Poincaré (or Lorentz) group which act on a very particular noncommutative version of Minkowski space called q-Minkowski space.

It has also been argued, for conceptual as well as practical, numerical reasons, that a lattice version of space-time or of space is quite satisfactory if one uses a random lattice structure or graph. The most widely used and successful modification of space-time is in fact what is called the lattice approximation. From this point of view the Lorentz group is a classical invariance group and is not valid at the microscopic level. Historically the first attempt to make a finite approximation to a curved manifold was due to Regge and this developed into what is now known as the Regge calculus. The idea is based on the fact that the Euler number of a surface can be expressed as an integral of the gaussian curvature. If one applies this to a flat cone with a smooth vertex then one finds a relation between the defect angle and the mean curvature of the vertex. The latter is encoded in the former. In recent years there has been a burst of activity in this direction, inspired by numerical and theoretical calculations of critical exponents of phase transitions on random surfaces. One chooses a random triangulation of a surface with triangles of constant fixed length, the lattice parameter. If a given point is the vertex of exactly six triangles then the curvature at the point is flat; if there are less than six the curvature is positive; it there are more than six the curvature is negative. Non-integer values of curvature appear through statistical fluctuation. Attempts have been made to generalize this idea to three dimensions using tetrahedra instead of triangles and indeed also to four dimensions, with euclidean signature. The main problem, apart from considerations of the physical relevance of a theory of euclidean gravity, is that of a proper identification of the curvature invariants as a combination of defect angles. On the other hand some authors have investigated random lattices from the point of view of noncommutative geometry. For an introduction to the lattice theory of gravity from these two different points of view we refer to the books by Ambjørn & Jonsson [4] and by Landi [77]. The work of Kaku [63] comes closest to bridging the gap. Compare also the loop-space approach to quantum gravity, for example in the monographs by Baez & Muniain [7] and by Gambini *et al.* [52].

Let $\mathcal{A}_{\bar{k}}$ be the algebra generated in some sense by the elements q^μ. We shall be here working on a formal level so that one can think of $\mathcal{A}_{\bar{k}}$ as an algebra of polynomials in the q^μ although we shall implicitly suppose that there are enough elements to generate smooth functions on space-time in the commutative limit. Since we have identified the generators as hermitian operators on some Hilbert space we can identify $\mathcal{A}_{\bar{k}}$ as a subalgebra of the algebra of all operators on the Hilbert space. We have added the subscript $\bar{k}$ to underline the dependence on this parameter but of course the commutation relations (1.1) do not determine the structure of $\mathcal{A}_{\bar{k}}$, We in fact conjecture that every possible gravitational field can be considered as the commutative limit of a noncommutative equivalent and that the latter is strongly restricted if not determined by the structure of the algebra $\mathcal{A}_{\bar{k}}$. We must have then a large number of algebras $\mathcal{A}_{\bar{k}}$ for each value of $\bar{k}$.

We mentioned above that the noncommutative structure gives rise to an ultraviolet cut-off. This idea has been developed by several authors [62], [84], [41] [70], [69], [23] since the original work of Snyder [119], [120]. It is the right-hand arrow of the diagram

$$\begin{array}{ccc} \mathcal{A}_{\bar{k}} & \Longleftarrow & \Omega^*(\mathcal{A}_{\bar{k}}) \\ \Downarrow & & \Uparrow \\ \text{Cut-off} & & \text{Gravity} \end{array} \tag{1.4}$$

We shall define and discuss it in Section 2. The top arrow is a mathematical triviality; the $\Omega^*(\mathcal{A}_{\bar{k}})$ is what gives a differential structure to the algebra. We shall define and discuss it in Section 4. We have argued elsewhere [85], not quite successfully, that each gravitational field is the unique 'shadow' in the limit $\bar{k} \to 0$ of some differential structure over some noncommutative algebra. This would define the left-hand arrow of the diagram. The composition of the three arrows is an expression of an old idea due to Pauli that perturbative ultraviolet divergences will one day be regularized by the gravitational field. For a recent review we refer to Garay [54]. The possibility we consider here is that the mechanism by which this works is through the introduction of noncommuting 'coordinates' such as the q^μ. A hand-waving argument can be given [87] which allows one to think of the noncommutative structure of space-time as being due to quantum fluctuations of the light-cone in ordinary 4-dimensional space-time. This relies on the existence of quantum gravitational fluctuations. A purely classical argument based on the formation of black-holes has been also given [41]. In both cases the classical gravitational field is to be considered as regularizing the ultraviolet divergences through the introduction of the noncommutative structure of space-time. This can be strengthened as the conjecture that the classical gravitational field and the noncommutative nature of space-time are two aspects of the same thing.

2 Ultraviolet Regularization

One example from which one can seek inspiration in looking for examples of noncommutative geometries is quantized phase space, which had been already studied from a noncommutative point of view by Dirac [40]. In particular it is instructive to consider the phase space of a particle in a plane: (q^1, q^2, p_1, p_2). In classical mechanics one has four commuting operators; in quantum mechanics one has the commutation relations

$$[q^1, p_1] = i\hbar, \qquad [q^2, p_2] = i\hbar. \tag{2.5}$$

The points of classical phase space have been replaced by 'Bohr cells' of area $2\pi\hbar$. Consider the divergent integral

$$I = \int \frac{dp_1 dp_2}{p^2}, \qquad p^2 = p_1^2 + p_2^2.$$

If one introduces a magnetic field B normal to the plane then the appropriately modified gauge-covariant momenta no longer commute:

$$[p_1, p_2] = i\hbar eB.$$

The points of momentum space have been replaced by 'Landau cells' of area $\hbar eB$. This serves in general as an infrared cut-off:

$$p^2 \gtrsim \hbar eB.$$

The noncommutative algebra generated by the (p_1, p_2) is of importance in the physics of the quantum Hall effect [117] and it has been studied in this respect from the point of view of noncommutative geometry [11]. If one were to replace the magnetic field by a gaussian curvature K, $eB \mapsto \hbar K$ then one would have the same effect; curvature in general acts as a mass.

In this example 'quantizing' position-space coordinates consists in replacing them by two operators which satisfy a commutation relation of the form

$$[q^1, q^2] = i\kbar q^{12}.$$

Ipso facto the points of position space are replaced by 'Planck cells' of area $2\pi\kbar$ and the integral I is completely regularized:

$$I \sim \log(\kbar K).$$

This vague idea can actually be implemented by explicit calculations [120], [41], [47], [69], [17], [98], [23], [73]. In general consider any $*$-algebra $\mathcal{A}$ with a trivial center, in some representation with a partial trace and let Δ be a linear operator on $\mathcal{A}$ with a set of eigenvectors $\phi_r \in \mathcal{A}$ and corresponding real eigenvalues λ_r:

$$\Delta\phi_r = \lambda_r\phi_r.$$

The parameter r here designates a point in some parameter space and we write the integral on this space as a sum over r. The corresponding classical action is

$$S = \mathrm{Tr}\,(\phi^* \Delta\phi), \qquad \phi \in \mathcal{A}. \tag{2.6}$$

The trace here must be defined in some representation of $\mathcal{A}$. We shall assume that with respect to this trace

$$\mathrm{Tr}\,(\phi_r^* \phi_s) = \delta_{rs} \tag{2.7}$$

and we define the Hilbert space $\mathcal{H} \subset \mathcal{A}$ of 1-particle states to be

$$\mathcal{H} = \{\phi = \sum_r a_r \phi_r \mid \sum_r |a_r|^2 < \infty\}.$$

As usual the a_r become operators when the field is quantized. For $f \in \mathcal{H}$ the completeness condition can be written as

$$\phi = \sum_r \phi_r \mathrm{Tr}\,(\phi_r^* \phi).$$

If we introduce the element

$$W = \sum_r \phi_r \otimes \phi_r^*$$

then the completeness condition can also be written

$$\mathrm{Tr}_2(W \cdot 1 \otimes \phi) = \phi \otimes 1.$$

The tensor product is here over the complex numbers and the subscript on the trace indicates that it is taken over the second factor. The element W is therefore the noncommutative generalization of the Dirac distribution in the commutative case; it is not an element of $\mathcal{H} \otimes \mathcal{H}$. We introduce also the element G defined by the formal sum

$$G = \sum \lambda_r^{-1} \phi_r \otimes \phi_r^*.$$

Since obviously $\Delta G = W$ this element generalizes the propagator corresponding to Δ. We wish to discuss the conditions under which the sum converges and G can be considered as a well-defined element of a weak closure of $\mathcal{H} \otimes \mathcal{H}$.

We shall restrict our attention to algebras which are generated by a set q^μ, $1 \leq \mu \leq n$, of n elements. We shall suppose that $\mathcal{A}$ is represented as an algebra of operators on a Hilbert space $L^2(V, \mu)$ and we fix an orthonormal basis $|i\rangle$. We can write then

$$q^\mu |i\rangle = \sum_j Q^\mu_{ji} |j\rangle$$

for some set of n matrices Q^μ_{ij}. If the algebra is commutative then $Q^\mu_{ij} = q^\mu_i \delta_{ij}$. As above, the symbol Σ here can represent a sum or an integral depending on the basis $|i\rangle$ it is convenient to choose. The index i belongs again to some parameter space which of course is not to be confused with the space to which r and s belong. The symbol δ_{ij} can represent therefore the Kronecker or Dirac delta.

Consider the differential d_u of the universal calculus, defined in Section 4. It is a map of $\mathcal{A}$ into $\mathcal{A}\otimes\mathcal{A}$ given by $d_u f = 1\otimes f - f\otimes 1$. We define the 'variation' δq^μ of the generator q^μ as

$$\delta q^\mu = \frac{1}{2} d_u q^\mu = \frac{1}{2}(1\otimes q^\mu - q^\mu \otimes 1). \tag{2.8}$$

We identify $q^\mu = q^\mu \otimes 1$ in the tensor product and we set $q^{\mu\prime} = 1\otimes q^\mu$. Thus we can write

$$\delta q^\mu = \frac{1}{2}(q^{\mu\prime} - q^\mu).$$

It follows from the commutation rules of the algebra that

$$[\delta q^\mu, \delta q^\nu] = \frac{1}{4} i\hbar (q^{\mu\nu}\otimes 1 + 1\otimes q^{\mu\nu}).$$

Suppose that a set of elements $\bar{q}^\mu$ of $\mathcal{A}\otimes\mathcal{A}$ can be found such that $\mathcal{A}\otimes\mathcal{A}$ is generated by the set $\{\bar{q}^\mu, \delta q^\mu\}$ and such that

$$[\bar{q}^\mu, \delta q^\nu] = 0. \tag{2.9}$$

Then we can write the tensor product $L^2(V,\mu)\otimes L^2(V,\mu)$ in the form

$$L^2(V,\mu)\otimes L^2(V,\mu) \simeq \mathcal{D}\otimes\mathcal{F} \tag{2.10}$$

where $\bar{q}^\mu$ acts on $\mathcal{D}$ and δq^μ on $\mathcal{F}$. We shall choose accordingly a basis

$$|\bar{\imath}, k\rangle = |\bar{\imath}\rangle_D \otimes |k\rangle_F$$

of $L^2(V,\mu)\otimes L^2(V,\mu)$. If $q^{\mu\nu}$ lies in the center of the algebra then the elements

$$\bar{q}^\mu = \frac{1}{2}(q^\mu + q^{\mu\prime})$$

are such that Equation (2.9) is satisfied. Further one has

$$q^\mu = \bar{q}^\mu - \delta q^\mu, \qquad q^{\mu\prime} = \bar{q}^\mu + \delta q^\mu$$

and with the obvious identifications

$$[\bar{q}^\mu, \bar{q}^\nu] = \frac{1}{2} i\hbar q^{\mu\nu}, \qquad [\delta q^\mu, \delta q^\nu] = \frac{1}{2} i\hbar q^{\mu\nu}. \tag{2.11}$$

The tensor product in the definition of G is now to be considered as a tensor product of a 'diagonal' algebra $\bar{\mathcal{A}}$, acting on $\mathcal{D}$ and a 'variation' $\delta\mathcal{A}$, acting on $\mathcal{F}$. That is, we rewrite

$$\mathcal{A}\otimes\mathcal{A} = \bar{\mathcal{A}}\otimes\delta\mathcal{A} \tag{2.12}$$

in accordance with (2.10). If (2.9) is not satisfied the factorization (2.10) can still be of interest if δq^μ acts only on $\mathcal{F}$. In general then $\bar{q}^\mu$ will act non-trivially on the complete tensor product $\mathcal{D}\otimes\mathcal{F}$. We shall suppose that the definition

(2.8) of δq^μ in terms of the tensor product coincides with the intuitive notion of the 'variation of a coordinate'. One can introduce a new differential calculus $(\bar{\Omega}^*(\mathcal{A}), \bar{d})$ defined by

$$\bar{d}\bar{q}^\mu = \delta q^\mu. \tag{2.13}$$

One would like this new calculus to be isomorphic to the original one if δq^μ and dq^μ are to be thought of as 'infinitesimal variations' [23].

Let $\mathcal{C}(M)$ be an algebra of functions on a space M. Let f be a map of M into itself and let f^* be the induced map of $\mathcal{C}(M)$ into itself. We set $\phi' = f^*(\phi)$ and define $\delta\phi = \phi' - \phi$. The ordinary propagator is a function of two points, an element of $\mathcal{C}(M) \otimes \mathcal{C}(M)$ and we are interested in the limit when the two points coincide. This limit must be taken with care since the partial derivative of a function after the limit and the limit of the derived function with respect to one of the variables are not in general equal. We are interested in the latter since the Laplace operator which defines the propagator acts only on one of the variables. If we set $\delta x = x' - x$ where $x' = f(x)$ then we can express the limit $\delta x \to 0$ as $\delta\phi \to 0$. We wish to study the element $G(q^\mu; q^{\nu\prime})$ of the tensor product $\mathcal{H} \otimes \mathcal{H}$ most particularly in the limit $q^{\mu\prime} \to q^\mu$. The q^μ are however fixed generators of the algebra and this limit must be defined otherwise. As a possible added complication, which will however not appear explicitly in the examples we shall consider, the generators q^μ are in general unbounded operators. We shall give a formal definition of the limit as a weak limit within the tensor product in terms of variations of the basis vectors $|i\rangle$. We shall use a tensor product which is not braided. We shall return to his assumption later.

Using the representation of $\mathcal{A}$ the propagator $G = G(q^\mu; q^{\nu\prime})$ can be expressed as a map

$$G: \ L^2(V,\mu) \otimes L^2(V,\mu) \to L^2(V,\mu) \otimes L^2(V,\mu).$$

It can be defined in terms of its (classical) matrix elements $\langle j, j'|G(q^\mu; q^{\nu\prime})|i, i'\rangle$. In the commutative limit $\hbar \to 0$ one would find

$$\langle j, j'|G(q^\mu; q^{\nu\prime})|i, i'\rangle \to G(q^\mu; q^{\nu\prime})\, \delta_{ij}\delta_{i'j'}$$

with

$$q^\mu|i\rangle = q_i^\mu|i\rangle, \qquad q^{\nu\prime}|i'\rangle = q_{i'}^{\nu\prime}|i'\rangle$$

and so, at least in a quasicommutative approximation, we can identify q^μ with a point $i \in V = \mathbb{R}^n$ and $q^{\mu\prime}$ with $i' \in V = \mathbb{R}^n$. We shall therefore represent graphically $G(q^\mu; q^{\mu\prime})$ as a line between i and i':

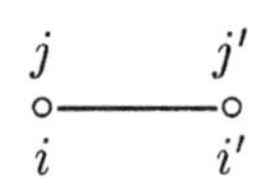

(2.14)

The extra pair of indices (j, j') is present because in general G acts as an operator on each end of the line. An ordinary propagator on a manifold diverges in the limit $q^{\mu\prime} \to q^\mu$. This limit can be redefined as the limit

$$|i'\rangle \to |i\rangle.$$

It makes sense now in the noncommutative case but it cannot be attained as we shall see below. We shall use therefore the identification (2.10) to express the limit as

$$|\bar{\imath}, k\rangle \to |\bar{\imath}, 0\rangle \equiv |\bar{\imath}\rangle. \tag{2.15}$$

In the graph (2.14) this means that the two ends of the line almost close to form a circle.

It is here that the representation, especially the representation of the tensor product, becomes of importance. We shall describe the second copy $\mathcal{F}$ of the Hilbert space using creation and annihilation operators. We choose then the basis $|k\rangle_F$ with $k \in \mathbb{Z}$. The states $|\bar{\imath}, 0\rangle$ are those in which collectively the operators δq^μ take their minimum value. In the language of quantum mechanics such a state is an example of a coherent state.

We introduce a set of n annihilation operators a_l with their adjoints a^*_m such that, as in quantum mechanics

$$[a_l \, , a^*_m] = \hbar\!\!\!\!- \delta_{lm}. \tag{2.16}$$

We shall see that each a_l annihilates and each a^*_l creates a unit of separation. The quantum mechanical analogue of this separation would be the energy of the harmonic oscillator. By analogy then we define a diagonal state to be a state annihilated by all the a_l. We define as usual the action of a_l on the diagonal basis element $|\bar{\imath}, 0\rangle \in \mathcal{D} \otimes \mathcal{F}$ by the condition $a_l |\bar{\imath}, 0\rangle = 0$ and we set recursively

$$a^*_l |\bar{\imath}, k_1, \dots, k_l, \dots k_n\rangle_F = \sqrt{\hbar\!\!\!\!-}\sqrt{k_l + 1}\, |\bar{\imath}, k_1, \dots, k_l + 1, \dots k_n\rangle_F.$$

The coincidence limit is attained on elements of $L^2(V, \mu) \otimes L^2(V, \mu)$ of the form $|\bar{\imath}, 0\rangle$.

The analogue of the integral I defined in the Introduction is defined then by the equation

$$\langle \bar{\jmath} | G(q^\mu; q^{\nu\prime}) | \bar{\imath} \rangle = \langle \bar{\jmath} | I(\hbar\!\!\!\!- \mu^2) | \bar{\imath} \rangle.$$

Here μ is a parameter in the operator Δ with the dimension of mass. In general $I(\hbar\!\!\!\!- \mu^2)$ is an operator acting on $\mathcal{D}$. In the example we shall consider however the space is homogeneous and it reduces to a constant. We can write then

$$\langle \bar{\jmath} | G(q^\mu; q^{\nu\prime}) | \bar{\imath} \rangle = I(\hbar\!\!\!\!- \mu^2)\, \langle \bar{\jmath} \,|\, \bar{\imath} \rangle.$$

To calculate $\langle \bar{\jmath} | G(q^\mu; q^{\mu\prime}) | \bar{\imath} \rangle$ we must express G in terms of the a_l and their adjoints. For this we write

$$\delta q^\mu = \sum_{l=1}^{n} (J^\mu_l a_l + J^{\mu *}_l a^*_l) \tag{2.17}$$

and from (2.11) we conclude that

$$\sum_{l=1}^{n} J^{[\mu}_l J^{\nu]*}_l = \frac{1}{2} i q^{\mu\nu}. \tag{2.18}$$

The J_l^μ appear here as the components of a symplectomorphism. They are fixed only to within a redefinition of the a_l and contain therefore $2n^2 + n$ free parameters. This is the number of elements of $GL(2n, \mathbb{R})$ which leave invariant the right-hand side of (2.18). If we interpret δq^μ as a 'string' joining two 'points' q^μ and $q^{\mu\prime}$ then each a_j creates a longitudinal displacement. They would correspond to the rigid longitudinal vibrational modes of the string. Since it requires no energy to separate two points the string tension would be zero.

If the differential calculus $(\bar{\Omega}^*(\mathcal{A}), \bar{d})$ defined in (2.13) has a frame $\bar{\theta}^\alpha = \bar{\theta}^\alpha_\lambda(\bar{q}^\mu)\, \bar{d}\bar{q}^\lambda$ then it would seem more appropriate to expand the variation in the form

$$\bar{\theta}^\alpha_\lambda(\bar{q}^\mu)\, \delta q^\lambda = \sum_{l=1}^{n} (j_l^\alpha a_l + j_l^{\alpha *} a_l^*). \tag{2.19}$$

We are motivated here by the desire to make δq^μ as similar as possible to the element dq^μ of the differential calculus. This would suggest, in particular, that the condition (2.9) is fulfilled only if the geometry is flat.

The 'non-local' modification we shall find in the propagator is to be associated not with the propagator but rather with the vertices at its end points. To see this we consider now the matrix elements

$$\begin{aligned} \langle j, j'|G(q^\mu; q^{\rho\prime})|i, i'\rangle\langle l', l|G(q^{\sigma\prime}; q^\nu)|k', k\rangle = \\ \langle j| \otimes \langle j'| \otimes \langle l'| \otimes \langle l|G \otimes G|i\rangle \otimes |i'\rangle \otimes |k'\rangle \otimes |k\rangle \end{aligned} \tag{2.20}$$

of the tensor product of two copies of the propagator, which we represent by the graph

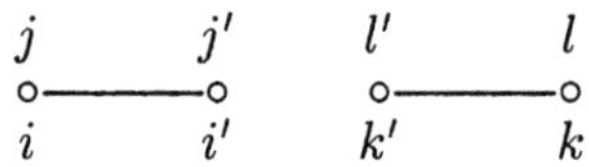

(2.21)

To form a vertex we must 'join' the 'point' k' to the 'point' i'. Following the prescription (2.15) this means that we replace the basis element

$$|i'\rangle \otimes |k'\rangle \in L^2(V, \mu) \otimes L^2(V, \mu)$$

by the basis element

$$|\bar{i}'\rangle = |\bar{i}', 0\rangle \in \mathcal{D} \otimes \mathcal{F}.$$

We are prompted then to introduce the projection

$$L^2(V, \mu) \otimes L^2(V, \mu) \otimes L^2(V, \mu) \otimes L^2(V, \mu) \xrightarrow{P} L^2(V, \mu) \otimes \mathcal{D} \otimes L^2(V, \mu)$$

defined by

$$P = \sum_{r, \bar{r}', s} |r, \bar{r}', s\rangle\langle r, \bar{r}', s|$$

and to define the propagator $G_2(q^\mu, q^{\rho'}, q^\nu)$ in terms of the matrix elements

$$\begin{aligned}
&\langle j, \bar{\jmath}', l|G_2|i, \bar{\imath}', k\rangle = \\
&\quad \sum_{r,\bar{r}',s} \langle j, \bar{\jmath}', l|G \otimes (1 \otimes 1)|r, \bar{r}', s\rangle \langle r, \bar{r}', s|(1 \otimes 1) \otimes G|i, \bar{\imath}', k\rangle = \\
&\qquad \sum_{r,\bar{r}',s} \langle j, \bar{\jmath}'|G \otimes 1|r, \bar{r}'\rangle\, \delta_{ls}\delta_{ri}\, \langle \bar{r}', s|1 \otimes G|\bar{\imath}', k\rangle = \\
&\qquad\quad \sum_{\bar{r}'} \langle j, \bar{\jmath}'|G \otimes 1|i, \bar{r}'\rangle \langle \bar{r}', l|1 \otimes G|\bar{\imath}', k\rangle
\end{aligned} \tag{2.22}$$

which we represent by the graph

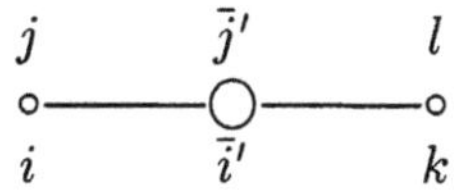

(2.23)

We could have also included the dummy multiplication index and written

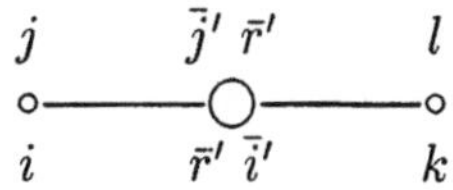

We have used the identifications

$$G \otimes G = G \otimes (1 \otimes 1) \cdot (1 \otimes 1) \otimes G$$

and the fact that

$$G \otimes G \in (\mathcal{A} \otimes \mathcal{A}) \otimes (\mathcal{A} \otimes \mathcal{A}) = \mathcal{A} \otimes (\mathcal{A} \otimes \mathcal{A}) \otimes \mathcal{A} = \mathcal{A} \otimes (\mathcal{D} \otimes \mathcal{F}) \otimes \mathcal{A}.$$

Since P projects $\mathcal{D} \otimes \mathcal{F}$ onto $\mathcal{D}$ we see that

$$G_2 \in \mathcal{A} \otimes \mathcal{D} \otimes \mathcal{A}.$$

In the commutative limit $k\!\!\!^- \to 0$ one would find

$$\langle j, \bar{\jmath}', l|G_2|i, \bar{\imath}', k\rangle \to G_2\, \delta_{ij}\delta_{i'j'}\delta_{kl}$$

on the left-hand side of (2.22) and

$$\langle j, \bar{\jmath}'|G \otimes 1|i, \bar{r}'\rangle \to G\, \delta_{ij}\delta_{r'j'}$$

on the right-hand side. One would normally choose as basis the eigenvectors of the position operator so that $q^\mu|i\rangle = q^\mu_i|i\rangle$ and one would normally drop the extra index on q^μ. The preceeding two limits would be written then respectively

$$\langle j, \bar{\jmath}', l|G_2|i, \bar{\imath}', k\rangle \to G_2(q^\mu, q^{\rho'}, q^\nu)$$

and

$$\langle j, \bar{\jmath}'|G \otimes 1|i, \bar{r}'\rangle \to G(q^\mu, q^{\rho'}).$$

The graph (2.23) in turn can be cut into the two graphs

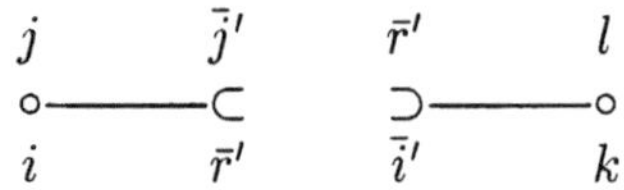

(2.24)

which represent respectively the factors

$$\langle j, \bar{j}' | G \otimes 1 | i, \bar{r}' \rangle, \qquad \langle \bar{r}', l | 1 \otimes G | \bar{i}', k \rangle.$$

We are prompted by this to introduce also the graph

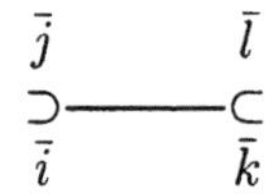

(2.25)

to represent the matrix elements

$$\langle \bar{j}, \bar{l} | 1 \otimes G \otimes 1 | \bar{i}, \bar{k} \rangle.$$

This is the propagator with 'fuzzy' vertices. It is obtained by joining (i, j) to (k, l) in the graph (2.23) and cutting it as in (2.24). To obtain an p-point vertex one would need a tensor product of $2p$ copies of the space $L^2(V, \mu)$. Of these p would describe the free end points of the lines and the remaining p would be rewritten as a factor of one copy of $\mathcal{D}$ to describe the 'mean position' of the vertex and $p - 1$ copies of $\mathcal{F}$ to describe the 'fuzz' [23].

As an example we consider a scalar field on the noncommutative flat plane. The noncommutative flat plane is the algebra $\mathcal{A}_{\hbar}$ generated by two hermitian elements $q^1 = x$ and $q^2 = y$ which satisfy the commutation relation $[x, y] = i\hbar$ and which has the associated differential calculus $\Omega^*(\mathcal{A}_{\hbar})$ given by $[q^\mu, dq^\nu] = 0$. The flatness is a consequence of the fact that dq^μ is a frame as we shall define this word in Section 4. If we introduce the two derivations

$$e_1 = -\frac{1}{i\hbar} \operatorname{ad} y, \qquad e_2 = \frac{1}{i\hbar} \operatorname{ad} x$$

then an appropriate generalization [89] of the Laplace operator Δ with mass μ is given by

$$\Delta = \Delta_{\hbar} + \mu^2, \qquad \Delta_{\hbar} = -(e_1^2 + e_2^2).$$

For each couple $(k_1, k_2) \in \mathbb{R}^2$ we introduce the unitary elements $u(k_1)$, $v(k_2) \in \mathcal{A}_{\hbar}$ defined by

$$u(k_1) = e^{ik_1 x}, \qquad v(k_2) = e^{ik_2 y}.$$

They satisfy the commutation relations

$$u(k_1) v(k_2) = q^{k_1 k_2 \hbar} v(k_2) u(k_1), \qquad q = e^{-i}.$$

A basis for the Hilbert space $\mathcal{H}$ is given by the eigenvectors

$$\phi_k = u(k_1) v(k_2), \qquad k = (k_1, k_2)$$

of Δ. The corresponding eigenvalues are

$$\lambda_k = k^2 + \mu^2, \qquad k^2 = k_1^2 + k_2^2.$$

The element G can be written then

$$G(x, y; x', y') = \frac{1}{(2\pi)^2} \int (k^2 + \mu^2)^{-1} \phi_k \otimes \phi_k^* \, dk, \qquad dk = dk_1 dk_2.$$

We must introduce a partial trace on $\mathcal{A}_{\hbar}$. This can be done only through a representation. The only properties which we shall need however are the identities

$$\mathrm{Tr}\,(u^*(k_1')u(k_1)) = 2\pi\delta(k_1' - k_1), \qquad \mathrm{Tr}\,(v^*(k_2')v(k_2)) = 2\pi\delta(k_2' - k_2).$$

That is:

$$\mathrm{Tr}\,(\phi_{k'}^* \phi_k) = (2\pi)^2 \delta^{(2)}(k' - k).$$

The commutation relations (2.11) become in this case

$$[\bar{x}, \bar{y}] = \frac{1}{2} i\hbar, \qquad [\delta x, \delta y] = \frac{1}{2} i\hbar. \tag{2.26}$$

As in (2.17) we write

$$\delta x = J^1 a + J^{1*} a^*, \qquad \delta y = J^2 a + J^{2*} a^*. \tag{2.27}$$

With (2.16) satisfied we have $J^{[1} J^{2]*} = \frac{1}{2} i q^{12}$. By a redefinition of a we can choose

$$J^1 = \frac{1}{2}, \qquad J^2 = \frac{1}{2i}, \qquad a = \delta x + i\delta y.$$

The freedom here is $SL(2, \mathbb{R})$, the symplectomorphism group in dimension 2. By a renormalization of $\hbar$ we can also choose $q^{12} = 1$.

We set $p = (p_1, p_2)$ and introduce the basis $|\bar{p}, k\rangle_F = |\bar{p}\rangle_D \otimes |k\rangle_F$ according to the prescription (2.10) of the previous section. We shall also re-express the tensor product according to (2.12) and drop the tensor-product symbol. We have then

$$u'^*(k_1)|\bar{p}'\rangle = e^{-ik_1 x'}|\bar{p}'\rangle = e^{-ik_1(\bar{x} + \delta x)}|\bar{p}'\rangle.$$

Since $\bar{x}$ and δx commute we can write this as

$$u'^*(k_1)|\bar{p}'\rangle = e^{-ik_1\bar{x}} e^{-ik_1(a + a^*)/2}|\bar{p}'\rangle.$$

Using the Baker-Campbell-Hausdorff formula

$$e^{\alpha a + \beta a^*} = e^{\beta a^*} e^{\alpha a} e^{\alpha\beta\hbar/2} = e^{\alpha a} e^{\beta a^*} e^{-\alpha\beta\hbar/2}$$

we find that

$$u'^*(k_1)|\bar{p}'\rangle = e^{-ik_1\bar{x}} e^{-k_1^2\hbar/8} e^{-ik_1 a^*/2}|\bar{p}'\rangle$$

and therefore

$$\phi_k'^* |\bar{p}'\rangle = e^{-ik_2 y'} e^{-ik_1 \bar{x}} e^{-k_1^2 \hbar/8} e^{-ik_1 a^*/2} |\bar{p}'\rangle$$

$$= e^{-ik_2 \bar{y}} e^{-ik_1 \bar{x}} e^{-\hbar k^2/8} e^{k_2 a^*/2} e^{-k_2 a/2} e^{-ik_1 a^*/2} |\bar{p}'\rangle$$

$$= e^{-ik_2 \bar{y}} e^{-ik_1 \bar{x}} e^{-\hbar k^2/8} e^{ik_1 k_2 \hbar/4} e^{(k_2 - ik_1) a^*/2} |\bar{p}'\rangle.$$

Similarly we find

$$\phi_k^* |\bar{p}\rangle = e^{-ik_2 \bar{y}} e^{-ik_1 \bar{x}} e^{-\hbar k^2/8} e^{ik_1 k_2 \hbar/4} e^{-(k_2 - ik_1) a^*/2} |\bar{p}\rangle.$$

From these last two equations we deduce that

$$\langle \bar{p}' | \phi_k \otimes \phi_k^* | \bar{p}\rangle = e^{-\hbar k^2/2} \langle \bar{p}' \mid \bar{p}\rangle. \tag{2.28}$$

The product here is the tensor product (2.12). Since the $\bar{\mathcal{A}}_\hbar$ factor reduces in fact to the identity, the product depends only on the second factor $\delta\mathcal{A}_\hbar$. We have dropped the prime on ϕ_k since the information is contained in the position in the tensor product.

The Fourier transform is the map

$$\tilde{\phi}(k) = \frac{1}{(2\pi)^2} \mathrm{Tr}\,(\phi_k^* \, \phi) \tag{2.29}$$

from $\mathcal{H}$ to the momentum space $L^2(\mathbb{R}^2, dk)$ and the map

$$\phi = \int \phi_k \, \tilde{\phi}(k) dk = \int e^{ik_2 y} e^{ik_1 x} e^{-ik_1 k_2 \hbar} \, \tilde{\phi}(k) dk \tag{2.30}$$

from $L^2(\mathbb{R}^2, dk)$ to $\mathcal{H}$. The Plancherel theorem is the completeness relation for the set of ϕ_k. We have the unitary map

$$\tilde{\phi}(l) = \frac{1}{(2\pi)^2} \mathrm{Tr}\left(\phi_l^* \int \tilde{\phi}(k) \, \phi_k \, dk\right)$$

from $L^2(\mathbb{R}^2, dk)$ onto itself and the unitary map

$$\phi \otimes 1 = \frac{1}{(2\pi)^2} \int \mathrm{Tr}_2 (\phi_k \otimes \phi_k^* \cdot 1 \otimes \phi)$$

of $1 \otimes \mathcal{H}$ onto $\mathcal{H} \otimes 1$. The Fourier transform defines the map

$$\tilde{\phi}(k, k') = \frac{1}{(2\pi)^4} \mathrm{Tr}\,(\phi_k^* \otimes \phi_{k'}^* \cdot \phi \otimes \phi) = \frac{1}{(2\pi)^4} \mathrm{Tr}\,(\phi_k^* \, \phi)\, \mathrm{Tr}\,(\phi_{k'}^* \, \phi)$$

from $\mathcal{H} \otimes \mathcal{H}$ to $L^2(\mathbb{R}^2, dk) \otimes L^2(\mathbb{R}^2, dk)$ and the map

$$\phi \otimes \phi = \int \phi_k \otimes \phi_{k'} \, \tilde{\phi}(k, k') dk dk'$$

from $L^2(\mathbb{R}^2, dk) \otimes L^2(\mathbb{R}^2, dk)$ to $\mathcal{H} \otimes \mathcal{H}$. If we write $\phi \otimes \phi = \bar{\phi} \otimes \delta\phi$ as in (2.12) then (2.28) states that the Fourier transform of the diagonal factor of $\phi_k \otimes \phi_{k'}$ is a constant function and that the projection onto the ground-state in $\mathcal{F}$ produces an exponential damping in momentum space.

We are now in a position to calculate the coincidence limit of the propagator. We have

$$\begin{aligned}\langle \bar{p}' | G(x, y; x', y') | \bar{p} \rangle &= \frac{1}{(2\pi^2)} \int (k^2 + \mu^2)^{-1} \langle \bar{p}' | \phi_k \otimes \phi_k^* | \bar{p} \rangle dk \\ &= \frac{1}{(2\pi^2)} \int \frac{e^{-\kbar k^2/2}}{k^2 + \mu^2} \langle \bar{p}' \mid \bar{p} \rangle dk.\end{aligned}$$

The Feynman rules here are the same as the commutative ones except for an extra factor $e^{-\kbar k^2/4}$ at each end of a propagator of momentum k to account for the projection onto the ground state in $\mathcal{F}$. We find then

$$\langle \bar{p} | G(x, y; x', y') | \bar{p} \rangle = I(\kbar \mu^2) \, \langle \bar{p} \mid \bar{p} \rangle$$

where $I(\kbar\mu^2)$ is given by the integral [126], [102], [41], [17], [23]

$$I(\kbar\mu^2) = \frac{1}{(2\pi)^2} \int \frac{e^{-\kbar k^2/2}}{k^2 + \mu^2} dk. \tag{2.31}$$

With a change of variable it can be written as

$$I(\kbar\mu^2) = \frac{1}{4\pi} \int_0^\infty \frac{e^{-x}}{x + \kbar\mu^2/2} dx = -\frac{1}{4\pi} e^{\kbar\mu^2/2} \mathrm{Ei}(-\kbar\mu^2/2),$$

where $\mathrm{Ei}(x)$ is the exponential-integral function. When $\kbar\mu^2 \to 0$ one finds

$$I(\kbar\mu^2) = \frac{1}{4\pi} \Big(-\log(\kbar\mu^2) + \log 2 - \gamma - \frac{1}{2} \kbar\mu^2 \log(\kbar\mu^2) + o(\kbar\mu^2) \Big)$$

and when $\kbar\mu^2 \to \infty$,

$$I(\kbar\mu^2) = \frac{1}{2\pi\kbar\mu^2} + o((\kbar\mu^2)^{-2}).$$

In this example the commutator q^{12} was a real constant and the gravitational field was trivial [37], [89]. Other examples, with non-vanishing curvature, have been considered [23].

3 Finite-Dimensional Algebras

Before giving the formal definition of a differential calculus and other geometric quantities it is important to have a few simple examples in mind to which one can refer. We mention here some finite-dimensional algebras over which noncommutative geometries have been constructed. They are all more or less based on the algebras M_n. The algebra M_n is 'smooth'; it has sufficient derivations

$$\mathrm{Der}(M_n) = \{X : M_n \to M_n \mid X(fg) = Xfg + fXg\}.$$

By 'sufficient' we mean that if $Xf = 0$ for all $X \in \mathrm{Der}(M_n)$ then $f \propto 1$. It is obvious that derivations are the natural noncommutative generalization of vector fields. An important difference is the fact that derivations in general do not form a left module over the algebra. That is, if $X \in \mathrm{Der}(\mathcal{A})$ and $f \in \mathcal{A}$ then in general $fX \notin \mathrm{Der}(\mathcal{A})$. This accounts for the fact that in noncommutative geometry the derivations play a relatively secondary role. This will become clear when we define a metric later. Consider the decomposition $\mathbb{C}^n = \mathbb{C}^m \oplus \mathbb{C}^{n-m}$ of the vector space $\mathbb{C}^n$ and the corresponding decomposition

$$M_n = \begin{pmatrix} M_m & M_n^{-\prime} \\ M_n^{-\prime\prime} & M_{n-m} \end{pmatrix} = M_n^+ \oplus M_n^-$$

of M_n, where $M_n^+ = M_m \times M_{n-m}$ and $M_n^- = M_n^{-\prime} \cup M_n^{-\prime\prime}$. We shall consider below the cases $n = 2$, $m = 1$ and $n = 3$, $m = 1$ [30], [34].

A second family of examples [83], [84], [85] which is of interest is given by the same algebra as above but considered as generated by 3 special matrices x^a which are a basis of an irreducible representation of the Lie algebra of SO_3:

$$M_n = M_n(x^a) = \{x^a \mid x^a = k\!\!\!{}^{-} r^{-1} J^a\}$$

where

$$[J_a, J_b] = i\epsilon_{abc}J^c, \qquad J_a J^a = (n^2 - 1)/4 = r^4/k\!\!\!{}^{-2}.$$

If we write the Casimir relation as $g_{ab}x^a x^b = r^2$ we find that

$$n \simeq \frac{4\pi r^2}{2\pi k\!\!\!{}^{-}}.$$

With the generators x^a and the associated derivations

$$e_a = \frac{1}{ik\!\!\!{}^{-}} \mathrm{ad}\, x_a$$

the algebra M_n describes the 'fuzzy sphere' we described in the Introduction. Graded extensions of these models have been also proposed [60], [58] and various field theories studied on them [55], [57, [59].

A third family of examples is given by the algebra M_n of the previous family but considered with a different set of generators:

$$M_n = \mathcal{A}_{l/n} = M_n(u, v) = \{(u, v) \mid u^n = v^n = 1\}$$

with

$$uv = qvu, \qquad q = e^{2\pi i\alpha}, \quad \alpha = l/n.$$

In these examples we set

$$n = \frac{(2\pi r)^2}{2\pi \hbar}.$$

Two representations [125] of $\mathcal{A}_{1/n}$ are given by the bases $\{|j\rangle_1\}$, $\{|k\rangle_2\} \in \mathbb{C}^n$ with the actions

$$u|j\rangle_1 = q^j|j\rangle_1, \qquad v|j\rangle_1 = |j+1\rangle_1,$$
$$u|k\rangle_2 = |k-1\rangle_2, \; v|k\rangle_2 = q^k|k\rangle_2$$

of u and v. There is a 'Fourier transformation' [116]

$$|j\rangle_1 = \frac{1}{\sqrt{n}}\sum_{l=0}^{n-1} q^{+jk}|k\rangle_2, \quad |k\rangle_2 = \frac{1}{\sqrt{n}}\sum_{j=0}^{n-1} q^{-jk}|j\rangle_1$$

between the two bases. One deduces immediately the relations given above for u and v. If we introduce hermitian matrices x and y by

$$x|j\rangle_1 = \frac{\hbar}{r}j|j\rangle_1, \qquad y|k\rangle_2 = \frac{\hbar}{r}k|k\rangle_2$$

we find that

$$u = e^{ix/r}, \; v = e^{iy/r}, \quad q = e^{i\hbar/r^2}.$$

Consider the derivations

$$e_1 = \frac{1}{i\hbar}\mathrm{ad}\, y, \quad e_2 = -\frac{1}{i\hbar}\mathrm{ad}\, x.$$

The action of e_1 and e_2 on the generators is given by

$$e_1 u = ir^{-1}u(1-nP_2), \; e_1 v = 0,$$
$$e_2 u = 0, \qquad\qquad e_2 v = ir^{-1}v(1-nP_1)$$

where

$$P_1 = |0\rangle_2\langle 0|, \quad P_2 = |n-1\rangle_1\langle n-1|.$$

Because of the projector terms one finds

$$e_1 u^n = 0, \quad e_2 v^n = 0, \quad [e_1, e_2] = 0.$$

We refer to the algebra $M_n(u,v)$ as the fuzzy torus. Recall that the 2-torus, with $\tilde{u} = e^{i\tilde{x}/r}$, $\tilde{v} = e^{i\tilde{y}/r}$ as generators of the algebra of smooth functions, has the two vector fields (derivations) $\tilde{e}_1 f = \partial_{\tilde{x}} f$, $e_2 f = \partial_{\tilde{y}} f$ with

$$\tilde{e}_1\tilde{u} = ir^{-1}\tilde{u}, \; \tilde{e}_1\tilde{v} = 0,$$
$$\tilde{e}_2\tilde{u} = 0, \qquad \tilde{e}_2\tilde{v} = ir^{-1}\tilde{v}.$$

If we replace the torus by a lattice approximation we can impose $\tilde{u}^n = 1$, $\tilde{v}^n = 1$ but the resulting algebra will have then no derivations. From these families of examples one sees that the same algebra can take on different aspects depending on the set of generators one uses and consequently on the set of derivations one considers as special.

4 Differential Calculi

Consider an associative algebra $\mathcal{A}$ and a graded algebra

$$\Omega^*(\mathcal{A}) = \bigoplus_{i\geq 0} \Omega^i(\mathcal{A}), \qquad \Omega^0(\mathcal{A}) = \mathcal{A}$$

which is a direct sum of a family of $\mathcal{A}$-bimodules; if the grading is a $\mathbb{Z}_{\not{k}}$-grading we write $\Omega^+(\mathcal{A}) = \mathcal{A}$. A differential d is a graded derivation of $\Omega^*(\mathcal{A})$ with $d^2 = 0$; if $\alpha \in \Omega^i(\mathcal{A})$ and $\beta \in \Omega^j(\mathcal{A})$ then $\alpha\beta \in \Omega^{i+j}(\mathcal{A})$ and $d(\alpha\beta) \in \Omega^{i+j+1}(\mathcal{A})$ with

$$d(\alpha\beta) = d\alpha\beta + (-1)^i \alpha d\beta.$$

A differential algebra is a graded algebra with a differential. One says that $\Omega^*(\mathcal{A})$ is a differential calculus over $\mathcal{A}$ and the elements of $\Omega^p(\mathcal{A})$ are known as p-forms.

Over any algebra $\mathcal{A}$, including the algebra of continuous functions on a compact manifold, one can define the universal calculus $\Omega^*_u(\mathcal{A})$. As usual $\Omega^0_u(\mathcal{A}) = \mathcal{A}$. The $\Omega^1_u(\mathcal{A}) \subset \mathcal{A} \otimes \mathcal{A}$ is defined to be the $\mathcal{A}$-bimodule generated by the map

$$\mathcal{A} \xrightarrow{d_u} \mathcal{A} \otimes \mathcal{A}$$

given on an element $f \in \mathcal{A}$ by

$$d_u f = 1 \otimes f - f \otimes 1.$$

We use the symbol f here to emphasis the role of the algebra as algebra of 'functions'. The algebra $\Omega^*_u(\mathcal{A})$ is the free algebra generated by the $\Omega^1(\mathcal{A})$. The map d_u has a unique extension as a differential to all of $\Omega^*_u(\mathcal{A})$.

There exists a construction of a differential calculus over $\mathcal{A}$ uniquely determined by the bimodule $\Omega^1(\mathcal{A})$. Define

$$\Omega^1_u(\mathcal{A}) \xrightarrow{\phi_1} \Omega^1(\mathcal{A})$$

by

$$\phi_1(d_u f) = df.$$

Because $d1 = 0$ the map is well defined. We have

$$\begin{array}{ccc} \mathcal{A} & \xrightarrow{d_u} & \Omega^1_u(\mathcal{A}) \\ \| & & \phi_1 \downarrow \\ \mathcal{A} & \xrightarrow{d} & \Omega^1(\mathcal{A}) \end{array}$$

and we can write

$$\Omega^1(\mathcal{A}) = \Omega^1_u(\mathcal{A}) / \operatorname{Ker} \phi_1.$$

Every bimodule of 1-forms can be in fact so expressed. There exists a construction [25], [27], [37], [89] which defines a differential calculus as the largest differential algebra consistent with the module structure of the 1-forms. In particular the map ϕ_1 can be extended to a map

$$\Omega^*_u(\mathcal{A}) \xrightarrow{\phi_*} \Omega^*(\mathcal{A})$$

uniquely defined by the bimodule $\Omega^1(\mathcal{A})$. We shall use mainly 1-forms but on occasion 2-forms and so we must mention the product

$$\Omega^1(\mathcal{A}) \otimes_{\mathcal{A}} \Omega^1(\mathcal{A}) \xrightarrow{\pi} \Omega^2(\mathcal{A}).$$

As an example let $\mathcal{A} = \mathcal{C}(V)$ and $\Omega^1(\mathcal{A}) \equiv \Omega^1(V)$. If $f \in \mathcal{A}$ then $d_u f$ is the function of 2 variables

$$d_u f(x, y) = f(y) - f(x).$$

Choose a local system $x^\lambda(x)$ of coordinates. The corresponding de Rham 1-form is $df = \partial_\lambda f dx^\lambda$. Expand the function $f(y)$ about the point x:

$$f(y) = f(x) + (x^\lambda(y) - x^\lambda(x))\partial_\lambda f + \cdots .$$

The map ϕ_1 is given by

$$\phi_1(x^\lambda(y) - x^\lambda(x)) = dx^\lambda.$$

It annihilates all $f(x, y) \in \Omega^1_u(\mathcal{A})$ of second order in $x - y$. One such form is $f d_u g - d_u g f$:

$$(f d_u g - d_u g f)(x, y) = -(f(y) - f(x))(g(y) - g(x)).$$

This does not vanish in $\Omega^1_u(\mathcal{A})$ but its image in $\Omega^1(\mathcal{A})$ under ϕ_1 is equal to zero

The de Rham differential calculus $\Omega^*(V)$ over a manifold V can be based on the above construction using the Dirac operator to define the differential. Consider in fact $D\!\!\!/\psi = i\gamma^\alpha D_\alpha \psi, \psi \in \mathcal{H}$, where

$$D\!\!\!/ = \begin{pmatrix} 0 & D^- \\ D^+ & 0 \end{pmatrix}, \qquad \mathcal{H} = \mathcal{H}^+ \oplus \mathcal{H}^-,$$

That is

$$D\!\!\!/\psi = D^+\psi^+ + D^-\psi^-, \qquad D^\pm\psi^\pm \in \mathcal{H}^\mp.$$

Introduce a moving frame e_α and its dual θ^α: $\theta^\alpha(e_\beta) = \delta^\alpha_\beta$. The e_α are derivations of the algebra $\mathcal{C}(V)$ of smooth functions on V and the θ^α are elements of $\Omega^1(V)$. For simplicity we assume that V is parallelizable. From the Leibniz rule

$$D\!\!\!/(f\psi) = (ie_\alpha f)\gamma^\alpha\psi + fD\!\!\!/\psi$$

we find that

$$e_\alpha f \gamma^\alpha = -i[D\!\!\!/, f].$$

Consider now the map $\gamma^\alpha \mapsto \theta^\alpha$ and write

$$\hat{d}f = e_\alpha f \theta^\alpha = -i[\not{D}, f].$$

If the commutator is taken to be graded we have

$$\hat{d}^2 f = -[\not{D}^2, f], \qquad \hat{d}^2 \neq 0. \tag{4.32}$$

The map $\gamma^\alpha \mapsto \theta^\alpha$ is not an algebra isomorphism since $\gamma^\alpha\gamma^\alpha + \gamma^\beta\gamma^\alpha = g^{\alpha\beta}$. This explains the presence of the Laplace operator $\not{D}^2$ on the right-hand side of (4.32) and it is the reason why $\hat{d}^2 \neq 0$. One can reproduce the de Rham forms using the construction outlined above. The set $(\mathcal{A}, \mathcal{H}, \not{D})$ is called a spectral triple [27]. It is conjectured that it characterizes the manifold V if the latter is compact [28], [72], [103]. We refer to Várilly [121] for an elementary discussion of the extra assumptions which one must make for this to be so.

Over all the finite-dimension algebras of the previous section one can construct differential calculi. As a first example write $\mathbb{C}^2 = \mathbb{C}^1 \oplus \mathbb{C}^1$ and decompose $M_2 = M_2^+ \oplus M_2^-$ as before. The commutative algebra M_2^+ is the algebra of functions on 2 points. Introduce a graded derivation $\hat{d}\alpha$ of $\alpha \in M_n$ by

$$\hat{d}\alpha = -[\eta, \alpha], \qquad \eta \in M_n^-.$$

The bracket is graded and η is antihermitian. We find that $\hat{d}\eta = -2\eta^2$ and that $\hat{d}^2\alpha = [\eta^2, \alpha]$. If we choose η such that $\eta^2 = -1$ then $d^2 = 0$. We set $\hat{d} \equiv d$. Then $\Omega_\eta^* = M_2$ is a differential calculus over M_2^+. Since for all p we have $\Omega_\eta^{2p} = M_2^+$ and $\Omega_\eta^{2p+1} = M_2^-$ we can identify

$$\Omega_\eta^* = \Omega_\eta^+ \oplus \Omega_\eta^-, \qquad \Omega_\eta^\pm = M_2^\pm$$

and consider the calculus as $\mathbb{Z}$-graded. Notice that

$$d\eta + \eta^2 = 1. \tag{4.33}$$

The spectral triple here is $(M_2^+, \mathbb{C}^2, i\eta)$. The differential calculus of this example is not based on derivations; it can however be considered [91] as a singular contraction of a 'smooth' differential calculus.

As a variation of the above example write $\mathbb{C}^3 = \mathbb{C}^2 \oplus \mathbb{C}^1$ and decompose

$$M_3 = M_3^+ \oplus M_3^-.$$

The algebra $M_3^+ = M_2 \times M_1$ can be considered as an algebra of functions on two points with an extra structure on one of them. Introduce a graded derivation $\hat{d}\alpha = -[\eta, \alpha]$ of $\alpha \in M_n$ with

$$\eta = \begin{pmatrix} 0 & 0 & a_1 \\ 0 & 0 & a_2 \\ -a_1^* & -a_2^* & 0 \end{pmatrix} \in M_n^-.$$

We have $\Omega^0_\eta = M_3^+$ and $\Omega^1_\eta = M_3^-$. It is not possible now to have $\hat{d}^2 = 0$. We define

$$\Omega^2_\eta = M_3^+/\mathrm{Im}\,\hat{d}^2 = M_1, \qquad \Omega^p_\eta = 0,\ p \geq 3.$$

The algebra Ω^*_η is a differential calculus over M_3^+. Notice that (4.33) is again satisfied. The spectral triple here is $(M_3^+, \mathbb{C}^3, i\eta)$.

As another set of examples we consider the algebra M_n with an antihermitian basis λ_a of SU_n and define the elements

$$\theta^a_u = \lambda_b \lambda^a d_u \lambda^b$$

in $\Omega^1_u(M_n)$. Then one can show that

$$\theta^a_u f = f\theta^a_u, \qquad f \in M_n.$$

We construct an algebra $\Omega^*(M_n)$ by imposing the relations

$$\theta^a\theta^b = -\theta^b\theta^a, \qquad \theta^a = \theta^a_u$$

and a differential d as the restriction of d_u. It is easily seen that

$$\Omega^1(M_n) \simeq \bigoplus_{i=1}^{n^2-1} M_n$$

and one can show that

$$d\theta^a = -\frac{1}{2}C^a{}_{bc}\theta^b\theta^c.$$

Introduce $\theta = -\lambda_a\theta^a$. Then one sees that

$$df = -[\theta, f]$$

and that

$$d\theta + \theta^2 = 0.$$

There is an obvious similarity between $\Omega^*(M_n)$ and the algebra of de Rham differential forms on the group SU_n. The spectral triple here is $(M_n, \mathbb{C}^n, i\theta)$. Differential calculi can be also constructed over M_n which are adapted to the bases (x^a) [83], [56], [13] and (u, v) [92] which were introduced in the previous section.

Almost all examples which have been studied in any detail belong to a special class of differential calculi which could be considered as the noncommutative generalization of a parallelizable manifold. The module $\Omega^1(\mathcal{A})$ is free as a left or right $\mathcal{A}$-module and has a special basis θ^a with

$$[f, \theta^a] = 0, \qquad 1 \leq a \leq d \tag{4.34}$$

which is dual to a set of derivations $e_a = \mathrm{ad}\,\lambda_a$:

$$df = e_a f\theta^a = [\lambda_a, f]\theta^a = -[\theta, f], \qquad \theta = -\lambda_a\theta^a.$$

We refer to θ^a as a frame or Stehbein. The 'Dirac operator' θ generates $\Omega^1(\mathcal{A})$ as a bimodule; it is not a free bimodule. The λ_a must satisfy the consistency condition [37], [89]

$$2\lambda_c\lambda_d P^{cd}{}_{ab} - \lambda_c F^c{}_{ab} - K_{ab} = 0. \tag{4.35}$$

The $P^{cd}{}_{ab}$ define the product in the algebra of forms:

$$\theta^a\theta^b = \pi(\theta^c \otimes \theta^d) = P^{ab}{}_{cd}\theta^c \otimes \theta^d.$$

The $F^c{}_{ab}$ are related to the 2-form $d\theta^a$:

$$d\theta^a = -\frac{1}{2}(F^a{}_{bc} - 2\lambda_e P^{(ae)}{}_{bc})\theta^b\theta^c.$$

The K_{ab} are related to the curvature of the 'Dirac operator':

$$d\theta + \theta^2 = \frac{1}{2}K_{ab}\theta^a\theta^b.$$

All the coefficients lie in the center $\mathcal{Z}(\mathcal{A})$ of the algebra. When

$$P^{ac}{}_{cd} = \frac{1}{2}(\delta^a_c\delta^b_d - \delta^a_d\delta^b_c)$$

the $F^c{}_{ab}$ are hermitian and the K_{ab} are anti-hermitian.

Using the derivations it is straightforward to impose a reality condition on the differential d:

$$(df)^*(e_a) = (df(e^\dagger_a))^*, \qquad e^\dagger_a f = (e_a f^*)^*.$$

It is easy to see that $e^\dagger_a = e_a$ if and only if $\lambda^*_a = -\lambda_a$. For general $f \in \mathcal{A}$ and $\xi \in \Omega^1(\mathcal{A})$ one has then

$$(f\xi)^* = \xi^* f^*, \qquad (\xi f)^* = f^*\xi^*.$$

There are $I^{ab}{}_{cd} \in \mathcal{Z}(\mathcal{A})$ such that

$$(\theta^a\theta^b)^* = \imath(\theta^a\theta^b) = I^{ab}{}_{cd}\theta^c\theta^d$$

and $J^{ab}{}_{cd} \in \mathcal{Z}(\mathcal{A})$ such that

$$(\theta^a \otimes \theta^b)^* = \jmath_2(\theta^a \otimes \theta^b) = J^{ab}{}_{cd}\theta^c \otimes \theta^d.$$

The reality condition is compatible with the product if

$$\pi \circ \jmath_2 = \imath \circ \pi.$$

It follows that $(\xi\eta)^* = -\eta^*\xi^*$. One finds also the relations

$$(f\xi\eta)^* = (\xi\eta)^* f^*, \qquad (f\xi \otimes \eta)^* = (\xi \otimes \eta)^* f^*.$$

If one has a representation [27] of the algebra and the differential calculus as von Neumann algebras then one can use the modular conjugation operator J to introduce a reality condition [28] under more general conditions.

5 Yang-Mills Connections

We recall that one of the possible definitions [75] of a connection in differential geometry is in terms of a covariant derivative, a map

$$D(dx^\lambda) = -\Gamma^\lambda_{\mu\nu} dx^\mu \otimes dx^\nu$$

from the module $\Omega^1(V)$ of de Rham 1-forms to $\Omega^1(V) \otimes \Omega^1(V)$. The $\Gamma^\lambda_{\mu\nu}$ are called Christoffel symbols. This definition can be readily carried over to noncommutative geometry [27]. We define a left connection (or Yang-Mills connection) on a left $\mathcal{A}$-module $\mathcal{H}$ as the map

$$\mathcal{H} \xrightarrow{D} \Omega^1(\mathcal{A}) \otimes_\mathcal{A} \mathcal{H}$$

with a left Leibniz rule

$$D(f\psi) = df \otimes \psi + fD\psi, \qquad f \in \mathcal{A}, \quad \psi \in \mathcal{H}.$$

It has an extension:

$$\Omega^*(\mathcal{A}) \otimes_\mathcal{A} \mathcal{H} \xrightarrow{D} \Omega^*(\mathcal{A}) \otimes_\mathcal{A} \mathcal{H}$$

given by

$$D(\alpha \otimes \psi) = d\alpha \otimes \psi + (-1)^p \alpha \otimes D\psi, \quad \alpha \in \Omega^p(\mathcal{A}).$$

We shall normally drop the tensor product symbol. In particular one verifies that

$$D^2(f\psi) = fD^2\psi.$$

This means that if we define the curvature as $\mathrm{Curv}(\psi) = D^2\psi$ then it is left linear.

As a first example consider M_3^+ with the differential calculus Ω^*_η and choose for $\mathcal{H}$ the bimodule M_3^+. A covariant derivative is given by

$$D_{(0)}\psi = -\eta\psi.$$

In fact one sees that

$$D_{(0)}(f\psi) = -\eta f\psi = -f\eta\psi + df\psi.$$

The most general D is necessarily of the form

$$D\psi = -\eta\psi - \psi\phi$$

where ϕ is a right-module morphism of $\mathcal{H}$. One can write $D\psi = d\psi + \omega\psi$ in terms of a 'connection form' ω which transforms as

$$\omega' = g^{-1}\omega g + g^{-1}dg, \quad g \in U_2 \times U_1.$$

In particular: $\eta' = \eta$; therefore

$$\omega = \eta + \phi, \qquad \phi' = g^{-1}\phi g.$$

The curvature is

$$\Omega = d\omega + \omega^2 = 1 + \phi^2 = 1 - |\phi|^2$$

and the analogue of the electromagnetic action is given by

$$V(\phi) = \frac{1}{4}\mathrm{Tr}\,(1 - |\phi|^2)^2.$$

We emphasize the fact that it is electromagnetism; the geometry has changed not the theory being studied. Because of the noncommutativity however the result often looks more like nonabelian Yang-Mills theory and so we rather refer to it as such.

As a second example consider the differential calculus $\Omega^*(M_n)$ over M_n introduced above and choose for $\mathcal{H}$ the bimodule M_n. A special covariant derivative is given by

$$D_{(0)}\psi = -\theta\psi$$

and the most general one is of the form

$$D\psi = -\theta\psi - \psi\phi.$$

One can again express D in terms of a 'connection form' which transforms as

$$\omega' = g^{-1}\omega g + g^{-1}dg, \quad g \in U_n.$$

In particular $\theta' = \theta$; therefore

$$\omega = \theta + \phi, \qquad \phi' = g^{-1}\phi g.$$

The curvature is

$$\Omega = d\omega + \omega^2 = \frac{1}{2}\Omega_{ab}\theta^a\theta^b$$

where

$$\Omega_{ab} = [\phi_a, \phi_b] - C^c{}_{ab}\,\phi_c.$$

The $C^c{}_{ab}$ is a sort of 'Christoffel symbol'; we shall see below in (6.39) that M_n with the present differential calculus is 'curved' as a geometry. The analogue of the electromagnetic action is

$$V(\phi) = \frac{1}{4}\mathrm{Tr}\,(\Omega_{ab}\Omega^{ab}).$$

Again, as above, this action describes 'electromagnetism' on a noncommutative 'space'. By radically changing the 'space' we have radically changed the aspect of a well-known theory.

6 Metrics and Linear Connections

Let $\mathcal{M}$ be an $\mathcal{A}$-bimodule and $\Omega^*(\mathcal{A})$ a differential calculus over some algebra $\mathcal{A}$. Consider a covariant derivative xcovariant derivative

$$\mathcal{M} \xrightarrow{D} \Omega^1(\mathcal{A}) \otimes_{\mathcal{A}} \mathcal{M}$$

as in the preceding section but with in addition a right Leibniz rule and a flip

$$\mathcal{M} \otimes_{\mathcal{A}} \Omega^1(\mathcal{A}) \xrightarrow{\sigma} \Omega^1(\mathcal{A}) \otimes_{\mathcal{A}} \mathcal{M}.$$

The right Leibniz rule is defined using the flip

$$D(\xi f) = \sigma(\xi \otimes df) + (D\xi)f. \tag{6.36}$$

The σ 'brings' d to the left where it belongs without changing the order of the factors. In general $\sigma^2 \neq 1$. The de Rham σ is necessarily of the form

$$\sigma(\xi \otimes \eta) = \eta \otimes \xi.$$

The flip is in all cases necessarily $\mathcal{A}$-bilinear. We define a bimodule $\mathcal{A}$-connection as the couple (D, σ). In the special case when $\mathcal{M} = \Omega^1(\mathcal{A})$ we speak of a linear connection [99], [43], [44], [67].

We define the torsion map

$$\Theta : \Omega^1(\mathcal{A}) \to \Omega^2(\mathcal{A})$$

by $\Theta = d - \pi \circ D$. It is left-linear and

$$\Theta(\xi)f - \Theta(\xi f) = \pi \circ (1 + \sigma)(\xi \otimes df).$$

We shall impose the condition

$$\pi \circ (\sigma + 1) = 0 \tag{6.37}$$

in order to assure that the torsion is bilinear. We shall find below other consequences of this condition.

We shall define a metric as a bilinear map

$$\Omega^1(\mathcal{A}) \otimes_{\mathcal{A}} \Omega^1(\mathcal{A}) \xrightarrow{g} \mathcal{A}$$

which satisfies the symmetry condition

$$g \propto g \circ \sigma. \tag{6.38}$$

This is a 'conservative' definition, a straightforward generalization of one of the possible definitions of a metric in ordinary differential geometry. The usual definition of a metric in the commutative case is a bilinear map

$$\mathcal{X} \otimes_{\mathcal{C}(V)} \mathcal{X} \xrightarrow{g} \mathcal{C}(V)$$

where $\mathcal{X}$ is the $\mathcal{C}(V)$-bimodule of vector fields on V. This definition is not suitable in the noncommutative case since the set of derivations of the algebra, which is the generalization of $\mathcal{X}$, has no natural structure as an $\mathcal{A}$-module. The linearity condition is equivalent to a locality condition for the metric; the length of a vector at a given point depends only on the value of the metric and the vector field at that point. In the noncommutative case bilinearity is the natural (and only possible) expression of locality. It would exclude, for example, a metric in ordinary geometry defined by a map of the form

$$g(\alpha, \beta)(x) = \int_V g_x(\alpha_x, \beta_y) G(x, y) dy.$$

Here $\alpha, \beta \in \Omega^1(V)$ and g_x is a metric on the tangent space at the point $x \in V$. The function $G(x, y)$ is an arbitrary smooth function of x and y and dy is the measure on V induced by the metric. Other definitions of a metric have been given, some of which are similar to that given above but which weaken the locality condition [15], [16] and one [31], [12] which defines a metric on the associated space of pure states.

Using σ one can also construct an extension

$$\mathcal{M} \otimes_{\mathcal{A}} \mathcal{M} \xrightarrow{D_2} \Omega^1(\mathcal{A}) \otimes_{\mathcal{A}} \mathcal{M} \otimes_{\mathcal{A}} \mathcal{M}$$

by

$$D_2(\xi \otimes \eta) = D\xi \otimes \eta + (\sigma \otimes 1) \circ (\xi \otimes D\eta).$$

The linear connection is said to be metric compatible if

$$(1 \otimes g) \circ D_2 = d \circ g.$$

This is a straightforward generalization of the usual definition of metric compatibility.

As an example consider the differential calculus Ω^*_η over M_3^+ [90]. Because of the bimodule identification

$$\Omega^1_\eta \otimes_{M_3^+} \Omega^1_\eta = M_3^+$$

we can conclude that $\sigma = \operatorname{diag}(\mu, \mu, -1)$ with $\mu \in \mathbb{C}$. Introduce the matrix

$$\eta_1 = \begin{pmatrix} 0 & 0 & a_1 \\ 0 & 0 & a_2 \\ 0 & 0 & 0 \end{pmatrix}$$

and define

$$\eta = \eta_1 - \eta_1^*, \quad \eta_{ij} = \eta_i \otimes \eta_j^*, \quad \zeta = \eta_1^* \otimes \eta_1.$$

Then

$$\sigma(\eta_{ij}) = \mu \eta_{ij}, \qquad \sigma(\zeta) = -1.$$

The unique bilinear metric is given by

$$g(\eta_{ij}) = \eta_i \eta_j^* \in M_2, \qquad g(\zeta) = -e \in M_1.$$

It is real on η_{ij} and imaginary on ζ. The unique covariant derivative is given by [90]

$$D\xi = -\eta \otimes \xi + \sigma(\xi \otimes \eta).$$

The torsion vanishes and the connection is metric compatible if $\mu = 1$.

If the geometry is parallelizable in the sense of Section 4 then the covariant derivative can be defined in terms of the basis:

$$D\theta^a = -\omega^a{}_{bc}\theta^b \otimes \theta^c, \qquad \omega^a{}_{bc} \in \mathcal{A},$$
$$\sigma(\theta^a \otimes \theta^b) = S^{ab}{}_{cd}\theta^c \otimes \theta^d, \quad S^{ab}{}_{cd} \in \mathcal{Z}(\mathcal{A}).$$

The metric must be of the form $g(\theta^a \otimes \theta^b) = g^{ab}$ with $g^{ab} \in \mathcal{Z}(\mathcal{A})$. A particular linear connection is given by

$$D\theta^a = -\theta \otimes \theta^a + \sigma(\theta^a \otimes \theta).$$

It is metric-compatible [37] if

$$\omega^a{}_{bd}g^{de} + \omega^e{}_{fg}S^{af}{}_{bh}g^{hg} = 0.$$

As an example consider M_n as algebra and introduce a metric by setting $g(\theta^a \otimes \theta^b) = g^{ab}$, the components of the SU_n Killing metric. The linear connection defined by

$$D\theta^a = -\omega^a{}_b \otimes \theta^b, \qquad \omega^a{}_b = -\frac{1}{2}C^a{}_{bc}\theta^c \tag{6.39}$$

has vanishing torsion and is compatible with the metric. With this connection the geometry of M_n looks like the invariant geometry of the group SU_n. Since the elements of the algebra commute with the frame θ^a, we can define D on all of $\Omega^*(M_n)$ using the left Leibniz rule. The map σ is necessarily given by

$$\sigma(\theta^a \otimes \theta^b) = \theta^b \otimes \theta^a. \tag{6.40}$$

It follows that D satisfies also the right Leibniz rule (6.36) and the metric satisfies the symmetry condition (6.38). We can suppose a general linear connection to be of the form

$$D\theta^a = -\omega^a{}_{bc}\,\theta^b \otimes \theta^c$$

with $\omega^a{}_{bc}$ arbitrary elements of M_n. Suppose that σ is given by (6.40). From the Leibniz rules we find that

$$0 = D([f, \theta^a]) = [f, D\theta^a]$$

and so the $\omega^a{}_{bc}$ must be all in the center of M_n. They are complex numbers. If we require that the torsion vanish then we have

$$\omega^a{}_{[bc]} = C^a{}_{bc}.$$

If we impose the condition that the connection be metric-compatible we find that

$$\omega^a{}_{(bc)} = 0.$$

The linear connection (6.39) is the unique torsion-free metric connection on $\Omega^1(M_n)$. If $\sigma^2 = 1$ then σ is necessarily given by (6.40).

The condition that D be real can be written

$$D\xi^* = (D\xi)^*, \qquad (\omega^a{}_{bc})^* = \omega^a{}_{de}(J^{de}{}_{bc})^*.$$

From the Leibniz rules and the equalities

$$(D(f\xi))^* = D((f\xi)^*) = D(\xi^* f^*)$$

for all f one finds the conditions [48]

$$(fD\xi)^* = (D\xi^*)f^*, \qquad (\xi \otimes \eta)^* = \sigma(\eta^* \otimes \xi^*).$$

We see than that the flip was necessary also to define a reality condition. The reality condition for the metric becomes

$$g((\xi \otimes \eta)^*) = (g(\xi \otimes \eta))^*, \qquad S^{ab}{}_{cd}g^{cd} = (g^{ba})^*.$$

Although it is not a completely satisfactory object in noncommutative geometry one can define the curvature as the map

$$\mathrm{Curv} : \Omega^1(\mathcal{A}) \longrightarrow \Omega^2(\mathcal{A}) \otimes_{\mathcal{A}} \Omega^1(\mathcal{A})$$

given by

$$\mathrm{Curv} = D^2 = \pi_{12} \circ D_2 \circ D.$$

The appropriate reality condition is the equality

$$\mathrm{Curv}(\xi^*) = (\mathrm{Curv}(\xi))^*.$$

We shall impose the stronger condition

$$D_2(\xi \otimes \eta)^* = (D_2(\xi \otimes \eta))^*.$$

There are $J^{abc}{}_{def} \in \mathcal{Z}(\mathcal{A})$ such that

$$(\theta^a \otimes \theta^b \otimes \theta^c)^* = J^{abc}{}_{def}\theta^d \otimes \theta^e \otimes \theta^f.$$

We find then that

$$J^{abc}{}_{def} = J^{ab}{}_{pq}J^{pc}{}_{dr}J^{qr}{}_{ef} = J^{bc}{}_{pq}J^{aq}{}_{rf}J^{rp}{}_{de}.$$

The second equality is the Yang-Baxter equation; it becomes the braid equation for the map σ:

$$\sigma_{12}\sigma_{23}\sigma_{12} = \sigma_{23}\sigma_{12}\sigma_{23}.$$

We have set here $\sigma_{12} = \sigma \otimes 1$ and $\sigma_{23} = 1 \otimes \sigma$. We refer to the litterature [94] for more details of these equations.

7 Infinite-Dimensional Models

An important family of infinite-dimensional models are based on the algebras $\mathbb{C}_q^n$ and $\mathbb{R}_q^n$ introduced by Faddeev *et al.* [45]. These are defined to be invariant under a generalization $SO_q(n)$ of the Lie group $SO(n,\mathbb{C})$ which is known as a quantum group [124], [95], [96]. The structure of $SO_q(n)$ is determined by a braid matrix $\hat{R}$ which has a decomposition

$$\hat{R} = qP_s - q^{-1}P_a + q^{1-n}P_t$$

with the P_s, P_a, P_t mutually orthogonal and

$$P_s + P_a + P_t = 1.$$

For example $P_t{}^{ij}{}_{kl} = (g^{mn}g_{mn})^{-1}g^{ij}g_{kl}$ and

$$g_{il}\,\hat{R}^{\pm 1 lh}{}_{jk} = \hat{R}^{\mp 1 hl}{}_{ij}\,g_{lk},$$
$$g^{il}\,\hat{R}^{\pm 1 jk}{}_{lh} = \hat{R}^{\mp 1 ij}{}_{hl}\,g^{lk}.$$

The g_{ij} is the q-deformed euclidean metric. The q-euclidean 'space' $\mathbb{C}_q^n$ has n generators x^i with

$$P_a{}^{ij}{}_{kl}x^kx^l = 0. \tag{7.41}$$

The real q-euclidean 'spaces' $\mathbb{R}_q^n$ are obtained by imposing $q \in \mathbb{R}^+$ and

$$(x^i)^* = x^jg_{ji}.$$

The 'length' squared $r^2 = g_{ij}x^ix^j = (x^i)^*x^i$ generates the center of $\mathbb{R}_q^n$. We shall extend $\mathbb{R}_q^n$ by adding the square rood r of r^2 as well as its inverse. r^{-1}. We shall also add an extra element Λ called the dilatator [100], [112], [93] which satisfies the conditions

$$x^i\Lambda = q\Lambda x^i, \qquad \Lambda^* = \Lambda^{-1}$$

The center is now trivial. We shall choose differential calculi with $d\Lambda = 0$. This is unsatisfactory from the point of view of the formalism which we described in Section 4 since we have an element which does not lie in the center and which has nevertheless a vanishing differential. This would be the equivalent in ordinary geometry of having a function which is not constant but such that all derivatives of it vanished. It would correspond to the geometry of a hyper-surface. We can consider the geometries described below as those of a 'hyper-surface' defined, very symbolically, by 'Λ = const.'.

There are two natural $SO_q(n)$-covariant differential calculi over $\mathbb{R}_q^n$ given by

$$x^i\xi^j = q\hat{R}^{ij}{}_{kl}\xi^kx^l$$

for $\xi^j \in \Omega^1(\mathbb{R}_q^n)$ and

$$x^i\bar{\xi}^j = q^{-1}\hat{R}^{-1 ij}{}_{kl}\bar{\xi}^kx^l$$

for $\bar\xi^j \in \bar\Omega^1(\mathbb{R}^n_q)$. There is no real calculus compatible with the coaction of the quantum group. One can extend the involution to $\Omega^1(\mathcal{A}) \oplus \bar\Omega^1(\mathcal{A})$ by setting

$$(\xi^i)^* = \bar\xi^j g_{ji}$$

There exists a frame [48] $(\theta^a, \bar\theta^a)$ with $(\theta^a)^* = \bar\theta^b g_{ba}$.

As an example consider $\mathbb{R}^1_q$. This 'space' has been studied from several points of view and has served as basis for several models [115], [114], [61], [46], [15], [16]. The algebra $\mathbb{R}^1_q$ consists of two generators x and Λ with $x\Lambda = q\Lambda x$. We choose x hermitian and $q \in (1, \infty)$ We introduce also an element y through the equation $x = q^y$. It follows that $\Lambda^{-1} y \Lambda = y + 1$. The differential calculus $\Omega^*(\mathbb{R}^1_q)$ is defined by the module structure

$$x dx = q dx x, \qquad dx \Lambda = q \Lambda dx.$$

If we introduce $z = q^{-1}(q-1) > 0$ and choose

$$\lambda_1 = -z^{-1}\Lambda$$

then we find that the calculus is defined by $e_1 = \mathrm{ad}\,\lambda_1$. Since Λ is unitary e_1 is not real. The second differential calculus $\bar\Omega^*(\mathbb{R}^1_q)$, with

$$x \bar{d}x = q^{-1} \bar{d}x x, \qquad \bar{d}x \Lambda = q \Lambda \bar{d}x$$

is based on the derivation $\bar{e}_1 = e_1^\dagger$.

The dual frames θ^1 and $\bar\theta^1$ are given by

$$\theta^1 = \theta^1_1 dx, \qquad \theta^1_1 = \Lambda^{-1} x^{-1},$$
$$\bar\theta^1 = \bar\theta^1_1 \bar{d}x, \qquad \bar\theta^1_1 = q^{-1} \Lambda x^{-1}.$$

Consider the element

$$\lambda_{R1} = (\lambda_1, \bar\lambda_1) = z^{-1}(-\Lambda, \Lambda^{-1})$$

of $\mathbb{R}^1_q \times \mathbb{R}^1_q$. The associated derivation $e_{R1} = \mathrm{ad}\,\lambda_{R1}$ is real. Using it one can construct a differential calculus

$$\Omega^*_R(\mathbb{R}^1_q) \subset \Omega^*(\mathbb{R}^1_q) \times \bar\Omega^*(\mathbb{R}^1_q)$$

whose structure is given by the relations

$$d_R \theta^1_R = 0, \quad (\theta^1_R)^2 = 0.$$

The forms θ^1, $\bar\theta^1$ and θ^1_R are all exact.

A rather straightforward calculation yields the result that there are two torsion-free connections [15], one of which is compatible with the unique local metric:

$$g(\theta^1_R \otimes \theta^1_R) = 1.$$

The flip is given by $\sigma_R = 1$ and the covariant derivatives are real.

One can represent [61], [46], [113], [15] $\mathbb{R}^1_q$ on a Hilbert space $\mathcal{R}_q = \{|k\rangle\}$ by

$$x|k\rangle = q^k|k\rangle, \qquad \Lambda|k\rangle = |k+1\rangle$$

The element y has then the representation

$$y|k\rangle = k|k\rangle$$

The representation can be extended to the differential calculi by setting

$$\theta^1 = 1, \quad \bar{\theta}^1 = 1, \quad \theta^1_R = 1.$$

The action of the two elements dx and $\bar{d}x$ is given then by

$$dx|k\rangle = q^{k+1}|k+1\rangle, \qquad \bar{d}x|k\rangle = q^k|k-1\rangle$$

and the real differential $d_R x$ can be represented by the operator

$$d_R x|k\rangle = q^k(q|k+1\rangle + \overline{|k-1\rangle})$$

We have placed a bar over the second copy of $\mathcal{R}_q$.

We have an interpretation of the metric in terms of observables since we have a representation of x and $d_R x$ on the Hilbert space $\mathcal{R}_q$. In this representation the distance s along the 'line' x is given by the expression

$$ds(k) = \|\sqrt{g'_{11}} d_R x(|k\rangle + \overline{|k\rangle})\| = \|\theta^1_R(|k\rangle + \overline{|k\rangle})\|.$$

We have used here

$$g'^{11} = g(d_R x \otimes d_R x) = (e_{R1} x)^2 g(\theta^1_R \otimes \theta^1_R).$$

We find that

$$ds(k) = \|\, |k\rangle + \overline{|k\rangle}\, \| = 1$$

The 'space' is discrete and the spacing between 'points' is uniform. One other, nonlocal, metric, associated to the second covariant derivative mentioned above, has been studied on $\mathbb{R}^1_q$ [115], [61], [46], [16]. It gives a more exotic structure to the lattice of points.

As a second example consider the algebra $\mathbb{R}^3_q$. It has three generators $x^i = (x^-, y, x^+)$. Introduce the parameter $h = \sqrt{q} - 1/\sqrt{q}$. The defining relations (7.41) can be written as

$$x^- y = q\, y x^-,$$

$$x^+ y = q^{-1} y x^+,$$

$$[x^+, x^-] = h y^2.$$

The metric matrix is given by $g_{ij} = g^{ij}$ with

$$g_{ij} = \begin{pmatrix} 0 & 0 & 1/\sqrt{q} \\ 0 & 1 & 0 \\ \sqrt{q} & 0 & 0 \end{pmatrix}.$$

Three linearly independent, hermitian generators (y^j) can be obtained as combinations of the x^i, given by

$$y^j = \Lambda^j_i x^i, \qquad \Lambda^j_i = \frac{1}{\sqrt{2}} \begin{pmatrix} 1 & 0 & \sqrt{q} \\ 0 & \sqrt{2} & 0 \\ i & 0 & -i\sqrt{q} \end{pmatrix}. \tag{7.42}$$

With respect to the new generators the metric is given by

$$g^{kl} = g^{ij}\Lambda^k_i \Lambda^l_j = \frac{1}{2}\begin{pmatrix} q+1 & 0 & i(q-1) \\ 0 & 2 & 0 \\ -i(q-1) & 0 & q+1 \end{pmatrix}.$$

It is hermitian but no longer real. In the limit $q \to 1$ one sees that $g^{kl} \to \delta^{kl}$. Except when considering the commutative limit it is more convenient to remain with the original 'coordinates' and a real metric.

The equations (4.34) admit to within a linear transformation a unique solution given by

$$\begin{aligned} \theta^- &= \Lambda^{-1} y^{-1} \xi^-, \\ \theta^0 &= \Lambda^{-1} r^{-1} (\sqrt{q}(q+1) y^{-1} x^+ \xi^- + \xi^0), \\ \theta^+ &= -\Lambda^{-1} r^{-2} (\sqrt{q} q (q+1) y^{-1} (x^+)^2 \xi^- + (q+1) x^+ \xi^0 - y \xi^+). \end{aligned}$$

An analogous expression can be found for the frame $\bar\theta^a$ of the differential calculus $\bar\Omega^*(\mathbb{R}^3_q)$. From the relations

$$[(\theta^a)^*, f^*] = -[\theta^a, f]^* = 0, \qquad f \in \mathbb{R}^3_q,$$

it follows that $(\theta^a)^*$ can be written in terms of $\bar\theta^b$. We choose the second frame so that the relation

$$(\theta^a)^* = \bar\theta^b g_{ba}$$

is satisfied. By direct calculation one finds that

$$P_t{}^{ab}{}_{cd} \theta^c \theta^d = 0, \qquad P_s{}^{ab}{}_{cd} \theta^c \theta^d = 0.$$

Therefore we can conclude that the coefficients $P^{ab}{}_{cd}$ of (4.35) are equal to the coefficients $P_{(a)}{}^{ab}{}_{cd}$ of (7.41):

$$P^{ab}{}_{cd} = P_{(a)}{}^{ab}{}_{cd}.$$

Consider the elements $\lambda_a \in \mathbb{R}^3_q$ with

$$\begin{aligned} \lambda_- &= +h^{-1} q \Lambda y^{-1} x^+, \\ \lambda_0 &= -h^{-1} \sqrt{q} \Lambda y^{-1} r, \\ \lambda_+ &= -h^{-1} \Lambda y^{-1} x^-. \end{aligned}$$

The θ^a are dual to the derivations $e_a = \mathrm{ad}\,\lambda_a$. The commutation relations of the λ_a are identical to those of x^i:

$$\begin{aligned}
\lambda_-\lambda_0 &= q\lambda_0\lambda_-, \\
\lambda_+\lambda_0 &= q^{-1}\lambda_0\lambda_+, \\
[\lambda_+, \lambda_-] &= h(\lambda_0)^2.
\end{aligned}$$

These equations can be rewritten more compactly in the form

$$P^{ab}{}_{cd}\lambda_a\lambda_b = 0.$$

This is the consistency relation (4.35) of the frame formalism with

$$C^a{}_{bc} = 0, \quad F_{ab} = 0.$$

In the commutative limit the frame becomes a moving frame in the sense of Cartan. We shall suppose that the constant Z of (1.3) is given by

$$Z = 1.$$

That is, we suppose that the generators of the algebra tend to their naive natural limit as (complex) coordinates on a real manifold and that the frame tends to the corresponding limit of a moving frame on this manifold. In the coordinates

$$(x, y, z) \equiv \tilde{y}^k = \lim_{q\to 1} y^k$$

one finds [49]

$$\begin{aligned}
\theta^1 &= (yr)^{-1}(rdx - xdr + izdr), \\
\theta^2 &= (yr)^{-1}(rdr - ixdz + izdx), \\
\theta^3 &= (yr)^{-1}(rdz - ixdr - zdr).
\end{aligned} \tag{7.43}$$

Although in the commutative limit the differential is real, we see that the frame is not. From (7.43) we find that in the commutative limit the metric is given by the line element

$$ds^2 = \sum_a (\theta^a)^2 = r^{-2}(dx^2 + dy^2 + dz^2). \tag{7.44}$$

If one uses spherical polar coordinates then one sees immediately that the Riemannian space is $S^2 \times \mathbb{R}$ with $\log r$ the preferred coordinate along the line. The radius of the sphere is equal to 1. It is often found in specific calculations in general relativity that it is more convenient to use a complex frame to calculate real curvature invariants. We can see however no property of the above frame which makes it particularly adapted to study the space $S^2 \times \mathbb{R}$. An interesting feature of this example is that there is a unique linear connection which is flat; it is equal to the Levi-Civita connection of a flat metric conformally equivalent to the one which we have found. The problem we are considering here lies in fact a little outside the range of the general theory because of the element Λ which is not in the center but which has nevertheless a vanishing differential. If one could in some way eliminate the Λ then r would lie in the center and any line element related to (7.44) by a conformal factor, function of r, would be local. This would of course include the flat metric.

As a final example we consider a noncommutative version of the Lobachevsky plane [79], defined as the set of points

$$V = \{(\tilde{x}, \tilde{y}) \in \mathbb{R}^2 \; \tilde{y} > 0\}.$$

A moving frame is given by

$$\theta^1 = \tilde{y}^{-1} d\tilde{x}, \qquad \theta^2 = \tilde{y}^{-1} d\tilde{y}.$$

Introduce the algebra $\mathcal{A}_h$ with hermitian generators (x, y) and relation

$$[x, y] = -2ihy.$$

A real frame is given by

$$\theta^1 = y^{-1} dx, \qquad \theta^2 = y^{-1} dy.$$

The structure of the differential calculus $\Omega^*(\mathcal{A})$ is given by

$$(\theta^1)^2 = 0, \qquad (\theta^2)^2 = 0, \qquad \theta^1\theta^2 + \theta^2\theta^1 = 0.$$

This algebra and differential calculus are invariant under the coaction of what is known as the Jordanian deformation [101], [76], [2], [3], [71], [22] $SL_h(2, \mathbb{C})$ of the special linear group SL_2. To within a normalization the unique metric is given by $g(\theta^a \otimes \theta^b) = g^{ab}$, where on the right-hand side are the components of the euclidean metric on $\mathbb{R}^2$. The unique torsion-free, metric-compatible linear connection is given by [71], [22]

$$D\theta^1 = \theta^1 \otimes \theta^2, \qquad D\theta^2 = -\theta^1 \otimes \theta^1.$$

The curvature map becomes

$$\mathrm{Curv}(\theta^1) = \theta^1\theta^2 \otimes \theta^2, \qquad \mathrm{Curv}(\theta^2) = -\theta^1\theta^2 \otimes \theta^1.$$

That is, it satisfies $R_{1212} = -1$, exactly as the commutative counterpart.

To construct a representation of $\mathcal{A}_h$ one can introduce (ξ, η) with $[\xi, \eta] = 2ih$ and write

$$x = \xi\eta - ih, \qquad y = \xi$$

A representation of ξ and η will yield a representation of $\mathcal{A}_h$. If one defines $\Lambda = e^{ix}$ and $q = e^{-2h}$ then $y\Lambda = q\Lambda y$, which defines $\mathbb{R}^1_q$ again but with another differential calculus.

There are many other infinite-dimensional algebras with associated differential calculi which have served as basis for exploring the possible applications of noncommutative geometry to physics. We mention in particular the noncommutative torus or rotation algebra [108], [109], [32], [27] which extends the noncommutative or fuzzy torus we defined in Section 3, the quantum plane [95], [96], [123], the quantum sphere [104] as well as 'quantum' deformations of Minkowski space [5], [105], [106], [14], [6], [74].

8 Gravity

The classical gravitational field is normally supposed to be described by a torsion-free, metric-compatible linear connection on a smooth manifold. One might suppose that it is possible to formulate a noncommutative theory of (classical/quantum) gravity by replacing the algebra of functions by a more general algebra and by choosing an appropriate differential calculus. We recall that the classical gravitational action is given by

$$S[g] = \mu_P^4 \Lambda_c + \mu_P^2 \int R.$$

In the noncommutative case there is a natural definition of the integral [32], [26], [27] but there does not seem to be a natural generalization of the Ricci scalar. One of the problems is the one we touched upon in Section 6: the natural generalization of the curvature form is in general not right-linear in the noncommutative case. The Ricci scalar then will not be local. One way of circumventing this problem is to consider classical gravity as an effective theory and the Einstein-Hilbert action as an induced action. There is an interesting theory of gravity, due to Sakharov [110], [111]. and popularized by Wheeler, called induced gravity, in which the gravitational field is a phenomenological coarse-graining of more fundamental fields. Flat Minkowski space-time is to be considered as a sort of perfect crystal and curvature as a manifestation of elastic tension, or possibly of defects, in this structure. A deformation in the crystal produces a variation in the vacuum energy which we perceive as gravitational energy. 'Gravitation is to particle physics as elasticity is to chemical physics: merely a statistical measure of residual energies.' The description of the gravitational field which we are attempting to formulate using noncommutative geometry is not far from this. We have noticed that the use of noncommuting coordinates is a convenient way of making a discrete structure like a lattice invariant under the action of a continuous group. In this sense what we would like to propose is a Lorentz-invariant version of Sakharov's crystal. Each coordinate can be separately measured and found to have a distribution of eigenvalues similar to the distribution of atoms in a crystal. The gravitational field is to be considered as a measure of the variation of this distribution just as elastic energy is a measure of the variation in the density of atoms in a crystal. The idea then is to identify the gravitational action with the quantum corrections to a classical field in a curved background. If $\Delta[g]$ is the operator which describes the propagation of a given mode in presence of a metric g then one finds that, with a cut-off Λ, the effective action is given by

$$\begin{aligned} \Gamma[g] \propto \operatorname{Tr} \log \Delta[g] \simeq \\ \Lambda^4 \mathrm{Vol}(V)[g] + \Lambda^2 S_1[g] + (\log \Lambda) S_2[g] + \cdots . \end{aligned}$$

If one identifies $\Lambda = \mu_P$ then one finds that $S_1[g]$ is the Einstein-Hilbert action. A problem with this is that it can be only properly defined on a compact manifold with a metric of euclidean signature and Wick rotation on a curved space-time is a rather delicate if not dubious procedure. Another problem with this theory, as indeed with the gravitational field in general, is that it predicts an extremely large cosmological constant. The expression Tr $\log \Delta[g]$ has a natural generalization to the noncommutative case [64], [1], [29], [19]. See also Example 7.3.5 of Madore [85].

We have defined gravity using a linear connection, which required the full bimodule structure of the $\mathcal{A}$-module of 1-forms. We argued that this was necessary to obtain a satisfactory definition of locality as well as a reality condition. It is possible to relax these requirements and define gravity as a Yang-Mills field [18], [118], [78], [20], [24], [51] or as a couple of left and right connections [35], [36]. If the algebra is commutative (but not an algebra of smooth functions) then to a certain extent all definitions coincide [38], [39], [77], [8].

One of the first, obvious applications of noncommutative geometry is as an alternative hidden structure of Kaluza-Klein theory [81], [82]. This means that one leaves space-time as it is and one modifies only the extra dimensions; one replaces their algebra of functions by a noncommutative algebra, usually of finite dimension to avoid the infinite tower of massive states of traditional Kaluza-Klein theory. Because of this restriction and because the extra dimensions are purely algebraic in nature the length scale associated with them can be arbitrary [86], indeed as large as the Compton wave length of a typical massive particle.

The algebra of Kaluza-Klein theory is therefore, for example, a product algebra of the form

$$\mathcal{A} = \mathcal{C}(V) \otimes M_n.$$

Normally V would be chosen to be a manifold of dimension four, but since much of the formalism which we shall outline is identical to that of the M(atrix)-theory of D-branes [10], [53], [33] we shall leave the dimension unspecified. We mention first electromagnetism and then gravity.

Let $\theta^i = (\theta^\alpha, \theta^a)$ be a frame over $\mathcal{A}$ with θ^α a moving frame on V. This means that we suppose that V is parallelizable. The matrix factor is also parallelizable with a differential calculus such that

$$\Omega^1(M_n) \simeq \bigoplus_1^d M_n$$

for some integer d. The interesting case is when

$$n \gg d.$$

We write $\Omega^1(\mathcal{A})$ as a direct sum

$$\Omega^1(\mathcal{A}) = \Omega^1_h \oplus \Omega^1_v,$$

with

$$\Omega^1_h = \Omega^1(V) \otimes M_n, \qquad \Omega^1_v = \mathcal{C}(V) \otimes \Omega^1(M_n).$$

The differential df of an element f of $\mathcal{A}$ is given by

$$df = d_h f + d_v f.$$

In terms of the frame θ^i we have as usual

$$d_h f = e_\alpha f \theta^\alpha, \qquad d_v f = e_a f \theta^a = -[\theta, f].$$

We introduce a torsion-free linear connection, compatible with the frame.

$$d\theta^i + \omega^i{}_j \theta^j = 0$$

with $\theta^\alpha + \omega^\alpha{}_\beta \theta^\beta = 0$, and $\omega^a{}_b = -\frac{1}{2} F^a{}_{bc} \theta^c$. Referring back to Section 4.2 we see that the electromagnetic action for the potential $\omega = A + (\theta + \phi)$ is given by

$$\begin{aligned} S[A, \phi] &= \int \mathcal{L}(A, \phi) = \frac{1}{4} \mathrm{Tr} \int \Omega_{ij} \Omega^{ij} \\ &= \frac{1}{4} \mathrm{Tr} \int F_{\alpha\beta} F^{\alpha\beta} + \frac{1}{2} \mathrm{Tr} \int D_\alpha \phi_a D^\alpha \phi^a - \int V(\phi) \end{aligned}$$

where [50], [97], [21], [42], [82]

$$V(\phi) = -\frac{1}{4} \mathrm{Tr} \, (\Omega_{ab} \Omega^{ab}).$$

As an example choose $d = 3$ and

$$F_{abc} = r^{-1} \epsilon_{abc}.$$

The hidden 'space' is a fuzzy sphere of radius r. The potential $V(\phi)$ vanishes when ϕ lies on a gauge orbit of a representation of SU_2. There are

$$p(n) \simeq \frac{e^{\pi\sqrt{2n/3}}}{4n\sqrt{3}}$$

such orbits.

If one replaces the matrix algebra by the algebra of the Connes-Lott model then one obtains a family of theories which includes the standard model of the electroweak interactions. It, and its extensions to include the strong interactions, have been extensively studied [30], [9], [65], [122], [80], [107]. For a recent review we refer to the Schladming lecture notes by Kastler [66].

For the simple models with a matrix extension one can use as gravitational action the Einstein-Hilbert action in 'dimension' $4 + d$, including possibly Gauss-Bonnet terms [82], [86], [85], [87], [68], [88].

References

[1] Ackermann, T., Tolksdorf, J. (1996) A generalized Lichnerowicz formula, the Wodzicki Residue and Gravity, J. Geom. Phys. **19** 143.

[2] Aghamohammadi, A. (1993) The two-parametric extension of h deformation of $GL(2)$ and the differential calculus on its quantum plane, Mod. Phys. Lett. A **8** 2607.

[3] Aghamohammadi, A., Khorrami, M., Shariati, A. (1995) h-deformation as a contraction of q-deformation, J. Phys. A: Math. Gen. **28** L225.

[4] Ambjørn, J., Jonsson, T. (1997) Quantum Geometry, a statistical field theory approach, Cambridge University Press.

[5] Aschieri, P., Castellani, L. (1996) Bicovariant calculus on twisted $ISO(N)$, Quantum Poincaré Group and Quantum Minkowski Space, Int. J. Mod. Phys. A **11** 4513.

[6] de Azcárraga, J.A., Kulish, P.P., Rodenas, F. (1997) Twisted h-spacetimes and invariant equations, Z. Physik C **76** 567.

[7] Baez, J.C., Muniain, J.P. (1994) Gauge Fields, Knots, and Gravity, (Series on Knots and Everything, Vol. 4) World Scientific.

[8] Balachandran, A.P., Bimonte, G., Landi, G., Lizzi, F., Teotonio-Sobrinho, P. (1998) Lattice Gauge Fields and Noncommutative Geometry, J. Geom. Phys. **24** 353.

[9] Balakrishna, B.S., Gürsey, F., Wali, K.C. (1991) Towards a Unified Treatment of Yang-Mills and Higgs Fields, Phys. Rev. **D44** 3313.

[10] Banks, T., Fischler, W., Shenker, S.H. Susskind, L. (1997) M Theory as a Matrix Model: a Conjecture, Phys. Rev. **D55** 5112.

[11] Bellissard, J., van Elst, A., Schulz-Baldes, H. (1994) The noncommutative geometry of the quantum Hall effect, J. Math. Phys. **35** 5373.

[12] Bimonte, G., Lizzi, F., Sparano, G. (1994) Distances on a Lattice from Noncommutative Geometry, Phys. Lett. **B341** 139.

[13] Carow-Watamura, U., Watamura, S. (1997) Chirality and Dirac Operator on Noncommutative Sphere, Comm. Math. Phys. **183** 365.

[14] Cerchiai, B.L., Wess, J. q-Deformed Minkowski Space based on a q-Lorentz Algebra, Munich Preprint LMU-TPW 98/02, `q-alg/9801104`.

[15] Cerchiai, B.L., Hinterding, R., Madore, J. Wess, J. The Geometry of a q-deformed Phase Space, Euro. J. of Phys. C (to appear).

[16] — A Calculus Based on a q-deformed Heisenberg Algebra, Euro. J. of Phys. C (to appear).

[17] Chaichian, M., Demichev, A., Prešnajder, P., Quantum Field Theory on Noncommutative Space-Times and the Persistence of Ultraviolet Divergences, Helsinki Preprint HIP-1998-77/TH, `hep-th/9812180`.

[18] Chamseddine, A.H., Felder, G., Fröhlich, J. (1993) Gravity in Non-Commutative Geometry, Comm. Math. Phys. **155** 205.

[19] Chamseddine, A.H., Connes, A. (1996) Universal Formula for Noncommutative Geometry Actions: Unification of Gravity and the Standard Model, Phys. Rev. Lett. **77** 4868.

[20] Chamseddine, A.H., Fröhlich, J., Grandjean, O. (1995) The gravitational sector in the Connes-Lott formulation of the standard model, J. Math. Phys. **36** 6255.

[21] Chapline, G., Manton, N.S. (1980) The Geometrical Significance of Certain Higgs Potentials: An approach to grand unification, Nucl. Phys. **B184** 391.

[22] Cho, S., Madore, J., Park, K.S. (1998) Non-commutative Geometry of the h-deformed Quantum Plane, J. Phys. A: Math. Gen. **31** 2639.

[23] Cho, S., Hinterding, R., Madore, J. Steinacker, H. Finite Field Theory on Noncommutative Geometries, Munich Preprint, LMU-TPW 99-06, hep-th/9903239.
[24] Chu, Chong-Sun, Ho, Pei-Ming, Zumino, B. Some Complex Quantum Manifolds and their Geometry, Lectures given at NATO Advanced Study Institute on Quantum Fields and Quantum Space Time, Cargese, France, 1996, hep-th/9608188.
[25] Connes, A. (1986) Non-Commutative Differential Geometry, Publications of the Inst. des Hautes Etudes Scientifiques **62** 257.
[26] — (1988) The Action Functional in Non-Commutative Geometry, Comm. Math. Phys. **117** 673.
[27] — (1994) Noncommutative Geometry, Plenum Press.
[28] — (1995) Noncommutative geometry and reality, J. Math. Phys. **36** 6194.
[29] — (1996) Gravity coupled with matter and the foundation of non-commutative geometry, Comm. Math. Phys. **192** 155.
[30] Connes, A., Lott, J. (1990) Particle Models and Noncommutative Geometry, in 'Recent Advances in Field Theory', Nucl. Phys. B (Proc. Suppl.) **18** 29.
[31] — (1992) The metric aspect of non-commutative geometry, Proceedings of the 1991 Cargèse Summer School, Plenum Press.
[32] Connes, A., Rieffel, M.A. (1987) Yang-Mills for non-commutative two-tori, Contemp. Math. **62** 237.
[33] Connes, A., Douglas, M.R., Schwarz, A. (1998) Noncommutative Geometry and Matrix Theory: Compactification on Tori, J. High Energy Phys. **02** 003.
[34] Coquereaux, R. (1989) Noncommutative geometry and theoretical physics, J. Geom. Phys. **6** 425.
[35] Cuntz, A., Quillen, D. (1995) Algebra extensions and nonsingularity, J. Amer. Math. Soc. **8** 251.
[36] Dabrowski, L., Hajac, P.M., Landi, G. Siniscalco, P. (1996) Metrics and pairs of left and right connections on bimodules, J. Math. Phys. **37** 4635.
Nucl. Phys.B3051988545.
[37] Dimakis, A., Madore, J. (1996) Differential Calculi and Linear Connections, J. Math. Phys. **37** 4647.
[38] Dimakis, A., Müller-Hoissen, F. (1993) A Noncommutative Differential Calculus and its Relation to Gauge theory and Gravitation, Int. J. Mod. Phys. A **3** 474.
[39] — (1994) Discrete Differential Calculus, Graphs, Topologies and Gauge Theory, J. Math. Phys. **35** 6703.
[40] Dirac, P.A.M. (1926) The Fundamental Equations of Quantum Mechanics, Proc. Roy. Soc. **A109** 642; (1926) On Quantum Algebras, Proc. Camb. Phil. Soc. **23** 412.
[41] Doplicher, S., Fredenhagen, K., Roberts, J.E. (1995) The Quantum Structure of Spacetime at the Planck Scale and Quantum Fields, Comm. Math. Phys. **172** 187.
[42] Dubois-Violette, M., Kerner, R., Madore, J. (1989) Gauge bosons in a noncommutative geometry, Phys. Lett. **B217** 485.
[43] Dubois-Violette, M., Madore, J., Masson, T. Mourad, J. (1995) Linear Connections on the Quantum Plane, Lett. Math. Phys. **35** 351.
[44] — (1996) On Curvature in Noncommutative Geometry, J. Math. Phys. **37** 4089.
[45] Faddeev, L.D., Reshetikhin, N.Y., Takhtajan, L.A. (1990) Quantization of Lie Groups and Lie Algebras, Leningrad Math. J. **1** 193.
[46] Fichtmüller, M., Lorek, A., Wess, J. (1996) q-Deformed Phase Space and its Lattice Structure, Z. Physik C **71** 533.
[47] Filk, T. (1996) Divergences in a field theory on quantum space, Phys. Lett. **B376** 53.

[48] Fiore, G., Madore, J. Leibniz Rules and Reality Conditions, Preprint 98-13, Dip. Matematica e Applicazioni, Università di Napoli, q-alg/9806071.

[49] — The Geometry of the Quantum Euclidean Space, Munich Preprint LMU-TPW 97-23, q-alg/9904027.

[50] Forgács, P., Manton, N.S. (1980) Space-Time Symmetries in Gauge Theories, Comm. Math. Phys. **72** 15.

[51] Fröhlich, J., Grandjean, O., Recknagel, A. Supersymmetric quantum theory, non-commutative geometry and gravitation, Lecture Notes, Les Houches 1995, hep-th/9706132.

[52] Gambini, R., Pullin, J., Ashtekar, A. (1996) Loops, Knots, Gauge Theories and Quantum Gravity, Cambridge University Press.

[53] Ganor, O.J., Ramgoolam, S., Taylor, W. (1997) Branes, Fluxes and Duality in M(atrix)-Theory, Nucl. Phys. **B492** 191.

[54] Garay, L.J. (1995) Quantum Gravity and Minimum Length, Int. J. Mod. Phys. A **10** 145.

[55] Grosse, H., Madore, J. (1992) A Noncommutative Version of the Schwinger Model, Phys. Lett. **B283** 218.

[56] Grosse, H., Prešnajder, P. (1993) The Construction of Noncommutative Manifolds Using Coherent States, Lett. Math. Phys. **28** 239.

[57] — (1995) The Dirac Operator on the Fuzzy Sphere, Lett. Math. Phys. **33** 171.

[58] Grosse, H., Reiter, G. The Fuzzy Supersphere, ESI Preprint, math-ph/9804013.

[59] Grosse, H., Klimčík, C., Prešnajder, P. (1996) Towards Finite Quantum Field Theory in noncommutative Geometry, Int. J. Theor. Phys. **35** 231.

[60] — (1997) Field Theory on a Supersymmetric Lattice, Comm. Math. Phys. **185** 155.

[61] Hebecker, A., Schreckenberg, S., Schwenk, J. Weich, W., Wess, J. (1994) Representations of a q-deformed Heisenberg algebra, Z. Physik C **64** 355.

[62] Hellund, E.J., Tanaka, K. (1954) Quantized Space-Time, Phys. Rev. **94** 192.

[63] Kaku, M. (1987) Generally Covariant Lattices, the Random Calculus, and the Strong Coupling Approach to the Renormalisation of Gravity, Quantum Field Theory and Quantum Statistics, Volume 2, I.A. Batalin, C.J. Isham and G.A. Vilkovisky (Ed.), Adam Hilger, Bristol.

[64] Kalau, W., Walze, M. (1995) Gravity, Non-Commutative Geometry and the Wodzicki Residue, J. Geom. Phys. **16** 327.

[65] Kastler, D. (1993) A detailed account of Alain Connes' version of the standard model in non-commutative geometry, Rev. Mod. Phys. **5** 477.

[66] — (1999) Noncommutative Geometry and Fundamental Physical Interactions, Proceedings of the 38. Internationale Universitätswochen für Kern- und Teilchenphysik, 'Geometrie und Quantenphysik', Schladming, Austria, Eds. H. Gausterer, H. Grosse, L. Pittner, Lecture Notes in Physics, Springer-Verlag.

[67] Kastler, D., Madore, J., Testard, D. (1997) Connections of bimodules in noncommutative geometry, Contemp. Math. **203** 159.

[68] Kehagias, A., Madore, J., Mourad, J. Zoupanos, G. (1995) Linear Connections in Extended Space-Time, J. Math. Phys. **36** 5855.

[69] Kempf, A., Mangano, G. (1997) Minimal Length Uncertainty Relation and Ultraviolet Regularization, Phys. Rev. **D55** 7909.

[70] Kempf, A., Mangano, G., Mann, R.B. (1995) Hilbert Space Representation of the Minimal Length Uncertainty Relation, Phys. Rev. **D52** 1108.

[71] Khorrami, M., Shariati, A., Aghamohammadi, A. (1997) $SL_h(2)$-Symmetric Torsionless Connections, Lett. Math. Phys. **40** 95.

[72] Krajewski, T. (1998) Classification of Finite Spectral Triples, J. Geom. Phys. **28** 1.

[73] Krajewski, T., Wulkenhaar, R. Perturbative Quantum Gauge Fields on the Noncommutative Torus, Marseille preprint CPT-99/P.3794, hep-th/9903187.

[74] Kosinski, P., Lukierski, J., Maslanka, P. Local $D = 4$ Field Theory on κ-Deformed Minkowski Space, Preprint, hep-th/9902037.

[75] Koszul, J.L. (1960) Lectures on Fibre Bundles and Differential Geometry, Tata Institute of Fundamental Research, Bombay.

[76] Kupershmit, B.A. (1992) The quantum group $GL_h(2)$, J. Phys. A: Math. Gen. **25** L1239.

[77] Landi, G. (1997) An Introduction to Noncommutative Spaces and their Geometries, Springer Lecture Notes, Springer-Verlag.

[78] Landi, G., Nguyen, Ai Viet, Wali, K.C. (1994) Gravity and electromagnetism in noncommutative geometry, Phys. Lett. **B326** 45.

[79] Leitenberger, F. (1996) Quantum Lobachevsky Planes, J. Math. Phys. **37** 3131.

[80] Lizzi, F., Mangano, G., Miele, G. Sparano, G. (1996) Constraints on Unified Gauge Theories from Noncommutative Geometry, Mod. Phys. Lett. A **11** 2561.

[81] Madore, J. (1989) Kaluza-Klein Aspects of Noncommutative Geometry, Proceedings of the XVII International Conference on Differential Geometric Methods in Theoretical Physics, Chester, August, 1988. World Scientific.

[82] — (1990) Modification of Kaluza-Klein Theory, Phys. Rev. **D41** 3709.

[83] — (1991) The Commutative Limit of a Matrix Geometry, J. Math. Phys. **32** 332; (1992) The Fuzzy Sphere, Class. and Quant. Grav. **9** 69.

[84] — (1992) Fuzzy Physics, Ann. Phys. **219** 187.

[85] — (1999) An Introduction to Noncommutative Differential Geometry and its Physical Applications, Cambridge University Press, 1995; 2nd Edition.

[86] Madore, J., Mourad, J. (1993) Algebraic-Kaluza-Klein Cosmology, Class. and Quant. Grav. **10** 2157.

[87] — (1995) On the Origin of Kaluza-Klein Structure, Phys. Lett. **B359** 43.

[88] — Noncommutative Kaluza-Klein Theory, Lecture given at the 5th Hellenic School and Workshops on Elementary Particle Physics, hep-th/9601169.

[89] — (1998) Quantum Space-Time and Classical Gravity, J. Math. Phys. **39** 423.

[90] Madore, J., Masson, T., Mourad, J. (1995) Linear Connections on Matrix Geometries, Class. and Quant. Grav. **12** 1429.

[91] Madore, J., Mourad, J., Sitarz, A. (1997) Deformations of Differential Calculi, Mod. Phys. Lett. A **12** 975.

[92] Madore, J., Saeger, L.A. (1998) Topology at the Planck Length, Class. and Quant. Grav. **15** 811.

[93] Majid, S. (1993) Braided Momentum in the q-Poincaré group, J. Math. Phys. **34** 2045

[94] Majid, S. (1995) Foundations of Quantum Group Theory, Cambridge University Press.

[95] Manin, Yu.I. (1988) Quantum Groups and Non-commutative Geometry, Les Publications du Centre de Recherche Mathématiques, Montréal.

[96] — (1989) Multiparametric Quantum Deformations of the General Linear Supergroup, Comm. Math. Phys. **123** 163.

[97] Manton, N.S. (1979) A New Six-Dimensional Approach to the Weinberg-Salam Model, Nucl. Phys. **B158** 141.

[98] Martín, C.P., Sánchez-Ruiz, D. The One-loop UV-Divergent Structure of $U(1)$-Yang-Mills Theory on Noncommutative $\mathbb{R}^4$, Madrid Preprint FT/UCM-15-99, hep-th/9903077.

[99] Mourad, J. (1995) Linear Connections in Non-Commutative Geometry, Class. and Quant. Grav. **12** 965.

[100] Ogievetsky, O., Schmidke, W.B., Wess, J. Zumino, B. (1992) q-deformed Poincaré algebra, Comm. Math. Phys. **150** 495-518.

[101] Ohn, C. (1992) A $*$-Product on $SL(2)$ and the Corresponding Nonstandard Quantum-$U(\mathfrak{sl}(2))$, Lett. Math. Phys. **25** 85.

[102] Pais, A., Uhlenbeck, G.E. (1950) On Field Theories with Non-Localized Action, Phys. Rev. **79** 145.

[103] Paschke, M,, Sitarz, A. (1998) Discrete spectral triples and their symmetries, J. Math. Phys. **39** 6191.

[104] Podleś, P. (1987) Quantum Spheres, Lett. Math. Phys. **47** 193.

[105] Podleś, P. (1996) Solutions of the Klein-Gordon and Dirac Equation on Quantum Minkowski Spaces, Comm. Math. Phys. **182** 569.

[106] Podleś, P., Woronowicz, S.L. (1996) On the Classification of Quantum Poincaré Groups, Comm. Math. Phys. **178** 61.

[107] Pris, I., Schücker, T. (1997) Noncommutative geometry beyond the standard model, J. Math. Phys. **38** 2255.

[108] Rieffel, M.A. (1980) Non-commutative Tori - A case study of Non-commutative Differentiable Manifolds, Contemp. Math. **105** 191.

[109] — (1981) C^*-algebras associated with irrational rotation, Pacific J. Math. **93** 415.

[110] Sakharov, A.D. (1967) Vacuum quantum fluctuations in curved space and the theory of gravitation, Doklady Akad. Nauk S.S.S.R. **177** 70.

[111] Sakharov, A.D. (1975) Spectral Density of Eigenvalues of the Wave Equation and Vacuum Polarization, Teor. i Mat. Fiz. **23** 178.

[112] Schlieker, M., Weich, W., Weixler, R. (1992) Inhomogeneous Quantum Groups, Z. Physik C **53** 79-82.

[113] Schmüdgen, K. Operator representations of a q-deformed Heisenberg algebra, Leibzig Preprint, q-alg/9805131.

[114] Schupp, P., Watts, P., Zumino, B. (1992) The 2-Dimensional Quantum Euclidean Algebra, Lett. Math. Phys. **24** 141.

[115] Schwenk, J., Wess, J. (1992) A q-deformed quantum mechanical toy model, Phys. Lett. **B291** 273.

[116] Schwinger, J. (1960) Unitary Operator Bases, Proc. Nat. Acad. Sci. **46** 570.

[117] Seiler, R. (1999) Geometric Properties of Transport in Quantum Hall Systems, Proceedings of the 38. Internationale Universitätswochen für Kern- und Teilchenphysik, 'Geometrie und Quantenphysik', Schladming, Austria, Eds. H. Gausterer, H. Grosse, L. Pittner, Lecture Notes in Physics, Springer-Verlag.

[118] Sitarz, A. (1994) Gravity from non-commutative geometry, Class. and Quant. Grav. **11** 2127.

[119] Snyder, H.S. (1947) Quantized Space-Time, Phys. Rev. **71** 38.

[120] — (1947) The Electromagnetic Field in Quantized Space-Time, Phys. Rev. **72** 68.

[121] Várilly, J.C. (1997) Introduction to Noncommutative Geometry, Proceedings of the European Mathematical Society Summer School on Noncommutative Geometry and Applications, Monsaraz and Lisbon.

[122] Várilly, J.C., Gracia-Bondía, J.M. (1993) Connes' noncommutative differential geometry and the Standard Model, J. Geom. Phys. **12** 223.

[123] Wess, J., Zumino, B. (1990) Covariant Differential Calculus on the Quantum Hyperplane, Nucl. Phys. B (Proc. Suppl.) **18** 302.

[124] Woronowicz, S.L. (1987) Twisted $SU(2)$ Group. An example of a Non-Commutative Differential Calculus, Publ. RIMS, Kyoto Univ. **23** 117; (1987) Compact Matrix Pseudogroups, Comm. Math. Phys. **111** 613.

[125] Weyl, H. (1950) The Theory of Groups and Quantum Mechanics, Dover, New York,

[126] Yukawa, H. (1949) On the radius of the Elementary Particle, Phys. Rev. **76** 300.

Geometric Properties of Transport in Quantum Hall Systems

Th. Richter and R. Seiler

Fachbereich Mathematik, TU-Berlin, Germany

1 Review

1.1 Introduction

In this first section, we present a short review of theoretical approaches to the quantum Hall effect. For an in depth coverage, we refer to the recent book D. J. Thouless (1998), as well as to M. Stone (1992).

Let us recall how a quantum Hall system in a laboratory looks like: a strong magnetic field runs perpendicular through a probe of a conductor or semiconductor, forming a two-dimensional system; this setup is typically realized as inversion layers in field effect transistors, formed at the interface between an insolator and a semiconductor under the influence of an electric field perpendicular to the interface. If the temperature of the system is near zero, the electrons are bound by a deep potential well, forming a two-dimensional system. We identify this inversion layer with the x-y plane, hence B is parallel to the z-axis.

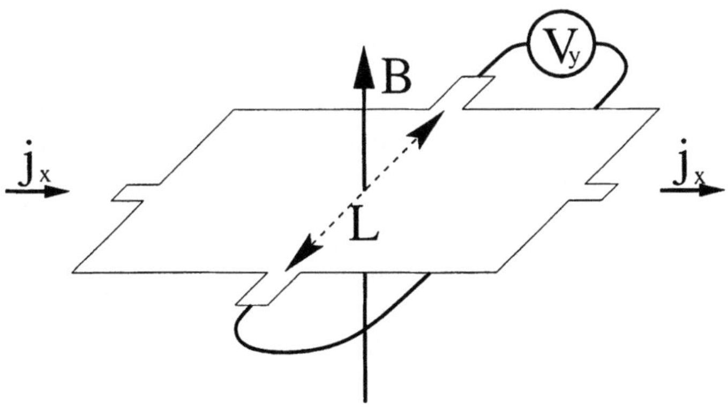

Fig. 1. The physical setup of the QHE

If we apply an external electric field E_y in the y-direction, the system will, due to the magnetic field, develop a current j_x in x direction, perpendicular to the magnetic field and the driving force E_y. The current j_x and E_y are, for small values of E_y, related by "Ohm's Law"

$$\begin{pmatrix} E_x \\ E_y \end{pmatrix} = \begin{pmatrix} \rho_{\mathrm{L}} & -\rho_{\mathrm{H}} \\ \rho_{\mathrm{H}} & \rho_{\mathrm{L}} \end{pmatrix} \begin{pmatrix} j_x \\ j_y \end{pmatrix} , \tag{1}$$

where we consider only the isotropic case for simplicity. Here ρ_{L} is the usual longitudinal resitance due to dissipative processes in the conductor, and ρ_{H} is the Hall resistance. The inverse of the resistance matrix is called the conductivity matrix[1] of the system, and the off-diagonal elements of it are the Hall conductance σ_{H}.

Even though this is how we want to consider the Hall effect mathematically, this is not how the concrete experiments are run; for practical reasons, one usually applies the current j_x and measures the potential difference V_y.

If we close now the system by two external loops connecting the opposite edges of the system we're able to relate the electric field E_y to the change of a first flux through the first handle and the current j_x to a mangetic flux through the second handle. Hence, the topology of the sytem in this model will be torus-like.

Another well-studied model of the Hall setup — and even the model first looked at by Laughlin to explain the quantum Hall effect — is that of a cylinder. This corresponds — with regard to the configuration put up in Fig. 2 — to an identification of opposite edges in x-direction, resulting in a cylinder geometry, with its related flux running in axis direction of the cylinder. The magnetic field perpendicular to through the surface is assumed to be constant.

Classically, the Hall resistance is expected to be proportional to the magnetic field, and this is just what was found by experimental physicists for the non-quantum mechanical Hall effect, say at room temperature (E. H. Hall (1879)). However, when K. von Klitzing, G. Dorda, M. Pepper (1980) applied a very strong magnetic field to a Hall system in a field effect transistor at very low temperature, they were puzzled by finding that the Hall conductivity of this system was indeed quantized: the Hall conductivity σ_H as a function of the magnetic field was not at all linear, but a step functions with plateaus of an unexpected precision of 10^{-8}, cf. Fig. 3. It was observed, too, that the longitudinal resistance ρ_{L} vanish for magnetic field giving rise to the plateaus. The conductivity σ_H is, in terms of natural units of e^2/h, an integer. This phenomen is called the "integral quantum Hall effect", and a first model of understanding it was presented by Laughlin, using the cylinder geometry system.

Later on, more experiments where run using a variety of systems, and in some of them plateaus of fractional conducitivity p/q were found. In most of these systems p and q are small integers, and q is usually an odd number (D. C. Tsui,

[1] The "conductivity" is the ratio of the current density to the electric field, whereas "conductance" is the ratio of current to voltage. The dimensions of "conductance" and "conductivity" are identical for two-dimensional systems, namely Ω^{-1}, so we no longer want to stress the difference between them.

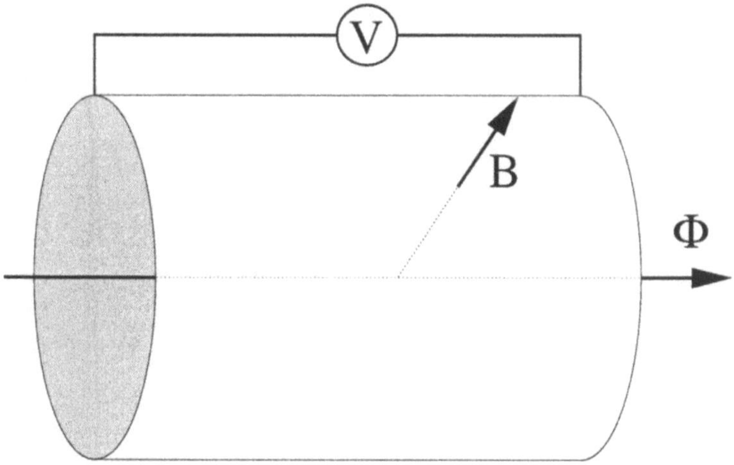

Fig. 2. The Laughlin model

H. L. Störmer, A. C. Gossard (1982), R. G. Clark et al. (1988), A. M. Chang, J. E. Cunningham (1989), J. A. Simmons et al. (1989)).

1.2 The Laughlin Argument

The first model for a quantum Hall system was invented by R. Laughlin (1981); it uses the cylinder geometry, as shown in Fig. 2, where the magnetic fields B points in normal direction of the cylinder. If the magnetic flux Φ through the cylinder changes in time by 2π, i.e. one flux quantum, there is a corresponding Hall current $I(t)$ in the direction of the cylinder axis. This can be easely seen in an effective one-particle theory: Laughlin consideres the usual isotropic effective-mass Hamiltonian

$$H = \frac{1}{2m^*}\left(\mathbf{p} - \frac{e}{c}\mathbf{A}\right)^2 + eV(y) \qquad V(y) = E_0 y \ , \tag{2}$$

where E_0 is the applied electric field and y is the coordinate in cylinder axis direction. Using the Landau gauge for $\mathbf{A}$, it is quite simple to calculate the eigenstates: They are — up to a phase factor — given by shifted harmonic oscillator eigenfunctions. The eigenstates are affected by a change of Φ only in the location of their centers, giving rise to a charge transport in cylinder axis direction. It is now easy to calculate the current I around the loop: it is given by the adiabatic derivative of the total energy of the system with respect to the magnetic flux. Due to the transport of states against the external electric field E_0, adjusting the flux goes along with an energy increase. One finds by direct computation:

$$I = n\frac{e^2 V}{h} \quad \Rightarrow \sigma_{\mathrm{H}} = n\frac{e^2}{h} \tag{3}$$

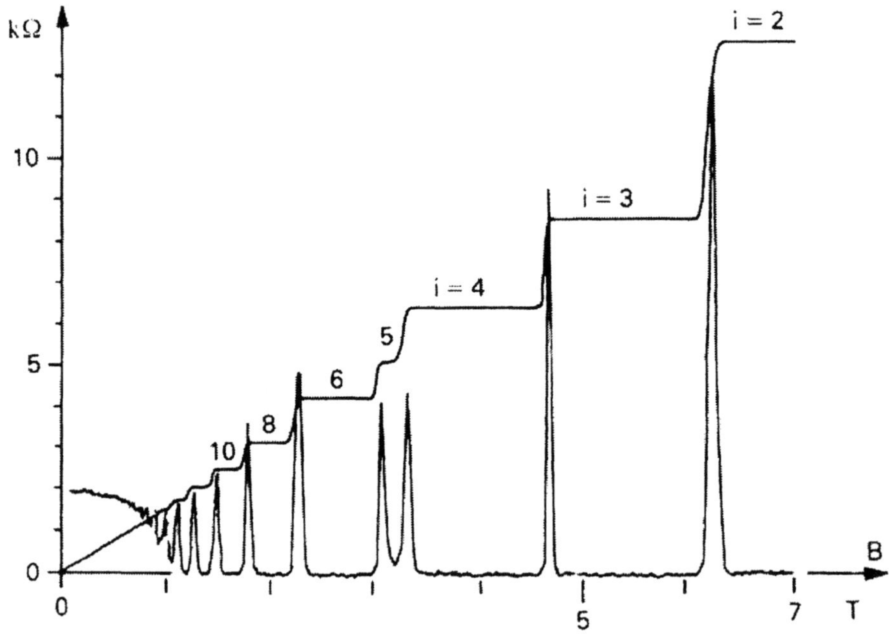

Fig. 3. The Hall resistance and the longitudinal resistance as a function of the magnetic field B

where n is the number of states transported from one edge of the system to the other under adiabatic change of Φ.

Laughlin does not discuss the dirty interacting system rigorously; we are therefore not following his arguments right here.

We close this section with a remark: If the Hall conductance σ_{H} is quantized in natural units of $(2\pi)^{-1}$, the charge Q transported by the Hall current I under an increase of the flux by 2π is an integer as well:

$$Q = \int_t I(t)dt = \sigma_{\mathrm{H}} \int \frac{d\Phi}{dt} dt = \frac{n}{2\pi} \int \frac{d\Phi}{dt} dt = n \ . \tag{4}$$

1.3 Thouless, Kohomoto, Nightingale and den Nijs

In 1982, Thouless, Kohomoto, Nightingale and den Nijs discovered a remarkable connection between the Hall resistence and a geometric object (D. J .Thouless, M. Kohmoto, P. Nightingale, M. den Nijs (1982)) which turned out to be a chern number. Their model is given by a one-electron Hamiltonian

$$H = \frac{1}{2} v^2 + W(x, y) \tag{5}$$

describing a two-dimensional electron gas. Here, v is the velocity operator

$$v = \frac{\hbar}{i}\mathrm{d} - A(x, y) \tag{6}$$

with the vector potential A due to an external homogenous magnetic field $B = \mathrm{curl}\, A$, perpendicular to the x-y plane, and W is a periodic background potential with lattice periods $q \cdot a$ and b. The magnetic flux through one lattice period is supposed to be rational, i.e. $\Phi := abB = p/q$, $p, q \in \mathbb{Z}$. Even though it is important to recognize that this defines intrinsically a torus geometry — by identifying the edges of the lattice sides — it is the torus geometry of the Brillouin zone that plays the eminent role in calculations.

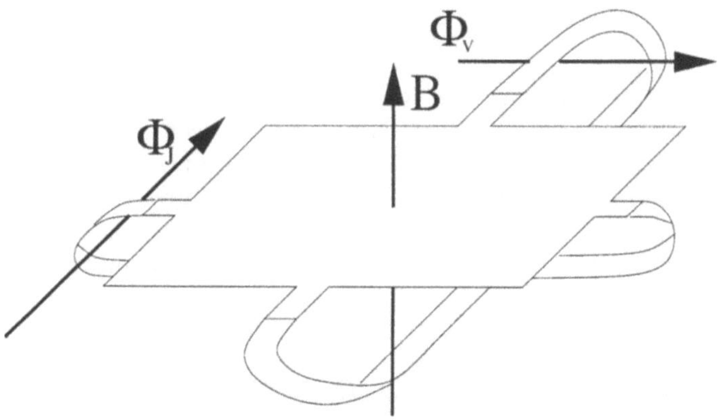

Fig. 4. The torus geometry of the physical space

Due to the external magnetic field, the ordinary momenta do no longer commute with H. They have to be replaced by the so-called quasi-momenta k_1, k_2 which are given by the phase factor relating the eigenfunctions of H at one edge compared to the same eigenfuntion taken at the opposite edge. More precisely, $\exp(ixk_1)$ and $\exp(iyk_2)$ are the eigenvalues of the so-called "magnetic shift operators" T_x and T_y, which move the eigenfunctions by one lattice period and multiply them with a phase. This phase has to be defined such that T_x and T_y *do* commute with H. Therefore, we can choose the eigenfunctions to satisfy the following "generalized Bloch condition", defining k_1 and k_2:

$$\begin{aligned} \psi_{k_1,k_2}(x + qa, y) &= \exp(2\pi i p y/b + i k_1 q a)\psi_{k_1,k_2}(x, y) \\ \psi_{k_1,k_2}(x, y + b) &= \exp(i k_2 b)\psi_{k_1,k_2}(x, y) \ . \end{aligned}$$

Mathematically speaking, this is the definition of a $U(1)$ line bundle family over the torus, parametrized by the quasi momenta; the Hamiltonian is a direct integral over the Brillouin zone.

The Hall conductivity is now calculated in terms of linear response theory, using the "Kubo Formula". It can be rewritten in terms of an integral over the eigenfunctions of the Hamiltonian H:

$$\sigma_H = \frac{ie^2}{2\pi h} \sum \int_{\mathcal{B}} < \partial_{k_1}\psi | \partial_{k_2}\psi > \; dk_1 \wedge dk_2 \qquad < \psi|\psi >= 1 \; , \tag{7}$$

where we integrate over the Brillouin zone $\mathcal{B}$ spanned by k_1 and k_2, and the sum has to be taken over the occupied electron sub-bands, i.e. all bands up to the Fermi-level.

This integral can be rewritten, by using the Stoke's Formula, as an integral over the boundary of the Brillouin zone:

$$\sigma_H = \frac{ie^2}{2\pi h} \int_{\partial\mathcal{B}} < \psi|\mathrm{d}\psi > - < \mathrm{d}\psi|\psi > \quad . \tag{8}$$

If the bands do not overlap, ψ is known to be a single-valued analytic function everywhere within the unit cell. It is now easely seen that this integral is just the change of the phase of the wave function around the unit cell, which has to be an integer. Hence, the Hall conductance is given in terms of a simple geometric entity, the phase difference of the eigenfunctions of the Hamiltonian. This winding number is well known in differential geometry, it is called the "Chern number" of the line bundle defined by the wave functions.

1.4 J. Avron, R. Seiler, Q. Niu, D. J. Thouless

In this model, interacting particles are considered (J. E. Avron, R. Seiler (1985), Q. Niu, D.J. Thouless (1987), J. E. Avron, R. Seiler, L. G. Yaffe (1987)). The configuration space is a compact domain in the two dimensional plain with two holes, with two Aharonov-Bohm fluxes Φ_1 and Φ_2 running through the holes, and again a strong magnetic field B perpendicular to the plane, see Fig. 1.4. Unlike the model discussed before, the Hamiltonian is a many-body Hamiltonian with an incompressibility condition, i.e. the spectrum is supposed to fulfill a gap condition. Apart from that, the Hamiltonian is rather general:

$$H(\Phi_1, \Phi_2) = \frac{1}{2} \sum_j (v_j - \Phi_1 a_1(x_j) - \Phi_2 a_2(x_j))^2 + \sum_{j<k} \frac{1}{|x_j - x_k|} + \sum_j W(x_j) \; ,$$

where v_j is the velocity operator of the j-th particle

$$v_j = \frac{\hbar}{i}\mathrm{d}_j - A(x_j) \tag{9}$$

with the vector potential A of the magnetic field $B = \operatorname{curl} A$ and W is a background potential. The coulombic particle-particle interaction could be replaced by any other potential with rather mild regularity conditions. The expressions

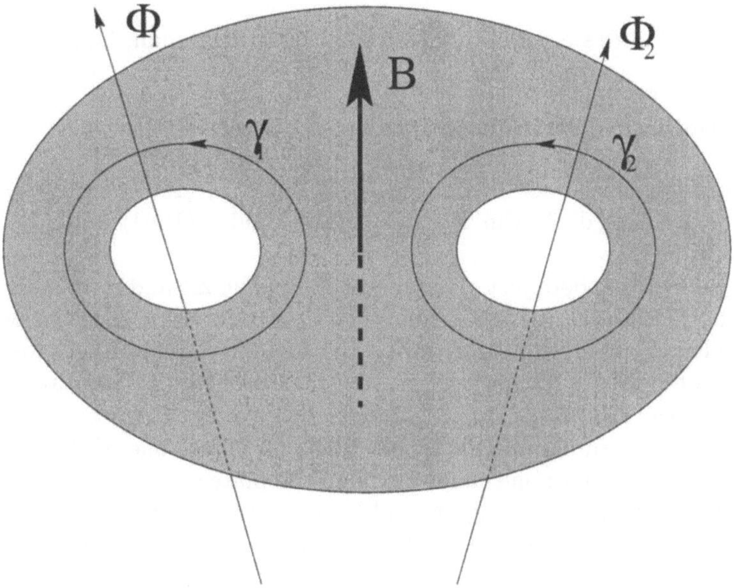

Fig. 5. The punctured plane model

$\Phi_l a_l(x_j)$ describe the vector potential of the Aharonov-Bohm fluxes as introduced above, requiring that the one-forms a_l are "dual" to the boundary of the domain, i.e.

$$\int_{\gamma_j} a_l = \delta_{l,j} \quad , \tag{10}$$

where γ_j is a closed loop around the j-th hole of the system.

The Hall voltage of this system is meant to be induced by the first flux, i.e. $V_H = \frac{d\Phi_1}{dt}$, using Faraday's law; the flux Φ_2, however, is generated by the current flowing around the second hole, relating the Hall conductance to the fluxes.

We will now argue in section 3 that the averaged Hall conductance of this system is given by the Chern number, up to an infinitely small error term in V_H, and hence an integer. The argument is — as we shall see — based on the adiabatic theorem of quantum mechanics. The basic reason why this theorem is relevant in this context is the following: in Ohm's law the limit $V_H \to 0$ is considered. Hence, the dependence on time due to $\partial_t \Phi_1$ is approximately zero, which is the so called "adiabatic limit".

In this setup one important condition is put in "by hand", namely the incompressibility of the system; or — in mathematical terms — the separation of the ground state energy of the system by gap from the rest of the spectrum. This assumption, which is expected to hold for quantum Hall systems, is however difficult to derive for many body systems. It can be analyzed in a satisfactory manner in an effective one-particle theory where the concept and mechanism of localization of states are well understood, cf. M. Aizenman, G. M. Graf (1998).

1.5 J. Bellissard, H. Schulz Baldes, A. Connes

A different approach was introduced by J. Bellisard (1987) and developed further by J. Bellissard, A. van Elst, H. Schulz-Baldes (1994) and M. Aizenman, G. M. Graf (1998).

The geometry is here just given by the $\mathbb{R}^2$ or the $\mathbb{Z}^2$ and a constant magnetic field B perpendicular to that plane. The approach is an effective one-particle theory, with the Hamiltonian given by the Landau Hamiltonian plus a random disorder potential W_ω. It is discussed by means of non-commutative geometry (cf. A. Connes (1994)).

The great advantage of this model is that it solves one of the difficulties of the Laughlin argument: one cannot explain the plateaus of the Hall conductivity as a function of the filling factor, and hence of the magnetic field, without the assumption of localized states in the spectral gaps of the unperturbed Landau Hamiltonian.

The starting point of the calculation is a generalized "Chern-Kubo formula", written by means of the projection P onto eigenfunctions of the Hamiltonian of energies below the Fermi level

$$\sigma_H(P) = \frac{e^2}{h}\frac{1}{2\pi i}\mathcal{T}\big(P[\partial_1 P, \partial_2 P]\big) =: \frac{e^2}{h}\mathrm{ch}(P) \ , \tag{11}$$

where $\mathcal{T}$ is the trace per unit volume. However, since the quasi-momenta k_1 and k_2 are no longer well-defined due to the disorder, the derivations with respect to k_1 resp. k_2 have to be replaced by their non-commutative counterparts, the commutator with the position operator X_j:

$$\partial_j A := -i[X_j, A] \ , \tag{12}$$

and — for the same reason — the trace can't be written in terms of an integral over the Brillouin zone anymore.

By a formula given by A. Connes (1985), the non-commutative Chern character is given by an average over the disorder with respect to a propability measure $\mathbf{P}$ on the disorder configuration space. To show that the Chern character — and hence the Hall conductance — is an integer, one has to compute that this average is the index of a Fredholm operator, namely

$$\mathrm{ch}(P_\omega) = \mathrm{index}\left(P_\omega \frac{X_1 + iX_2}{|X_1 + iX_2|}\bigg|_{\mathrm{range}\, P_\omega} \right) \ , \tag{13}$$

where X_i are again the position operators. This formula holds whenever the states at the Fermi level are dynamically localized.

The required calculations have been greatly simplified later on in J. E. Avron, R. Seiler, B. Simon (1994a), avoiding the language of non-commutative geometry completely. Note that the operator U defined by multiplication with

$$u := \frac{X_1 + iX_2}{|X_1 + iX_2|} \tag{14}$$

is the gauge transformation related to unit flux tube piercing the $\mathbb{R}^2$ at the origin. The above Fredholm index compares therefore the projections P and $Q := UPU^{-1}$. We therefore define the so called "relative index" of P and Q by the Fredholm index above

$$\operatorname{index}(P, Q) := \operatorname{index}(PU|_{\text{range } P}) \quad \text{where } Q := UPU^{-1} \ . \tag{15}$$

This relative index can be easely computed if the difference $P - Q$ is in one of the Schatten ideals $\mathcal{S}^{(2n-1)}$, i.e. $(P - Q)^{2n-1}$ is of trace class. Then

$$\operatorname{index}(P, Q) = \operatorname{tr}\, (P - Q)^{2m-1} \quad \text{for all } m \geq n \ . \tag{16}$$

Especially, this expression does not depend on m provided m is large enough to make $(P - Q)^{2m-1}$ trace class.

Physically speaking, the relative index compares the dimensions of the kernels of P and Q. It can be seen that, by adding a flux tube, some eigenstates of H are "driven to infinity". For example, taking for P and Q the ground state projections, the kernels of P and Q should "differ by some states" and their relative index is therefore an integer, counting the "deficiency". In particular, this integer is one in the perturbed Landau Hamiltonian case.

1.6 J. Fröhlich, Q. Niu, X. G. Wen, A. Zee

Another approach to the quantum Hall effect is that of using methods of quantum field theory and Chern-Simons theory. This setup has been used by several authors, in particular by X. G. Wen (1989), J. Fröhlich, T. Kerler (1991) and X. G. Wen, A. Zee (1992). We shall, however, only scratch on this threory and show — using an argument by J. Fröhlich — that abelian Chern-Simons theory appears quite naturally in this context.

If one writes the Ohm Hall law for one of the observed plateaus where the longitudinal resistances vanishes, one finds for the current $\mathbf{j}$ and the electric field $\mathbf{E}$

$$\mathbf{j} = \sigma_H \cdot \epsilon \mathbf{E} \qquad \epsilon = \begin{pmatrix} 0 & 1 \\ -1 & 0 \end{pmatrix} \ , \tag{17}$$

where $\sigma_H \in \mathbb{R}$ is the Hall conductivity. Additionally, we make use of the following more fundamental laws: for first the charge conservation

$$\frac{\partial j^0}{\partial t} + \operatorname{div} \mathbf{j} = 0 \ , \tag{18}$$

where j^0 is the charge density, and Faraday's induction law

$$\frac{\partial \mathbf{B}}{\partial t} + \operatorname{curl} \mathbf{E} = 0 \qquad \mathbf{B} = \begin{pmatrix} 0 \\ 0 \\ B \end{pmatrix} . \tag{19}$$

By combining these three equations, one finds simply by integration

$$j^0 = \sigma_H B . \tag{20}$$

It is now more convenient to write these equations in terms of the field tensor of (2 + 1) dimensional electrodynamics

$$F := \begin{pmatrix} 0 & E_x & E_y \\ -E_x & 0 & -B \\ -E_y & B & 0 \end{pmatrix} \tag{21}$$

to obtain the following reformulation of the equations above:

$$J := \begin{pmatrix} j^0 \\ \mathbf{j} \end{pmatrix} = -\sigma_H F \text{ from (20)} \tag{22}$$

$$\mathrm{d}J = 0 \text{ charge conservation} \tag{23}$$

$$\mathrm{d}F = 0 \text{ induction law} \tag{24}$$

If the confinement domain Ω of the system is contractible, the last two equations can be integrated by introducing two one-forms a and b such that

$$J = \mathrm{d}b \qquad F = \mathrm{d}a . \tag{25}$$

Rewriting (22) in terms of a and b yields

$$\mathrm{d}b = -\sigma_H \mathrm{d}a . \tag{26}$$

This equation is the Euler-Lagrange equation derived from an action principle of a Chern-Simons type action S_Λ on the space-time domain $\Lambda := \Omega \times \mathbb{R}$, varied with respect to the dymanical variable b:

$$S_\Lambda = \frac{1}{4\pi\sigma_H} \int_\Lambda b \wedge \mathrm{d}b + \frac{1}{2\pi} \int_\Lambda a \wedge \mathrm{d}b + \frac{1}{2\pi} \Gamma_{\partial\Lambda} , \tag{27}$$

where the last term is a boundary term arising to make the equation gauge-invariant, related to the edge-currents mentioned above.

This action principle allows now an obvious quantization using Feynman path integrals. A close investigation of this quantization results in the observation that the quantum Hall conductivity σ_H must be a rational number. The required computations can be carried out explicitly in case the domain Ω is a disk: the term $\Gamma_{\partial\Lambda}$ is the generating functional of connected Green's functions of a chiral $U(1)$ current circulating around the boundary $\partial\Omega$ of the system. Using the requirements that the total action S_Λ is gauge-invariant and that every localizable excitation of finite energy and charge ±1 obeys Fermi statistics, one finds that

$$\sigma_H = \sum_{i,j=1}^{N} (K^{-1})_{ij} \ , \tag{28}$$

where N is an integer — the number of chiral currents — and

$$K_{i,i} \in 2\mathbb{Z} + 1 \text{ for all } i, \quad \text{and} \quad K_{i,j} \in \mathbb{Z} \text{ for all } i, j \ . \tag{29}$$

Hence, σ_H is necessarely a rational number. Moreover, for $N = 1$ one has

$$\sigma_H = \frac{1}{2l+1} \qquad l \in \mathbb{N}_0 \ , \tag{30}$$

i.e. fractions with odd denominator.

An alternative approach would start again from the action integral S_Λ and would use results of topological Chern-Simons theory.

2 Adiabatics

2.1 The Adiabatic Setup

We aim now at the adiabatic theorem of quantum mechanics, following the article J. E. Avron, R. Seiler, L. G. Yaffe (1987). Even though the theorem itself is rather old — its first formulation goes back to Born and Fock (M. Born, V. Fock (1928)) — its proper formulation was found years later by T. Kato (1950) in the context of pertubation theory of linear operators.

The general setup is that of non-relativistic quantum mechanics. One consideres explicitly time dependent quantum systems, whose dynamics are given in terms of a time-dependent Hamiltonian $H(\cdot)$. Furthermore, we introduce a time scale T. Hence, including this time scale, the Schrödinger equation looks like this:

$$i\partial_t \Psi_T(t) = H(t/T)\Psi_T(t) \tag{31}$$

We're now interested at the limit $T \to \infty$, hence in the limit of "infinitely slow" change of the Hamiltonian. This is called the "adiabatic limit" of the system.

To formulate our adiabatic theorem, some more assumptions have to be made: first, we require that the Hamiltonian $H(s)$ is continuously differentiable in s in the strong sense. Furthermore, we assume that the spectrum of the Hamiltonian $H(s)$ — where $s := t/T$ is the external parameter — is separated by a gap such that the size of the gap is uniformely bounded from below[2]. Hence, we may define the projection $P(s)$ onto one part of the spectrum, as separated by the gap. Our third assumption is that this projection is of finite rank.

[2] Some more recent adiabatic theorems work without this condition. It is sufficient to have, for example, an embedded eigenvalue in a continous spectrum of H. However, there is no control of the error in terms of T anymore, cf. J. E. Avron, A. Elgart (1998).

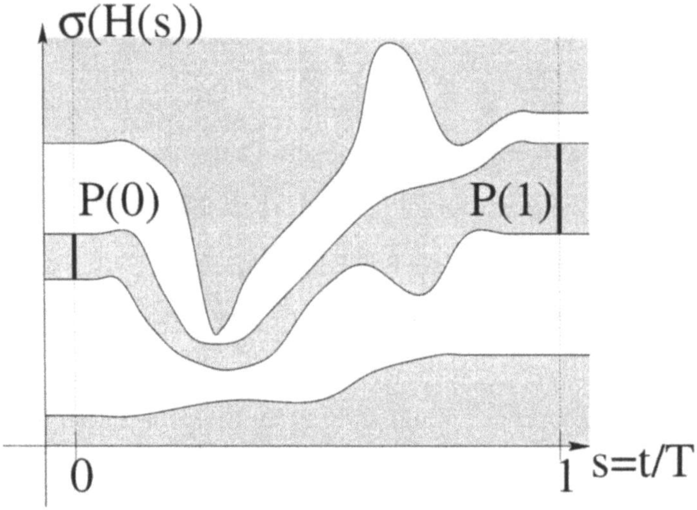

Fig. 6. The gap condition

If we start now the time evolution with a state within this part of the spectrum, i.e. $\Psi_T(0) \in P(0)$, the adiabatic theorem tells us that the state $\Psi_T(t)$ for a later time t of order T is still within this part of the spectrum up to a small error term, which is controlled by the time-scale T and the width of the gap. Morally, very little of the state Ψ "leaks out" to different parts of the spectrum, where the size of the gap defines a typical time scale since energy is related to time by Planck's constant.

2.2 Kato's Equation

If we want to make this statement more precise, we somehow have to compare the real, physical time evolution by some kind of ideal evolution that does not "leak" at all. Hence, the unitary time evolution operator $U_{\mathrm{AD}}(s)$ of this dynamics — called the adiabatic dynamics for that reason — would have to map $P(0)$ into $P(s)$, or would have to fulfill the following intertwining condition

$$\begin{aligned} & P(s) = U_{\mathrm{AD}}(s)P(0)U_{\mathrm{AD}}^{-1}(s) \\ \Longleftrightarrow \quad & U_{\mathrm{AD}}(s)P(0) = P(s)U_{\mathrm{AD}}(s) \ . \end{aligned} \tag{32}$$

For example, if we put

$$H_{\mathrm{Kato}}(s) := \frac{i}{T}[\dot{P}(s), P(s)] \ , \tag{33}$$

where the dot denotes the derivative with respect to the parameter s, we easely find that a solution of the Schrödinger equation of this Hamiltonian

$$i\partial_s\Psi(s) = i[\dot{P}(s), P(s)]\Psi(s)$$
$$\Psi(0) \in P(0)$$

fulfills indeed $\Psi(s) \in P(s)$. This is straightforewards to calculate and uses not much more than just

$$\begin{aligned} P^2(s) = P(s) \qquad P^{\perp}(s)P(s) = P(s)P^{\perp}(s) = 0 \\ \Rightarrow \quad \dot{P}(s) = P(s)\dot{P}(s)P^{\perp}(s) + P^{\perp}(s)\dot{P}(s)P(s) \ , \end{aligned} \tag{34}$$

where $P^{\perp}(s) = \mathbb{1} - P(s)$ is the projection onto the orthogonal complement of the image of $P(s)$.

It turns out that there is a complete family of Hamiltonians whose time evolution fulfills the intertwining property. For example, we may add any operator that commutes with $H(s)$[3]. Hence, another choice of a generator for the adiabatic time evolution would be

$$H_{\mathrm{A}}(s) := H(s) + \frac{i}{T}[\dot{P}(s), P(s)] \tag{35}$$

The importance of this choice for H amongst the family of Hamiltonians fulfilling the intertwining property is that the dynamics given in terms of H_{A} is "closest" — in some sense — to the true physical dynamics of the system.

The first choice, however, can be given a nice geometric interpretation as well. With a little algebra, check that a solution of the Schrödinger equation of Kato's Hamiltonian with $\Psi(0) \in P(0)$ can be written in another nice way, namely by

$$P\mathrm{d}\Psi = 0 \tag{36}$$

where d denotes the exterial differentiation with respect to the parameter s. It is easely checked that the operator $\nabla := P\mathrm{d}$ defined in this way fulfills all axioms of a connection - a well studied object in differential geometry which is the nearest-possible analogon of exterior derivation in curved space:

1. The operator ∇ is $\mathbb{C}$-linear, i.e. for all λ and $\mu \in \mathbb{C}$, we have

$$\nabla(\lambda\Psi + \mu\chi) = \lambda\nabla\Psi + \mu\nabla\chi \tag{37}$$

2. It fulfills the Leibnitz identity

$$\nabla f\Psi = df\Psi + f\nabla\Psi \quad f \in \mathbf{C}^{\infty} \tag{38}$$

[3] or even any operator that commutes with $P(s)$, even though this is usually not considered.

This connection acts as an operator in a vector bundle which is defined by projecting out the sub-bundle $P(s)$ from a trivial $\mathbf{L}^2$-bundle over the parameter space of H. Even though this connection looks absolutely simple - just taking the derivative and brute-force projection down to the bundle where we want to have its image - this construction is more natural than it might seem to. The reader should be reminded of the Levi-Civita connection of the tangent bundle of an embedded surface in $\mathbb{R}^3$ which works the like, but looks more complicated in local coordinates.

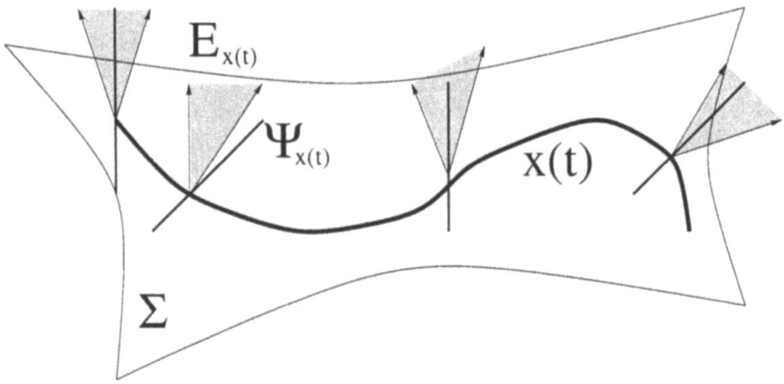

Fig. 7. The problem of a horizontal lift

Hence, solving the adiabatic evolution $Pd\Psi = \nabla\Psi = 0$ is nothing but parallel transport of the vector $\Psi(s)$ along the curve described by s in the parameter space of the Hamiltonian H, within the bundle defined by $P(s)$, or in more modern language, of finding a "horizontal lift" of Ψ along the curve the Hamiltonian describes in parameter space. The adiabatic time-evolution $U_{\mathrm{AD}}(s)$ is the "explicit" solution of this differential equation and hence the operator that performs the parallel transport.

2.3 The Adiabatic Theorem

In terms of the notation introduced above, we're now able to formulate the adiabatic theorem. It's basic contents, however, can be summarized as follows: *In the adiabatic limit, quantum mechanics becomes geometric.*

Theorem 1 *Let $H(s)$ be a (smooth) one-parameter family of Hamiltonians such that the gap-condition holds uniformly in s. Let $U_T(s)$ the physical time evolution, parametrized in the rescaled time $s = t/T$, i.e. let $U_T(s)$ be the solution of*

$$i\partial_s U_T(s) = T \cdot H(s) U_T(s) \ . \tag{39}$$

Let $P(s)$ be the projection onto the states below the gap. Comparing the projection and its time evolution, we have

$$U_T(s)P(0)U_T^{-1}(s) = P(s) + O(T^{-1}) \ . \tag{40}$$

The size of the error term depends on the size of the gap and on the time scale T.

Moreover, if 0 and s are not in the support of $\partial_s H(s)$, we get a much better error estimate:

$$U_T(s)P(0)U_T^{-1}(s) = P(s) + O(T^{-\infty}) \ , \tag{41}$$

i.e. the error term is of infinitesimal order.

We don't want to give a full proof of this theorem, but prefer to sketch the general idea. For first, define the "wave operator":

$$\Omega(s) := U_{\mathrm{AD}}^{-1} U_T(s) \tag{42}$$

It measures — as in scattering theory — the difference between the "ideal"[4] time evolution U_{AD} and the physical time evolution U_T. Using the adiabatic time evolution, we see that $\Omega(s)$ is the solution of the "Volterra" integral equation:

$$\Omega(s) = \mathbb{1} - \int_0^s K(s')\Omega(s')ds' \quad \text{with} \tag{43}$$

$$K(s) := U_{\mathrm{AD}}^{-1}(s)[\dot{P}(s), P(s)]U_{\mathrm{AD}}(s) \tag{44}$$

This integral equation can be used to start an iteration, by putting

$$\Omega_0(s) := \mathbb{1} \quad \text{as first approximation} \tag{45}$$

$$\text{and} \quad \Omega_j(s) := -\int_0^s K(s')\Omega_{j-1}(s')ds' \ . \tag{46}$$

By using these definitions, we can check now that

$$\Omega(s) - \sum_{j=0}^{N} \Omega_j(s) = O(T^{-N}) \ . \tag{47}$$

The proof of this statement builds on the following key-lemma, using mainly integration by parts:

Lemma 2 *Let $R(s,z)$ be the resolvent of $H(s)$ and define for a bounded operator $X(s)$, continously differentiable in s in the strong sense, the "twiddle operation" by*

$$\tilde{X}(s) := -\frac{1}{2\pi i}\int_\Gamma R(s,z)X(s)R(s,z)dz \tag{48}$$

where Γ is a path in the complex plane enclosing the part of the spectrum $P(s)$ projects onto. Note that, due to the gap condition, this integral is well defined.

[4] In scattering theory, U_{AD} would be the time evolution without the scattering potential.

Let $Y(s)$ be another bounded operator family, again continously differentiable in the strong sense. Then, the following equation holds:

$$
\begin{aligned}
&P^{\perp}(0)\int_0^t U_{\mathrm{AD}}^{-1}(s)X(s)U_{\mathrm{AD}}(s)P(0)Y(s)ds = \\
&= \frac{i}{T}P(0)^{\perp}\Big[U_{\mathrm{AD}}^{-1}(s)\tilde{X}(s)U_{\mathrm{AD}}(s)P(0)Y(s)\Big|_0^t \\
&\quad - \int_0^t U_{\mathrm{AD}}^{-1}(s)\dot{\tilde{X}}(s)U_{\mathrm{AD}}(s)P(0)Y(s)ds \\
&\quad - \int_0^t U_{\mathrm{AD}}^{-1}(s)\tilde{X}(s)U_{\mathrm{AD}}(s)P(0)\dot{Y}(s)ds\Big]
\end{aligned} \tag{49}
$$

This lemma is now applied to the operators $X(s) := K(s)$ and $Y(s) = \Omega(s)$. The additonal $P(0)$ in the left-hand side of (49) is for free due to (34). Analyzing the resulting right-hand side shows

$$
||\Omega_{j+1}|| < \frac{C}{T}\sup_{0<s<t}\left(||\Omega_j(s)||, ||\Omega_j'(s)||\right) \ . \tag{50}
$$

Inserting now $\Omega_j'(s) = -K(s)\Omega_j(s)$ and using that $\Omega_j(s)$ and K are bounded reveals

$$
||\Omega_{j+1}|| < \frac{C}{T}\sup_{0<s<t}\left(||\Omega_j(s)||, ||\Omega_{j-1}(s)||\right) \ , \tag{51}
$$

which is enough to prove the claim.

2.4 Adiabatic Curvature, and Applications

As we've seen, the adiabatic time evolution is mainly "geometric". Solving the adiabatic equation is equivalent to integrating the connection Pd, or finding the parallel transport of the wave function along the curve of $H(s)$ in parameter space.

To give an application for this machinery, let us look at the torus geometry system introduced before: the Hamiltonian is parametrized by the two magnetic fluxes ϕ_1 and ϕ_2 through the handles of the torus system. This parameter space forms — by using gauge equivalence — itself a torus, namely

$$
\Phi := \mathbb{R}^2/(2\pi\mathbb{Z}^2) \ , \tag{52}
$$

which is called the "flux torus".

Furthermore, let $P(\phi_1, \phi_2)$ be the projection onto the ground state of the Hamiltonian $H(\phi_1, \phi_2)$. By this construction, we get a vector bundle $E \xrightarrow{\pi} \Phi$ over the flux torus whose fibre is the image of $P(\phi_1, \phi_2)$, i.e. we define this bundle in terms of a projection as sub-bundle of the trivial bundle $\mathbf{L}^2 \times \Phi$. We equip this bundle with the natural connection $\nabla = P\mathrm{d}$, describing the adiabatic transport. This is all we need to calculate an important bundle invariant, the (first) Chern number. It is given by

$$c_1(E) := \frac{1}{2\pi i}\int_{\Phi} \mathrm{tr}\ \nabla^2 \tag{53}$$

$$= \frac{1}{2\pi i}\int_{\Phi} \mathrm{tr}\ P(\mathrm{d}P)\wedge(\mathrm{d}P)P\ , \tag{54}$$

where the trace has to be taken over the fibre.

3 Chern Number Approach

In this section, we want to show how the chern number defined in the last section relates to the transport coefficients of quantum Hall systems.

3.1 The QHE for Interacting Fermion Systems

We start — as an example — with the following Hamiltonian by J. E. Avron, R. Seiler (1985) as already mentioned in the introduction, cf. Fig. 1.4.

It is the model Hamiltonian for an interacting fermion system in a compact configuration space, which is by definition a subset of $\mathbb{R}^2$ with two holes. On the boundary, we impose Dirichlet conditions. A constant magnetic field B runs through the plane, and two magnetic fluxes ϕ_1 and ϕ_2 flow through the holes of the domain. The Hamiltonian is defined by

$$H(\phi_1,\phi_2) = \frac{1}{2}\sum_{i=1}^{N}(v_i - \phi_1 a_1(x) - \phi_2 a_2(x))^2 + \sum_{i<j}^{N}\frac{1}{|x_i - x_j|} + \sum_{i=1}^{N} W(x_i)\ . \tag{55}$$

The operator $v_i := (-i\mathrm{d}+A)$ is the velocity operator, A is the vector potential of the external magnetic field, i.e $\mathrm{d}A = Bdx\wedge dy$ and W is a background potential. The terms $\phi_1 a_1$ resp. $\phi_2 a_2$ describe the fluxes, where a_i is a closed one-form fulfilling

$$\int_{\gamma_j} a_i = \delta^i_j\ . \tag{56}$$

The loop γ_j encircles the j-th hole of the plane. We furthermore assume that the gap condition holds.

The Hall voltage is applied by making ϕ_1, and hence H, explicitly time dependent; therefore, it is given by $V_{\mathrm{H}} = \dot{\phi}_1$. Hence, the adiabatic limit of slow time dependence is now the limit of small voltages V_{H}. To apply the adiabatic theorem, we select a "switch" function for ϕ_1: the flux remains turned off for negative times, is then adiabatically increased by one flux unit with slope V_{H} and is then again held constant. We furthermore introduce the rescaled time $s = \frac{t}{T} = V_{\mathrm{H}}t$. In this time scale, the Schrödinger equation reads:

$$i\partial_s U(s,\phi_2) = \frac{1}{V_{\mathrm{H}}}H(\phi_1(s),\phi_2)U(s,\phi_2) \tag{57}$$

$$U(0,\phi_2) = \mathbb{1} \tag{58}$$

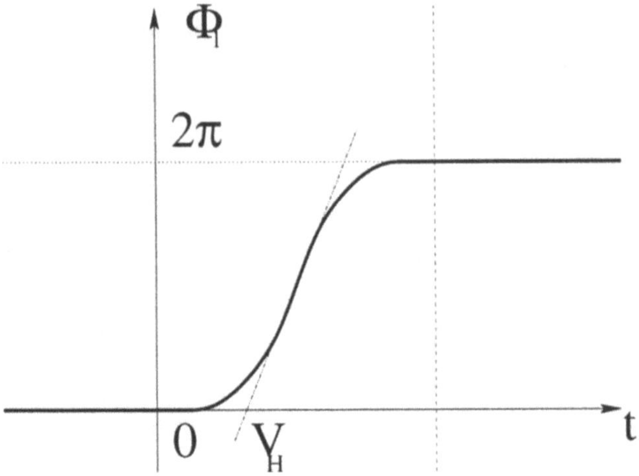

Fig. 8. A "switch function"

Since $\phi_1(s)$ is a monotonically increasing function of s, we may make a variable transform and use ϕ_1 as independent variable instead of s. In a slight abuse of notation, we write now $U(\phi_1, \phi_2)$ instead of $U(s(\phi_1), \phi_2)$ etc., and consider in the following all quantities as functions of Φ_1 and Φ_2.

Let us now denote the projection onto the ground state of H by $P(\phi_1, \phi_2)$, as before, and the physical, time evolved state by $\rho(\phi_1, \phi_2)$

$$\rho(\phi_1, \phi_2) := \frac{1}{q}\hat{P}(\phi_1, \phi_2) := \frac{1}{q}U(\phi_1, \phi_2)P(0, \phi_2)U^{-1}(\phi_1, \phi_2) \tag{59}$$

$$\rho(0, \phi_2) := \frac{1}{q}P(0, \phi_2) \quad q := \text{tr } P(0, \phi_2) \ . \tag{60}$$

Furthermore, we denote its energy expectation value by

$$E(\phi_1, \phi_2) := \frac{1}{q}\text{tr } \hat{P}(\phi_1, \phi_2)H(\phi_1, \phi_2) \ . \tag{61}$$

Since H is periodic in ϕ_2 up to a gauge transformation, E is periodic in ϕ_2. The current is now given by the expectation of the current operator, the derivative of H by ϕ_2:

$$j_x(\phi_1, \phi_2) := \text{tr } \rho(\phi_1, \phi_2)\frac{\partial H(\phi_1, \phi_2)}{\partial \phi_2} \tag{62}$$

$$= \frac{i}{q}\frac{\partial \Phi_1}{\partial s}\partial_{\Phi_1}\text{tr } \hat{P}(\phi_1, \phi_2)U^{-1}(\phi_1, \phi_2)\partial_{\phi_2}U(\phi_1, \phi_2) \ . \tag{63}$$

This expression can be rewritten in terms of the so called "persistent-current" formula:

$$j_x \mathrm{d}dt \wedge \mathrm{d}\phi_2 = \frac{\partial E}{\partial_{\phi_2}}(\phi_1, \phi_2)\mathrm{d}t \wedge \mathrm{d}\phi_2 + \frac{i}{q}\text{tr } \hat{P}(\mathrm{d}\hat{P}) \wedge (\mathrm{d}\hat{P})\hat{P}\ (\phi_1, \phi_2) \tag{64}$$

The last term looks very like the adiabatic curvature term: it is of order V_{H} and hence vanishes linearly in the adiabatic limit. The first term, however, can be shown to persist in the limit, i.e. is of order $O(1)$. However, since it is periodic, this term will vanish if we integrate this equation over ϕ_2 for calculating an averaged transport.

To make the last term the curvature, we need to replace the physical projection $\hat{P}(\phi_1, \phi_2)$ by the adiabatically transported projection $P(\phi_1, \phi_2)$. According to the adiabatic theorem, this can be done up to a small error in powers of the time scale, or — equivalently — in powers of the voltage V_{H}.

We calculate now the ϕ_2-averaged current transport $\overline{Q}$ when switching on ϕ_1 as described above. This yields:

$$\overline{Q} := \frac{1}{2\pi} \int_0^{2\pi} \int_0^{2\pi} j_x \, \mathrm{d}\phi_1 \wedge \mathrm{d}\phi_2 \tag{65}$$

$$= \frac{i}{2\pi q} \int_{S^1 \times S^1} \mathrm{tr}\, \hat{P}(\mathrm{d}\hat{P}) \wedge (\mathrm{d}\hat{P})\hat{P} \tag{66}$$

because the first term is cancled by periodicity. By means of the adiabatic theorem, we may now replace $\hat{P}$ by P and obtain the desired result

$$\bar{Q} = \frac{1}{q}\frac{i}{2\pi} \int \mathrm{tr}\, P(\mathrm{d}P) \wedge (\mathrm{d}P)P + O(V_{\mathrm{H}}^{\infty}) \ , \tag{67}$$

namely, that the averaged charge transport is given by the first chern number of the ground state bundle of H.

3.2 Fluctuations, and Quillen's Formula

Besides the interpretation as curvature, are we able to calculate tr $P(dP) \wedge (dP)P$ more explictly? Moreover, since the above formula speaks only about the average of this expression, what about the fluctuations of the trace? They can be calculated in a different model, indeed (J. E. Avron, R. Seiler, P.G. Zograf (1994)).

The base of the vector bundle defined by P is here a (two-dimensional) Riemann surface Σ of genus g, with g magnetic fluxes through the handles of the surface and a "constant" magnetic field; since Σ is by construction a curved space, we have to be make clear what we mean by this:

In terms of the complex local coordinates z, the surface comes with a conformal metric $ds^2 = \rho(z, \bar{z}) dz\, d\bar{z}$. This metric defines naturally the volume form of the surface, namely $\sigma = \frac{i}{2}\rho dz \wedge d\bar{z}$, and the Hodge-star operator $\star$. Since we can identify magnetic fields with two-forms, we call a magnetic field "constant" if it is a constant multiple of the volume form. To allow a geometric interpretation of the system, we furthermore impose "Dirac quantization", the integral of the magnetic field two-form over the surface area is $2\pi i$ times an integer:

$$\int_{\Sigma} B = 2\pi i f \tag{68}$$

This ensures that we may later on interpret our wave-function as sections in a $U(1)$-bundle over the surface Σ.

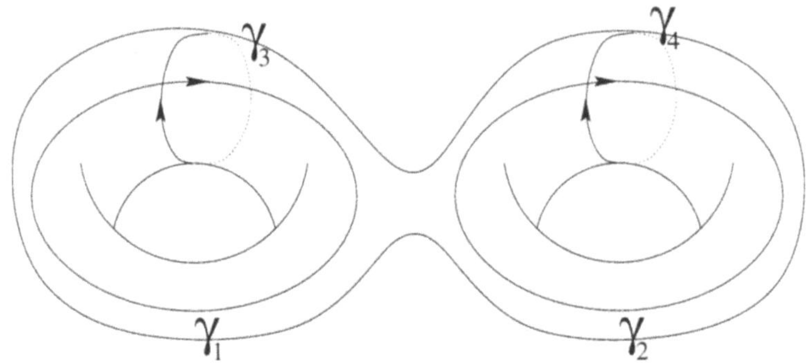

Fig. 9. The fundamental cycles of a Riemann surface

To introduce the fluxes, we make use of the DeRham theorem: it guarantees the existence of a basis of $2g$ real harmonic[5] one-forms dual to the fundamental cycles $\gamma_1, \ldots, \gamma_{2g}$, enclosing pairwise the handles of the surface.

$$\int_{\gamma_k} \omega_i = \delta_{i,k} \quad \omega_i \in \Omega^1(\Sigma) \tag{69}$$

A little calculation shows that each (real) vector A potential giving rise to the same magnetic field B, i.e. $dA = B$, can be written in the following way:

$$A = A_0 + \sum_{j=1}^{2g} \omega_j \phi^j + \mathfrak{g}^{-1} \mathrm{d}\mathfrak{g}. \tag{70}$$

Here, ϕ^1 to ϕ^{2g} are $2g$ magnetic Aharonov-Bohm fluxes through the handles of the surface, defined modulo $2\pi\mathbb{Z}$, and $\mathfrak{g} \in \mathbf{C}^\infty(\Sigma)$ is a gauge-transformation. A_0 is an arbitrary "origin" in this space with $dA_0 = B$. Hence, the space of vector potentials modulo gauge-transformations forms a $2g$ affine torus $\Phi = \mathbb{R}^{2g}/(2\pi\mathbb{Z}^{2g})$, parametrized by the magnetic fluxes, therefore called the "flux torus"[6]

This space can be given a natural symplectic and a natural Riemannian structure by

$$\Omega := \sum_{j=1}^{g} \mathrm{d}\phi^j \wedge \mathrm{d}\phi^{j+g} \tag{71}$$

[5] i.e. closed and co-closed, $d\omega = d \star \omega = 0$.

[6] This is in some sense the dual of the "Jacobian" of the base manifold, i.e. the Teichmüller space of holomorphic line bundles on Σ. Instead of deforming the bundle, we parametrized the connections $\nabla = -id + A$, leaving the $U(1)$ bundle fixed.

$$G_{i,j} := \int_{\Sigma} \omega_i \wedge \star\omega_j \ , \tag{72}$$

where $\star$ denotes the Hodge-star operator. Moreover, we may introduce an almost complex structure on this ad-hoc real manifold. For that, denote that the tangent space of Φ is naturally parametrized by harmonic one-forms, and we may act on them by the *real linear* hodge star from the cotangent bundle of Σ. If we denote this operation by J, we obviously have an almost complex structure since

$$J^2 = \star\star = -\mathbb{1} \tag{73}$$

for one-forms. It turns out that this almost-complex structure is integrable and hence a complex structure. Moreover, the flux torus Φ is kählerian, i.e. we have by straighforeward calculation

$$G(X,Y) = \Omega(JX,Y) \tag{74}$$

for tangent vector fields X and $Y \in \mathrm{T}_A\Phi$.

We consider now the following one-particle Hamiltonian on the base space Σ:

$$H(\phi) = (-i\mathrm{d} + A(\phi))^*(-i\mathrm{d} + A(\phi)) = 4D^*(\phi)D(\phi) + B \ , \tag{75}$$

where $A(\phi)$ is the vector potential parametrized by the flux torus as given by (70), B is the magnetic field and D is the Dirac operator - or in more geometric languague, the 0, 1-part of the connection $-i\mathrm{d} + A$. It is now well-known that the ground-state of H is given by the kernel of D. Moreover, we can compute the dimension of the kernel for large magnetic fields:

$$\frac{1}{2\pi i}\int_{\Sigma} B = f \geq 2g - 1 \ , \tag{76}$$

using the Riemann-Roch index formula for the operator D:

$$\operatorname{index} D = 1 - g + f \quad \Rightarrow \quad \dim\ker D = f - g + 1, \tag{77}$$

independent of the fluxes ϕ. This, and the compactness of the torus Φ, guarantees the existance of an energy gap, as required for the application of the adiabatic theorem.

As before, let $P(\phi)$ the projection onto the ground-state of $H(\phi)$. Quillen's local index formula (D. Quillen (1985)) states now that the adiabatic curvature

$$\sigma(\phi) := \operatorname{tr} P(\mathrm{d}P) \wedge (\mathrm{d}P)P \tag{78}$$

splits into two parts: one constant part given by the geometry of the system, and one fluctuating part which integrates out by taking the average:

$$\sigma(\phi) = 2\pi i\Omega + \frac{i}{2}\mathrm{d}J\mathrm{d}\log\det(D^*D) \tag{79}$$

The right-hand side determinant is the zeta-regularized determinant of elliptic operators, and d is the exterior differentiation with respect to the fluxes ϕ.

3.3 Quantum Viscosity

We present now an application for the chern-number approach which is not directly related to the quantum Hall conductivity, but to so called "quantum viscosity", cf. J. E. Avron, R. Seiler, P.G. Zograf (1995). The model presented here was later on generalized by P. Lévay (1995).

Let us review some basics from classical continuum mechanics: if we deform a macroscopic body by acting on it with an exernal force, a small region within that body gets moved from the point $x \in \mathbb{R}^3$ to $x' = x + u$. The vector field u is called the "distortion field" of the movement. Its differential splits into an antisymmetric part which describes just an infinitesimal rotation of the system, and a symmetric part which is called the "strain tensor":

$$u_{\alpha,\beta} := \frac{\partial u_\alpha}{\partial x^\beta} + \frac{\partial u_\beta}{\partial x^\alpha} \tag{80}$$

The internal forces of the body are described by another tensor, the "stress tensor" $\sigma_{\alpha,\beta}$. The force F acting on an internal cut with normal n is given by

$$F_\alpha = \sum_\beta \sigma_{\alpha,\beta} n_\beta \ . \tag{81}$$

For the limit of small strain rates the stress of a fluid depends linearly on the strain and on its first time derivative, the strain rate:

$$\sigma_{\alpha,\beta} = \sum_{\gamma,\delta} \lambda_{\alpha,\beta,\gamma,\delta} u_{\gamma,\delta} - \sum_{\gamma,\delta} \eta_{\alpha,\beta,\gamma,\delta} \dot{u}_{\gamma,\delta} \tag{82}$$

The coefficient $\lambda_{\alpha,\beta,\gamma,\delta}$ is called the "elastic modulus tensor", $\eta_{\alpha,\beta,\gamma,\delta}$ the "viscosity tensor".

For a newtonian fluid, the tensor $\sigma_{\alpha,\beta}$ is symmetric and hence the viscosity is symmetric in both index pairs:

$$\eta_{\alpha,\beta,\gamma,\delta} = \eta_{\beta,\alpha,\gamma,\delta} = \eta_{\alpha,\beta,\delta,\gamma} \ . \tag{83}$$

With respect to the index permutation $\alpha, \beta, \gamma, \delta \to \gamma, \delta, \alpha, \beta$, the viscosity splits into an symmetric part associated to dissipation and an antisymmetric part describing non-dissipative response

$$\begin{gathered} \eta = \eta^S + \eta^A \\ \eta^S_{\alpha,\beta,\gamma,\delta} = \eta^S_{\gamma,\delta,\alpha,\beta} \qquad \eta^A_{\alpha,\beta,\gamma,\delta} = -\eta^A_{\gamma,\delta,\alpha,\beta} \end{gathered} \tag{84}$$

One usually assumes the antisymmetric part to vanish because of no compelling evidence to think otherwise.

Quantum fluids, however, can have a grounds state which is separated by a finite gap from the rest of the spectrum; such a fluid will have a non-dissipative response with $\eta^S = 0$ at zero temperature, whereas η^A may or may not vanish. For example, time reversal symmetry will cause $\eta^A = 0$ due to the Onsager relation, but a sytem with broken time symmetry — as the Hall fluid with a full Landau level — will have $\eta^A \neq 0$ in general.

We consider now a two dimensional quantum Hall fluid on a torus $\mathcal{T} = \mathbb{R}^2/\mathbb{Z}^2$ with a flat metric, and use the Landau Hamiltonian to describe the kinetic energy of the system. Instead of deforming the base space and chosing the euclidian metric of the $\mathbb{R}^2$, we keep the fundamental domain fixed and deform instead the metric g of this torus such that the volume $V = \sqrt{\det g}$ does not change. The space of these flat metrics on tori is parametrized by one complex variable $\tau = \tau_1 + i\tau_2$:

$$g(V,\tau) = \frac{V}{\tau_2}(dx^2 + 2\tau_1 dxdy + |\tau|^2 dy^2) \ , \tag{85}$$

where it is enough to consider τ in the fundamental domain of $SL(2,\mathbb{Z})$ because all other choices are obtained by simply choosing a different base in the lattice $\mathbb{Z}^2$. This domain is a two-sphere with two conical points and one puncture, and the analog of the "flux torus" of the previous section, cf. Fig. 3.3.

The Hamiltonian with respect to this metric, with Aharonov-Bohm gauge fields ϕ_1, ϕ_2 and a constant magnetic field B perpendicular to the torus is given by

$$H(V,\tau,\phi) = \frac{1}{V\tau_2}\left(|\tau|^2 D_x^2 - \tau_1(D_x D_y + D_y D_x) + D_y^2\right) \ , \tag{86}$$

where the Dirac operators D_x and D_y are

$$\begin{aligned} D_x &= -i\partial_x + 2\pi(By + \phi_1 + B/2) \\ D_y &= -i\partial_y + 2\pi(\phi_2 + B/2) \ . \end{aligned} \tag{87}$$

We furthermore require that $B \in \mathbb{Z}$ is an integer and impose the usual magnetic translation boundary conditions:

$$\psi(x+1,y) = \psi(x,y) \quad \psi(x,y+1) = \mathrm{e}^{-2\pi i Bx}\psi(x,y) \tag{88}$$

The stress operator is now, by the principle of virtual work, the derivation of H with respect to the strain

$$\sigma_{\alpha,\beta} = -\frac{1}{V}\frac{\partial H}{\partial u_{\alpha,\beta}} \tag{89}$$

Adiabatic deformation gives the quantum version of (82), which is the analog of the persistent-current equation (64):

$$\left\langle \frac{\partial H}{\partial u_{\alpha,\beta}} \right\rangle = \frac{\partial E}{\partial u_{\alpha,\beta}} + \sum_{\gamma,\delta} \Omega_{\alpha,\beta,\gamma,\delta}\dot{u}_{\gamma,\delta} + O(T^{-1}) \ , \tag{90}$$

where T is again the adiabatic time scale parameter, E is the expectation of the energy and Ω is the adiabatic curvature, which plays now the role of the non-dissipative viscosity. For homogeneous fluids, the viscosity and the curvature are related by

$$\Omega_{\alpha,\beta,\gamma,\delta} = V\eta^A_{\alpha,\beta,\gamma,\delta} \ . \tag{91}$$

Luckely, the ground states of the Hamiltonian can be written down explicitly in terms of theta-functions, so it's not hard to calculate the adiabatic curvature: one gets, as far as deformation is concerned:

$$\Omega = B\frac{d\tau_1 \wedge d\tau_2}{4\tau_2^2} \tag{92}$$

The physical interpretation of this formula is the following: consider a two dimensional Hall fluid on a surface of a cylinder. Compressing it in the radial or axial direction results in a twist rate of the left boundary circle relative to the right circle. And vice versa: a shear of the two boundary circles results in a compression rate in the radial and a stretching rate in the axial direction.

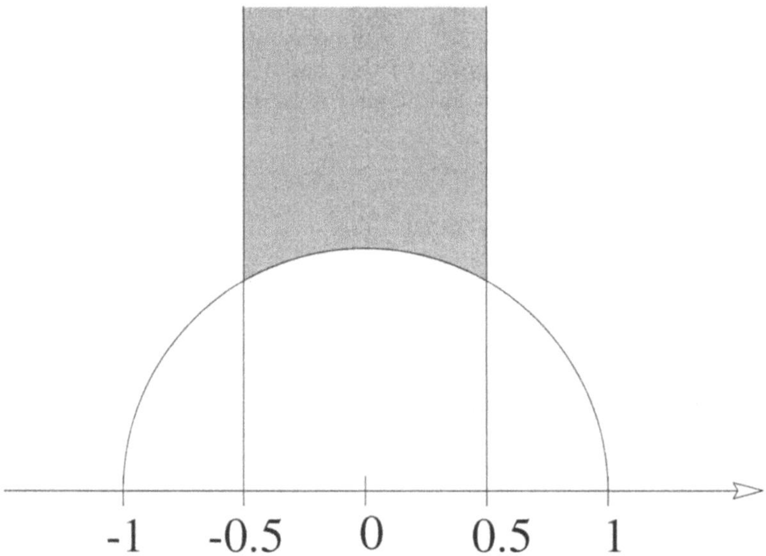

Fig. 10. The modulus space of complex tori

Similary to what we did in the Hall conductance setup, we may average the curvature over the moduli space, i.e. the fundamental domain $\mathcal{F}$ of $SL(2,\mathbb{Z})$. Calculating this number, we find:

$$<\Omega>=\frac{1}{2\pi}\int_{\mathcal{F}} B\frac{d\tau_1 \wedge d\tau_2}{4\tau_2^2} = \frac{B}{24} \ . \tag{93}$$

Though this is not an integer in general — because the parameter space is not a smooth compact manifold here — this is still a topogical invariant.

4 Index Approach, Bulk and Edge

We will discuss now another interpretation of the quantum Hall conductance, namely that of an index.

4.1 The Algebra of Two Projectors

Before we're aiming at defining an index, we first have to have a close look at the algebra generated by two orthogonal projections on a Hilbert space, following J. E. Avron, R. Seiler, B. Simon (1994a), J. E. Avron, R. Seiler, B. Simon (1994b) and S. Borac (1995).

The setup is as follows: we start with two orthogonal projections P and Q on a Hilbert space $\mathcal{H}$, and the algebra $\mathcal{R}$ generated by this two projections. Even though this looks like a very simple object, it has a surprisingly rich structure, as it allows the introduction of "differential calculus" and the definition of an index. Let's begin with a closer analysis of this algebra:

Defining the operators A and B by

$$A := P - Q \quad B := \mathbb{1} - P - Q, \tag{94}$$

we see that both A and B are selfadjoint, and generate $\mathcal{R}$ as well. Following J. E. Avron, R. Seiler, B. Simon (1994b), we call B the "Kato-dual" to A. It's easely checked that they fulfill the following algebraic relations:

$$A^2 + B^2 = 1 \quad \{A, B\} := AB + BA = 0 \tag{95}$$

From this we see that A^2 and B^2, and with them $|A|$ and $|B|$, commute with both A and B, and are therefore central in $\mathcal{R}$. We can now define an "operator valued angle" θ between the projections P and Q by

$$\theta := \arcsin |A| \tag{96}$$

The eigenprojections of A and B for the eigenvalues 0, +1 and −1 play an eminent role in what follows: we denote them by $E_A(0)$, $E_B(1)$ etc. Their images can be expressed in the intersection of the ranges of P and Q and their orthogonal complements $P^\perp$ and $Q^\perp$. We have, for example,

$$\begin{aligned} \operatorname{range} E_A(-1) &= \operatorname{range} P^\perp \cap \operatorname{range} Q \\ &= \operatorname{range} E_A(1) = \operatorname{range} P \cap \operatorname{range} Q^\perp \\ \operatorname{range} E_B(-1) &= \operatorname{range} P \cap \operatorname{range} Q \\ \operatorname{range} E_B(1) &= \operatorname{range} P^\perp \cap \operatorname{range} Q^\perp \ . \end{aligned} \tag{97}$$

If we define now the projection E as the sum $E_A(-1)+E_A(1)+E_B(-1)+E_B(1)$, we can check that this is just the projection onto the maximal abelian subalgebra $\mathcal{R}E$ within $\mathcal{R}$. The algebra $\mathcal{R}$ splits therefore into an abelian part $\mathcal{R}E$ and a completely non-commutative part $\mathcal{R}(\mathbb{1} - E)$ where the commutator of P and Q has trivial kernel. It can be proven that this completely non-commutative part is a type I_2 v. Neumann algebra, and its center is the v. Neumann algebra generated by the angle operator θ.

We define now the unitaries U and V by the polar decomposition of A and B:

$$A := U|A| \qquad B := V|B| \tag{98}$$

$$\Rightarrow \quad [P, Q] = -UV|A||B| \tag{99}$$

One can check that, using these definitions, we may write every element $T \in \mathcal{R}(\mathbb{1} - E)$ as

$$T = c_0(\theta)\mathbb{1} + c_1(\theta)U + c_2(\theta)V + c_3(\theta)(-iUV) \tag{100}$$

where $c_0, \ldots, c_3$ are central.

If we write $\mathbb{1}$, U, V and $-iUV$ in terms of the self-adjoint matrix units of the algebra

$$\begin{aligned} E_{11} &= \frac{\mathbb{1} - iUV}{2} \qquad & E_{12} &= U\frac{\mathbb{1} + iUV}{2} \\ E_{21} &= U\frac{\mathbb{1} - iUV}{2} \qquad & E_{22} &= \frac{1 + iUV}{2} \end{aligned} \tag{101}$$

we see that these unitaries are "morally" the Pauli matrices.

4.2 First Order Calculus on $\mathcal{A} := \mathcal{R}(1 - E)$

Recall that the one-forms $\Omega(\Sigma)$ as sections in the cotangent bundle over a smooth manifold Σ form a bimodule over the algebra $\mathbf{C}^\infty(\Sigma)$. The external derivative

$$\mathrm{d} : \mathbf{C}^\infty(\Sigma) \to \Omega(\Sigma) \tag{102}$$

is a linear map satisfying the Leibniz rule

$$\mathrm{d}(fg) = (\mathrm{d}f)g + f(\mathrm{d}g) \quad \text{for all} \quad f, g \in \mathbf{C}^\infty(\Sigma) \tag{103}$$

This algebraic setting can be used to define the concept of first order differential calculus for an arbitrary (unital) algebra $\mathcal{A}$:

Definition 3 *A triple $(\mathcal{A}, \Omega, \mathrm{d})$ with $\mathcal{A}$ a unital algebra, Ω a bi-module of $\mathcal{A}$ and d a linear map $\mathcal{A} \to \Omega$ is said to be a first order differential calculus over $\mathcal{A}$ if d fulfills the Leibniz rule*

$$\mathrm{d}(ab) = (\mathrm{d}a)b + a(\mathrm{d}b) \quad \textit{for all} \quad a, b \in \mathcal{A} \tag{104}$$

and further, if any $\rho \in \Omega$ is a finite sum of the form

$$\rho = \sum_k a_k \mathrm{d}b_k \qquad a_k, b_k \in \mathcal{A} \tag{105}$$

Any first order differential calculus is given by the following construction, up to isomorphism: let I be the sub-bimodule of $\mathcal{A} \otimes \mathcal{A}$ given by the kernel of the multiplication:

$$I := \text{kern}(m : \mathcal{A} \otimes \mathcal{A} \to \mathcal{A}) \tag{106}$$

Then every Ω is, up to bi-module isomorphism, given by $\Omega = I/N$ where N is a sub-bimodule of I and $\mathrm{d} = \pi \circ \mathcal{D}$ where $\pi : I \to \Omega$ is the canonical projection map and $\mathcal{D}$ is the linear map

$$\begin{aligned} \mathcal{D} : \mathcal{A} &\to I \\ x &\mapsto \mathbb{1} \otimes x - x \otimes \mathbb{1} \ . \end{aligned} \tag{107}$$

Therefore, the choice of a differential calculus is equivalent to the choice of the sub-bimodule $N \subset I$.

We can now understand the classical differential calculus of commutative algebras in a second way, namely by defining it by taking for N the bimodule generated by the image of the map

$$\text{range}\left((\text{id} + \tau)|_I : I \to I \right) \tag{108}$$

where τ is the twist map:

$$\begin{aligned} \tau : \mathcal{A} \otimes \mathcal{A} &\to \mathcal{A} \otimes \mathcal{A} \\ x \otimes y &\mapsto y \otimes x \end{aligned} \tag{109}$$

Since $\mathcal{A}$ is here commutative, the multiplication is "τ-commutative", i.e. $m = m\tau$.

This concept can be generalized to non-commutative algebras, where τ gets replaced by the "Yang-Baxter operator"

$$R : \mathcal{A} \otimes \mathcal{A} \to \mathcal{A} \otimes \mathcal{A}. \tag{110}$$

with the properties

$$\begin{aligned} R(a \otimes \mathbb{1}) &= \mathbb{1} \otimes a \\ R(\mathbb{1} \otimes a) &= a \otimes \mathbb{1} \\ R(m \otimes \text{id}) &= (\text{id} \otimes m) R_1 R_2 \\ R(\text{id} \otimes m) &= (m \otimes \text{id}) R_2 R_1 \ , \end{aligned} \tag{111}$$

where R_1 resp. R_2 denotes the action of R onto the first two resp. last two factors of the product $\mathcal{A} \otimes \mathcal{A} \otimes \mathcal{A}$.

The sub-bimodule N generated by

$$\text{range}\left((\text{id} + R)|_I : I \to I \right) \tag{112}$$

forms then a differential calculus on $\mathcal{A}$ in the sense above.

The algebra $\mathcal{A} := \mathcal{R}(1-E)$ can be equipped with such a structure by

$$\begin{aligned} R : C_1 \otimes C_2 &\mapsto C_2 \otimes C_1 \\ U \otimes C &\mapsto C \otimes U \qquad \text{and vice versa} \\ U \otimes V &\mapsto -V \otimes U \\ V \otimes U &\mapsto -U \otimes V \ , \end{aligned} \tag{113}$$

where C, C_1 and C_2 are central elements. We find that $\mathcal{A}$ is R-commutative, i.e. $m = mR$ and furthermore, that in the differential calculus defined by R:

$$d(UV) = U(dV) - V(dU) \tag{114}$$

Moreover, $R^2 = \text{id}$.

4.3 The Index of a Pair of Projections

Let P and Q be orthogonal projections on a separable Hilbert space $\mathcal{H}$, as in the previous section. We say that the pair (P,Q) is "fredholm" if the map

$$C := QP : \text{range}\, P \to \text{range}\, Q \tag{115}$$

is a Fredholm operator. We call the index of this map the relative index of the pair (P,Q), written $\text{index}(P,Q)$. Using the notation from above, it can be shown that

$$\begin{aligned} \text{index}(P,Q) &= \dim E_A(1) - \dim E_A(-1) \\ &= \dim \text{kern}(P - Q - \mathbb{1}) - \dim \text{kern}(Q - P - \mathbb{1}) \ . \end{aligned} \tag{116}$$

The following relations for the index are not unexpected:

$$\text{index}(Q,P) = -\text{index}(P,Q) \tag{117}$$

$$\text{index}(UPU^{-1}, UQU^{-1}) = \text{index}(P,Q) \tag{118}$$

$$\text{index}(P,R) = \text{index}(P,Q) + \text{index}(Q,R) \tag{119}$$

for all orthogonal projections R such that either $Q-R$ or $P-Q$ is compact and all unitaries U.

Moreover, we can prove a very convenient formula for the index in case $P-Q$ is in one of the trace class ideals $\mathcal{I}_{2n+1}$, i.e. $(P-Q)^{2n+1}$ is trace class. We have then:

$$\text{index}(P,Q) = \text{tr}\ (P-Q)^{2m+1} \quad \text{for all } m \geq \text{n}. \tag{120}$$

The proof of this theorem is not too hard, using only the algebraic relations of $A = P - Q$ and its Kato-dual $B = P - Q^{\perp}$. For first, check that the spectrum of A without the points $+1$ and -1 is invariant under reflection:

$$\begin{aligned} Ae_n &= \lambda_n e_n \qquad \lambda_n \notin \{-1,+1\} \\ \Rightarrow \quad A(Be_n) &= -B(Ae_n) = -\lambda_n(Be_n) \\ \text{furthermore} \quad Be_n = 0 &\Rightarrow A^2 e_n = e_n \ \Rightarrow \ Ae_n = \pm e_n \ , \end{aligned}$$

which shows that $Be_n \neq 0$ and hence the claim. If we denote now the multiplicity of the eigenvalue λ by m_λ, we have by Lidskii's theorem

$$\begin{aligned}\operatorname{tr}(P-Q)^{2m+1} &= \sum \lambda^{2m+1} m_\lambda = \sum_{\lambda>0} \lambda^{2m+1}(m_\lambda - m_{-\lambda}) \\ &= m_1 - m_{-1} = \operatorname{index}(P,Q) \ .\end{aligned}$$

Even though this proof does not hold if P and Q are not selfadjoint, the theorem remains true in this more general case.

4.4 Index Approach to the QHE

We consider now an application of the index approach to a quantum Hall system, following J. E. Avron, R. Seiler, B. Simon (1994a). Our model describes non relativistic, non interacting fermions in the punctured plane $\mathbb{C} \setminus \{a\}$, $a \in \mathbb{C}$, with random impurities. As always, a constant magnetic field B perpendicular to the plane acts on the particles. The one-particle Hamiltonian of this system reads

$$H_\omega := \frac{1}{2}(p-A)^2 + W_\omega = H_0 + W_\omega, \tag{121}$$

where $p = -id$ is the momentum operator, A is the vector potential with curl $A = B$ and W_ω is a random potential.

We introduce now the magnetic translation operators $T(a)$ by requiring

$$T(a)f(z) = e^{i\theta(B,a,z)} f(z-a) \qquad [H_0, T(a)] = 0 \quad (a, z \in \mathbb{C}) \tag{122}$$

where $\theta(B,a,z)$ is a phase factor. Note that, due to the magnetic field, the ordinary translation operators $\theta = 0$ do no longer commute with H_0. We require furthermore that the translations act ergodically on the propability space.

We add now adiabatically one magnetic flux unit through the point a, i.e. we consider the time-dependent Hamiltonian

$$H_\omega(t) := \frac{1}{2}(p - A + \Lambda(t) d\phi_a(z))^2 + W_\omega \tag{123}$$

$$\text{where} \qquad \phi_a(z) = \frac{z-a}{|z-a|} \ . \tag{124}$$

The function ϕ models the additional magnetic flux, and $\Lambda(t)$ is a "switch" function, 0 for negative t and monotonically increasing towards 1 for $t \to \infty$. Obviously, we have $H_\omega(-\infty) = H_\omega$. Since the Hamiltonian with one additional flux unit piercing at a is gauge equivalent to that without the flux, we have in the adiabatic limit, i.e. the limit $\frac{d}{dt}\Lambda(t) \to 0$, that

$$H_\omega(+\infty) = U_a H_\omega U_a^{-1} \qquad u_a(z) = \frac{z-a}{|z-a|} \tag{125}$$

where U_a is the gauge transformation acting by multiplication with u_a.

We fix a Fermi level E_F and define P_ω to be the projection onto all eigenstates of H_ω of energies below this level. It can be seen that — by turning on the flux adiabatically — we drive the states outwards to infinity. The number of states transported, i.e. the "charge deficiency" is given by the relative index of the projection P_ω and the related projection of $H_\omega(\infty)$, namely $U_a P_\omega U_a^{-1}$:

$$Q_\omega^D = \mathrm{index}(P_\omega, U_a P_\omega U_a^{-1}) \quad \omega\text{-almost sure} \tag{126}$$

This charge deficiency is ω-almost sure identically to the charge deficiency Q^D of the non-probabilistic system without the random potential W_ω. We have the following theorem:

Let H be a Schrödinger operator on the domain $\mathbb{C}$ with a finite gap in its spectrum and let P be a spectral projection of H onto all eigenstates below the gap. Furthermore, let P have an integral kernel $p(z_1, z_2)$ which is jointly continous in z_1 and $z_2 \in \mathbb{C}$ and decays away from the diagonal, which is essentially a localization condition:

$$|p(z_1, z_2)| \leq \frac{C}{1 + |z_1 - z_2|^{2+\epsilon}} \quad (C > 0, \epsilon > 0) \tag{127}$$

These conditions are, for example, fulfilled by the Landau-Hamiltonian H_0.

Let U be the unitary operator which acts by multiplication with a function $u(z)$ of modulus one, which is differentiable away from a single point $a \in \mathbb{C}$. We assume furthermore that there are constants $C_1 > 0$ and $C_2 > 0$ such that

$$\left|1 - \frac{u(z_1 + z_2)}{u(z_2)}\right| \leq C_1 \frac{|z_1|}{|z_2|} \qquad \text{for all} \quad \frac{|z_1|}{|z_2|} \leq C_2 \tag{128}$$

In our example, it is easy to check that the flux tube

$$u_a(z) = \frac{z - a}{|z - a|},$$

fulfills these conditions.

Under these hypothesis, one can prove that $(P - UPU^{-1}) \in \mathcal{I}_3$ and therefore

$$Q^D = \mathrm{index}(P, UPU^{-1}) = \mathrm{tr}\,(P - UPU^{-1})^3 \tag{129}$$

The right hand side can be written more explicitly, using the integral kernels for P and the explicit form of U:

$$Q^D = \int p(z_1, z_2) p(z_2, z_3) p(z_3, z_1) \left(1 - \frac{u(z_1)}{u(z_2)}\right) \left(1 - \frac{u(z_2)}{u(z_3)}\right) \left(1 - \frac{u(z_3)}{u(z_1)}\right) dz_1 dz_2 dz_3 \tag{130}$$

If we furthermore assume that the projection P is covariant, i.e. there exists a (unitary) gauge transformation $T(a)$ acting by multiplication by a phase and a shift of the argument on the integral kernel $p(z_1, z_2)$:

$$p(z_1, z_2) = \mathrm{e}^{i\theta(a, z_1)} p(z_1 - a, z_2 - a) \mathrm{e}^{-i\theta(a, z_2)} \tag{131}$$

we can evaluate this integral:

$$Q^D = 2\pi i N(U) \int p(0, z_1) p(z_1, z_2) p(z_2, 0)(z_1 \wedge z_2) dz_1 dz_2 \tag{132}$$

The number $N(U)$ is the winding number of the unitary $u(z)$ around the point a, and

$$z_1 \wedge z_2 := \mathrm{Re}(z_1)\mathrm{Im}(z_2) - \mathrm{Im}(z_1)\mathrm{Re}(z_2) \tag{133}$$

is twice the area of the triangle spanned by z_1, z_2 and the origin.

In our case, P is of course covariant, the required gauge transformation is given by the magnetic translation operators $T(a)$, and obviously $N(U) = 1$. It is remarkable that in the case of this simple flux tube the required calculations can be performed explicitly and more or less boil down to the computation of the area of triangles.

4.5 Edge vs. Bulk

All models for the QHE presented so far but the quantum field theoretic approach do not take the boundary of the sample into account, even though several authors focus on the importance of states localized near the edge of the model and their interplay with the states in the bulk of the probe. We're now going to present a more suitable model (E. Akkermans, J. E. Avron, R. Narevich, R. Seiler (1998)):

To keep things as simple as possible, we consider again the cylinder symmetry system of Laughlin, i.e. the configuration space is $\Sigma = [0, L]_x \times S^1_y$. In addition to that we have a constant magnetic field B perpendicular to the cylinder surface and a gauge flux Φ in direction of the cylinder axis. The critical point is now *to define* what bulk and edge states should be and how to keep them apart. Therefore, we're going to introduce chiral boundary conditions, closely related to that of Atiyah, Patodi and Singer.

The question what bulk and edge means has an obvious answer in classical mechanics: if we consider a classical charged particle in a "billiard" under the influence of a constant magnetic field, it will rotate — say — in clockwise direction as long as it doesn't touch the boundary, but it will in average rotate counter-clockwise if it hits the boundary. Therefore, we call a state $\Psi \in \mathbf{L}^2(\Sigma)$ a bulk state, if

$$\begin{aligned} &\int_0^{2\pi} \overline{\psi(0,y)} v_y(0) \psi(0,y) dy > 0 \quad \text{and} \\ &\int_0^{2\pi} \overline{\psi(L,y)} v_y(L) \psi(L,y) dy < 0 \ , \end{aligned} \tag{134}$$

where $v_y(x) := -i\partial_y - Bx - \Phi/(2\pi)$ is the velocity operator in y-direction taken at x. Hence, a state is in the bulk if the expectation of its y-velocity is positive at the left hand side, and negative on the right hand side. It is now obvious that the $\mathbf{L}^2(\Sigma)$ splits into three parts which we call "left edge" — the first expectation is negative — "bulk" and "right edge" — the second expectation is positive.

The definition of the Hamiltonian is crucial: let $D(\Phi)$ be the Dirac operator

$$D(\Phi) := \partial_x - v_y = \bar{\partial} + Bx + \frac{\Phi}{2\pi} \tag{135}$$

and define the energy functional by the following quadratic form on $\mathbf{C}^\infty(\Sigma)$:

$$\begin{aligned} E(\psi) := & \langle D\psi, D\psi \rangle \\ & - \int_0^{2\pi} \overline{P_{\mathrm{LE}}\psi}(0,y)(v_y P_{\mathrm{LE}}\psi)(0,y)dy \\ & + \int_0^{2\pi} \overline{P_{\mathrm{RE}}\psi}(L,y)(v_y P_{\mathrm{RE}}\psi)(L,y)dy \ , \end{aligned} \tag{136}$$

where P_{LE} and P_{RE} are the projections onto the left-edge resp. right-edge part of $\mathbf{L}^2(\Sigma)$. Both boundary integrals are positive, hence we added an energy penalty to states living marginaly.

Since E is now a positive, symmetric quadratic form, it defines a selfadjoint Hamiltonian H by

$$E(\psi) = \langle \psi, H\psi \rangle \tag{137}$$

The operator H is formally given by D^*D, but the energy penalty of the edge states is now encoded in spectral boundary conditions defining the domain of H. A core for its domain is given by

$$\begin{aligned} D(H) \supset \Big\{ \psi \in \mathbf{C}^\infty(\Sigma) \Big| \\ & (DP_{\mathrm{LE}}^\perp\psi)(0,y) = 0 \wedge (DP_{RE}^\perp\psi)(L,y) = 0 \\ \wedge \quad & (\partial_x P_{LE}\psi)(0,y) = 0 \wedge (\partial_x P_{\mathrm{RE}}\psi)(L,y) = 0 \Big\} \ . \end{aligned} \tag{138}$$

These boundary conditions look almost like the spectral boundary conditions considered by Atiah, Patodi and Singer, except that we obtained a Neumann type boundary condition on the right resp. left edge part of the Hilbert space whereas APS consider Dirichlet boundary conditions there.

The physical relevance of these boundary conditions becomes even more clear if we use the cylinder symmetry of the system and Fourier-transform it:

$$\begin{aligned} \mathcal{F} : \ell^2(\mathbb{Z}) \otimes \mathbf{L}^2([0,L]) & \to \mathbf{L}^2(\Sigma) \\ \{\psi_m(x)\}_{m\in\mathbb{Z}} & \mapsto \sum_{m\in\mathbb{Z}} \psi_m(x) e^{imy} \end{aligned} \tag{139}$$

The Fourier-decomposed operator is, hence, an operator valued matrix, which is diagonal due to the rotation symmetry. It is just a harmonic oscillator centered at $\rho(m)$:

$$(\mathcal{F}^{-1}H\mathcal{F})_{m,m'} = \delta_{m,m'}\left(-\frac{d^2}{dx^2} + (m - Bx - \frac{\Phi}{2\pi})^2 - B\right) \tag{140}$$

$$=: \delta_{m,m'}h(\rho) = \delta_{m,m'}\left(-\frac{d^2}{dx^2} + B^2\big(x - \rho(m)\big)^2 - B\right)$$

$$\text{where} \quad \rho(m) := \frac{2\pi m - \Phi}{2\pi B} \tag{141}$$

The advantage of this approach is that we have now a very simple characterization for left edge, bulk and right edge, namely

$$\mathbf{L}^2(\Sigma) \overset{\mathcal{F}}{\cong} \tag{142}$$

$$\bigoplus_{m,\rho(m)<0} e^{imy}\mathbf{L}^2([0,L]) \quad \bigoplus_{m,\rho(m)\in[0,L]} e^{imy}\mathbf{L}^2([0,L]) \quad \bigoplus_{m,\rho(m)>L} e^{imy}\mathbf{L}^2([0,L]) \ ,$$

where the summands are left-edge, bulk and right-edge part, respectively.

The boundary conditions for this one-dimensional problem are now very simple:

$$\begin{aligned} \text{left edge} \quad \rho < 0 &: \partial_x\psi_m(0) = 0 \wedge \big(\partial_x + (x-\rho)\big)\psi_m(L) = 0 \\ \text{bulk} \quad \rho \in [0,L] &: \big(\partial_x + (x-\rho)\big)\psi_m(0) = 0 \wedge \big(\partial_x + (x-\rho)\big)\psi_m(L) = 0 \\ \text{right edge} \quad \rho > L &: \big(\partial_x + (x-\rho)\big)\psi_m(0) = 0 \wedge \partial_x\psi_m(L) = 0 \end{aligned} \tag{143}$$

Since the kernel of $h(\rho)$ is the kernel of $\big(\partial_x + (x-\rho)\big)$, these kernel eigenfunctions fulfill the bulk boundary conditions automatically and are therefore identified as bulk ground states. Hence, the bulk ground states are identical to those of the Landau Hamiltonian in the infinte plane: gaussians, centered at ρ, and localized in the interiour of the cylinder.

As we adjust Φ, these states will get moved on the cylinder. Note that for ρ equal to 0 or L, the Gaussians have a horizontal tangent at the edge and are therefore both bulk, and edge states. This shows that the spectrum of $h(\rho)$ depends continously on ρ, unlike for the APS conditions where can be shown to jump at the transition from edge to bulk.

However, if we increase or decrease ρ further into the edge, the eigenfunctions will look more complicated and their energy will increase. In particular, for the lowest edge branch one has in the limit $\rho \nearrow 0$ or $\rho \searrow L$ a finite unique sound velocity for the chiral edge currents:

$$\left.\frac{\partial E}{\partial \rho}\right|_{\rho\nearrow 0} = \sqrt{\frac{B}{\pi}} \tag{144}$$

We return now to our starting point, the argument Laughlin presented in his first paper: consider the second quantization of this Hamiltonian for non-interacting fermion particles with the Fermi level set to zero, the bulk ground state energy: the multi-particle ground state wave function is given by "filling up" the ground state of the one-particle system:

$$\Psi := \psi_0 \wedge \psi_1 \wedge \ldots \wedge \psi_{M-1} \tag{145}$$

were the degeneracy M of the one-particle ground state is $M = B \cdot L$ and the wave function ψ_k is a Gaussian centered at k/B for $\Phi = 0$. If we increase now Φ by 2π, the leftmost state becomes an left-edge state and all other states move just one step to the left. The number of states transported by the increase of Φ by one flux unit, i.e. 2π, is therefore one and the Hall conductivity is one.

Following Laughlin in his second paper, we can build a simple model for the fractional quantum hall effect as well. If we consider the many-body Laughlin wave-function

$$\psi_l(z_1, \ldots, z_k) := \prod_{1 \leq i < k \leq N} (\mathrm{e}^{z_i} - \mathrm{e}^{z_k})^3 \exp\left(-B/2 \cdot \sum_{i=1}^{N} \left(x_i - \frac{\Phi}{2\pi B}\right)^2\right) \tag{146}$$

where $z_i := x_i + iy_i$, we find for $\Phi = 0$ exactly $N = B \cdot L/3$ states and obtain in this way a filling factor of one third. It is easy to check that it requires three flux units to move the Laughlin states one step to the left, hence the Hall conductivity is now $1/3$ and therefore fractional. The degeneracy of such states is threefold; hence, it is of no surprise that the conductance is an integer divided by 3, see eq. 67.

References

E. Abrahams, P.W. Anderson, D. Licciardello, T. Ramakrishnan (1979): Scaling Theory of Localization: Absense of Quantum Diffusion in Two Dimensions. Phys. Rev. Lett. **42**, 673

M. Aizenman, G. M. Graf (1998): Localization bounds for an electron gas. J. Phys. A **31** 6783

E. Akkermans, J. E. Avron, R. Narevich, R. Seiler (1998): Boundary conditions for Bulk and Edge States in Quantum Hall Systems. European J. Phys. B **1**, 117

M. F. Atiyah, V. K. Patodi, I. M. Singer (1975): Spectral asymmetry and Riemannian Geometry I. Math. Proc. Camb. Phil. Soc. **77**, 43

J. E. Avron, A. Elgart (1998): The Adiabatic Theorem of Quantum Mechanics. Private communications, to appear in Comm. Math. Phys. The paper can be obtained at `http://xxx.lanl.gov/abs/math-ph/9805022`.

J. E. Avron, R. Seiler (1985): Quantization of the Hall conductance for general multi-particle Schrödinger Hamiltonians. Phys. Rev. Lett. **54** 259

J. E. Avron, R. Seiler, L. G. Yaffe (1987): Adiabatic theorems and applications to the quantum Hall effect. Comm. Math. Phys. **110** 33

J. E. Avron, R. Seiler, B. Simon (1994a): Charge deficiency, charge transport and comparison of dimension. Comm. Math. Phys. **159**, 399

J. E. Avron, R. Seiler, B. Simon (1994b): The index of a pair of projections. J. of Funct. Anal. **120** 220

J. E. Avron, R. Seiler, P. G. Zograf: Adiabatic Quantum Transport: Quantization and Fluctuations. Phys. Rev. Lett. **73** No. 24, 3255

J. E. Avron, R. Seiler, P. G. Zograf: Viscosity of Quantum Hall Fluids. Phys. Rev. Lett. **75** No. 4 697

J. Bellisard in: Proceedings of the Bad Schandau conference on localization, Eds. Ziesche and Weller. Teubner-Verlag

J. Bellissard, A. van Elst, H. Schulz-Baldes (1994): The non commutative goemetry of the quantum Hall effect. J. Math. Phys. **35**, 5373

M. V. Berry (1984): Quantal phase factors accompanying adiabatic changes. Proc. R. Soc. Lond. A **392** 45

S. Borac (1995): On the Algebra Generated by Two Projections. J. Math. Phys. **36**, No. 2, 863

M. Born, V. Fock (1928): Beweis des Adiabatensatzes. Z. Phys. **51** 165-169

A. M. Chang, J. E. Cunningham (1989): Solid State Comm. **72** 652

R. G. Clark et al. (1988): Experimental Determinatino of Fractional Charge e/q for Quasiparticle Excitations in the Fractional Quantum Hall Effect. Phys. Rev. Lett. **60** 1747

A. Connes (1985): Introduction to non commutative differential geometry. Publ. IHES **62**, 257

A. Connes (1994): Noncommutative Geometry. Academic Press, London.

J. Fröhlich, T. Kerler (1991): Universality in quantum Hall systems. Nulc. Phys. B **354** 369

J. Fröhlich, U. Studer (1992): Gauge invariance in non-relativistic many-body theory. In: Mathematical Physics X. K. Schmüdgen Eds. Springer, Berlin, Heidelberg, New York.

E. H. Hall (1879): On a new action of the Magnet on Electric Currents: Amer. J. Math. **2** 287

B. I. Halperin (1982): Quantized Hall conductance, current-carrying edge states, and the existance of extended states in a two dimensional disordered potential, Phys. Rev. B **25** 2185

B. I. Halperin, P. A. Lee, N. Read (1993): Phys. Rev. B **47**, 7312

T. Kato (1950): On the adiabatic theorem of quantum mechanics. J. Phys. Soc. J., **5**, 435

T. Kato (1966): Pertubation theory of linear operators. Springer, Berlin, Heidelberg, New York.

M. Klein, R. Seiler (1990): Power law corrections to the Kubo formula vanish in quantum Hall systems. Comm. Math. Phys. **128** 141

K. von Klitzing, G. Dorda, M. Pepper (1980): Realization of a resistance standard based on fundamental constants, Phys. Rev. Lett. **45** 494

M. Komoto (1985): Topological invariants and the quantization of the Hall conductance. Ann. Phys. **160** 343-354

R. Laughlin (1981): Quantized Hall conductivity in two dimensions, Phys. Rev. B **23** 5652

R. Laughlin (1983): Anomalous quantum Hall effect: an incompressible quantum fluid with fractionally charged excitations. Phys. Rev. Lett **50** 1395

R. Laughlin (1987) in: R. E. Prange and S. M. Girvin, Eds.: The Quantum Hall Effect Springer, Berlin, Heidelberg, New York.

P. Lévay (1995): Berry Phases for Landau Hamiltonians on deformed tori. J. Math. Phys. **36** No. 6

Q. Niu, D.J. Thouless, Y. S. Wu (1985): Quantum Hall conductance as a topological invariant. Phys. Rev. B **31** 3372
Q. Niu, D.J. Thouless (1987): Quantum Hall effect with realistic boundary conditions. Phys. Rev. B **35** 2188
D. Quillen (1985): Determinants of Cauchy-Riemann operators on Riemann surfaces. Funkts. Anal. Prilozh. **19**, 37
B. Simon (1979): Trace ideals and their applications. Cambridge Univ. Press
B. Simon (1971): Hamiltonians defined as quadratic forms. Princeton Univ. Press.
J. A. Simmons et al. (1989): Phys. Rev. Lett. **63** 1731
M. Stone: Ann. Phys. **207** 38
M. Stone, Ed.: The Quantum Hall Effect World Scientific, Singapore.
D. J .Thouless, M. Kohmoto, P. Nightingale, M. den Nijs (1982): Quantized Hall Conductance in a Two-Dimensional Periodic Potential Phys. Rev. Lett. **49**, 40
D. J. Thouless (1994): Topological interpretations of quantum Hall conductance J. Math. Phys. **35** No. 10, 5362
D. J. Thouless (1998): Topological Quantum Numbers in Nonrelativistic Physics World Scientific, Singapore.
D. C. Tsui, H. L. Störmer, A. C. Gossard (1982): Two-Dimensional Magnetotransport in the Extreme Quantum Limit. Phys Rev. Lett. **48** 1559
X. G. Wen (1989): Vacuum degeneracy of chiral spin states in compactified space. Phys. Rev. B **40** 7387
X. G. Wen, A. Zee (1992): Classification of Abelian quantum Hall states and matrix formulation of topological fluids. Phys. Rev. B **46** 2290

q-Deformed Heisenberg Algebras

Julius Wess

Sektion Physik der Ludwig-Maximilians-Universität
Theresienstr. 37, D-80333 München
and
Max-Planck-Institut für Physik
(Werner-Heisenberg-Institut)
Föhringer Ring 6, D-80805 München

Introduction

This lecture consists of two sections. In section 1 we consider the simplest version of a q-deformed Heisenberg algebra as an example of a noncommutative structure. We first derive a calculus entirely based on the algebra and then formulate laws of physics based on this calculus. Then we realize that an interpretation of these laws is only possible if we study representations of the algebra and adopt the quantum mechanical scheme. It turns out that observables like position or momentum have discrete eigenvalues and thus space gets a lattice-like structure.

In section 2 we study a framework for higher dimensional noncommutative spaces based on quantum groups. The Poincaré-Birkhoff-Witt property and conjugation properties play an essential role there. In these spaces derivatives are introduced and based on these derivatives a q-deformed Heisenberg algebra can be constructed.

1 q-Deformed Heisenberg Algebra in One Dimension

1.1 A calculus based on an algebra

We try to develop a formal calculus entirely based on an algebra. In this lecture we consider the q-deformed Heisenberg algebra as an example and define derivatives and an integral on purely algebraic grounds. The functions which we differentiate and integrate are elements of a subalgebra - we shall call them fields. The integral is the inverse image of the derivative - thus it is an indefinite integral. For the derivatives a Leibniz rule can be found quite naturally and this leads to a rule for partial integration.

The algebra

As a model for the noncommutative structure that arises from quantum group considerations we are going to study the q-deformed Heisenberg algebra.

$$q^{\frac{1}{2}}xp - q^{-\frac{1}{2}}px = i\Lambda,\ \Lambda p = qp\Lambda, \quad \Lambda x = q^{-1}x\Lambda \\ q \in \mathbf{R}, \quad q \neq 0 \tag{1.1}$$

To motivate this algebra we start with a few heuristic arguments. In the context of quantum groups it is natural to introduce q-commutators as we did in the first equation of (1.1). The element x of the algebra will be identified with the observable for position in space, the element p with the canonical conjugate observable, usually called momentum. Observables have to be represented by selfadjoint linear operators in a Hilbert space. This will guarantee real eigenvalues and a complete set of orthogonal eigenvectors.

For this reason we require already at the level of the algebra an antilinear involution that will be identified with the conjugation operation of linear operators in Hilbert space. At the algebraic level we use bar and at the operator level star to denote this involution. Algebraic selfadjoint elements will have to be represented by selfadjoint operators in Hilbert space.

As a first step we extend the algebra (1.1) by conjugate elements $\overline{x}$, $\overline{p}$, $\overline{\Lambda}$ and find from (1.1) ($\overline{q} = q$):

$$q^{\frac{1}{2}}\overline{p}\,\overline{x} - q^{-\frac{1}{2}}\overline{x}\,\overline{p} = -i\overline{\Lambda} \tag{1.2}$$

Then for reasons explained above we demand:

$$\overline{x} = x, \quad \overline{p} = p \tag{1.3}$$

With this assumption follows from (1.2) and (1.1) for $q \neq 1$:

$$\begin{aligned} px &= i\lambda^{-1}(q^{-\frac{1}{2}}\Lambda - q^{\frac{1}{2}}\overline{\Lambda}) \\ xp &= i\lambda^{-1}(q^{\frac{1}{2}}\Lambda - q^{-\frac{1}{2}}\overline{\Lambda}), \quad \lambda = q - \frac{1}{q} \end{aligned} \tag{1.4}$$

For Λ selfadjoint this leads to $xp + px = 0$. This is too strong a relation which we do not admit as it does not allow a smooth transition for $q \to 1$. In great generality we can demand Λ to be the product of a unitary and a hermitean element. The simplest choice is for Λ to be unitary. Thus we extend the algebra by Λ^{-1} and demand

$$\overline{\Lambda} = \Lambda^{-1} \tag{1.5}$$

We want to have x^{-1} as an observable as well, thus we extend the algebra once more, this time by x^{-1}.

The following ordered monomials form a basis of the algebra:

$$x^m \Lambda^n, \quad m, n \in \mathbf{Z} \tag{1.6}$$

The element p can be expressed in this basis

$$p = i\lambda^{-1}x^{-1}(q^{\frac{1}{2}}\Lambda - q^{-\frac{1}{2}}\overline{\Lambda}) \tag{1.7}$$

This follows from (1.4).

To summarize, we study the associative algebra over the complex numbers freely generated by the elements p, x, Λ, x^{-1}, Λ^{-1} and their conjugates. This algebra is to be divided by the ideal generated by the relations (1.1),(1.3) and (1.5) and those that follow for x^{-1} and Λ^{-1}.

Fields and derivatives:

At the algebraic level we define a field f as an element of the subalgebra generated by x and x^{-1}, then completed by formal power series.

$$f(x) \in \Big[[x, x^{-1}]\Big] \equiv \mathcal{A}_x \tag{1.8}$$

A derivative we define as a map of $\mathcal{A}_x$ into $\mathcal{A}_x$, as we are going to explain now. From the algebra follows:

$$pf(x) = g(x)p - iq^{\frac{1}{2}}h(x)\Lambda \tag{1.9}$$

where $g(x)$ and $h(x)$ can be computed using (1.1). The derivative is defined as follows:

$$\nabla : \mathcal{A}_x \to \mathcal{A}_x, \quad \nabla f(x) = h(x) \tag{1.10}$$

The monomials x^m, $m \in \mathbf{Z}$ form a basis of $\mathcal{A}_x$. On these elements the derivative acts as follows:

$$\nabla x^m = [m]x^{m-1}, \quad [m] = \frac{q^m - q^{-m}}{q - q^{-1}} \tag{1.11}$$

We see that the element x^{-1} is not in the image of ∇. The ∇ map also has a kernel, the constants:

$$\nabla c = 0, \quad c \in \mathbf{C} \tag{1.12}$$

In a similar way we can define the maps L and L^{-1} from $\mathcal{A}_x$ onto $\mathcal{A}_x$. We start from the algebraic relation

$$\Lambda f(x) = j(x)\Lambda, \quad \Lambda^{-1} f(x) = k(x)\Lambda^{-1} \tag{1.13}$$

and define

$$\begin{aligned} L : \mathcal{A}_x \to \mathcal{A}_x,\ & Lf(x) = j(x) \\ L^{-1} : \mathcal{A}_x \to \mathcal{A}_x,\ & L^{-1} f(x) = k(x) \end{aligned} \tag{1.14}$$

For the x-basis we obtain:

$$Lx^m = q^{-m}x^m, \quad L^{-1}x^m = q^m x^m \tag{1.15}$$

The elements x, x^{-1} of the algebra $\mathcal{A}_x$ define a map $\mathcal{A}_x \to \mathcal{A}_x$ in a natural way.
These maps form an algebra

$$\begin{aligned} Lx = q^{-1}xL, \quad & L\nabla = q\nabla L, \\ q^{\frac{1}{2}}x\nabla - q^{-\frac{1}{2}}\nabla x &= -q^{-\frac{1}{2}}L \end{aligned} \tag{1.16}$$

homomorphic to the algebra (1.1) with the identification:

$$L \sim \Lambda, \quad x \sim x, \quad -iq^{\frac{1}{2}}\nabla \sim p \tag{1.17}$$

We do not consider the complex extension of the algebra (1.16) and do not define a bar operation on L and ∇. Nevertheless, it can be verified directly from the definition of L, L^{-1} and ∇ that

$$\begin{aligned} \nabla &= \lambda^{-1}x^{-1}(L^{-1} - L) \\ \nabla x^m &= \frac{1}{\lambda}(q^m - q^{-m})x^{m-1} = [m]x^{m-1} \end{aligned} \tag{1.18}$$

which agrees with (1.11).

Leibniz rule:

For usual functions we know the Leibniz rule:

$$\partial fg = (\partial f)g + f(\partial g) \tag{1.19}$$

There is a Leibniz rule for the derivative ∇ as well. It can be obtained from (1.18) if we know how L and L^{-1} acts on the product of fields. We compute this action by taking products of elements in the x-basis:

$$\begin{aligned} Lx^n x^m &= Lx^{m+n} = q^{-(m+n)}x^{m+n} \\ &= q^{-m}x^m q^{-n}x^n = (Lx^m)(Lx^n) \end{aligned} \tag{1.20}$$

Similar for L^{-1}:

$$L^{-1}x^n x^m = (L^{-1}x^m)(L^{-1}x^n) \tag{1.21}$$

This leads to the rule for the product of arbitrary elements of $\mathcal{A}_x$:

$$\begin{aligned} Lfg &= (Lf)(Lg) \\ L^{-1}fg &= (L^{-1}f)(L^{-1}g) \end{aligned} \tag{1.22}$$

For the maps x, x^{-1} we have the obvious rule:

$$\begin{aligned} x\ fg &= (xf)g \equiv f(xg) \\ x^{-1}\ fg &= (x^{-1}f)g \equiv f(x^{-1}g) \end{aligned} \tag{1.23}$$

These formulas can be used to obtain the Leibniz rule for ∇, using (1.18):

$$\begin{aligned} \nabla\ fg &= \lambda^{-1}x^{-1}(L^{-1} - L)\ fg \\ &= \lambda^{-1}x^{-1}\Big((L^{-1}f)((L^{-1}g) - (Lf)(Lg)\Big) \end{aligned} \tag{1.24}$$

Now we form a derivative on f for the first and on g for the second term:

$$\begin{aligned} \nabla\ fg = &\Big(\lambda^{-1}x^{-1}(L^{-1} - L)f\Big)(L^{-1}g) + \lambda^{-1}(x^{-1}Lf)(L^{-1}g) \\ &+(Lf)\lambda^{-1}x^{-1}(L^{-1} - L)g - \lambda^{-1}(Lf)(x^{-1}L^{-1})g \end{aligned} \tag{1.25}$$

The result is:

$$\nabla\ fg = (\nabla f)(L^{-1}g) + (Lf)(\nabla g) \tag{1.26}$$

The role of f and g can be exchanged, f and g commute. In a similar way that led from (1.24) to (1.25) we could have formed the derivative on g for the first term and on f for the second term of (1.24). The result then is:

$$\nabla fg = (\nabla f)(Lg) + (L^{-1}f)(\nabla g) \tag{1.27}$$

The expressions (1.26) and (1.27) are identical due to the identity in (1.23).

There is also a q-version of Green' s theorem. We compute:

$$\begin{aligned} \nabla(\nabla f)(L^{-1}g) &= (\nabla^2 f)g + (L^{-1}\nabla f)(\nabla L^{-1}g) \\ \nabla(L^{-1}f)(\nabla g) &= (\nabla L^{-1}f)(L^{-1}\nabla g) + f(\nabla^2 g) \end{aligned} \tag{1.28}$$

The two versions of the Leibniz rule (1.26) and (1.27) have been used. We subtract the two equations and obtain Green' s theorem:

$$(\nabla^2 f)(g) - (f)(\nabla^2 g) = \nabla\Big((\nabla f)(L^{-1}g) - (L^{-1}f)(\nabla g)\Big) \tag{1.29}$$

The indefinite integral:

We define the indefinite integral over a field as the inverse image of the derivative (1.10) and (1.11). We know that x^{-1} is not in the range of ∇ and that the constants are in the kernel of ∇.

$$\int^x x^n = \frac{1}{[n+1]}x^{n+1} + c, \quad n \in \mathbf{Z},\ n \neq -1 \tag{1.30}$$

We can also use formulas (1.18) to invert ∇:

$$\nabla^{-1} = \lambda\frac{1}{L^{-1} - L}x \tag{1.31}$$

This map is not defined on x^{-1}. We show that it reproduces (1.30) with $c = 0$:

$$\begin{aligned} \nabla^{-1}x^n &= \lambda\frac{1}{L^{-1} - L}x^{n+1} \\ &= \frac{\lambda}{q^{n+1} - q^{-n-1}}x^{n+1} \\ &= \frac{1}{[n+1]}x^{n+1} \end{aligned} \tag{1.32}$$

The map ∇^{-1} is now defined on any field that when expanded in the x^n basis does not have an x^{-1} term.

$$\begin{aligned} \nabla^{-1}f(x) &= \lambda\sum_{\nu=0}^{\infty} L^{2\nu}Lxf(x) \\ &= -\lambda\sum_{\nu=0}^{\infty} L^{-2\nu}L^{-1}xf(x) \end{aligned} \tag{1.33}$$

We shall use the first or second expansion depending on what series converges.
As an example:

For $n \geq 0$

$$\nabla^{-1}x^n = \lambda \sum_{\nu=0}^{\infty} L^{2\nu} L x^{n+1} \tag{1.34}$$
$$= \lambda \sum_{\nu=0}^{\infty} q^{-(2\nu+1)(n+1)} x^{n+1}$$

This sum converges for $q > 1$ and we obtain (1.32) from (1.34).

For $n < -1$ we find that the second expansion of (1.33) converges and gives again the result(1.32).

From the very definition of the integral follows:

$$\int^x \nabla f = f + c \tag{1.35}$$

This can be combined with the Leibniz rule (1.26) or (1.27) to give a formula for partial integration:

$$\int^x \nabla fg = fg + c = \int^x (\nabla f)(L^{-1}g) + \int^x (Lf)(\nabla g) \tag{1.36}$$

or

$$\int^x \nabla fg = fg + c = \int^x (\nabla f)(Lg) + \int^x (L^{-1}f)(\nabla g) \tag{1.37}$$

1.2 Field equations in a purely algebraic context

Based on the calculus that we have developed in the previous section we introduce field equations that define the time development of fields. For this purpose we have to enlarge the algebra by a central element, the time. Fields now depend on x and t.

We will consider the Schroedinger equation and the Klein-Gordon equation and demonstrate that there are continuity equations for a charge density and for an energy momentum density.

It is also possible to separate space and time dependence to obtain the time independent Schroedinger equation. To interpet it as an eigenvalue equation a Hilbert space for the solutions has to be defined. This cannot be done on algebraic grounds only.

Schroedinger equation:

This is the equation of motion that governs the time dependence of a quantum mechanical system.

To define it we extend the algebra $\mathcal{A}_x$ by the time variable t, in our case it will be a central element and $\bar{t} = t$. We call the extended algebra $\mathcal{A}_{x,t}$.

The Schroedinger equation acts on a field as follows:

$$i\frac{\partial}{\partial t}\psi(x,t) = \Big(-\frac{1}{2m}\nabla^2 + V(x)\Big)\psi(x,t) \qquad (1.38)$$

$$\psi \in \mathcal{A}_{x,t}, \quad V \in \mathcal{A}_x, \quad \overline{V} = V$$

We show that there is a continuity equation for

$$\rho(x,t) = \overline{\psi}(x,t)\psi(x,t) \in \mathcal{A}_{x,t} \qquad (1.39)$$

The continuity equation can be written in the form

$$\frac{\partial\rho}{\partial t} + \nabla j = 0 \qquad (1.40)$$

This is a consequence of the Schroedinger equation (1.38) and its conjugate equation. To find the conjugate equation we consider $\nabla\overline{\psi}$ and $\overline{\nabla\psi}$ as elements of $\mathcal{A}_{x,t}$, where conjugation is defined. We find $\nabla\overline{\psi} = \overline{\nabla\psi}$ and therefore

$$-i\frac{\partial}{\partial t}\overline{\psi} = (-\frac{1}{2m}\nabla^2 + V)\overline{\psi} \qquad (1.41)$$

This yields

$$\frac{\partial\rho}{\partial t} = \frac{i}{2m}\left\{\overline{\psi}(\nabla^2\psi) - (\nabla^2\overline{\psi})\psi\right\} \qquad (1.42)$$

Now we can use Green's theorem (1.29) and we obtain:

$$j = -\frac{i}{2m}L^{-1}\left\{\overline{\psi}(L\nabla\psi) - (L\nabla\overline{\psi})\psi\right\} \qquad (1.43)$$

The time independent Schroedinger equation can be obtained from (1.38) by separation:

$$\psi(x,t) = \varphi(t)U(x) \qquad (1.44)$$

The usual argument leads to:

$$i\frac{\partial}{\partial t}\varphi(t) = E\varphi(t), \qquad (1.45)$$

$$\left(-\frac{1}{2m}\nabla^2 + V\right)U(x) = EU(x), \quad E \in \mathbf{C}$$

This is as far as we can get using the algebra only. To find meaningful solutions of (1.45) we have to define a linear space with a norm, a Hilbert space, and the solutions $U(x)$ will have to be elements of this space.

There is also a continuity equation for the energy momentum density:

$$\mathcal{H} = \frac{1}{2mq}(\nabla L^{-1}\overline{\psi})(\nabla L^{-1}\psi) + V\overline{\psi}\psi \tag{1.46}$$

$$\pi = \frac{1}{2m}\left((\nabla\overline{\psi})(L^{-1}\dot{\psi}) + (L^{-1}\dot{\overline{\psi}})\nabla\psi\right) \tag{1.47}$$

It follows from (1.38) and (1.40) that

$$\frac{d}{dt}\mathcal{H} - \nabla\pi = 0 \tag{1.48}$$

Both conservation laws will follow from a variational principle that we shall formulate as soon as we know how to define a definite integral. In the meantime it is left as an exercise to prove (1.46).

Klein-Gordon equation:

We define

$$\frac{\partial^2}{\partial t^2}\phi - \nabla^2\phi + m^2\phi = 0, \qquad \phi \in \mathcal{A}_{x,t} \tag{1.49}$$

For complex ϕ (1.49) and its conjugate lead to current conservation:

$$\rho = \dot{\overline{\phi}}\phi - \overline{\phi}\dot{\phi}, \quad j = -(\nabla\overline{\phi})(L^{-1}\phi) + (L^{-1}\overline{\phi})(\nabla\phi) \tag{1.50}$$

$$\dot{\rho} + \nabla j = 0 \tag{1.51}$$

We have used Green's theorem (1.29). The verification of (1.50) as well as the verification of energy momentum conservation is again left as an exercise.

$$\mathcal{H} = \dot{\overline{\phi}}\dot{\phi} + \frac{1}{q}(\nabla L^{-1}\overline{\phi})(\nabla L^{-1}\phi) + m^2\overline{\phi}\phi \tag{1.52}$$

$$\pi = (\nabla\overline{\phi})(L^{-1}\dot{\phi}) + (L^{-1}\dot{\overline{\phi}})(\nabla\phi) \tag{1.53}$$

$$\dot{\mathcal{H}} - \nabla\pi = 0 \tag{1.54}$$

1.3 Gauge theories in a purely algebraic context

It is possible to develop a covariant gauge theory starting from fields as defined in the previous chapter. These fields are supposed to have well-defined properties under a gauge transformation. Using the concepts of connection and vielbein covariant derivatives can be defined. These covariant derivatives form an algebra and a curvature arises from this algebra in a natural way.

Finally we show how an exterior calculus can be set up that opens the way to differential geometry on the algebra.

Covariant derivatives:

We assume that $\psi(x,t)$ spans a representation of a compact gauge group. Let T_l be the generators of the group in this representation and let $\alpha(x,t)$ be Lie algebra-valued

$$\alpha(x,t) = \sum_l g\alpha_l(x,t)T_l \qquad (1.55)$$
$$\alpha_l(x,t) \in \mathcal{A}_{x,t}$$

We have introduced a coupling constant g, ($g \in \mathbf{C}$).

The field $\psi(x,t)$ is supposed to transform as follows:

$$\psi'(x,t) = e^{i\alpha(x,t)}\psi(x,t) \qquad (1.56)$$

A covariant derivative is a derivative such that

$$(\mathcal{D}\psi)' = e^{i\alpha(x,t)}(\mathcal{D}\psi) \qquad (1.57)$$

and $\mathcal{D} = \nabla$ for $g = 0$.

We make the Ansatz:

$$\mathcal{D}_x\psi = E(\nabla + \phi)\psi \qquad (1.58)$$

Here we have introduced the connection ϕ and the Vielbein E. For $g = 0$, ϕ has to be zero and E has to be the unit matrix.

We aim at a transformation law for E and ϕ such that

$$E'(\nabla + \phi')e^{i\alpha}\psi = e^{i\alpha}E(\nabla + \phi)\psi \qquad (1.59)$$

From the Leibniz rule (1.26) for ∇ follows:

$$\nabla\psi' = (\nabla e^{i\alpha})L\psi + (L^{-1}e^{i\alpha})\nabla\psi \qquad (1.60)$$

Thus (1.59) will be satisfied if:

$$E' = e^{i\alpha}E(L^{-1}e^{-i\alpha}) \qquad (1.61)$$
$$\phi' = (L^{-1}e^{i\alpha})\phi(e^{-i\alpha}) - (\nabla e^{i\alpha})(Le^{-i\alpha})L$$

From the inhomogeneous terms in the transformation law of ϕ in (1.61) we see that ϕ has to be L-valued.

$$\phi = g\varphi L \qquad (1.62)$$

From (1.61) follows

$$g\varphi' = (L^{-1}e^{i\alpha})g\varphi(Le^{-i\alpha}) - (\nabla e^{i\alpha})(Le^{-i\alpha}) \qquad (1.63)$$

To further analyze (1.63) we use the expression (1.18) for ∇

$$(\nabla e^{i\alpha})(Le^{-i\alpha}) = \lambda^{-1}x^{-1}\Big((L^{-1} - L)e^{i\alpha}\Big)(Le^{-i\alpha}) \qquad (1.64)$$
$$= \lambda^{-1}x^{-1}\left[(L^{-1}e^{i\alpha})(Le^{-i\alpha}) - 1\right]$$

This suggests to rewrite (1.63) as follows:

$$g\varphi' - \lambda^{-1}x^{-1} = (L^{-1}e^{i\alpha})\left[g\varphi - \lambda^{-1}x^{-1}\right](Le^{-i\alpha}) \qquad (1.65)$$

From the transformation law of E (1.61) now follows that the object

$$E(g\varphi - \lambda^{-1}x^{-1})(LE) = g\chi \qquad (1.66)$$

transforms homogeneously

$$\chi' = e^{i\alpha}\chi e^{-i\alpha} \qquad (1.67)$$

Thus χ can be chosen to be proportional to the unit matrix or to be Lie algebra-valued.

$$\chi = \chi_0 1 \quad + \quad \sum_l \chi_l T_l \qquad (1.68)$$

We are now going to show that $\chi_l = 0$ if we demand a covariant version of (1.18). For this purpose we first have to find a covariant version of L:

$$(\mathcal{L}\psi)' = e^{i\alpha}\mathcal{L}\psi \quad , \quad \mathcal{L} = L \quad for \quad g = 0 \qquad (1.69)$$

We try the Ansatz:

$$\mathcal{L} = \tilde{E}L \qquad (1.70)$$

with a new version of a vielbein $\tilde{E}$. From (1.69) follows

$$\tilde{E}' = e^{i\alpha}\tilde{E}(Le^{-i\alpha}) \qquad (1.71)$$

It is natural to identify $\tilde{E}$ with (LE^{-1}) because both have the same transformation property:

$$\tilde{E} = (LE^{-1}) \qquad (1.72)$$

For $\mathcal{L}$ and its inverse we found:

$$\mathcal{L} = (LE^{-1})L = LE^{-1}, \quad \mathcal{L}^{-1} = EL^{-1} \qquad (1.73)$$

$\mathcal{L}^{-1}$ transforms covariant as well.

Now we postulate:

$$\mathcal{D}_x = \lambda^{-1}x^{-1}(\mathcal{L}^{-1} - \mathcal{L}) \qquad (1.74)$$

In more detail:

$$\begin{aligned}\mathcal{D}_x &= \lambda^{-1}x^{-1}\left\{E(L^{-1}-L)+\left(E-(LE^{-1})\right)L\right\} \\ &= E\nabla + \lambda^{-1}x^{-1}E\left(1-(E^{-1})(LE^{-1})\right)L\end{aligned} \tag{1.75}$$

If we compare this with (1.58) we find:

$$g\varphi = \lambda^{-1}x^{-1}\left(1-E^{-1}(LE^{-1})\right) \tag{1.76}$$

The connection is entirely expressed in terms of the vielbein.

Now we show that with this choice of φ the covariant derivative of the vielbein vanishes.

Let us start with a field H that transforms like E:

$$H' = e^{i\alpha}H(L^{-1}e^{-i\alpha}) \tag{1.77}$$

We apply $\mathcal{L}$ and $\mathcal{L}^{-1}$ to this object:

$$\begin{aligned}\mathcal{L}H &= (LE^{-1})(LH)E \\ \mathcal{L}^{-1}H &= E(L^{-1}H)(L^{-1}E^{-1})\end{aligned} \tag{1.78}$$

This we insert into (1.73) to obtain:

$$\mathcal{D}_xH = \lambda^{-1}x^{-1}\{E(L^{-1}H)(L^{-1}E^{-1}) - (LE^{-1})(LH)E\} \tag{1.79}$$

If in this formula we substitute E for H we obtain:

$$\mathcal{D}_xE = 0 \tag{1.80}$$

For the covariant time derivative we follow the standard construction

$$\mathcal{D}_t = (\partial_t + \omega)\psi, \quad \omega = \sum_l \omega_l(x,t)T_l \tag{1.81}$$

The transformation property of ω is:

$$\omega' = e^{i\alpha}\omega e^{-i\alpha} + e^{i\alpha}\partial_t e^{-i\alpha} \tag{1.82}$$

Curvature:

The covariant derivatives have an algebraic structure as well. With the choice (1.76) for the connection it follows from (1.74) that

$$\mathcal{L}\mathcal{D}_x\psi = q\mathcal{D}_x\mathcal{L}\psi, \ \ \mathcal{L}^{-1}\mathcal{D}_x\psi = q^{-1}\mathcal{D}_x\mathcal{L}^{-1}\psi \tag{1.83}$$

$$(\mathcal{L}\mathcal{D}_t - \mathcal{D}_t\mathcal{L})\psi = \mathcal{L}T\psi(\mathcal{L}^{-1}\mathcal{D}_t - \mathcal{D}_t\mathcal{L}^{-1})\psi = -T\mathcal{L}^{-1}\psi$$

T is a tensor quantity:

$$T = (\partial_t E)E^{-1} - (E)(L^{-1}\omega)E^{-1} + \omega \tag{1.84}$$

such that

$$T' = e^{i\alpha}Te^{-i\alpha} \tag{1.85}$$

The commutator of $\mathcal{D}_t$ and $\mathcal{D}_x$ is easy to compute from (1.74) and (1.83). We find:

$$(\mathcal{D}_t\mathcal{D}_x - \mathcal{D}_x\mathcal{D}_t)\psi = T\mathcal{D}_x\psi + \lambda^{-1}x^{-1}\{\mathcal{L}T + T\mathcal{L}^{-1}\}\psi \tag{1.86}$$

To avoid the $\lambda^{-1}x^{-1}$ factor we can write this term also in the form:

$$\lambda^{-1}x^{-1}\left\{\mathcal{L}T + T\mathcal{L}^{-1}\right\} = EF(LE)\mathcal{L} \tag{1.87}$$

with

$$F = g\partial_t\varphi - \nabla\omega + (L^{-1}\omega)g\varphi - g\varphi(L\omega) \tag{1.88}$$

and

$$E'F'(LE') = e^{i\alpha}EFLEe^{-i\alpha} \tag{1.89}$$

The tensor T plays the role of a curvature.

Leibniz rule for covariant derivatives:

We want to learn how covariant derivatives act on products of representations. To distinguish the representations we are going to use indices. Thus ψ and χ are two representations such that

$$\psi'_\alpha = (e^{i\alpha})_\alpha{}^\beta\psi_\beta, \quad \chi'_a = (e^{i\alpha})_a{}^b\chi_b \tag{1.90}$$

Repeated indices are to be summed. The Vielbein is a matrix object $E_\alpha{}^\beta$ or $E_a{}^b$, depending on what representation it acts on.

$$\begin{aligned} E'^{\ \beta}_{\alpha} &= (e^{i\alpha})_\alpha{}^\sigma E_\sigma{}^\rho(L^{-1}e^{-i\alpha})_\rho{}^\beta \\ E'^{\ b}_{a} &= (e^{i\alpha})_a{}^s E_s{}^r(L^{-1}e^{-i\alpha})_r{}^b \end{aligned} \tag{1.91}$$

On the product of representations it acts as follows:

$$E_{\alpha a}{}^{\beta b}\psi_\beta\chi_b = E_\alpha{}^\beta E_a{}^b\psi_\beta\chi_b \tag{1.92}$$

This gives us a chance to obtain the vielbein starting from fundamental representations. It is left to show that such a construction leads to a unique vielbein for each representation.

To obtain the Leibniz rule we start with a scalar:

$$f' = f \tag{1.93}$$

In this case the derivatives are the covariant derivatives:

$$\mathcal{D}_x f = \nabla f, \quad \mathcal{D}_t f = \frac{\partial}{\partial t} f, \quad \mathcal{L} f = L f, \quad \mathcal{L}^{-1} f = L^{-1} f \tag{1.94}$$

If we combine this representation with ψ we obtain

$$\begin{aligned}(\mathcal{L} f \psi)_\alpha &= (LE^{-1})_\alpha{}^\beta (L f \psi)_\beta = (Lf)(LE^{-1})_\alpha{}^\beta (L\psi)_\beta \\ &= \mathcal{L} f (\mathcal{L}\psi)_\alpha\end{aligned} \tag{1.95}$$

As a second example we treat the scalar obtained from a covariant and a contravariant representation.

$$\psi' = e^{i\alpha}\psi \quad , \quad \chi' = \chi - e^{i\alpha} \tag{1.96}$$

such that

$$\chi'\psi' = \chi\psi \tag{1.97}$$

We take the covariant action of L on $\chi\psi$:

$$\begin{aligned}\mathcal{L}(\chi^\alpha \psi_\alpha) &= L(\chi^\alpha \psi_\alpha) = (L\chi^\alpha)(L\psi_\alpha) \\ &= (L\chi^\sigma)(LE)_\sigma{}^\alpha (LE^{-1})_\alpha{}^\beta (L\psi)_\beta \\ &= (\mathcal{L}\chi)^\alpha (\mathcal{L}\psi)_\alpha\end{aligned} \tag{1.98}$$

Encouraged by this result we continue with the product of two arbitrary representations:

$$\begin{aligned}\mathcal{L}(\psi\chi)_{\alpha a} &= (LE^{-1})_\alpha{}^\beta{}_a{}^b (L\psi_\beta)(L\chi_b) = (LE^{-1})_\alpha{}^\beta (LE^{-1})_a{}^b (L\psi)_\beta (L\chi)_b \\ &= (\mathcal{L}\psi)_\alpha (\mathcal{L}\chi)_a\end{aligned} \tag{1.99}$$

This is the Leibniz rule for the covariant version of L. It is obvious that the same holds for $\mathcal{L}^{-1}$:

$$\mathcal{L}^{-1}\psi_\alpha \chi_a = (\mathcal{L}^{-1}\psi)_\alpha (\mathcal{L}^{-1}\chi)_a \tag{1.100}$$

The covariant derivative $\mathcal{D}_x$ can be obtained from $\mathcal{L}$ and $\mathcal{L}^{-1}$ according to the same argument that led to the Leibniz rule for the derivative ∇ (1.26), we can now show that:

$$\mathcal{D}_x(\psi\chi) = (\mathcal{D}_x\psi)(\mathcal{L}\chi) + (\mathcal{L}^{-1}\psi)(\mathcal{D}_x\chi) \tag{1.101}$$

or

$$\mathcal{D}_x(\psi\chi) = (\mathcal{D}_x\psi)(\mathcal{L}^{-1}\chi) + (\mathcal{L}\psi)(\mathcal{D}_x\chi) \tag{1.102}$$

This is the Leibniz rule for covariant derivatives.

The Leibniz rule (1.100), (1.101) and (1.102) allow us to drop the field ψ in the equations (1.83) and (1.87) and write them as algebra relations.

$$\begin{aligned} \mathcal{L}\mathcal{D}_x = q\mathcal{D}_x\mathcal{L}, \quad \mathcal{L}^{-1}\mathcal{D}_x = q^{-1}\mathcal{D}_x\mathcal{L}^{-1}\mathcal{L}\mathcal{D}_t - \mathcal{D}_t\mathcal{L} = \mathcal{L}T \\ \mathcal{L}^{-1}\mathcal{D}_t - \mathcal{D}_t\mathcal{L}^{-1} = -T\mathcal{L}^{-1}\mathcal{D}_t\mathcal{D}_x - \mathcal{D}_x\mathcal{D}_t = T\mathcal{D}_x + EF(LE)\mathcal{L} \end{aligned} \tag{1.103}$$

T plays the role of an independent tensor that is a function of the vielbein and the connection ω. It is defined in equation (1.84). F depends on T and is defined in equation (1.87).

In a later chapter we shall see that the vielbein can be expressed in terms of a connection A_μ in the usual definition of a covariant derivative.

Exterior Derivative:

To define differentials we have to extend the algebra by an element dx. We assume for the moment that there is an algebraic relation that allows us to order x and dx:

$$dxf(x) = a[f(x)]dx, \quad a : \mathcal{A}_x \to \mathcal{A}_x \tag{1.104}$$

The map a has to be an automorphism of the algebra $\mathcal{A}_x$

$$a[f(x)g(x)] = a[f(x)]a[g(x)] \tag{1.105}$$

This can be seen as follows

$$\begin{aligned} dx\ f \times g &= a[f \times g]dx \\ &= a[f]dxg = a[f]a[g]dx \end{aligned} \tag{1.106}$$

The exterior derivative d can be defined as follows:

$$d\,x = dx, \quad d = dx\nabla, \quad d^2 = 0 \tag{1.107}$$

From (1.11) follows:

$$d\,x^m = [m]dxx^{m-1} \tag{1.108}$$

This tells us how d acts on any field $f(x) \in \mathcal{A}_x$.

To derive a Leibniz rule for the exterior derivative we use the Leibniz rule for the derivative (1.26)

$$d\,fg = dx\{(\nabla f)(L^{-1}g) + (Lf)\nabla g\} \tag{1.109}$$

The relation (1.36) allows us to obtain a Leibniz rule for d:

$$d\,fg = (df)(L^{-1}g) + a[Lf]dg \tag{1.110}$$

The simplest choice is:

$$a[f] = f, \quad dx\,x = x\,dx \tag{1.111}$$

Then we obtain from (1.108)

$$d\,fg = (df)(L^{-1}g) + (Lf)dg \tag{1.112}$$

We could have used the Leibniz rule (1.110), then we would have obtained:

$$d\,fg = dx\{(\nabla f)(Lg) + (L^{-1}f)\nabla g\} \tag{1.113}$$

For $a(f)$ defined by (1.111) this yields:

$$d\,fg = (df)(Lg) + (L^{-1}f)dg \tag{1.114}$$

A different choice for $a[f]$ could be

$$a[f] = L^p f, \quad p \in \mathbf{Z} \tag{1.115}$$

This satisfies (1.105).

The Leibniz rule (1.110) then yields

$$d\,fg = (df)(L^{-1}g) + (L^{p+1}f)(dg) \tag{1.116}$$

or

$$d\,fg = (df)(Lg) + (L^{p-1}f)(dg) \tag{1.117}$$

1.4 q-Fourier transformations

In the next chapter we will study representations of the algebra (1.1). As mentioned before we are interested in "good" representations where the coordinates and the momenta can be diagonalized. We also want to have explicit formulas for the change of the coordinate bases to the momentum bases. This is the q-Fourier transformation and it turns out that the relevant transition functions are the well known q-functions $cos_q x$ and $sin_q x$. We are going to discuss these functions now. It should be mentioned that in addition to quantum groups there is another reservoir of detailed mathematical knowledge - these are the basic hypergeometric series, special example: $sin_q(x)$ and $cos_q(x)$.

The q-Fourier transformation is based on the q-deformed *cosine* and *sine* functions that are defined as follows:

$$\begin{aligned} \cos_q(x) &= \sum_{k=0}^{\infty}(-1)^k \frac{x^{2k}}{[2k]!}\frac{q^{-k}}{\lambda^{2k}} \\ \sin_q(x) &= \sum_{k=0}^{\infty}(-1)^k \frac{x^{2k+1}}{[2k+1]!}\frac{q^{k+1}}{\lambda^{2k+1}} \end{aligned} \tag{1.118}$$

the symbol $[n]$ was defined in (1.11), and $[n]!$ stands for:

$$[n]! = [n]\,[n-1]\,\cdots\,[1] \tag{1.119}$$

$$[n] = \frac{q^n - q^{-n}}{q - \frac{1}{q}} = q^{n-1} + q^{n-3} + \ \ldots\ + q^{-n+3} + q^{-n+1}$$

These cos_q and sin_q functions are solutions of the equations:

$$\begin{aligned} \frac{1}{x}\left(\sin_q(x) - \sin_q(q^{-2}x)\right) &= \cos_q(x) \\ \frac{1}{x}\left(\cos_q(x) - \cos_q(q^{-2}x)\right) &= -q^{-2}\sin_q(q^{-2}x) \end{aligned} \tag{1.120}$$

The $cos_q(x)$ and $sin_q(x)$ functions (1.118) are determined by these equations up to an overall normalization. To prove this is straightforward, as an example we verify the first of the equations (1.120)

$$\frac{1}{x}\left(\sin_q(x) - \sin_q(q^{-2}x)\right) = \sum_{k=0}^{\infty}(-1)^k \frac{x^{2kx}}{[2k+1]!}\frac{q^{k+1}}{\lambda^{2k+1}}(1 - q^{-2(2k+1)}) \tag{1.121}$$

but

$$1 - q^{-2(2k+1)} = \frac{\lambda}{q}\, q^{-2k}[2k+1] \tag{1.122}$$

and (1.120) follows. The relations (1.120) are the analogon to the property of the usual cos and sin functions that the derivative of $sin(cos)$ is $cos(-sin)$.

The most important property of the cos_q and sin_q functions is that they each form a complete and orthogonal set of functions. This allows us to formulate the q-Fourier theorem for functions $g(q^{2n})$ that are defined on the q-lattice points $q^{2n}, \quad n \in \mathbf{Z}$.

$$\begin{aligned} \tilde{g}_c(q^{2\nu}) &= N_q \sum_{n=-\infty}^{\infty} q^{2n}\cos_q(q^{2(\nu+n)})\, g(q^{2n}) \\ g(q^{2n}) &= N_q \sum_{\nu=-\infty}^{\infty} q^{2\nu}\cos_q(q^{2(\nu+n)})\, \tilde{g}_c(q^{2\nu}) \end{aligned} \tag{1.123}$$

and:

$$\sum_{n=-\infty}^{\infty} q^{2n}\left|g(q^{2n})\right|^2 = \sum_{\nu=-\infty}^{\infty} q^{2\nu}\left|\tilde{g}_c(q^{2\nu})\right|^2 \tag{1.124}$$

The normalization constant N_q can be calculated:

$$N_q = \prod_{\nu=0}^{\infty} \left(\frac{1 - q^{-2(2\nu+1)}}{1 - q^{-4(\nu+1)}} \right), \quad q > 1 \tag{1.125}$$

Another transformation is obtained by using sin_q instead of cos_q:

$$\tilde{g}_s(q^{2\nu}) = N_q \sum_{n=-\infty}^{\infty} q^{2n} \sin_q(q^{2(\nu+n)})\, g(q^{2n}) \tag{1.126}$$

$$g(q^{2n}) = N_q \sum_{\nu=-\infty}^{\infty} q^{2\nu} \sin_q(q^{2(\nu+n)})\, \tilde{g}_s(q^{2\nu})$$

and

$$\sum_{n=-\infty}^{\infty} q^{2n} \left|g(q^{2n})\right|^2 = \sum_{\nu=-\infty}^{\infty} q^{2\nu} \left|\tilde{g}_s(q^{2\nu})\right|^2 \tag{1.127}$$

That sin_q and cos_q individually form a complete set of functions can be understood because the range of $x = q^{2n}$ is restricted to $x \geq 0$.

The orthogonality and completeness property can be stated as follows:

$$N_q^2 \sum_{\nu=-\infty}^{\infty} q^{2\nu} \cos_q(q^{2(n+\nu)}) \cos_q(q^{2(m+\nu)}) = q^{-2n}\, \delta_{n,m} \tag{1.128}$$

$$N_q^2 \sum_{\nu=-\infty}^{\infty} q^{2\nu} \sin_q(q^{2(n+\nu)}) \sin_q(q^{2(m+\nu)}) = q^{-2n}\, \delta_{n,m}$$

That the role of n and ν can be exchanged to get the completeness (orthogonality) relations from the orthogonality (completeness) relations is obvious.

There is also a deformation of the relation $cos^2 + sin^2 = 1$. It is:

$$\cos_q(x) \cos_q(qx) + q^{-1} \sin_q(x) \sin_q(q^{-1}x) = 1 \tag{1.129}$$

the general term in the cos_q product is:

$$\cos_q(x) \cos_q(qx) = \sum_{l,k=0}^{\infty} \left(\frac{x}{\lambda}\right)^{2(k+l)} (-1)^{k+l} \frac{q^{l-k}}{[2k]!\,[2l]!} \tag{1.130}$$

and in the sin_q product it is:

$$q^{-1} \sin_q(x) \sin_q(q^{-1}x) = \sum_{l,k=0}^{\infty} \left(\frac{x}{\lambda}\right)^{2(k+l+1)} (-1)^{k+l} \frac{q^{l-k}}{[2k+1]!\,[2l+1]!} \tag{1.131}$$

The terms of (1.129) that add up to a fixed power $2n$ of x are

$$(-1)^n \left(\frac{x}{\lambda}\right)^{2n} \frac{1}{[2n]!} \sum_{k=0}^{2n} (-1)^k q^{n-k} \frac{[2n]!}{[k]![2n-k]!} \tag{1.132}$$

For $n = 0$ this is one, for $n = 1$ it is:

$$-\left(\frac{x}{\lambda}\right)^2 \frac{1}{[2]} \left\{ q - [2] + \frac{1}{q} \right\} = 0 \tag{1.133}$$

For the general term (1.129) was proved by J.Schwenk.

In the formulas for the Fourier transformations (1.123) and (1.126) only even powers of q enter whereas (1.129) connects even powers to odd powers. From (1.128) we see that $cos_q(q^{2n})$ and $sin_q(q^{2n})$ have to tend to zero for $n \to \infty$. For (1.129) to hold $cos_q(q^{2n+1})$ and $sin_q(q^{2n+1})$ have to diverge for $n \to \infty$. This behaviour is illustrated by Fig.1. and Fig.2. They show that $cos_q(x)$ and $sin_q(x)$ are functions that diverge for $x \to \infty$, they are not functions of L^2. However, the points $x = q^{2n}$ are close to the zeros of $cos_q(x)$ and $sin_q(x)$ and for $n \to \infty$ tend to these zeros such that the sum in (1.128) is convergent.

We can also consider $cos_q(x)$ and $sin_q(x)$ as a field, i.e. as an element of the algebra $\mathcal{A}_x$. Then we can apply ∇ to it. We use (1.18) as an expression for ∇:

$$\nabla cos_q(kx) = \frac{1}{\lambda}\frac{1}{x} \left\{ \cos_q(qkx) - \cos_q(q^{-1}kx) \right\} \tag{1.134}$$

Now we use (1.120) for the variable $y = qkx$ and we obtain

$$\nabla cos_q(kx) = -k \frac{1}{q\lambda} \sin_q(q^{-1}kx) \tag{1.135}$$

The same can be done for $sin_q(kx)$:

$$\nabla sin_q(kx) = k \frac{q}{\lambda} \cos_q(qkx) \tag{1.136}$$

This shows that $cos_q(kx)$ and $sin_q(kx)$ are eigenfunctions of ∇^2:

$$\begin{aligned} \nabla^2 \cos_q(kx) &= -\frac{k^2}{q\lambda^2} \cos_q(kx) \\ \nabla^2 \sin_q(kx) &= -\frac{k^2 q}{\lambda^2} \sin_q(kx) \end{aligned} \tag{1.137}$$

We have found eigenfunctions for the free Schroedinger equation (1.45) ($V = 0$). To really give a meaning to them we have to know how to define a Hilbert space for the solutions.

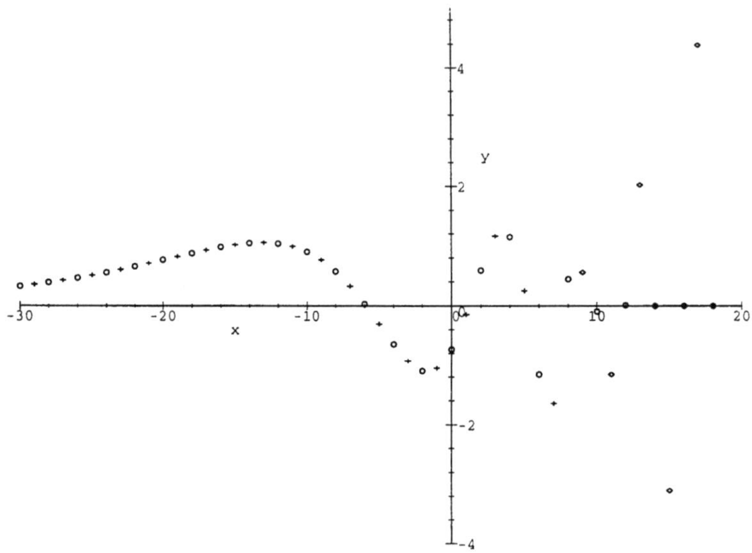

Fig. 1. $sin_q(q^n)$ for $q = 1.1$. Crosses (circles) indicate odd (even) n. For $n > 8$ a logarithmic y-scale was used.

1.5 Representations

In the first three chapters we have developed a formalism that is entirely based on the algebra. For a physical interpretation we have to relate this formalism to real numbers - real numbers being the result of measurements. This can be done by studying representations of the algebra and adopting the interpretation scheme of quantum mechanics. Thus we have to aim at representations where selfadjoint elements of the algebra are represented by selfadjoint linear operators in a Hilbert space. We want to diagonalize these operators, they should have real eigenvalues and the eigenvectors should form a basis in a Hilbert space. We shall call these representations "good" representations.

The Hilbert space $\mathcal{H}_s^\sigma$:

We first represent the algebra

$$x\Lambda = q\Lambda x, \quad \bar{x} = x, \quad \bar{\Lambda} = \Lambda^{-1} \tag{1.138}$$

From what was said above we can assume x to be diagonal. From (1.138) follows that with any eigenvalue s of the linear operator x there will be the eigenvalues

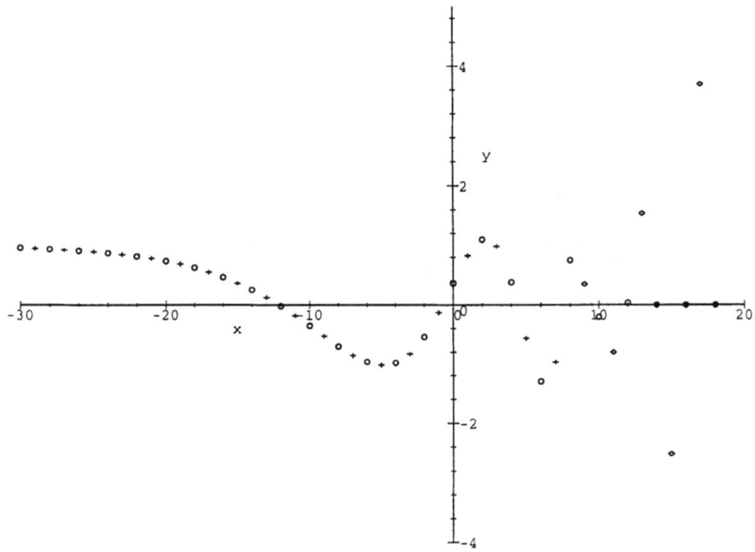

Fig. 2. $cos_q(q^n)$ for $q = 1.1$. Crosses (circles) indicate odd (even) n. For $n > 8$ a logarithmic y-scale was used.

$q^n s, \quad n \in \mathbf{Z}$. As s is the eigenvalue of a selfadjoint operator it is real. We shall restrict s to be positive and use $-s$ for negative eigenvalues. In the following we will assume $q > 1$.

We start with the following eigenvectors and eigenvalues:

$$x|n,\sigma\rangle^s = \sigma s q^n |n,\sigma\rangle^s \qquad (1.139)$$
$$n \in \mathbf{Z}, \quad \sigma = \pi_D 1, \quad 1 \leq s < q$$

The algebra (1.138) is represented on the states with σ and s fixed:

$$\Lambda|n,\sigma\rangle^s = |n+1,\sigma\rangle^s \qquad (1.140)$$

A possible phase is absorbed in the definition of the states. These states are supposed to form an orthonormal basis in a Hilbert space, which we will call $\mathcal{H}_s^\sigma$.

$$|n,\sigma\rangle^s \in \mathcal{H}_s^\sigma \qquad (1.141)$$
$${}^s\langle m,\sigma|n,\sigma\rangle^s = \delta_{n,m}$$

This makes Λ, defined by (1.140), a unitary linear operator.

To obtain a representation of the algebra (1.1) we use formula (1.7) to represent p:

$$p|n,\sigma\rangle^s = i\lambda^{-1}\frac{\sigma}{s}q^{-n}\left\{q^{-\frac{1}{2}}|n+1,\sigma\rangle^s - q^{\frac{1}{2}}|n-1,\sigma\rangle^s\right\} \qquad (1.142)$$

This defines the action of p on the states of the bases (1.141). A direct calculation shows that the linear operator p defined by (1.142) indeed satisfies the algebra (1.1) and that p is hermitean.

$$\overline{\langle n+1p|n\rangle} = \rangle np|n+1\rangle = -i\lambda^{-1}q^{-n-\frac{1}{2}} \qquad (1.143)$$

However, p is not selfadjoint. To show this we assume p to be selfadjoint, i.e. diagonizable with real eigenvalues and eigenstates that form a basis of the Hilbert space (1.141). At the same time the algebra (1.1) should be represented.

$$p|p_0\rangle = p_0|p_0\rangle,\ |p_0\rangle = \sum_n c_n^{p_0}|n,\sigma\rangle^s \qquad (1.144)$$

From the algebra follows that $\Lambda|p_0\rangle$ and $\Lambda^{-1}|p_0\rangle$ are orthogonal to $|p_0\rangle$ because they belong to different eigenvalues of p.

$$p\Lambda|p_0\rangle = \frac{1}{q}p_0\Lambda|p_0\rangle \qquad (1.145)$$
$$p\Lambda^{-1}|p_0\rangle = qp_0\Lambda^{-1}|p_0\rangle$$

We conclude

$$\langle p_0\Lambda|p_0\rangle = \rangle p_0\Lambda^{-1}|p_0\rangle = 0 \qquad (1.146)$$

for every eigenvalue p_0 of p.
From the algebra (1.1) follows:

$$\langle p_0\left(q^{\frac{1}{2}}xp - q^{-\frac{1}{2}}px\right)|p_0\rangle\rangle = p_0(q^{\frac{1}{2}} - q^{-\frac{1}{2}})\langle p_0x|p_0\rangle\rangle = 0 \qquad (1.147)$$

If there is a $p_0 \neq 0$ we find

$$\langle p_0x|p_0\rangle\rangle = 0 \qquad (1.148)$$

Now we use the representation of $|p_0\rangle$ in the $|n,\sigma\rangle^s$ basis to conclude

$$\sigma\sum_n q^n\left|c_n^{p_0}\right|^2 = 0 \qquad (1.149)$$

For fixed σ this can only be true if all the c_n^{po} vanish. A clear contradiction.

We have shown that p as defined by (1.142) is not a selfadjoint linear operator. It does, however, satisfy the algebra (1.1). We are now going to show that it has selfadjoint extensions. We show that there is a basis in $\mathcal{H}_s^\sigma$ where p is diagonal and has real eigenvalues. We define the states:

$$|\tau p_\nu, \sigma\rangle_I^s = \frac{1}{\sqrt{2}} N_q \sum_{n=-\infty}^{\infty} q^{n+\nu} \Big\{ \cos_q(q^{2(n+\nu)})|2n, \sigma\rangle^s + \tau i \sin_q(q^{2(n+\nu)})|2n+1, \sigma\rangle^s \Big\} \tag{1.150}$$

$$\tau = \pi_D 1$$

First we prove orthogonality with the help of (1.128):

$$\begin{aligned} {}_I^s\langle \tau' p_\mu, \sigma \tau p_\nu, \sigma\rangle_I^s &= \frac{1}{2} N_q^2 \sum_{n=-\infty}^{\infty} q^{2n+\nu+\mu} \Big(\cos_q(q^{2(n+\nu)}) \cos_q(q^{2(n+\mu)}) \\ &\qquad + \tau\tau' \sin_q(q^{2(n+\nu)}) \sin_q(q^{2(n+\mu)}) \Big) \\ &= \frac{1}{2} \delta_{\nu\mu}(1 + \tau\tau') = \delta_{\nu\mu}\delta_{\tau\tau'} \end{aligned} \tag{1.151}$$

Next we prove completeness. First we form a linear combination of the states (1.150) to be able to use (1.128) again:

$$N_q \sum_n q^{n+\nu} \cos_q(q^{2(n+\nu)})|2n, \sigma\rangle^s = \frac{1}{\sqrt{2}} \left\{ |p_\nu, \sigma\rangle_I^s + |-p_\nu, \sigma\rangle_I^s \right\} \tag{1.152}$$

and find

$$N_q \sum_\nu q^{m+\nu} \cos_q(q^{2(m+\nu)}) \frac{1}{\sqrt{2}} \left\{ |p_\nu, \sigma\rangle_I^s + |-p_\nu, \sigma\rangle_I^s \right\} = |2m, \sigma\rangle^s \tag{1.153}$$

and similarly:

$$N_q \sum_\nu q^{m+\nu} \cos_q(q^{2(m+\nu)}) \frac{(-i)}{\sqrt{2}} \left\{ |p_\nu, \sigma\rangle_I^s - |-p_\nu, \sigma\rangle_I^s \right\} = |2m+1, \sigma\rangle^s \tag{1.154}$$

Thus the states of (1.150) form a basis. We finally show that they are eigenstates of p. We have to use (1.120) and we obtain

$$p|\tau p_\nu, \sigma\rangle_I^s = \frac{1}{s\lambda q^{\frac{1}{2}}} \sigma\tau q^{2\nu} |\tau p_\nu, \sigma\rangle_I^s \tag{1.155}$$

these are eigenstates of p with positive and negative eigenvalues but with eigenvalues spaced by q^2. From the algebra we expect a spacing by q as was shown in (1.145). We conclude that the selfadjoint extension of p does not represent the algebra (1.1).

We can apply Λ to the state (1.150) and we obtain:

$$\begin{aligned}\Lambda|\tau p_\nu,\sigma\rangle_I^s &= \frac{1}{\sqrt{2}}N_q \sum_{n=-\infty}^{\infty} q^{n+\nu}\Big\{\cos_q(q^{2(n+\nu)})|2n+1,\sigma\rangle^s \\ &\qquad + i\tau q^{-1}\sin_q(q^{2(n+\nu-1)})|2n,\sigma\rangle^s\Big\} \\ &= |\tau p_\nu,\sigma\rangle_{II}^s \end{aligned} \tag{1.156}$$

This defines the states $|\tau p_\nu,\sigma\rangle_{II}^s$. They again form a basis and are eigenstates of p

$$p|\tau p_\nu,\sigma\rangle_{II}^s = \frac{1}{s\lambda q^{\frac{1}{2}}}\sigma\tau q^{2\nu-1}|\tau p_\nu,\sigma\rangle_{II}^s \tag{1.157}$$

This can be shown in the same way as for the states (1.150). These states define a different selfadjoint extension of p. Now the eigenvalues differ by a factor q from the previous eigenvalues.

We note that the sign of the eigenvalues depends on $\sigma\tau$ and it is positive and negative for σ positive as well as for σ negative. This suggests to start from reducible representations $\mathcal{H}_s^+ \oplus \mathcal{H}_s^-$ and to extend p in the reducible representations. We define

$$\begin{aligned}|\tau p_\nu\rangle_I^s &= \frac{1}{\sqrt{2}}\{|\tau p_\nu,+\rangle_I^s + |-\tau p_\nu,-\rangle_I^s\} \\ &= \frac{1}{2}N_q \sum_{n=-\infty}^{\infty} q^{n+\nu}\Big\{\cos_q(q^{2(n+\nu)})(|2n,+\rangle^s + |2n,-\rangle^s) \\ &\qquad + i\tau \sin_q(q^{2(n+\nu)})(|2n+1,+\rangle^s - |2n+1,-\rangle^s)\Big\}\end{aligned} \tag{1.158}$$

These are eigenstates of p:

$$p|\tau p_\nu\rangle_I^s = \frac{1}{s\lambda q^{\frac{1}{2}}}\tau q^{2\nu}|\tau p_\nu\rangle_I^s \tag{1.159}$$

They are orthogonal because the $\sigma = +1$ states are orthogonal to the $\sigma = -1$ states. By the same analysis as for (1.153), (1.154) we obtain the states

$$\begin{aligned}&\frac{1}{\sqrt{2}}\{|2m,+\rangle^s + |2m,-\rangle^s)\} \\ \text{and}\qquad &\frac{1}{\sqrt{2}}\{|2m+1,+\rangle^s - |2m+1,-\rangle^s)\}\end{aligned} \tag{1.160}$$

To obtain all the states in $\mathcal{H}_s^+ \oplus \mathcal{H}_s^-$ we add the states

$$\begin{aligned}|\tau p_\nu\rangle_{II}^s &= \frac{1}{\sqrt{2}}\{|\tau p_\nu, +\rangle_{II}^s + |-\tau p_\nu, -\rangle_{II}^s\} \\ &= \frac{1}{2}N_q \sum_{n=-\infty}^{\infty} q^{n+\nu}\Big\{\cos_q(q^{2(n+\nu)})(|2n+1,+\rangle^s + |2n+1,-\rangle^s) \\ &\qquad + i\tau q^{-1}\sin_q(q^{2(n+\nu-1)})(|2n,+\rangle^s - |2n,-\rangle^s)\Big\}\end{aligned} \tag{1.161}$$

They are eigenstates of p:

$$p|\tau p_\nu\rangle_{II}^s = \frac{1}{s\lambda q^{\frac{1}{2}}}\tau q^{2\nu-1}|\tau p_\nu\rangle_{II}^s \tag{1.162}$$

They allow us to obtain the states

$$\begin{aligned}&\frac{1}{\sqrt{2}}(|2m,+\rangle^s - |2m,-\rangle^s)) \\ \text{and} \qquad &\frac{1}{\sqrt{2}}(|2m+1,+\rangle^s + |2m+1,-\rangle^s))\end{aligned} \tag{1.163}$$

by a Fourier transformation.

The states defined in (1.158) and (1.161) form a complete set of states. The states defined in (1.158) are orthogonal by themselves as well as the states defined in (1.161). It remains to be shown that the states defined in (1.158) are orthogonal to the states defined in (1.161). But this is obvious because

$$\{{}^s\langle n,+ + {}^s\langle***n,-\}\{|m,+\rangle^s - |m,-\rangle^s\} = 0 \tag{1.164}$$

The states $|\tau p_\nu\rangle_I^s$ and $|\tau p_\nu\rangle_{II}^s$ form a basis in $\mathcal{H}_s^+ \oplus \mathcal{H}_s^-$, they are eigenstates of p:

$$\begin{aligned}p|\tau p_\nu\rangle_I^s &= \frac{1}{s\lambda q^{\frac{1}{2}}}q^{2\nu}|\tau p_\nu\rangle_I^s \\ p|\tau p_\nu\rangle_{II}^s &= \frac{1}{s\lambda q^{\frac{1}{2}}}q^{2\nu-1}|\tau p_\nu\rangle_{II}^s\end{aligned} \tag{1.165}$$

and Λ is defined on them and it is unitary.

$$\begin{aligned}\Lambda|\tau p_\nu\rangle_I^s &= |\tau p_\nu\rangle_{II}^s \\ \Lambda|\tau p_\nu\rangle_{II}^s &= |\tau p_{\nu-1}\rangle_I^s\end{aligned} \tag{1.166}$$

A dynamics can now be formulated by defining a Hamilton operator in terms of p, x and Λ. The simplest Hamiltonian is

$$H = \frac{1}{2}p^2 \tag{1.167}$$

We have calculated its eigenvalues and its eigenfunctions. From the analysis of the $cos_q(x)$ and $sin_q(x)$ functions and from Fig.1. and Fig.2. we know that these states get more squeezed with increasing energy.

1.6 The definite integral and the Hilbert space $L^2{}_q$

The definite integral

Within a given representation of the algebra (1.1) a definite integral can be defined. We consider the representation where s of eqn (1.139) is one $s = 1$ and $\sigma = +1$. The states we shall denote by $|n, +\rangle$.

We define the definite integral as the difference of matrix elements of the indefinite integral (1.30):

$$\int_N^M f(x) = \langle +, M \int^x f(x)\, |M, +\rangle - \rangle +, N \int^x f(x)\, |N, +\rangle \tag{1.168}$$

From (1.35) follows:

$$\int_N^M \nabla f = \langle +, M\, f\, |M, +\rangle - \langle +, N\, f\, |N, +\rangle \tag{1.169}$$

We can use (1.168) for monomials to integrate power series in x:

$$\int_N^M x^n = \frac{1}{[n+1]} \left(q^{M(n+1)} - q^{N(n+1)} \right) \tag{1.170}$$

or we can use (1.33) where we defined ∇^{-1} and take the appropriate matrix elements

$$\begin{aligned} \langle +, M \int^x f(x)\, |M, +\rangle &= \langle +, M\, \nabla^{-1} f(x)\, |M, +\rangle \\ &= \lambda \sum_{\nu=0}^{\infty} \langle +, M\, L^{2\nu} Lx\, f(x) |M, +\rangle \end{aligned} \tag{1.171}$$

The action of L can be replaced by the operator Λ:

$$\langle +, M \int^x f\, |M, +\rangle = \lambda \sum_{\nu=0}^{\infty} \langle +, M \Lambda^{2\nu} Lx f(x) \Lambda^{-2\nu} |M, +\rangle \tag{1.172}$$

and Λ can now act on the states:

$$\langle +, M \int^x f |M, +\rangle = \lambda \sum_{\nu=0}^{\infty} \langle +, M - 2\nu Lx f(x) |M - 2\nu, +\rangle \tag{1.173}$$

This suggests treating the matrix elements for even and odd values of μ separately:

$$\langle +, 2M \int^x f|2M, +\rangle = \lambda \sum_{\mu=-\infty}^{M} \langle +, 2\mu Lx f(x)|2\mu, +\rangle \tag{1.174}$$

$$\langle +, 2M+1 \int^x f|2M+1, +\rangle = \lambda \sum_{\mu=-\infty}^{M} \langle +, 2\mu+1 Lx f(x)|2\mu+1, +\rangle$$

The definite integral (1.168) from even (odd) N to even (odd) M now finds its natural form:

$$\int_{2N}^{2M} f(x) = \lambda \sum_{\mu=N+1}^{M} \langle +, 2\mu Lx f(x)|2\mu, +\rangle \tag{1.175}$$

$$\int_{2N+1}^{2M+1} f(x) = \lambda \sum_{\mu=N+1}^{M} \langle +, 2\mu+1 Lx f(x)|2\mu+1, +\rangle$$

These are Riemannian sums for the Riemannian integral.

With the formula (1.174) it is possible to integrate x^{-1} as well:

$$\int_{2N}^{2M} \frac{1}{x} = \lambda(M-N) \tag{1.176}$$

If we take the limit $q \to 1$ and define $\overline{x} = q^{2M}$ and $\underline{x} = q^{2N}$ (1.176) approaches the formula

$$\int_{\underline{x}}^{\overline{x}} \frac{1}{x} dx = \ln \overline{x} - \ln \underline{x} \tag{1.177}$$

We now take the limit $N \to -\infty$, $M \to \infty$:

$$\int_{-\infty}^{2M} f(x) = \lambda \sum_{\mu=-\infty}^{M} \langle +, 2\mu Lx f(x)|2\mu, +\rangle \tag{1.178}$$

$$\int_{2M}^{\infty} f(x) = \lambda \sum_{\mu=M+1}^{\infty} \langle +, 2\mu Lx f(x)|2\mu, +\rangle$$

and

$$\int_{-\infty}^{\infty} f(x) = \lambda \sum_{\mu=-\infty}^{\infty} \langle +, 2\mu Lx f(x)|2\mu, +\rangle \tag{1.179}$$

The factor x in the matrix element allows the sum to converge for $N \to -\infty$ for fields that do not vanish at $x = 0$.

Similar formulas are obtained for μ, M odd.

The case $\sigma = -1$ can be treated analogously. From the discussions on selfadjoint operators in the previous chapter we know that $\sigma = +1$ and $\sigma = -1$ should be considered simultaneously.

The Hilbert space $L^2{}_q$

The integral (1.179) can be used to define a scalar product for fields. We shall assume that the integral exists for even and odd values of μ and for $\sigma = +1$ and $\sigma = -1$.

We define the scalar product for $L^2{}_q$:

$$(\chi, \psi) = \int \overline{\chi}\psi = \frac{\lambda}{2} \sum_{\substack{\mu=-\infty \\ \sigma=+1,-1}}^{\infty} \sigma\langle\sigma, \mu Lx\, \overline{\chi}\, \psi|\mu, \sigma\rangle \tag{1.180}$$

For fields that vanish at $x = \pi_D\infty$ we conclude from (1.169) that

$$\int \nabla(\overline{\chi}\,\psi) = 0 \tag{1.181}$$

This leads to a formula for partial integration:

$$\int (\nabla\overline{\chi})(L\,\psi) + \int (L^{-1}\overline{\chi})(\nabla\psi) = 0 \tag{1.182}$$

$$\int (\nabla\overline{\chi})(L^{-1}\psi) + \int (L\,\overline{\chi})(\nabla\psi) = 0$$

From Green's theorem follows:

$$\int (\nabla^2\overline{\chi})\,\psi = \int \overline{\chi}\,(\nabla^2\psi) \tag{1.183}$$

This shows that ∇^2 is a hermitean operator in $L^2{}_q$.

Eigenfunctions of the operator have been obtained in (1.137) These are the functions $cos_q(kx)$ and $sin_q(kx)$. From the discussion of these functions in chapter four we know that kx has to be an even power of q for the functions to be in $L^2{}_q$. We therefore split $L^2{}_q$ in four subspaces with μ even or μ odd and $\sigma = +1$ or $\sigma = -1$.

$$L^2{}_q = \mathcal{H}^{\mu\ even}_{\sigma=+1} \oplus \mathcal{H}^{\mu\ odd}_{\sigma=+1} \oplus \mathcal{H}^{\mu\ even}_{\sigma=-1} \oplus \mathcal{H}^{\mu\ odd}_{\sigma=-1} \tag{1.184}$$

The functions

$$N_q \left(\frac{2q^{\frac{1}{2}}}{\lambda}\right)^{\frac{1}{2}} q^k \cos_q(xq^{2k+1}) \tag{1.185}$$

form an orthogonal basis in $\mathcal{H}^{\mu\ even}_{\sigma=+1}$. This follows from (1.128) and the definition of the scalar product (1.180) has to be remembered. The function (1.185) belongs to the eigenvalue:

$$\nabla^2 \cos_q(xq^{2k+1}) = -\frac{1}{\lambda^2} q^{4k+1} \cos_q(xq^{2k+1}) \tag{1.186}$$

We now give a table for the various eigenfunctions:

	Function	Eigenvalue
$\mathcal{H}^{even}_{\sigma=+1}$:	$N_q\sqrt{\frac{2q}{\lambda}}q^k \cos_q(xq^{2k+1})$	$-\frac{1}{\lambda^2}q^{4k+1}$
	$N_q\sqrt{\frac{2q}{\lambda}}q^k \sin_q(xq^{2k+1})$	$-\frac{1}{\lambda^2}q^{4k+3}$
$\mathcal{H}^{odd}_{\sigma=+1}$:	$N_q\sqrt{\frac{2q}{\lambda}}q^k \cos_q(xq^{2k})$	$-\frac{1}{\lambda^2}q^{4k-1}$
	$N_q\sqrt{\frac{2q}{\lambda}}q^k \sin_q(xq^{2k})$	$-\frac{1}{\lambda^2}q^{4k+1}$

For $\sigma = -1$ we obtain the same table. The cos_q functions as well as the sin_q functions form an orthonormal basis in the respective subspaces of ${L^2}_q$. The set of eigenfunctions is overcomplete. We conclude that ∇^2 is hermitean but not selfadjoint. On any of the bases defined above a selfadjoint extension of the operator ∇^2 can be defined. The set of eigenvalues depends on the representation.

1.7 Variational principle

Euler-Lagrange equations:

We assume that there is a Lagrangian that depends on the fields ψ, $\nabla L^{-1}\psi$ and $\dot{\psi}$. An action is defined

$$W = \int_{t_1}^{t_2} dt \int \mathcal{L}(\dot{\psi}, \psi, \nabla L^{-1}\psi) \tag{1.187}$$

In this chapter the integrals over the algebra are taken from $-\infty$ to $+\infty$ if not specified differently.

The action should be stable under the variation of the fields as they are kept fixed at t_1 and t_2

$$\delta\psi(t_1) = \delta\psi(t_2) = 0 \tag{1.188}$$

The variation of the action is:

$$\begin{aligned} \delta W &= \int_{t_1}^{t_2} dt \int \left\{ \frac{\partial \mathcal{L}}{\partial \dot{\psi}} \delta\dot{\psi} + \frac{\partial \mathcal{L}}{\partial \psi} \delta\psi + \frac{\partial \mathcal{L}}{\partial \nabla L^{-1}\psi} \delta \nabla L^{-1}\psi \right\} \\ &= \int_{t_1}^{t_2} dt \int \left\{ -\frac{d}{dt}\frac{\partial \mathcal{L}}{\partial \dot{\psi}} - \nabla L \frac{\partial \mathcal{L}}{\partial \nabla L^{-1}\psi} + \frac{\partial \mathcal{L}}{\partial \psi} \right\} \delta\psi \end{aligned} \tag{1.189}$$

The boundary terms do not contribute because of (1.188) and ψ and its variation are supposed to vanish at infinity. The partial integration for ∇L follows from (1.27) when we use it for $(Lf)(L^{-1}g)$:

$$\nabla\left((Lf)(L^{-1}g)\right) = (\nabla Lf)(g) + f(\nabla L^{-1}g) \tag{1.190}$$

We obtain the Euler-Lagrange equations:

$$\frac{\partial \mathcal{L}}{\partial \psi} = \frac{d}{dt}\frac{\partial \mathcal{L}}{\partial \dot{\psi}} + \nabla L \frac{\partial \mathcal{L}}{\partial \nabla L^{-1}\psi} \tag{1.191}$$

The Schroedinger equation (1.38) can be obtained from the following Lagrangian. The variations of ψ and $\overline{\psi}$ are independent:

$$W = \int_{t_1}^{t_2} dt \int \left\{ i\overline{\psi}\frac{\partial}{\partial t}\psi - \frac{1}{2mq}(\nabla L^{-1}\overline{\psi})(\nabla L^{-1}\psi) - \overline{\psi}V\psi \right\} \tag{1.192}$$

Noether theorem:

We study the variation of the action under an arbitrary variation of the fields and the time:

$$\begin{aligned} t' &= t + \tau \\ \psi'(t') &= \psi(t) + \Delta\psi \end{aligned} \tag{1.193}$$

The variation of the action is defined as usual:

$$\Delta W = \int_{t_1'}^{t_2'} dt' \int_{2N}^{2M} \mathcal{L}(\dot{\psi}'(t'), \psi'(t'), \nabla L^{-1}\psi'(t')) - \int_{t_1}^{t_2} dt \int \mathcal{L} \tag{1.194}$$

First we change the t' variable of integration to t.

$$dt' = dt(1+\dot{\tau})$$

$$\begin{aligned} &\int_{t_1'}^{t_2'} dt' \int_{2N}^{2M} \mathcal{L}(\dot{\psi}'(t'), \psi'(t'), \nabla L^{-1}\psi'(t')) \\ &= \int_{t_1}^{t_2} dt(1+\dot{\tau}) \int_{2N}^{2M} \mathcal{L}(\dot{\psi}'(t+\tau)), \psi'(t+\tau), \nabla L^{-1}\psi'(t+\tau)) \end{aligned} \tag{1.195}$$

Next we make a Taylor expansion of $\mathcal{L}$ in time and obtain:

$$\begin{aligned} \delta W &= \int_{t_1}^{t_2} \int_{2N}^{2M} \frac{d}{dt}\tau\mathcal{L} \\ &+ \int_{t_1}^{t_2} dt \int_{2N}^{2M} \mathcal{L}(\dot{\psi}'(t), \psi'(t), \nabla L^{-1}\psi'(t)) - \mathcal{L} \end{aligned} \tag{1.196}$$

The variation of ψ at the same time we call $\delta\psi$:

$$\delta\psi = \Delta\psi - \tau\dot{\psi} \tag{1.197}$$

and we obtain in the same way as we derived the Euler-Lagrange equations:

$$\begin{aligned}\Delta W = &\int_{t_1}^{t_2} dt \frac{d}{dt}\tau \int_{2N}^{2M} \mathcal{L} \\ &+ \int_{t_1}^{t_2} dt \int_{2N}^{2M} \left\{ -\frac{d}{dt}\frac{\partial \mathcal{L}}{\partial \dot{\psi}} - \nabla L \frac{\partial \mathcal{L}}{\partial \nabla L^{-1}\psi} + \frac{\partial \mathcal{L}}{\partial \psi} \right\} \delta\psi \\ &+ \int_{t_1}^{t_2} dt \int_{2N}^{2M} \left[\frac{d}{dt}\left\{ \frac{\partial \mathcal{L}}{\partial \dot{\psi}} \delta\psi \right\} + \nabla \left\{ \left(L \frac{\partial \mathcal{L}}{\partial \nabla L^{-1}\psi} \right) (L^{-1}\delta\psi) \right\} \right]\end{aligned} \tag{1.198}$$

As a consequence of the Euler-Lagrange equation we find

$$\begin{aligned}\Delta W = &\int_{t_1}^{t_2} dt \frac{d}{dt}\tau \int_{2N}^{2M} \left(\mathcal{L} - \frac{\partial \mathcal{L}}{\partial \dot{\psi}} \dot{\psi} \right) \\ &- \int_{t_1}^{t_2} dt \tau \int_{2N}^{2M} \nabla \left(\left(L \frac{\partial \mathcal{L}}{\partial \nabla L^{-1}\psi} \right) (L^{-1}\dot{\psi}) \right) \\ &+ \int_{t_1}^{t_2} dt \int_{2N}^{2M} \left\{ \frac{d}{dt}\left(\frac{\partial \mathcal{L}}{\partial \dot{\psi}} \Delta\psi \right) + \nabla \left(\left(L \frac{\partial \mathcal{L}}{\partial \nabla L^{-1}\psi} \right) L^{-1}\Delta\psi \right) \right\}\end{aligned} \tag{1.199}$$

If we know that $\Delta W = 0$ because (1.195) are symmetry transformations of the action we obtain the Noether theorem

$$\begin{aligned}&\frac{d}{dt}\left\{ \tau \left(\mathcal{L} - \frac{\partial \mathcal{L}}{\partial \dot{\psi}} \dot{\psi} \right) + \frac{\partial \mathcal{L}}{\partial \dot{\psi}} \Delta\psi \right\} \\ &+ \nabla \left\{ \left(L \frac{\partial \mathcal{L}}{\partial \nabla L^{-1}\psi} \right) \left(L^{-1}\Delta\psi - \tau L^{-1}\dot{\psi} \right) \right\} = 0\end{aligned} \tag{1.200}$$

This equation holds for the densities because the limits of the integrations are arbitrary.

Charge conservation:

The Lagrangian (1.192) is invariant under phase transformations:

$$\psi' = e^{i\alpha}\psi, \quad \tau = 0, \quad \Delta\psi = \delta\psi = i\alpha\psi \tag{1.201}$$

For the charge density we find from (1.200) and (1.192)

$$\begin{aligned}-\alpha\rho = \frac{\partial \mathcal{L}}{\partial \dot{\psi}} \delta\psi + \frac{\partial \mathcal{L}}{\partial \dot{\overline{\psi}}} \delta\overline{\psi} &= -\alpha\overline{\psi}\psi \\ &= -\alpha\rho\end{aligned} \tag{1.202}$$

For the current density

$$\left(L\frac{\partial\mathcal{L}}{\partial\nabla L^{-1}\psi}\right)(L^{-1}\delta\psi)+\left(L\frac{\partial\mathcal{L}}{\partial\nabla L^{-1}\overline{\psi}}\right)(L^{-1}\Delta\overline{\psi})$$
$$=-\alpha\frac{1}{2im}L^{-1}\left\{\overline{\psi}(L\nabla\psi)-(L\nabla\overline{\psi})\psi\right\} \quad (1.203)$$
$$=-\alpha j$$

This agrees with the definition of ρ and j in eqn (1.50).

Energy conservation:
The action (1.192) is invariant under time translations:

$$t'=t+\tau_0 \quad (1.204)$$
$$\Delta\psi=0$$

From (1.200) we find the energy density and the density of the energy flow:

$$\mathcal{H}=\frac{1}{2mq}(\nabla L^{-1}\overline{\psi})(\nabla L^{-1}\psi)+V\overline{\psi}\psi \quad (1.205)$$
$$\pi=\frac{1}{2m}\left\{(\nabla\overline{\psi})L^{-1}\dot{\psi}+(L^{-1}\dot{\overline{\psi}}\nabla\psi\right\}$$

1.8 The Hilbert space $\mathcal{L}^2{}_q$

It is easy to find a representation of the algebra (1.1) in the Hilbert space L^2 of square integrable functions. The usual quantum mechanical operators in this space will be denoted with a hat:

$$[\hat{x},\hat{p}]=i,\ \overline{\hat{x}}=\hat{x},\ \overline{\hat{p}}=\hat{p} \quad (1.206)$$

A further operator that we will use frequently is:

$$\hat{z}=-\frac{i}{2}(\hat{x}\hat{p}+\hat{p}\hat{x}),\ \overline{\hat{z}}=-\hat{z} \quad (1.207)$$

It has the commutation property:

$$\hat{x}\hat{z}=(\hat{z}+1)\hat{x},\ \hat{p}\hat{z}=(\hat{z}-1)\hat{p} \quad (1.208)$$

An immediate consequence of (1.208) is that the algebra (1.138) is represented by

$$x=\hat{x},\ \Lambda=q^{\hat{z}} \quad (1.209)$$

x is selfadjoint and Λ is unitary.

The element p can be expressed by the formula (1.7)

$$p = i\lambda^{-1}\left(q^{-\frac{1}{2}}q^{\hat{z}} - q^{\frac{1}{2}}q^{-\hat{z}}\right)\hat{x}^{-1} \tag{1.210}$$

That this represents the algebra (1.1) can be easily verified with the help of (1.206). It is a consequence of the algebraic properties of $\hat{x}$ and $\hat{p}$ only. Any unitary transformation of $\hat{x}$ and $\hat{p}$ will not change this property.

A short calculation going through the steps:

$$\hat{z} = -i\hat{x}\hat{p} - \frac{1}{2}, \; i\hat{x}^{-1} = \left(\hat{z} - \frac{1}{2}\right)^{-1}\hat{p} \tag{1.211}$$

leads to the following representation for p:

$$p = \frac{[\hat{z} - \frac{1}{2}]}{\hat{z} - \frac{1}{2}}\hat{p} \tag{1.212}$$

where we use the symbol $[A]$

$$[A] = \lambda^{-1}(q^A - q^{-A}) \tag{1.213}$$

for any A.

It can be shown that p as defined in (1.212) is hermitean but not a selfadjoint operator in L^2.

We now show that it is possible to define a Hilbert space where p as defined in (1.210) is selfadjoint. We start from functions $f(\hat{x}) \in C^\infty$. They form the algebra of C^∞ functions which we denote by F.

The set of functions $f(\hat{x})$ that vanish at $\hat{x} = \pi_D x_0 q^n$, $n \in \mathbf{Z}$ forms an ideal in F. We shall denote this ideal by $F^{x_0}_{\pi_D q^n}$ and introduce the factor space:

$$\mathcal{F}^q_{x_0} = F/(F^{x_0}_{+q^n} \cap F^{x_0}_{-q^n}) \tag{1.214}$$

We will show that the operators x, Λ and p as defined by (1.209) and (1.210) act on this factor space, that it is possible to introduce a norm on $\mathcal{F}^q_{x_0}$, defining a Hilbert space this way and that x and p are represented by selfadjoint operators in this Hilbert space and that Λ is unitary.

We have to show that x and Λ leave the ideals $F_{\pi_D q^n}$ invariant. For x this is obvious, to show it for Λ we use (1.138) and find

$$\Lambda f(\hat{x}) = q^{-\frac{1}{2}} f(\frac{1}{q}\hat{x}) \tag{1.215}$$

Here we consider Λ to be represented by (1.209) acting on $f \in F$. Thus Λ transforms the zeros of an element of F_{q^n} into the zeros of the transformed element and this leaves the ideal invariant. The operator p as represented by (1.210) or (1.212) is entirely expressed in terms of x and Λ and thus leaves the ideal invariant as well.

The algebra (1.1) is now represented on $\mathcal{F}^q_{x_0}$.

Next we show that it is possible to define a scalar product on $\mathcal{F}^q_{x_0}$. Guided by (1.180) we define:

$$(g, f) = \lambda \sum_{\substack{n=-\infty \\ \sigma=+1,-1}}^{\infty} \sigma q^n \, \overline{g}(q^n x_0) f(q^n x_0) \tag{1.216}$$

This is the well known Jackson integral. It defines a norm on $\mathcal{F}^q_{x_0}$ because the scalar product (1.216) for a function of $F^{x_0}_{\pi_D q^n}$ and any function of F is zero.

The operators x, Λ and p are now linear operators in this Hilbert space. For x it is obvious that it is selfadjoint as it is diagonal. Let us show that Λ is unitary:

$$\Lambda f(x_0 q^n) = q^{-\frac{1}{2}} f(x_0 q^{n-1}), \; \Lambda^{-1} f(x_0 q^n) = q^{\frac{1}{2}} f(x_0 q^{n+1}) \tag{1.217}$$

$$\begin{aligned}(g, \Lambda f) &= \lambda \sum_{\substack{n=-\infty \\ \sigma=+1,-1}}^{\infty} \sigma q^n \, \overline{g}(q^n x_0) f(q^{n-1} x_0) q^{-\frac{1}{2}} \\ &= \lambda \sum_{\substack{n=-\infty \\ \sigma=+1,-1}}^{\infty} \sigma q^{n+1} \, \overline{g}(q^{n+1} x_0) f(q^n x_0) q^{-\frac{1}{2}} \\ &= \lambda \sum_{\substack{n=-\infty \\ \sigma=+1,-1}}^{\infty} \sigma q^n (\Lambda^{-1} \overline{g})(q^n x_0) f(q^n x_0) \\ &= (\Lambda^{-1} g, f)\end{aligned} \tag{1.218}$$

We now turn to the operator p. We use the expression (1.7) for p:

$$p = i\lambda^{-1} x^{-1} \left\{ q^{\frac{1}{2}} \Lambda - q^{-\frac{1}{2}} \Lambda^{-1} \right\} \tag{1.219}$$

and have it act on elements of $\mathcal{F}$. Using (1.217) we find:

$$pf(x) = i\lambda^{-1} x^{-1} \left\{ f(\frac{1}{q} x) - f(qx) \right\} \tag{1.220}$$

The elements $f \in F$ are considered as representants of the cosets in $\mathcal{F}^q_{x_0}$ such that (1.219) is derived by acting with the differential operator p on elements of F.

We know that $sin_q(x)$ and $cos_q(x)$ solve the eigenvalue equations of p as defined in (1.225). this is just equation (1.120). This equation links even (odd) powers of q in the argument of $cos_q(x)$ or $sin_q(x)$ to odd (even) powers. But we know from chapter 4 that the behaviour of $cos_q(q^{2n+1})$ and $sin_q(q^{2n+1})$ for $n \to \infty$ does not allow these functions to be in L^2 or in the Hilbert space defined with the norm (1.216).

To show that p has a selfadjoint extension we shall draw from our experience in chapter five. There it was essential to split states with eigenvalues belonging to even and odd powers of q. In chapter six (1.184) we have made the same experience.

We shall start from four ideals in F: $F_{+q^{2n}}$, $F_{+q^{2n+1}}$, $F_{-q^{2n}}$ and $F_{-q^{2n+1}}$ For x_0 we again choose $x_0 = 1$ and drop it in the notation. The respective factor spaces we denote by $\mathcal{F}_+^{2n}$, $\mathcal{F}_+^{2n+1}$, $\mathcal{F}_-^{2n}$, and $\mathcal{F}_-^{2n+1}$.

The intersection of the ideals leads to the union of the factor spaces and we identify $\mathcal{F}_1^q$ with this union.

The operator x acts on each of these factor spaces individually. The operator Λ maps $\mathcal{F}_{+(-)}^{2n}$ onto $\mathcal{F}_{+(-)}^{2n+1}$ and vice versa. This follows from (1.217) which shows that the ideals $F_{\pi_D q^{2n}}(F_{\pi_D q^{2n+1}})$ are mapped to the ideals $F_{\pi_D q^{2n+1}}(F_{\pi_D q^{2n}})$.

We shall distinguish functions $f \in F$ as representants of $\mathcal{F}_{\pi_D}^{2n}$ and $\mathcal{F}_{\pi_D}^{2n+1}$:

$$f^g \in \mathcal{F}_{\pi_D}^{2n}, \; f^u \in \mathcal{F}_{\pi_D}^{2n+1} \tag{1.221}$$

Thus we have to read (1.216) as follows:

$$\Lambda f^{g(u)}(x) = \Lambda f(x) = q^{-\frac{1}{2}} f(\frac{1}{q}x) = q^{-\frac{1}{2}} f^{u(g)}(\frac{1}{q}x) \tag{1.222}$$

The same way we have to read (1.220):

$$p\, f^{g(u)}(x) = i\lambda^{-1}x^{-1}\left\{ f^{u(g)}(\frac{1}{q}x) - f^{u(g)}(qx)\right\} \tag{1.223}$$

To show that p has a selfadjoint extension we construct a basis where p is diagonal. Eqns. (1.222) and (1.120) show that $cos_q(x)$ and $sin_q(x)$ will play the role of transition functions.

We first choose specific representatives in the spaces $\mathcal{F}_+^{2n}$ and $\mathcal{F}_+^{2n+1}$

$$\begin{aligned} C_+^g(q^{2n})^{(\nu)} &\sim \cos_q(xq^{2\nu}) \\ C_+^u(q^{2n+1})^{(\nu)} &\sim \cos_q(xq^{2\nu+1}) \\ S_+^g(q^{2n})^{(\nu)} &\sim \sin_q(xq^{2\nu}) \\ S_+^u(q^{2n+1})^{(\nu)} &\sim \sin_q(xq^{2\nu+1}) \end{aligned} \tag{1.224}$$

We calculate according to the rule (1.223):

$$\begin{aligned} p\,C_+^{g(\nu)} &\sim p\cos_q(xq^{2\nu}) = i\lambda^{-1}x^{-1}\left\{\cos_q(xq^{2\nu-1}) - \cos_q(xq^{2\nu+1})\right\} \\ &= i\lambda^{-1}\frac{1}{q}\sin_q(q^{-1}x) \\ &= i\lambda^{-1}q^{2\nu-1}\sin_q(xq^{2\nu-1}) \\ &= i\lambda^{-1}q^{2\nu-1}S_+^{u(\nu-1)} \end{aligned} \tag{1.225}$$

A similar calculation shows that:

$$p\,S_+^{u(\nu-1)} = -i\lambda^{-1}q^{2\nu}C_+^{g(\nu)} \tag{1.226}$$

The map p can now be diagonalized in the space $\mathcal{F}_+^{2n} \cup \mathcal{F}_+^{2n+1}$. A set of eigenvectors are:

$$\mathcal{C}_+^{g(\nu)} \mp iq^{-\frac{1}{2}}\mathcal{S}_+^{u(\nu-1)} \in \mathcal{F}_+^{2n} \cup \mathcal{F}_+^{2n+1} \tag{1.227}$$

They belong to the eigenvalues:

$$p\left\{\mathcal{C}_+^{g(\nu)} \mp iq^{-\frac{1}{2}}\mathcal{S}_+^{u(\nu-1)}\right\} = \pi_D \frac{1}{\lambda q^{\frac{1}{2}}} q^{2\nu} \left\{\mathcal{C}_+^{g(\nu)} \mp iq^{-\frac{1}{2}}\mathcal{S}_+^{u(\nu-1)}\right\} \tag{1.228}$$

When we compare (1.228) with (1.155) we know how to continue. It is exactly the same analysis as in chapter five that we have to repeat to construct a basis that consists of eigenstates of p.

We now turn our attention to the operator p^2. As a differential operator it acts on C^∞ functions. Starting from the representation (1.212) for p a short calculation yields:

$$\begin{aligned} p^2 &= \frac{4}{\lambda^2}\hat{p}\,\frac{q+q^{-1}-2\cos(\hat{x}\hat{p}+\hat{p}\hat{x})h}{1+(\hat{x}\hat{p}+\hat{p}\hat{x})}\,\hat{p} \\ &= \hat{p}^2 + O(h) \\ q &= e^h. \end{aligned} \tag{1.229}$$

We can consider h as a coupling constant and expand p^2 at $\hat{p}^2$. Thus we could consider a Hamiltonian:

$$H = \frac{1}{2}p^2 = \frac{1}{2}\hat{p}^2 + O(h) \tag{1.230}$$

We would find that H is not a selfadjoint operator in L^2. It is, however, a selfadjoint operator on the space $\mathcal{F}^q$ equipped with the norm (1.216). Its eigenvalues are $\frac{1}{2q\lambda}q^{4\nu}$ and $\frac{1}{2q\lambda}q^{4\nu+2}$. The eigenstates are the same as the eigenstates of p.

A final remark on the representation of p and Λ on the space $\mathcal{F}^q$. Both can be viewed as a linear mapping of $\mathcal{F}^q$ into $\mathcal{F}^q$ in very much the same way as a vector field is a linear mapping of C^∞ into C^∞. The derivative property of a vector field is changed, however. The operator Λ acts group like:

$$\Lambda fg = (\Lambda f)(\Lambda g) \tag{1.231}$$

and p acts with the Leibniz rule:

$$pfg = (pf)(\Lambda g) + (\Lambda^{-1})(pg) \tag{1.232}$$

They form an algebra:

$$(\Lambda p - qp\Lambda)f = 0 \tag{1.233}$$

1.9 Gauge theories on the factor spaces

The usual gauge transformations on wave functions $\psi(\hat{x}, t)$

$$\begin{aligned} \psi'(\hat{x}, t) &= e^{i\alpha(\hat{x},t)}\psi(\hat{x}, t) \\ \alpha(\hat{x}, t) &= g\sum_l \alpha_l(\hat{x}, t)T_l \end{aligned} \tag{1.234}$$

define a gauge transformation on $\mathcal{F}^q$ as well. We want to show that there are covariant expressions for x, Λ and p that reduce to x, Λ and p foı $g = 0$.

The operator x by itself is obviously covariant and acts on $\mathcal{F}^q$.

A covariant expression of Λ is obtained by replacing the canonical momentum in (1.209) by the covariant momentum

$$\hat{p} \to \hat{p} - gA^l(\hat{x}, t)T_l \tag{1.235}$$

where $A^l T_l$ transforms in the usual way:

$$A'^l T_l = e^{i\alpha} A^l T_l e^{-i\alpha} - \frac{i}{g} e^{i\alpha}\, \partial\, e^{-i\alpha} \tag{1.236}$$

This leads to the expression

$$\tilde{\Lambda} = q^{-i\hat{x}(\hat{p}-yA^l T_l)-\frac{1}{2}} \tag{1.237}$$

It is certainly true that

$$\widetilde{\Lambda}' = e^{i\alpha(\hat{x},t)} \Lambda\, e^{-i\alpha(\hat{x},t)} \tag{1.238}$$

because it is true for any power of the covariant derivative $\hat{p} - gA^l T_l$. Thus:

$$\widetilde{\Lambda}'\psi'(\hat{x}, t) = e^{i\alpha(\hat{x},t)}\widetilde{\Lambda}\psi(\hat{x}, t) \tag{1.239}$$

We have to show that $\widetilde{\Lambda}$ is defined on $\mathcal{F}^q$. This will be the case if $\widetilde{\Lambda}$ leaves the ideals $F_{\pi_D q^n}$ invariant. We adapt the notation of (1.73) and define:

$$\begin{aligned} E^{-1} &= q^{i\hat{x}\hat{p}+\frac{1}{2}}\widetilde{\Lambda} = q^{i\hat{x}\hat{p}} q^{i\hat{x}(\hat{p}-gA^l T_l)} \\ E &= q^{i\hat{x}(\hat{p}-gA^l T_l)} q^{-i\hat{x}\hat{p}} \end{aligned} \tag{1.240}$$

As a consequence of the Baker-Campbell-Hausdorff formula E can be written in the form

$$E = q^{ia_l T^l} \tag{1.241}$$

where a is a functional of A and its space derivatives. Thus E acts on $\mathcal{F}_q$.

E was defined such that

$$\widetilde{\Lambda} = \Lambda E^{-1}(\hat{x}) \tag{1.242}$$

This shows that $\tilde{\Lambda}$ acts on $\mathcal{F}_q$ as E^{-1} and Λ do so.

Formula (1.239) can be used to compute a in (1.240) explicitely. We make use of the time dependence in $A(\hat{x}, t)$ and compute the expression $E\frac{\partial}{\partial t}E^{-1}$ first with E in the form (1.240) and then with E in the form (1.239) and compare the result. We demonstrate this for the abelian case:

$$E\frac{\partial}{\partial t}E^{-1} = \frac{\partial}{\partial t} - i\frac{\partial}{\partial t}ah, \quad q = e^h \tag{1.243}$$

Now the second way:

$$\begin{aligned} E\frac{\partial}{\partial t}E^{-1} &= q^{i\hat{x}(\hat{p}-gA)}\frac{\partial}{\partial t}q^{-i\hat{x}(\hat{p}-gA)} \qquad (1.244)\\ &= \frac{\partial}{\partial t} + [-ig\hat{x}Ah, \frac{\partial}{\partial t}] + \frac{1}{2}\left[i\hat{x}(\hat{p}-gA)h, \left[-ig\hat{x}Ah, \frac{\partial}{\partial t}\right]\right] + \ldots \\ &= \frac{\partial}{\partial t} + igh\hat{x}\dot{A} + \frac{1}{2}h^2 g\hat{x}\frac{\partial}{\partial \hat{x}}(ig\hat{x}\dot{A}) + \ldots \\ &\quad + \frac{1}{n!}h^n(\hat{x}\frac{\partial}{\partial \hat{x}})^{n-1}(ig\hat{x}\dot{A}) + \ldots \end{aligned}$$

This can be written in a more compact way:

$$E\frac{\partial}{\partial t}E^{-1} = \frac{\partial}{\partial t} + \frac{q^{\hat{x}\frac{\partial}{\partial \hat{x}}} - 1}{\hat{x}\frac{\partial}{\partial \hat{x}}}ig\frac{\partial}{\partial t}\hat{x}A \tag{1.245}$$

A comparison of (1.244) and (1.242) shows that

$$a = -g\frac{q^{\hat{x}\frac{\partial}{\partial \hat{x}}} - 1}{h\hat{x}\frac{\partial}{\partial \hat{x}}}\hat{x}A \tag{1.246}$$

In an expansion in $\hat{x}\frac{\partial}{\partial \hat{x}}$ of (1.245) there is never a negative power. Another way of writing (1.245) is:

$$\begin{aligned} a &= -g\frac{q^{-\frac{1}{2}}\Lambda^{-1} - 1}{h\hat{x}\frac{\partial}{\partial \hat{x}}}\hat{x}A \qquad (1.247)\\ &= -gh^{-1}(q^{-\frac{1}{2}}\Lambda^{-1} - 1)\partial_x^{-1}A(\hat{x}, t) \end{aligned}$$

In this form it is obvious that a has the right transformation property such that E transforms as in (1.71).

We show the transformation law in the non-abelian case for E as well:

$$E(\hat{x}) = q^{i\hat{x}(\hat{p}-gA^l(\hat{x})T_l)}q^{-i\hat{x}\hat{p}} \tag{1.248}$$

$$\begin{aligned} E'(\hat{x}) &= q^{i\hat{x}(\hat{p}-gA'^l(\hat{x})T_l)}q^{-i\hat{x}\hat{p}} \qquad (1.249)\\ &= e^{i\alpha(\hat{x})}q^{i\hat{x}(\hat{p}-gA^l(\hat{x})T_l)}e^{-i\alpha(\hat{x})}q^{-i\hat{x}\hat{p}} \\ &= e^{i\alpha(\hat{x})}E(\hat{x})e^{-i\alpha(q\hat{x})} \end{aligned}$$

This agrees with (1.61).

It is now obvious how to define the covariant version of p

$$\tilde{p} = i\lambda^{-1}\hat{x}^{-1}(q^{\frac{1}{2}}\tilde{\Lambda} - q^{-\frac{1}{2}}\tilde{\Lambda}^{-1}) \tag{1.250}$$

This is covariant under gauge transformations, it has the property that $\tilde{p} \to p$ for $g \to 0$ and it is defined on $\mathcal{F}_q$.

The gauge covariant differential operators $\tilde{\Lambda}$, $\tilde{p}$ and $\mathcal{D}_t$ acting on time dependent elements of $\mathcal{F}_q$ form an algebra. This is the algebra (1.103). We can now express the tensor T through the vector potential $A^1 = A$, $A^0 = -ig\omega$

$$T = -E(\partial_t E^{-1}) - igA^0 + igE(\Lambda^{-1}A^0)E^{-1} \tag{1.251}$$

The first expression we have calculated in (1.244) for an abelian gauge group. We now compute T for an abelian gauge group:

$$T = \frac{1}{\hat{x}\frac{\partial}{\partial\hat{x}}} ig\left\{(\Lambda^{-1} - 1)\hat{x}\left(-\frac{\partial}{\partial t}A^1 + \frac{\partial}{\partial\hat{x}}A^0\right)\right\} \tag{1.252}$$

We see that the curvature T is a functional of the usual curvature F:

$$F = \frac{\partial}{\partial\hat{x}}A^0 - \frac{\partial}{\partial t}A^1 \tag{1.253}$$

and its space derivatives

$$T = ig(\Lambda^{-1} - 1)\partial_x^{-1}F \tag{1.254}$$

For the non-abelian case the calculations are more involved and as I have not done them I have to leave them to the reader.

References

AA Gaspar, M. Rahman, *Basic Hypergeometric Series*, Cambridge University Press, Cambridge, 1990

AB Schmuedgen, *Unbounded Operator Algebras and Representation Theory*, Berin, 1990.

AC N. Lebedev, *Special Functions and their Appliations*, Prentice Hall, 1965.

AD Schwenk, J. Wess, *A q-deformed Quantum Mechanical Toy Model*, Phys. Lett. **B** **291**:273–277, 1992.

AE Hebecker, S. Schreckenberg, J. Schwenk, W. Weich, J. Wess, *Representations of a q-deformed Heisenberg Algebra*, Z. Phys. **C** **64**:355–359, 1994.

AF Fichtmüller, A. Lorek, J. Wess, *q-Deformed phase space and its lattice structure.* Z. Phys. **C** **71**: 533 1996.

AF L. Cerchiai, R. Hinterding, J. Madore, J. Wess, *A calculus based on a q-deformed Heisenberg Algebra*, Eur. Phys. J. **C** **8**: 547, 1999.

AH Schwenk, *Fourier transformations for the q-deformed Heisenberg Algebra*, Varenna 1994, *Quantum Groups and their Applications in Physics* IOS Press 1996.

AI H. Koornwinder, R. F. Swarttouw, *On q-Analogues of the Fourier and Hankel Transforms*, Trans. AMS 333, 445- 461, 1992.
AJ Schmuedgen, *Operator Representations of a q-deformed Heisenberg Algebra*, Eur. Phys. J. **C 5**: 553–566, 1998.
AK Lorek, A. Ruffing, J Wess, *A q-deformation of the Harmonic Oscillator*, Z.Phys. **C 74**:369, 1997.
AL J. Macfarlane, *On q-analogues of the quantum harmonic oscillator and quantum group $SU_q(2)$*, J. Phys. **A 22**: 4581 1989.
AM C. Biedenharn, *The quantum group $SU_q(2)$ and a q-analogue of the boson operators*, J. Phys. **A 22**: L837, 1989.
AN Dick, A. Pollok-Narayanan, H. Steinacker, J. Wess, *Convergent Perturbation Theory for a q-deformed Anharmonic Oscillator*, Eur. Phys. J.**C 6**: 701, 1999.
AO Lorek, J. Wess, *Dynamical symmetries in q-deformed quantum mechanics*, Z.Phys. **C 67**:671, 1995.
AP Kempf, S. Majid, *A q-deformation of the Harmonic Oscillator*, J. Math. Phys. **35**:6802, 1994.
AQ L. Cerchiai, R. Hinterding, J. Madore, J. Wess, *The geometry of a q-deformed phase space*, Eur. Phys. J. **C 8**:533, 1999.
AR Hinterding, J. Wess, *q-deformed Hermite Polynomials in q-Quantum Mechanics*, Eur. Phys. J. **C 6**: 183, 1999.

2 q-Deformed Heisenberg Algebra in n Dimensions

2.1 $SL_q(2)$, Quantum groups and the R-matrix.

Let me first present a simple example of a quantum group, $SL_q(2)$. This will allow me to exhibit the mathematical structure of quantum groups and at the same time I can demonstrate why a physicist might get interested in such an object.

Take a two by two matrix with entries a, b, c, d:

$$T = \begin{pmatrix} a & b \\ c & d \end{pmatrix} \tag{2.255}$$

If the entries are complex numbers or real numbers and if the determinant is not zero T will be an element of $GL(2, C)$ or $GL(2, R)$.

For $GL_q(2)$ we demand that the entries a, b, c, d are elements of an algebra $\mathcal{A}$ which we define as follows: We take the free associative algebra generated by $1, a, b, c, d$ and divide by the ideal generated by the following relations:

$$\begin{aligned} ab &= qba \\ ac &= qca \\ ad &= da + \lambda bc \\ bc &= cb \\ bd &= qdb \\ cd &= qdc \end{aligned} \tag{2.256}$$

q is a complex number, $q \in \mathbb{C}, q \neq 0$ and $\lambda = q - q^{-1}$.
In the algebra $\mathcal{A}$ we allow formal power series.
A direct calculation shows that

$$det_q T = ad - qbc \tag{2.257}$$

is central, it commutes with a, b, c and d. If $det_q T \neq 0$ we call T an element of $GL_q(2)$, if $det_q T = 1$, T will be an element of $SL_q(2)$.

So far for the definition of $SL_q(2)$. It looks quite arbitrary; by what follows, however, it will become clear that the relations (2.256) have been chosen carefully such that the algebra $\mathcal{A}$ has some very nontrivial properties.

The relations (2.256) allow an ordering of the elements a, b, c, d. We could decide to write any monomial of degree n as a sum of monomials $a^{n_1} b^{n_2} c^{n_3} d^{n_4}$ with $n = n_1 + n_2 + n_3 + n_4$. We could have chosen any other ordering, e.g. $b^{m_1} a^{m_2} d^{m_3} c^{m_4}, \quad n = m_1 + m_2 + m_3 + m_4$. Moreover, the monomials in a given order are a basis for polynomials of fixed degree (Poincaré-Birkhoff-Witt).

The relations depend on a parameter q and for $q = 1$ the algebra becomes commutative. In this sense we call the quantum group $GL_q(2)$ a q deformation of $GL(2, C)$.

That the Poincaré-Birkhoff-Witt property is far from being trivial can be demonstrated by the following example. Consider an algebra freely generated by two elements x, y and divided by the ideal generated by the relation $yx = xy + x^2 + y^2$. If we now try to order $y^2 x$ in the xy ordering we find $x^3 + y^3 + x^2 y + xy^2 = 0$ The polynomials of third degree in the x, y ordering are not independent. This algebra does not have the Poincaré-Birkhoff-Witt property.[1]

That our algebra $\mathcal{A}$ has the Poincaré-Birkhoff-Witt property follows from the fact that it can be formulated with the help of an $\hat{R}$ matrix and that this matrix satisfies the quantum Yang Baxter equation.

Let me introduce the concept of an $\hat{R}$ matrix. The relation (2.256) can be written in the form

$$\hat{R}^{ij}{}_{kl} T^k{}_r T^l{}_s = T^i{}_k T^j{}_l \hat{R}^{kl}{}_{rs} \tag{2.258}$$

The indices take the value one and two, repeated indices are to be summed, $T^i{}_j$ stands for a, b, c, d in an obvious assignment and $\hat{R}$ is the following 4 by 4 matrix

$$\hat{R} = \begin{pmatrix} q & 0 & 0 & 0 \\ 0 & \lambda & 1 & 0 \\ 0 & 1 & 0 & 0 \\ 0 & 0 & 0 & q \end{pmatrix} \tag{2.259}$$

The rows and columns of $\hat{R}$ are labelled by 11, 12, 21 and 22.

As an example:

$$\hat{R}^{12}{}_{ij} T^i{}_2 T^j{}_2 = T^1{}_i T^2{}_j \hat{R}^{ij}{}_{22}$$

becomes

$$\lambda T^1{}_2 T^2{}_2 + T^2{}_2 T^1{}_2 = q T^1{}_2 T^2{}_2$$

[1] This example has been shown to me by Phung Ho Hai (thesis).

or

$$\lambda bd + db = qbd \rightarrow bd = qdb$$

The relations (2.258) are called $\hat{R}TT$ relations. They are 16 relations that reduce to the 6 relations of eq. (2.256). This of course is due to particular properties of the $\hat{R}$ matrix (2.259). We could start from an arbitrary $\hat{R}$ matrix, but then the $\hat{R}TT$ relations might have $T = 1$ and 0 as the only solutions. If on the other hand we take $\hat{R}$ to be the unit matrix $\hat{R}^{ij}_{kl} = \delta^i_k \delta^j_l$, there would be no relation for $abcd$. If $\hat{R}^{ij}_{kl} = \delta^i_l \delta^j_k$ all elements of the T matrix would commute. The art is to find an $\hat{R}$ matrix that guarantees the Poincaré Birkhoff Witt property, in this case for polynomials in a, b, c, d.

In any case, the existence of an $\hat{R}$ matrix has far-reaching consequences. E.g. from the $\hat{R}TT$ relations follows:

$$\begin{aligned} &\hat{R}^{i_1 i_2}{}_{j_1 j_2}(T^{j_1}{}_r \otimes T^r{}_{k_1}) \cdot (T^{j_2}{}_s \otimes T^s{}_{k_2}) \\ &= \hat{R}^{i_1 i_2}{}_{j_1 j_2} T^{j_1}{}_r T^{j_2}{}_s \otimes T^r{}_{k_1} T^s{}_{k_2} \\ &= T^{i_1}{}_{j_1} T^{i_2}{}_{j_2} \hat{R}^{j_1 j_2}{}_{rs} \otimes T^r{}_{k_1} T^s{}_{k_2} \\ &= (T^{i_1}{}_r \otimes T^r{}_{l_1}) \cdot (T^{i_2}{}_s \otimes T^s{}_{l_2}) \hat{R}^{l_1 l_2}{}_{k_1 k_2} \end{aligned} \tag{2.260}$$

This shows that we can define a co-product:

$$\Delta T^j{}_l = T^j{}_r \otimes T^r{}_l \tag{2.261}$$

It is compatible with the $\hat{R}TT$ relations:

$$\hat{R}^{ij}{}_{kl} \Delta T^k{}_r \Delta T^l{}_s = \Delta T^i{}_k \Delta T^j{}_l \hat{R}^{kl}{}_{rs} \tag{2.262}$$

This co-multiplication is an essential ingredient of a Hopf algebra. Another one is the existence of an inverse (antipode). We have already seen that $det_q T$ is central. We can enlarge the algebra by the inverse of $det_q T$. Then it is easy to see that

$$T^{-1} = \frac{1}{det_q T} \begin{pmatrix} d, & -\frac{1}{q} b \\ -qc, & a \end{pmatrix} \tag{2.263}$$

is the inverse matrix of T.

We have now learned that a quantum group is a Hopf algebra, it is a q-deformation of a group and it has an $\hat{R}$ matrix associated with it.

We continue by studying the $\hat{R}$ matrix (2.259) in more detail. It is an easy exercise to verify that it satisfies the characteristic equation:

$$(\hat{R} - q)(\hat{R} + \frac{1}{q}) = 0 \tag{2.264}$$

The matrix $\hat{R}$ has q and $-\frac{1}{q}$ as eigenvalues. The projectors that project on the respective eigenspaces follow from the characteristic equation:

$$\mathrm{A} = -\frac{\mathrm{q}}{1 + \mathrm{q}^2}(\hat{\mathrm{R}} - \mathrm{q}) \qquad \mathrm{S} = \frac{\mathrm{q}}{1 + \mathrm{q}^2}(\hat{\mathrm{R}} + \frac{1}{\mathrm{q}}) \tag{2.265}$$

A is a deformation of an antisymmetrizxer and S of a symmetrizer. The normalization is such that:

$$A^2 = A, \quad S^2 = S, \quad AS = SA = 0. \quad 1 = S + A, \quad \hat{R} = qS - \frac{1}{q}A \tag{2.266}$$

The $\hat{R}$ matrix approach can be generalized to n dimensions. The n^2 by n^2 $\hat{R}$ matrix for the quantum group $GL_q(n)$ is:

$$\hat{R}^{ji}_{kl} = \delta^i_k \delta^j_l [1 + (q-1)\delta^{ij}] + (q - \frac{1}{q})\theta(i-j)\delta^j_k \delta^i_l \tag{2.267}$$

where $\theta(i-j) = 1 \quad$ for $i > j$ and $\theta(i-j) = 0 \quad$ for $i \leq j$. The $\hat{R}TT$ relations (2.258) now refer to an $n \times n$ matrix T. There are n^4 relations for the n^2 entries of T. It can be shown that the thus defined T matrix has the Poincaré-Birkhoff-Witt property. Comultiplication is defined the same way as by (2.261) and the $\hat{R}$ matrix satisfies the same characteristic equation (2.265). This leads to the projectors A and S in the n-dimensional case as well.

The $\hat{R}$-matrices (2.259, 2.267) are symmetric:

$$\hat{R}^{ab}{}_{cd} = \hat{R}^{cd}{}_{ab} \tag{2.268}$$

In such a case, the transposed matrix $\tilde{T}$:

$$\tilde{T}^a{}_b = T^b{}_a \tag{2.269}$$

will also satisfy the RTT relations (2.258)

$$\begin{aligned} \tilde{T}^a{}_c \tilde{T}^b{}_d \hat{R}^{cd}{}_{ef} &= T^c{}_a T^d{}_b \hat{R}^{ef}{}_{cd} \\ = T^e{}_i T^f{}_j \hat{R}^{ij}{}_{ab} &= \hat{R}^{ab}{}_{ij} \tilde{T}^i{}_e \tilde{T}^j{}_f \end{aligned} \tag{2.270}$$

Therefore $\tilde{T} \in GL_q(n)$, if $T \in GL_q(n)$. It then has an inverse $\tilde{T}^{-1}$:

$$\tilde{T}^{-1\,a}{}_i \tilde{T}^i{}_b = \delta^a_b, \qquad \tilde{T}^a{}_i \tilde{T}^{-1\,i}{}_b = \delta^a_b \tag{2.271}$$

In general, (for $q \neq 1$), however, $\tilde{T}^{-1} \neq \widetilde{T^{-1}}$. For $n = 2$:

$$\tilde{T}^{-1} = \frac{1}{det_q T}\begin{pmatrix} d & -\frac{1}{q}c \\ -qb & a \end{pmatrix} \quad , \widetilde{T^{-1}} = \frac{1}{det_q T}\begin{pmatrix} d & -qc \\ -\frac{1}{q}b & a \end{pmatrix}. \tag{2.272}$$

$$\begin{pmatrix} q^2 & 0 \\ 0 & 1 \end{pmatrix} \begin{pmatrix} d & -\frac{1}{q}c \\ -qb & a \end{pmatrix} \begin{pmatrix} \frac{1}{q^2} & 0 \\ 0 & 1 \end{pmatrix} = \begin{pmatrix} d & -qc \\ -\frac{1}{q}b & a \end{pmatrix} \tag{2.273}$$

2.2 Quantum planes

After having defined the algebraic structure of a quantum group we are interested in its comodules. Such comodules will be called quantum planes.

Let the matrix T satisfy $\hat{R}TT$ relations of the type (2.258) with some general matrix $\hat{R}$ and let T co-act on a n-component "vector" as follows:

$$\omega(x^i) = T^i{}_k \otimes x^k \tag{2.274}$$

This defines a contravariant vector.

We ask for an algebraic structure of the quantum plane that is compatible with the co-action (2.274). From the $\hat{R}TT$ relations follows that for any polynomical $\mathcal{P}(\hat{R})$ it is true that:

$$\mathcal{P}(\hat{R})^{ij}{}_{kl}T^k{}_rT^l{}_s = T^i{}_kT^j{}_l\mathcal{P}(\hat{R})^{kl}{}_{rs} \tag{2.275}$$

As a consequence, any relation of the type

$$\mathcal{P}(\hat{R})^{ij}{}_{kl}x^kx^l = 0 \tag{2.276}$$

implies

$$\mathcal{P}(\hat{R})^{ij}{}_{kl}\omega(x^k)\omega(x^l) = 0 \tag{2.277}$$

We shall say that the relation (2.276) is covariant.

A natural definition of an algebraic structure is:

$$P_A^{ij}{}_{kl}x^kx^l = 0 \tag{2.278}$$

if P_A, the "antisymmetrizer", can be obtained as a polynomial in the $\hat{R}$ matrix. This then generalizes the property that the coordinates of an ordinary space commute and (1.42) is covariant.

For $GL_q(n)$ the most general polynomial of $\hat{R}$ is of degree one and from (1.10) we know that (2.278) becomes:

$$x^ix^j = \frac{1}{q}\hat{R}^{ij}{}_{kl}x^kx^l \tag{2.279}$$

In two dimensions, this reduces to the condition:

$$x^1x^2 = qx^2x^1 \tag{2.280}$$

The relations (2.279) can be generalized to the case of two or more copies of quantum planes, for instance (x^1, x^2) and (y^1, y^2). The relations

$$x^iy^j = \frac{\kappa}{q}\hat{R}^{ij}{}_{kl}y^kx^l \tag{2.281}$$

are consistent, i.e. they have the Poincaré -Birkhoff-Witt property for arbitrary $\kappa, \kappa \neq 0$ and they are covariant. For $n = 2$ the relations (2.281) are

$$x^1y^1 = \kappa y^1x^1$$

$$x^1y^2 = \frac{\kappa}{q}y^2x^1 + \frac{\kappa}{q}\lambda y^1x^2 \tag{2.282}$$
$$x^2y^1 = \frac{\kappa}{q}y^1x^2$$
$$x^2y^2 = \kappa y^2x^2$$

Consistency can be checked explicitely:

$$x^1(y^1y^2 - qy^2y^1) = \kappa y^1x^1y^2 - q\kappa(\frac{1}{q}y^2x^1 + \frac{1}{q}\lambda y^1x^2)y^1 = \kappa^2(\frac{1}{q}y^1y^2 - y^2y^1)x^1 \tag{2.283}$$

Taking x^1 through the y relations does not give rise to new relations. Similar for all third order relations and then, by induction, we conclude that (2.282) does not generate new higher order relations and therefore the Poincaré -Birkhoff-Witt property holds.

We can do this more systematically if we start from a general $\hat{R}$ matrix as we did at the beginning of this chapter. We consider three copies of quantum planes, x, y and z. Covariant relations are:

$$xy = \hat{R}yx, \quad yz = \hat{R}zy, \quad xz = \hat{R}zx \tag{2.284}$$

(indices as in (2.281)). We demand that a reordering of xyz to zyx should give the same result, independent of the way we achieve this reordering. There are two "independent" ways to do it. The first one starts by changing first xy:

$$xyz \to yxz \to yzx \to zyx,$$

the second one by changing first yz:

$$xyz \to xzy \to zxy \to zyx.$$

The result should be the same. This leads to an equation on the $\hat{R}$ matrix that is called Quantum Yang-Baxter equation. It can easily be formulated by introducing n^3 by n^3 matrices:

$$\hat{R}_{12}{}^{(i)}{}_{(j)} = \hat{R}^{i_1i_2}{}_{j_1j_2}\delta^{i_3}{}_{j_3} \tag{2.285}$$

and

$$\hat{R}_{23}{}^{(i)}{}_{(j)} = \delta^{i_1}{}_{j_1}\hat{R}^{i_2i_3}{}_{j_2j_3} \tag{2.286}$$

The Yang-Baxter equation that follows from the independence of the reordering then is:

$$\hat{R}_{12}\hat{R}_{23}\hat{R}_{12} = \hat{R}_{23}\hat{R}_{12}\hat{R}_{23} \tag{2.287}$$

These are n^6 equations for n^4 independent entries of the $\hat{R}$-matrix. It can be checked that the $\hat{R}$-matrices (2.259) and (2.267) do satisfy the Yang-Baxter equation.

A direct consequence of (2.287) is

$$\mathcal{P}(\hat{R}_{12})\hat{R}_{23}\hat{R}_{12} = \hat{R}_{23}\hat{R}_{12}\mathcal{P}(\hat{R}_{23}) \tag{2.288}$$

where $\mathcal{P}(\hat{R}_{12})$ is any polynomial in $\hat{R}_{12}$, e.g. a projector P_{12} or $\hat{R}_{12}{}^{-1}$.

We now use (2.287) to discuss relations of the type (2.283) more systematically:

$$\begin{aligned} P_{A23}xyy &= \frac{\kappa}{q}P_{A23}\hat{R}_{12}yxy \\ &= \frac{\kappa^2}{q^2}P_{A23}\hat{R}_{12}\hat{R}_{23}yyx \\ &= \frac{\kappa^2}{q^2}\hat{R}_{12}\hat{R}_{23}P_{A12}yyx. \end{aligned} \tag{2.289}$$

Thus the relation (2.281) is consistent with the xx and yy relations (2.276) if $\hat{R}$ satisfies the Yang-Baxter equation.

For the $\hat{R}$-matrix (2.287) becomes

$$\hat{R}^{ab}{}_{\alpha\beta}\hat{R}^{\beta c}{}_{\gamma t}\hat{R}^{\alpha\gamma}{}_{rs} = \hat{R}^{bc}{}_{\alpha\beta}\hat{R}^{a\alpha}{}_{r\gamma}\hat{R}^{\gamma\beta}{}_{st} \tag{2.290}$$

It is interesting to note that this relation expresses the fact that the $n \times n$ matrices $t^{\alpha}{}_{\gamma}$

$$(t^{\alpha}{}_{\gamma})_{ab} = \hat{R}^{a\alpha}{}_{\gamma b} \tag{2.291}$$

- where a, b label the n rows and n columns respectively and α, γ label n^2 matrices - represent a solution of the $\hat{R}TT$ relations. With the notation (2.291) eqn. (2.290) takes the form

$$t^{b}{}_{\alpha}t^{c}{}_{\gamma}\hat{R}^{\alpha\gamma}{}_{rs} = R^{bc}{}_{\alpha\beta}t^{\alpha}{}_{r}t^{\beta}{}_{s} \tag{2.292}$$

where the ts are multiplied matrixwise.

The inverse of $\hat{R}_{12}$ or $\hat{R}_{23}$ is $\hat{R}^{-1}{}_{12}$ or $\hat{R}^{-1}{}_{23}$. With $\hat{R}$ the matrix $\hat{R}^{-1}$ will satisfy the Yang-Baxter equation as well. Thus we could have used $\hat{R}^{-1}$ in (2.281) as well.

We now try to define a covariant transformation law:

$$\omega(y_i) = S^{l}{}_{i} \otimes y_l \tag{2.293}$$

such that

$$\omega(y_l x^l) = 1 \otimes y_l x^l \tag{2.294}$$

This leads to

$$S = T^{-1} \tag{2.295}$$

Thus a covariant quantum plane exists if we can define T^{-1}. Covariant relations for this quantum plane are of the form where P is a projector

$$y_a y_b P^{ba}_{lm} = 0 \tag{2.296}$$

Note the position of the indices.

$$\begin{aligned}\omega(y_a y_b)P^{ba}{}_{lm} &= T^{-1}{}^r{}_a T^{-1}{}^s{}_b \otimes y_r y_s P^{ba}{}_{lm} \\ &= P^{sr}{}_{ab} T^{-1}{}^b{}_m T^{-1}{}^a{}_l \otimes y_r y_s = 0\end{aligned} \tag{2.297}$$

The second step follows from the RTT relations (2.258).

It is possible to define covariant and consistent relations between the covariant and contravariant quantum planes. We start from the Ansatz:

$$x^r y_l = \Gamma^{mr}_{sl} y_m x^s \tag{2.298}$$

Covariance means that

$$\begin{aligned}\omega(x^r y_l) &= T^r{}_s S^m{}_l \otimes x^s y_m = T^r{}_s S^m{}_l \otimes \Gamma^{ns}_{tm} y_n x^t \\ &= \Gamma^{mr}_{sl} S^n{}_m T^s{}_t \otimes y_n x^t\end{aligned} \tag{2.299}$$

or:

$$\begin{aligned}T^r{}_a S^b{}_l \Gamma^{ca}_{db} &= \Gamma^{sr}_{tl} S^c{}_s T^t{}_d \\ T^a{}_b T^c{}_d \Gamma^{bd}_{sr} &= \Gamma^{ac}_{db} T^d{}_s T^b{}_r\end{aligned} \tag{2.300}$$

This is the condition for covariance. Now we have to prove consistency.

$$\begin{aligned}0 = P^{ab}{}_{cd} x^c x^d y_l &= P^{ab}{}_{cd} \Gamma^{ed}_{fl} x^c y_e x^f \\ &= P^{ab}{}_{cd} \Gamma^{ed}_{fl} \Gamma^{kc}_{me} y_k x^m x^f \\ &= \Gamma^{ka}_{\alpha\beta} \Gamma^{\beta c}_{\gamma l} P^{\alpha\gamma}{}_{mf} y_k x^m x^f\end{aligned} \tag{2.301}$$

For the last step we have to use the Yang-Baxter equation (2.287) for Γ and that P is a polynomial in Γ.

Thus Γ has to be a solution of the Yang-Baxter equation such that the ΓTT relations hold for Ts defined by the $\hat{R}TT$ relations (2.258). In general there might be several such solutions. Any such solution Γ can be decomposed into projectors on the invariant subspaces - this follows from covariance. We know that with $\hat{R}$, $\hat{R}^{-1}$ will always be a solution as well.

If we use $\Gamma = q\hat{R}^{-1}$ we find that $y_l x^l$ is central, it commutes with all x^s and y_r.

Another way to define a covariant transformation law is:

$$\omega(\hat{y}_l) = \hat{S}^k_l \otimes \hat{y}_k \tag{2.302}$$

and require that:

$$\omega(x^l \hat{y}_l) = 1 \otimes x^l \hat{y}_l \tag{2.303}$$

This leads to:

$$\hat{S} = \widetilde{\tilde{T}^{-1}} \tag{2.304}$$

with the notation of (2.269).

A similar analysis as above shows that

$$\hat{y}_l x^k = \Gamma^{kr}_{ls} x^s \hat{y}_r \tag{2.305}$$

is consistent and covariant. Γ again might be any solution of the Yang-Baxter equation that can be decomposed into covariant projectors. Compare the position of the matrices in (2.305) and (2.298).

$$\hat{y}_a \hat{y}_b P^{ba}{}_{lm} = 0 \tag{2.306}$$

is covariant again.

If we reorder x and y by (2.298) we can relate (2.295) and (2.304). If χ is the inverse matrix of Γ

$$\chi^{ar}_{bs} \Gamma^{ms}_{lr} = \delta^a_l \delta^m_b \tag{2.307}$$

we find:

$$y_a x^b = \chi^{br}_{al} x^l y_r \tag{2.308}$$

and, therefore

$$y_a x^a = \chi^{ar}_{al} x^l y_r = x^l \hat{y}_l \tag{2.309}$$

Thus

$$\hat{y}_l = \chi^{ar}_{al} y_r, \qquad \chi^r_l = \chi^{ar}_{al} \tag{2.310}$$

has to transform with $\hat{S}$ as in (2.302). A direct calculation for $SL_q(2)$ shows that this is consistent with (2.272). We have

$$\widetilde{T^{-1}} = \chi^{-1} T^{-1} \chi \tag{2.311}$$

The change from (2.293) to (2.302) can be achieved by a linear change of the basis in the quantum plane.

If, for $SL_q(2)$ we start from $\Gamma = \hat{R}$ in (2.298) we find from (2.310) that in (2.305) the respective Γ is $\Gamma = \hat{R}^{-1}$. Thus by a direct computation from

$$x^r y_l = \hat{R}^{mr}_{sl} y_m x^s \tag{2.312}$$

it follows

$$\hat{y}_a x^b = (\hat{R}^{-1})^{bc}_{ad} x^a \hat{y}_c \tag{2.313}$$

2.3 Quantum derivatives

Another algebraic structure on co-modules is obtained by generalizing the Leibniz rule of derivatives:

$$\frac{\partial}{\partial x^i} x^j = \delta_i^j + x^j \frac{\partial}{\partial x^i} \tag{2.314}$$

We demand that the algebra generated by the elements of the quantum plane algebra x^i and the derivatives ∂_i, divided by proper ideals, has the Poincaré-Birkhoff-Witt property.In addition we shall show that there is an exterior differential calculus based on these quantum derivatives.

We start with an Ansatz:

$$\partial_i x^j = \delta_i^j + C_{il}^{jk} x_k \partial^l \tag{2.315}$$

It allows arbitrary coefficients C_{il}^{jk} that should take care of the noncommutativity of the space. Covariance and the PBW requirement will determine the $n^2 \times n^2$ matrix C to a large extent.

Covariance will be achieved by requiring ∂_i to transform covariantly:

$$\omega(\partial_i) = \hat{S}^l{}_i \otimes \partial_l . \tag{2.316}$$

This leads to certain conditions on C which we derive now:

$$\begin{aligned} \omega(\partial_i x^j) &= \hat{S}^l{}_i T^j{}_k \otimes \partial_l x^k \\ &= \hat{S}^l{}_i T^j{}_k \otimes \{\delta_l^k + C_{ln}^{km} x^n \partial_m\} \end{aligned} \tag{2.317}$$

$$\omega(\delta_i^j + C_{il}^{jk} x^l \partial_k) = \delta_i^j + C_{il}^{jk} T^l{}_n S^m{}_k \otimes x^n \partial_m$$

Equating these expressions yields:

$$\hat{S}^l{}_i T^j{}_k C_{ln}^{km} = C_{il}^{jk} T^l{}_n \hat{S}^m{}_k \quad \text{and} \quad \hat{S}^j{}_a T^b{}_j = \delta_a{}^b, \quad T^j{}_a \hat{S}^b{}_j = \delta_a{}^b \tag{2.318}$$

Thus $\hat{S}$ is as in (2.302). As a further consequence of (2.318):

$$C_{cd}^{ab} T^c{}_r T^d{}_s = T^a{}_c T^b{}_d C_{rs}^{cd} \tag{2.319}$$

Eqn. (2.319) can be satisfied if C is a polynomial in $\hat{R}$. To guarantee covariance such a polynomial has to be a combination of projectors:

$$C = \sum_{\text{all projectors}} c_l P_l \tag{2.320}$$

Next we demand consistency with the algebraic relations for the quantum plane:

$$P_A{}^{ab}{}_{cd} x^c x^d = 0 \tag{2.321}$$

P_A is the antisymmetrizer. We compute

$$\partial_i P_A{}^{ab}{}_{cd} x^c x^d = (P_A{}^{ab}{}_{ij} + P_A{}^{ab}{}_{cd} C^{cd}{}_{ij}) x^j + P_A{}^{ab}{}_{cd} C_{im}^{cr} C_{rn}^{ds} x^m x^n \partial_s \tag{2.322}$$

Consistency requires that this expression should be zero. There are two equations to be satisfied. The first one is:

$$P_A + P_A C = 0 \tag{2.323}$$

From (2.323) follows:

$$C = -P_A + \sum_{\text{symmetric projectors}} c_l P_l \tag{2.324}$$

The second equation that we obtain is more difficult to analyse. We realize that the last term in eqn (2.322) can be written as follows

$$(P_{A12} C_{23} C_{12})^{\alpha_1 \alpha_2 \alpha_3}{}_{\beta_1 \beta_2 \beta_3} x^{\beta_2} x^{\beta_3} \partial_{\alpha_3}$$

Here we use the notation of (2.285), (2.286). If we manage to carry P_A to the right hand side as a P_{A23}, then P_A would act on: $x^{\beta_2} x^{\beta_3}$ and give zero. This hints at the structure of a Yang-Baxter equation:

$$P_{A12} C_{23} C_{12} = C_{23} C_{12} P_{A23} \tag{2.325}$$

If C is a solution of the Yang-Baxter equation:

$$C_{12} C_{23} C_{12} = C_{23} C_{12} C_{23} \tag{2.326}$$

then it would be true that for any polynomial $\mathcal{P}(C)$ we would have:

$$\mathcal{P}(C)_{12} C_{23} C_{12} = C_{23} C_{12} \mathcal{P}(C)_{23}. \tag{2.327}$$

We conclude that if C has the structure given in eqn(2.324) and if coefficients c_l can be found such that C satisfies the Yang-Baxter equation and if all the $c_l \neq -1$ (so that we can write the projector P_A as a polynomial in C) then the derivative defined in (2.315) is consistent with the quantum plane relations (2.321). This is the case for $SL_q(n)$. There we have eqn (2.324):

$$\hat{R} = qS - \frac{1}{q} A, \quad \hat{R}^{-1} = \frac{1}{q} S - qA \tag{2.328}$$

and therefore:

$$C = q\hat{R} \quad \text{or} \quad C^{-1} = \frac{1}{q} \hat{R}^{-1}. \tag{2.329}$$

There are two solutions that have the desired properties.

We conclude that

$$\partial_i x^j = \delta_i^j + q \hat{R}^{jk}{}_{il} x^l \partial_k \tag{2.330}$$

or

$$\hat{\partial}_i x^j = \delta_i^j + \frac{1}{q} \hat{R}^{-1jk}{}_{il} x^l \hat{\partial}_k \tag{2.331}$$

are two possibilities to define a covariant derivative on a quantum plane consistent with the defining relations of $SL_q(n)$ quantum planes.

In general, if there are consistent and covariant derivatives, i.e. if we can find a matrix $C^{ab}{}_{cd}$ that satisfies the Yang-Baxter equation and that has the decomposition in projectors (2.324), then

$$C^{-1} = -P_A + \sum_{\text{sym.}} c_l^{-1} P_l \tag{2.332}$$

has the same properties and we have two possibilities to define covariant derivatives.

To complete the algebra we have to find the $\partial\partial$ relations. To this end we compute:

$$\partial_j \partial_i x^a x^b = \delta_i^a \delta_j^b + C_{ij}^{ab} + \dots$$

The unit $\delta_i^a \delta_j^b$ has a projector decomposition:

$$1 = P_A + \sum_{\text{sym.}} P_l \tag{2.333}$$

Combining this with (2.324) we find

$$\partial_j \partial_i x^a x^b = \sum_{\text{sym.}} (c_l + 1) P_l{}^{ab}{}_{ij} + \dots$$

and we conclude that it would be consistent to demand

$$\partial_j \partial_i P_A{}^{ij}{}_{rs} = 0. \tag{2.334}$$

This is a covariant condition and we can show by an argument similar to the one that followed eqn (2.322) that (2.334) is consistent with (2.330), including all the terms that have been indicated by dots.

We summarize

$$P_A{}^{ab}{}_{cd} x^c x^d = 0 \tag{2.335}$$

$$\partial_a \partial_b P_A{}^{ba}{}_{cd} = 0$$

$$\partial_a x^b = \delta_a^b + C_{ad}^{bc} x^d \partial_c$$

and

$$C = -P_A + \sum_{\text{sym.}} c_l P_l \tag{2.336}$$

$$C_{12} C_{23} C_{12} = C_{23} C_{12} C_{23}$$

It defines a covariant and consistent algebra. The matrix C can be replaced by C^{-1} and we again obtain a covariant and consistent algebra.

$$\hat{\partial}_a \hat{\partial}_b P_A{}^{ba}{}_{cd} = 0 \tag{2.337}$$

$$\hat{\partial}_a x^b = \delta_a^b + C^{-1bc}{}_{ad} x^d \hat{\partial}_c$$

We can consider the algebra generated by x, ∂ and $\hat{\partial}$. A similar argument that led to (2.334) shows that

$$\hat{\partial}_a \partial_b = C^{cd}{}_{ba} \partial_d \hat{\partial}_c \tag{2.338}$$

is a consistent and covariant condition.

Let me list the relations for the algebra based on the $\hat{R}$ matrix (2.319), i.e. for $SL_q(2)$:

$$x^i x^j = \frac{1}{q} \hat{R}^{ij}{}_{kl} x^k x^l : \tag{2.339}$$
$$x^1 x^2 = q x^2 x^1$$

$$\partial_i x^j = \delta_i^j + q \hat{R}^{jk}_{il} x^l \partial_k : \tag{2.340}$$
$$\partial_1 x^1 = 1 + q^2 x^1 \partial_1 + q\lambda x^2 \partial_2$$
$$\partial_1 x^2 = q x^2 \partial_1$$
$$\partial_2 x^1 = q x^1 \partial_2$$
$$\partial_2 x^2 = 1 + q^2 x^2 \partial_2$$

$$\partial_a \partial_b = \frac{1}{q} \partial_c \partial_d \hat{R}^{dc}{}_{ba} : \tag{2.341}$$
$$\partial_1 \partial_2 = \frac{1}{q} \partial_2 \partial_1$$

$$\hat{\partial}_i x^j = \delta_i^j + \frac{1}{q} \hat{R}^{-1jk}{}_{il} x^k \hat{\partial}_l : \tag{2.342}$$
$$\hat{\partial}_1 x^1 = 1 + \frac{1}{q^2} x^1 \hat{\partial}_1$$
$$\hat{\partial}_1 x^2 = \frac{1}{q} x^2 \hat{\partial}_1$$
$$\hat{\partial}_2 x^1 = \frac{1}{q} x^1 \hat{\partial}_2$$
$$\hat{\partial}_2 x^2 = 1 + \frac{1}{q^2} x^2 \hat{\partial}_2 - \frac{\lambda}{q} x^1 \hat{\partial}_1$$

$$\hat{\partial}_a \hat{\partial}_b = \frac{1}{q} \hat{\partial}_c \hat{\partial}_d \hat{R}^{dc}{}_{ba} : \tag{2.343}$$

$$\hat{\partial}_1 \hat{\partial}_2 = \frac{1}{q} \hat{\partial}_2 \hat{\partial}_1$$
$$\hat{\partial}_a \partial_b = q \hat{R}^{cd}{}_{ba} \partial_d \hat{\partial}_c : \tag{2.344}$$
$$\hat{\partial}_1 \partial_1 = q^2 \partial_1 \hat{\partial}_1$$

$$\hat{\partial}_1 \partial_2 = q \partial_2 \hat{\partial}_1$$
$$\hat{\partial}_2 \partial_1 = q \partial_1 \hat{\partial}_2 + \lambda q \partial_2 \hat{\partial}_1$$
$$\hat{\partial}_2 \partial_2 = q^2 \partial_2 \hat{\partial}_2$$

There is an exterior differential calculus based on these quantum derivatives. We introduce differentials and as a generalization of the anticommutativity of ordinary differentials we demand:

$$P_S{}^{ab}{}_{cd} dx^c dx^d = 0 \tag{2.345}$$

and as usual:

$$d = dx^i \partial_i \tag{2.346}$$

$$d^2 = 0, \quad d(fg) = (df)g + fdg, \quad ddx^i = -dx^i d \tag{2.347}$$

We can make use of (2.345) and (2.346) to find relations for x and dx. We start with an Ansatz:

$$x^i dx^j = O^{ij}_{ks} dx^k x^s \tag{2.348}$$

Acting with d on this equation yields:

$$\left[dx^i dx^j = -O^{ij}_{ks} dx^k dx^s \right] \tag{2.349}$$

We combine this with (2.345) and obtain

$$\left[(1+O) = \sum_{\text{sym.}} c_l P_l \right] \tag{2.350}$$

and in turn:

$$O = -P_A + \sum_{\text{sym.}} (c_l - 1) P_l \tag{2.351}$$

Next we evaluate the equation:

$$d \cdot x^i = dx^i + x^i d$$
$$dx^l \partial_l x^i = dx^i + x^i dx^l \partial_l$$
$$dx^l C^{ir}{}_{ls} x^s \partial_r = O^{ir}{}_{ls} dx^l x^s \partial_r \tag{2.352}$$

We conclude:

$$O^{ir}{}_{ls} = C^{ir}{}_{ls} \tag{2.353}$$

This is consistent with (2.351).

We could also have derived the exterior algebra:

$$P_A{}^{ij}{}_{kl} x^k x^l = 0 \tag{2.354}$$
$$P_S{}^{ij}{}_{kl} dx^k dx^l = 0$$
$$x^i dx^j = C^{ij}{}_{kl} dx^k x^l$$

from consistency arguments and the properties of the differential as given by (2.347). The derivatives and their properties then would have followed from (2.346). For our purpose, however, the existence of derivatives and their properties is more essential. That there is an exterior calculus with all the properties (2.347) - especially the unchanged Leibniz rule - is a pleasant surprise.

We can now deal with the entire algebra generated by x^i, dx^j, ∂_l and divided by the respective ideals. For this purpose the dx^l, ∂_j relations have to be specified.

We again start with an Ansatz:

$$\partial_j dx^i = D^{ik}{}_{jl} dx^l \partial_k \tag{2.355}$$

and we multiply this equation by x^r from the right:

$$\begin{aligned} \partial_j dx^i x^r &= D^{ik}{}_{jl} dx^l \partial_k x^r \\ C^{-1ir}{}_{st} \partial_j x^s dx^t &= D^{ik}{}_{jl} dx^l \partial_k x^r \end{aligned}$$

This equation splits into two parts:

$$C^{-1ir}{}_{jt} dx^t = D^{ir}{}_{jt} dx^t$$

and

$$C^{-1ir}{}_{st} C^{sa}{}_{jb} x^b \partial_a dx^t = D^{ik}{}_{jl} C^{la}{}_{kb} dx^l x^b \partial_a$$

From the first equation we conclude:

$$D^{ir}{}_{jt} = C^{-1ir}{}_{jt} \tag{2.356}$$

The second equation is true because C satisfies the Yang-Baxter equation. For the x, ∂ and dx algebra to be consistent it is now necessary that C satisfies the Yang-Baxter equation. The result is:

$$\partial_j dx^i = C^{-1ik}{}_{jl} dx^l \partial_k \tag{2.357}$$

It should be noted that whereas x and d have simple commutation properties ($dx^a = (dx^a) + x^a d$), this is not true for d and ∂_i.

We compute:

$$\partial_i d = \partial_i dx^l \partial_l = C^{-1la}{}_{ib} dx^b \partial_a \partial_l \tag{2.358}$$

The antisymmetric projector of C does not contribute to (2.358). The result, however, depends on C via the symmetric projectors. For $SL_q(n)$ it can be evaluated explicitely:

$$C^{-1} = q^{-1}\hat{R}^{-1} = -A + q^{-2}S = q^{-2}1 - (1+q^{-2})A \tag{2.359}$$

and we obtain for $SL_q(n)$:

$$\partial_i d = q^{-2} d\partial_i \tag{2.360}$$

To complete the list of explicit relations for $SL_2(2)$ we finish this chapter by giving the xdx relations:

$$\begin{aligned} dx^1 x^1 &= q^2 x^1 dx^1 \\ dx^1 x^2 &= q x^2 dx^1 + (q^2-1) x^1 dx^2 \\ dx^2 x^1 &= q x^1 dx^2 \\ dx^2 x^2 &= q^2 x^2 dx^2 \end{aligned} \tag{2.361}$$

2.4 Conjugation

Let us finally enrich the algebraic structure by adding a conjugation. For a physical interpretation this will be essential because such an interpretation will rest on Hilbert space representations of the algebra and observables will have to be identified with essentially self-adjoint operators in Hilbert space. The conjugation defined here is a purely algebraic operation to start with but later has to be identified with mapping to the adjoint operator in Hilbert space. We introduce conjugate variables as new independent elements of the algebra:

$$\overline{x^i} \equiv \overline{x}_i \quad , \quad \overline{x^i x^j} = \overline{x}_j \overline{x}_i \tag{2.362}$$

The lower index of $\overline{x}_i$ does not mean that they transform covariantly as defined by (2.293) or (2.302). The transformation law follows from (2.274):

$$\omega(\overline{x}_i) = \overline{T^i{}_k} \otimes \overline{x}_k = \overline{T}^k{}_i \otimes \overline{x}_k \tag{2.363}$$

We have defined:

$$\overline{T^i{}_k} \equiv \overline{T}^k{}_i \tag{2.364}$$

From the $\hat{R}TT$ relations (2.258) follows by conjugating:

$$\hat{R}^{+sr}{}_{lk} \overline{T}^l{}_j \overline{T}^k{}_i = \overline{T}^s{}_l \overline{T}^r{}_k \hat{R}^{+lk}{}_{ji} \tag{2.365}$$

$\hat{R}^{+sr}$ is defined as follows:

$$\hat{R}^{+sr}{}_{lk} = \overline{\hat{R}^{kl}{}_{rs}} \tag{2.366}$$

$\hat{R}^+$ satisfies the Quantum Yang-Baxter equation if $\hat{R}$ does. This can be verified by direct calculation.

For the $\hat{R}$ matrix (2.259), (2.267) and real $q(\overline{q} = q)$ we find:

$$\hat{R}^{+sr}{}_{lk} = \hat{R}^{rs}{}_{lk} \tag{2.367}$$

$\hat{R}^+$ in this case is obtained from $\hat{R}$ by a symmilarity transformation and has the same eigenvalues.

The $\overline{x}\,\overline{x}$ relations are obtained by conjugating (2.279):

$$\overline{x}_j \overline{x}_i = \frac{1}{q} \hat{R}^{kl}{}_{ij} \overline{x}_l \overline{x}_k \tag{2.368}$$

This is exactly of the same type as the relation (2.296) with $P = A$ of (2.265). A consistent $x\overline{x}$ relation is therefore obtained from (2.298) by $\Gamma = q\hat{R}^{-1}$. We choose this possibility to have $\overline{x}_i x^i$ central:

$$x^i \overline{x}_j = q \hat{R}^{-1li}{}_{kj} \overline{x}_l x^k \tag{2.369}$$

This is consistent with conjugation as well.

We demand covariance of this relation with respect to (2.274) and (2.363).

$$\begin{aligned}\omega(x^i\overline{x}_j) &= T^i{}_l\overline{T}^k{}_j x^l\overline{x}_k \\ &= qT^i{}_l\overline{T}^k{}_j(\hat{R}^{-1})^{rl}{}_{sk}\overline{x}_r x^s \\ &= q(\hat{R}^{-1})^{li}{}_{kj}\overline{T}^r{}_l T^k{}_s\overline{x}_r x^s\end{aligned} \tag{2.370}$$

This leads to:

$$T^i{}_l\overline{T}^k{}_j(\hat{R}^{-1})^{rl}{}_{sk} = (\hat{R}^{-1})^{li}{}_{kj}\overline{T}^r{}_l T^k{}_s \tag{2.371}$$

These are the $\hat{R}T\overline{T}$ relations. All this could be put together to one big quantum plane of $2n$ elements $(x^i, \overline{x}_j)$ and one big $(2n)^2 \times (2n)^2 \hat{R}$ matrix satisfying the corresponding Yang-Baxter equation. What we arrive at is the complex quantum plane generated by $x^i, \overline{x}_j$ and divided by the ideals generated by the relations (2.279), (2.368) and (2.369). The quantum group represented by T and $\overline{T}$ is $GL_q(2,C)$ or, for $det_q T = 1$, $SL_q(2,C)$.

A generalization to several quantum planes is possible starting from (2.281).

$$\begin{aligned}x^i y^j &= \frac{\kappa}{q}\hat{R}^{ij}{}_{kl}y^k x^l \\ \overline{y}_j\overline{x}_i &= \frac{\kappa}{q}\hat{R}^{kl}{}_{ij}\overline{x}_l\overline{y}_k \\ x^i\overline{y}_j &= \kappa q\hat{R}^{-1}{}^{li}_{kj}\bar{y}_l x^k \\ y^j\bar{x}_i &= \kappa q\hat{R}^{-1}{}^{kj}_{li}\bar{x}_k y^l\end{aligned} \tag{2.372}$$

For $n = 2$ the explicit $x\bar{y}$ relations are (for $\kappa = 1$):

$$\begin{aligned}x^1\bar{y}_1 &= \bar{y}_1 x^1 - q\lambda\bar{y}_2 x^1 \\ x^1\bar{y}_2 &= q\bar{y}_2 x^1 \\ x^2\bar{y}_1 &= q\bar{y}_1 x^2 \\ x^2\bar{y}_2 &= \bar{y}_2 x^2\end{aligned} \tag{2.373}$$

The $\bar{y}\bar{x}$ relations follow from (2.282) by conjugating.

For the entries of $T, \bar{T}$, as they are defined by (2.255) and its conjugate the $T, \bar{T}$ relations are:

$$\begin{aligned}a\bar{a} &= \bar{a}a - q\lambda\bar{c}c \\ a\bar{b} &= \frac{1}{q}\bar{b}a - \lambda\bar{d}c \\ a\bar{c} &= q\bar{c}a \\ a\bar{d} &= \bar{d}a \\ b\bar{a} &= \frac{1}{q}\bar{a}b - \lambda\bar{c}d \\ b\bar{b} &= \bar{b}b + q\lambda(\bar{a}a - \bar{d}d - q\lambda\bar{c}c)\end{aligned} \tag{2.374}$$

$$
\begin{aligned}
b\bar{c} &= \bar{c}b \\
b\bar{d} &= q\bar{d}b + \lambda q^2 \bar{c}a \\
c\bar{a} &= q\bar{a}c \\
c\bar{b} &= \bar{b}c \\
c\bar{c} &= \bar{c}c \\
c\bar{d} &= \frac{1}{q}\bar{d}c \\
\\
d\bar{a} &= \bar{a}d \\
d\bar{b} &= q\bar{b}d + \lambda q^2 \bar{c}a \\
d\bar{c} &= \frac{1}{q}\bar{c}d \\
d\bar{d} &= \bar{d}d + \lambda q\bar{c}c
\end{aligned}
$$

A comparison of (2.363) with (2.293) shows that it is possible to identify $\overline{T}$ with T^{-1}. This then defines the quantum group $U_q(n)$ or for $det_q T = 1$ the quantum group $SU_q(n)$.

For $n = 2$,

$$
T = \begin{pmatrix} a & b \\ c & d \end{pmatrix} \quad , \quad T^{-1} = \begin{pmatrix} d & \frac{-1}{q}b \\ -qc & a \end{pmatrix} \tag{2.375}
$$

we find

$$
\begin{aligned}
\overline{a} &= d, \overline{b} = -qc \\
\overline{d} &= a, \overline{c} = -\frac{1}{q}b
\end{aligned} \tag{2.376}
$$

It can be verfied directly that (2.376) is consistent with (2.374).

To extend conjugation to an algebra with derivatives we start with differentials. From (2.345) we have

$$
x^c dx^d = q\hat{R}^{cd}{}_{ab} dx^a x^b \tag{2.377}
$$

and therefore

$$
dx^c dx^d = -q\hat{R}^{cd}{}_{ab} dx^a dx^b \tag{2.378}
$$

Conjugation leads to

$$
\overline{dx}_d \overline{x}_c = q\hat{R}^{ab}{}_{cd} \overline{x}_b \overline{dx}_a \tag{2.379}
$$

and

$$
\overline{dx}_d \overline{x}_d = -q\hat{R}^{ab}{}_{cd} \overline{dx}_b \overline{dx}_a \tag{2.380}
$$

From (2.372) we conclude:

$$
x^i \overline{dx}_j = q\hat{R}^{-1}{}^{li}_{kj} \overline{dx}_l x^k \tag{2.381}
$$

and

$$
dx^i \overline{dx}_j = -q\hat{R}^{-1}{}^{li}_{kj} \overline{dx}_l dx^k \tag{2.382}
$$

2.5 q-Deformed Heisenberg algebra

The canonical commutation relations are at the basis of a quantum mechanical system:

$$[\hat{x}, \hat{p}] = \hat{x}\hat{p} - \hat{p}\hat{x} = i \tag{2.383}$$

The elements of this algebra are supposed to be selfadjoint

$$\overline{\hat{x}} = \hat{x} \quad , \quad \overline{\hat{p}} = \hat{p} \tag{2.384}$$

A physical system is defined through a representation of this algebra in a Hilbert space where selfadjoint elements of the algebra have to be represented by (essentially) selfadjoint linear operators.

In quantum mechanics the elements of (2.383) are represented in the Hilbert space of square-integrable functions by:

$$\hat{x} = x \quad , \quad \hat{p} = -i\frac{\partial}{\partial x} \tag{2.385}$$

Starting from the algebra (2.383), the spectrum of the linear operator $\hat{x}$ can be interpreted as the manifold on which the physical system lives - i.e. the configuration space.

In quantum mechanics it is $\mathbb{R}_{\not{k}}$ ($\curvearrowleft \in \mathbb{R}_{\not{k}}$). The element $\hat{p}$ is a differential operator on this manifold.

We shall change the algebra (2.383) in accord with quantum group considerations. It is natural to assume that $\hat{x}$ is an element of a quantum plane and to relate $\hat{p}$ to a derivative in such a plane. The simplest example that we can consider is suggested by the last equation of (2.341). With an obvious change in notation we study the algebra

$$\partial x = 1 + qx\partial \tag{2.386}$$

More precisely, we study the free algebra generated by the elements x and ∂ and divided by the ideal generated by (2.386).

If we assume x to be selfadjoint

$$\overline{x} = x \tag{2.387}$$

we see that this cannot be the case for $i\partial$ because we find from (2.386) that:

$$\overline{\partial} x = -\frac{1}{q} + \frac{1}{q}x\overline{\partial} \tag{2.388}$$

In general $\overline{\partial}$ will be related to $\hat{\partial}$ rather than to ∂. (See the second equation of (2.342)). Thus we could study the algebra generated by x, ∂ and $\overline{\partial}$ and divide by (2.386), (2.388) and an ideal generated by $\partial\overline{\partial}$ relations. These can be found by a similar argument that led to the $\partial\hat{\partial}$ relations (2.338). We compute from (2.386) and (2.388) $\partial\overline{\partial}x$ and $\overline{\partial}\partial x$ and find that

$$\overline{\partial}\partial = q\partial\overline{\partial} \tag{2.389}$$

is consistent with these calculations. If we now try to define an operator $\hat{p}$ by $\hat{p} = -\frac{i}{2}(\partial - \overline{\partial})$ we find that the $x, \hat{p}$ relations do not close. The real part of ∂ has to be introduced as well. Thus our Heisenberg algebra would have one space and two momentum operators - a system that will hardly find a physical interpretation.

It turns out that $\overline{\partial}$ can be related to ∂ and x in a nonlinear way. This relation involves the scaling operator Λ:

$$\begin{aligned} \Lambda &\equiv q^{\frac{1}{2}}\left(1+(q-1)x\partial\right) \\ \Lambda x &= qx\Lambda \\ \Lambda\partial &= q^{-1}\partial\Lambda \end{aligned} \tag{2.390}$$

The scaling property follows from (2.386).

We now define

$$\tilde{\partial} = -q^{-\frac{1}{2}}\Lambda^{-1}\partial \tag{2.391}$$

Λ^{-1} is defined by an expansion in $(q-1)$. We find

$$\begin{aligned} \tilde{\partial}x &= -\tfrac{1}{q} + \tfrac{1}{q}x\tilde{\partial} \\ \tilde{\partial}\partial &= q\partial\tilde{\partial} \end{aligned} \tag{2.392}$$

Comparing this with (2.388) and (2.389) it follows from (2.391) and (2.392) that conjugation in the x, ∂ algebra can be defined by

$$\overline{x} = x \quad , \quad \overline{\partial} = -q^{-\frac{1}{2}}\Lambda^{-1}\partial \tag{2.393}$$

Conjugating Λ and using (2.393) shows that

$$\overline{\Lambda} = \Lambda^{-1} \tag{2.394}$$

Λ is a unitary element of the algebra, this justifies the factor $q^{\frac{1}{2}}$ in the definition of Λ.

The existence of a scaling operator Λ and the definition of the conjugation (2.393) seems to be very specific for the x, ∂ algebra (2.386). It is however generic in the sense that a scaling operator and a definition of conjugation based on it can be found for all the quantum planes defined by $SO_q(n)$ and $SO_q(1,n)$.

The definition of the q-deformed Heisenberg algebra will now be based on the definition of the momentum:

$$p = -\frac{i}{2}(\partial - \overline{\partial}) \tag{2.395}$$

It is selfadjoint. From the x, ∂ algebra and the definition of $\overline{\partial}$ follows

$$\begin{aligned} &q^{\frac{1}{2}}xp - q^{-\frac{1}{2}}px = i\Lambda^{-1} \\ &\Lambda x = qx\Lambda \qquad \Lambda p = q^{-1}p\Lambda \end{aligned} \tag{2.396}$$

and

$$\overline{p} = p \quad , \quad \overline{x} = x \quad , \quad \overline{\Lambda} = \Lambda^{-1} \tag{2.397}$$

These algebraic relations can be verified in the x, ∂ representation where the ordered x, ∂ monomials form a basis. We shall take (2.396) and (2.397) as the defining relations for the q-deformed Heisenbergalgebra without making further reference to its x, ∂ representation.

2.6 The q-deformed Lie algebra $\mathbf{sl_q(2)}$

The algebra $sl_q(2)$ is the dual object to the quantum group $SL_q(2)$. In this chapter we first define this algebra without referring to the quantum group and then study its representations.

The algebra is defined as follows:

$$\begin{aligned} \frac{1}{q}T^+T^- - qT^-T^+ &= T^3 \\ q^2T^3T^+ - \frac{1}{q^2}T^+T^3 &= (q+\frac{1}{q})T^+ \\ \frac{1}{q^2}T^3T^- - q^2T^-T^3 &= -(q+\frac{1}{q})T^- \end{aligned} \tag{2.398}$$

It is often convenient to introduce the "group like" element τ:

$$\begin{aligned} \tau &= 1 - \lambda T^3, \qquad \lambda = q - \frac{1}{q} \\ T^3 &= \lambda^{-1}(1-\tau) \end{aligned} \tag{2.399}$$

From (2.398) follows:

$$\begin{aligned} \tau T^+ &= \frac{1}{q^4}T^+\tau \\ \tau T^- &= q^4 T^- \\ \frac{1}{q}T^+T^- - qT^-T^+ &= \frac{1}{\lambda}(\tau - 1) \end{aligned} \tag{2.400}$$

The algebra (2.398) has a Casimir operator

$$\mathbf{T}^2 = q^2(T^-T^+ + \frac{1}{\lambda^2})\tau^{-1/2} + \frac{1}{\lambda^2}(\tau^{1/2} - 1 - q^2) \tag{2.401}$$

$\mathbf{T}^2$ commutes with T^+, T^- and T^3. The constant term has been added to give the usual Casimir operator in the limit $q \to 1$.

The algebra (2.398) allows for a coproduct:

$$\begin{aligned} \Delta(T^3) &= T^3 \otimes 1 + \tau \otimes T^3 \\ \Delta(T^{\pi_D}) &= T^{\pi_D} \otimes 1 + \tau^{1/2} \otimes T^{\pi_D} \end{aligned} \tag{2.402}$$

This follows from its construction as the dual object of the Hopf algebra $SL_q(2)$ or it can be verified directly that:

$$\begin{aligned}
&\frac{1}{q}\Delta(T^+)\Delta(T^-) - q\Delta(T^-)\Delta(T^+) = \Delta(T^3) \\
&q^2\Delta(T^3)\Delta(T^+) - \frac{1}{q^2}\Delta(T^+)\Delta(T^3) = (q+\frac{1}{q})\Delta(T^+) \\
&\frac{1}{q^2}\Delta(T^3)\Delta(T^-) - q^2\Delta(T^-)\Delta(T^3) = -(q+\frac{1}{q})\Delta(T^-)
\end{aligned} \tag{2.403}$$

To complete the definition of $sl_q(2)$ as a Hopf algebra we add the definition of the co-unit:

$$\varepsilon(T) = 0 \tag{2.404}$$

and the antipode:

$$\begin{aligned}
S(T^{\pi_D}) &= -T^{\pi_D}\tau^{-1} \\
S(T^3) &= -T^3\tau^{-1}
\end{aligned} \tag{2.405}$$

For the Hopf algebra $su_q(2)$ conjugation properties have to be assigned. It is easy to see that

$$\overline{T}^3 = T^3, \quad \overline{T}^+ = \frac{1}{q^2}T^-, \quad \overline{T}^- = q^2T^+ \tag{2.406}$$

is compatible with all previous relations.

With the very same methods with which the representations of angular momentum are constructed in quantum mechanics we can construct representations of $su_q(2)$. The representations are characterized by the eigenvalue of the Casimir operator and the states in a representation by the eigenvalues of T^3.

We first define the q number

$$[n] = \frac{q^n - q^{-n}}{q - q^{-1}} \tag{2.407}$$

and list the non-vanishing matrix elements:

$$\begin{aligned}
\mathbf{T}^2|j,m> &= q[j][j+1]|j,m> \\
T^3|j,m> &= q^{-2m}[2m]|j,m> \\
T^+|j,m> &= q^{-m-\frac{3}{2}}\sqrt{[j+m+1][j-m]}|j,m+1> \\
T^-|j,m> &= q^{-m+\frac{3}{2}}\sqrt{[j+m][j-m+1]}|j,m-1> \\
\tau|j,m> &= q^{-4m}|j,m>
\end{aligned} \tag{2.408}$$

The representation characterized by j is $2j+1$ dimensional, and it is easy to see that in the limit $q \to 1$ it becomes the representation of the usual angular momentum j^i:

$$q \to 1 : T^3 \to 2j_3, \quad T^{\pi_D} \to j^{\pi_D} \tag{2.409}$$

such that

$$\begin{aligned} j^+j^- - j^-j^+ &= 2j_3 \\ j_3j^+ - j^+j_3 &= j^+ \\ j^-j_3 - j_3j^- &= j^- \end{aligned} \tag{2.410}$$

It is remarkable that the non-vanishing matrix elements are exactly the same as for the representations of the undeformed angular momentum. This suggests that the T matrices are in the enveloping algebra of the j matrices. For the representations of the algebra (2.410) we know that

$$\begin{aligned} j_3|j,m> &= m|j,m> \\ j^+|j,m> &= \sqrt{(j+m+1)(j-m)}|j,m+1> \\ j^-|j,m> &= \sqrt{(j+m)(j-m+1)}|j,m-1> \end{aligned} \tag{2.411}$$

or

$$\begin{aligned} |j,m+1>&= \frac{1}{\sqrt{j(j+1)-m^2-m}}j^+|j,m>= \frac{1}{\sqrt{\mathbf{j}^2-j_3^2+j_3}}j^+|j,m> \\ |j,m-1>&= \frac{1}{\sqrt{j(j+1)-m^2+m}}j^-|j,m>= \frac{1}{\sqrt{\mathbf{j}^2-j_3^2-j_3}}j^-|j,m> \end{aligned} \tag{2.412}$$

In the last step we replaced the numbers s, m by operators.

Comparing (2.411) with (2.408) we find that

$$\begin{aligned} T^3 &= q^{-2j_3}[2j_3] = \frac{1}{\lambda}(1-q^{-4j_3}) \\ T^+ &= \frac{1}{\lambda}q^{-(j_3+\frac{1}{2})}\sqrt{\frac{q^{\sqrt{1+4\mathbf{j}^2}}+q^{-\sqrt{1+4\mathbf{j}^2}}-q^{-2j_3+1}-q^{2j_3-1}}{\mathbf{j}^2-j_3^2+j_3}}\,j^+ \\ T^- &= \frac{1}{\lambda}q^{-(j_3-\frac{1}{2})}\sqrt{\frac{q^{\sqrt{1+4\mathbf{j}^2}}+q^{-\sqrt{1+4\mathbf{j}^2}}-q^{-2j_3-1}-q^{2j_3+1}}{\mathbf{j}^2-j_3^2-j_3}}\,j^- \\ \tau &= q^{-4j_3} \end{aligned} \tag{2.413}$$

have the same matrix elements (2.408). That they satisfy the same algebra (2.398) can be verified by a direct calculation. The eqns (2.410) should be used in the form

$$(j_3-1)j^+ = j^+j_3 \qquad \text{or} \qquad j_3j^+ = j^+(j_3+1)$$

and

$$j^+j^- = \mathbf{j}^2 - j_3^2 + j_3, \quad j^-j^+ = \mathbf{j}^2 - j_3^2 - j_3$$

We calculate:

$$\begin{aligned}
T^+T^- &= \frac{1}{\lambda^2} q^{-2j_3+1} \frac{q^{\sqrt{1+4\mathbf{j}^2}} + q^{-\sqrt{1+4\mathbf{j}^2}} - q^{-2j_3+1} - q^{2j_3-1}}{\mathbf{j}^2 - j_3^2 + j_3} j^+ j^- \\
T^-T^+ &= \frac{1}{\lambda^2} q^{-2j_3-1} \frac{q^{\sqrt{1+4\mathbf{j}^2}} + q^{-\sqrt{1+4\mathbf{j}^2}} - q^{-2j_3+1} - q^{2j_3-1}}{\mathbf{j}^2 - j_3^2 - j_3} j^- j^+ \\
\frac{1}{q} T^+T^- - qT^-T^+ &= \frac{1}{\lambda^2} q^{-2j_3} \left(q^{\sqrt{1+4\mathbf{j}^2}} + q^{-\sqrt{1+4\mathbf{j}^2}} - q^{-2j_3+1} - q^{2j_3-1} \right) \\
&= \frac{1}{\lambda^2} q^{-2j_3} (q^{2j_3} - q^{-2j_3})(q - \frac{1}{q}) = \frac{1}{\lambda}(1 - q^{-4j_3}) \\
&= T^3
\end{aligned} \tag{2.414}$$

This should serve as an example.

For $j = 1/2$ and $j = 1$ we show the representations explicitly:

$$\begin{aligned}
T^+|\frac{1}{2}, -\frac{1}{2}> &= q^{-1}|\frac{1}{2}, \frac{1}{2}> \\
T^+|\frac{1}{2}, \frac{1}{2}> &= 0 \\
T^-|\frac{1}{2}, -\frac{1}{2}> &= 0 \\
T^-|\frac{1}{2}, \frac{1}{2}> &= q|\frac{1}{2}, -\frac{1}{2}> \\
T^3|\frac{1}{2}, -\frac{1}{2}> &= -q|\frac{1}{2}, -\frac{1}{2}> \\
T^3|\frac{1}{2}, \frac{1}{2}> &= q^{-1}|\frac{1}{2}, \frac{1}{2}>
\end{aligned} \tag{2.415}$$

$$\begin{aligned}
T^+|1, -1> &= \frac{1}{q}\sqrt{1+q^2}|1, 0> \\
T^+|1, 0> &= \frac{1}{q^2}\sqrt{1+q^2}|1, 1> \\
T^+|1, 1> &= 0 \\
T^-|1, -1> &= 0 \\
T^-|1, 0> &= q\sqrt{1+q^2}|1, -1> \\
T^-|1, 1> &= \sqrt{1+q^2}|1, 0> \\
T^3|1, -1> &= -q(1+q^2)|1, -1> \\
T^3|1, 0> &= 0 \\
T^3|1, 1> &= \frac{1}{q}(1 + \frac{1}{q^2})|1, 1>
\end{aligned} \tag{2.416}$$

We can identify the vectors of a representation with elements of a quantum plane. The elements of a quantum plane can be multiplied. This product we identify with the tensor product of the representations to obtain its transformation properties. From the comultiplication (2.402) we find the rule how the products transform:

$$\begin{aligned} T^3 x\cdots &= (\Delta(T^3)x\cdots) = (T^3 x)\cdots + (\tau x)T^3\cdots \\ T^{\pi_D} x\cdots &= (\Delta(T^{\pi_D})x\cdots) = (T^{\pi_D}x)\cdots + (\tau^{1/2}x)T^{\pi_D}\cdots \end{aligned} \tag{2.417}$$

the dots indicate the additional factors more explicitely for $j = 1/2$. We identify

$$|\frac{1}{2},-\frac{1}{2}>= x^1, \quad |\frac{1}{2},\frac{1}{2}>= x^2 \tag{2.418}$$

and obtain from (2.415)

$$\begin{aligned} T^3x^1 &= q^2x^1T^3 - qx^1 \\ T^3x^2 &= q^{-2}x^2T^3 + q^{-1}x^2 \\ T^+x^1 &= qx^1T^+ + q^{-1}x^2 \\ T^+x^2 &= q^{-1}x^2T^+ \\ T^-x^1 &= qx^1T^- \\ T^-x^2 &= q^{-1}x^2T^- + qx^1 \end{aligned} \tag{2.419}$$

Now we ask what are the algebraic relations on the variables x^i that are compatible with 2.334.

$$\begin{aligned} T^+x^1x^2 &= (qx^1T^+ + \tfrac{1}{q}x^2)x^2 = x^1x^2T^+ + \tfrac{1}{q}x^2x^2 \\ T^+x^2x^1 &= \tfrac{1}{q}x^2T^+x^1 = x^2x^1T^+ + \tfrac{1}{q^2}x^2x^2 \\ T^+(x^1x^2 - qx^2x^1) &= (x^1x^2 - qx^2x^1)T^+ \end{aligned} \tag{2.420}$$

The relation

$$x^1x^2 = qx^2x^1 \tag{2.421}$$

is compatible with 2.334. This can be verified for all the Ts.

We have discovered the relation 2.280 for the two-dimensional quantum plane. This is not surprising, as we know that this plane is covariant under $SU_q(2)$ We could have started from 2.280 and asked for all linear transformations of the type (2.417) that are compatible with 2.280. This way we would have found (2.417) and from there the "multiplication" rule (2.398) and the comultiplication (2.402). It is all linked via the $\hat{R}$ matrix.

From 2.362 and (2.406) we can deduce the action of the Ts on the conjugation plane:

$$\begin{aligned} T^3\overline{x}_1 &= \frac{1}{q^2}\overline{x}_1T^3 + \frac{1}{q}\overline{x}_1 \\ T^3\overline{x}_2 &= q^2\overline{x}_2T^3 - q\overline{x}_2 \\ T^+\overline{x}_1 &= \frac{1}{q}\overline{x}_1T^+ \end{aligned} \tag{2.422}$$

$$
\begin{aligned}
T^+\overline{x}_2 &= q\overline{x}_2 T^+ - \overline{x}_1 \\
T^-\overline{x}_1 &= \frac{1}{q}\overline{x}_1 T^- - \overline{x}_2 \\
T^-\overline{x}_2 &= q\overline{x}_2 T^-
\end{aligned}
$$

2.7 q-deformed Euclidean space in three dimensions

The $\hat{R}$-matrix for the fundamental representation of a quantum group contains all the information on the $\hat{R}$-matrices for the other representations. One way to extract this information is to construct quantum space variables as products of the quantum space variables of the fundamental representation. Let us demonstrate this for $SU_q(2)$.

We start from the relations (2.284) and the $\hat{R}$-matrix (2.259). We take four copies of the quantum planes x, y, u and v such that

$$
\begin{aligned}
xu &= \frac{1}{q}\hat{R}ux, \quad & xy &= \frac{1}{q}\hat{R}vx \\
yu &= \frac{1}{q}\hat{R}uy, \quad & yv &= \frac{1}{q}\hat{R}vy
\end{aligned} \tag{2.423}
$$

Then we consider "bispinors":

$$
xy \sim X, \quad uv \sim \widetilde{X} \tag{2.424}
$$

We can use (2.419) to single out those components that transform like a three-vector corresponding to (2.416):

$$
X^- = x^1 y^1, \quad X^0 = \frac{1}{\sqrt{1+q^2}}(x^1 y^2 + q x^2 y^1), \quad X^+ = x^2 y^2 \tag{2.425}
$$

and the same for $\tilde{X}$ with xy replaced by uv.

The relations (2.424) are sufficient to compute the 9×9 $\hat{\mathcal{R}}$-matrix:

$$
X\widetilde{X} = \hat{\mathcal{R}}\widetilde{X}X \tag{2.426}
$$

The relations are consistent with (2.406) and the reality condition on the quantum space:

$$
\overline{X^0} = X^0, \quad \overline{X^+} = -qX^- \tag{2.427}
$$

This makes X^0, X^+, X^- to be quantum space variable of $SO_q(3)$.

The $\hat{\mathcal{R}}$-matrix will satisfy the Yang-Baxter equation if $\hat{R}$ satisfies it.

The $\hat{\mathcal{R}}$-matrix, calculated that way, has three different eigenvalues $1(5), -q^{-4}(3)$ and $q^{-6}(1)$. The number in bracket is the multiplicity of the respective eigenspaces and corresponds to the $j=2$, $j=1$ and $j=0$ representations.

As a consequence, $\hat{\mathcal{R}}$ satisfies the characteristic equation:

$$(\hat{\mathcal{R}} - 1)(\hat{\mathcal{R}} + \frac{1}{q^4})(\hat{\mathcal{R}} - \frac{1}{q^6}) = 0 \tag{2.428}$$

which in turn gives the projectors on the irreducible subspaces:

$$\begin{aligned} P_1 &= \frac{q^{12}}{(1+q^2)(1-q^6)}(\hat{\mathcal{R}} - 1)(\hat{\mathcal{R}} + \frac{1}{q^6}) \\ P_3 &= \frac{q^{10}}{(1+q^2)(1+q^4)}(\hat{\mathcal{R}} - 1)(\hat{\mathcal{R}} - \frac{1}{q^4}) \\ P_5 &= \frac{q^{10}}{(q^4+1)(q^6-1)}(\hat{\mathcal{R}} + \frac{1}{q^4})(\hat{\mathcal{R}} - \frac{1}{q^6}) \end{aligned} \tag{2.429}$$

The projectors contain the Clebsch-Gordon coefficients for the reduction of the $3 \otimes 3$ representation of $SO_q(3)$. Thus the Clebsch-Gordon coefficients can be derived from the $\hat{\mathcal{R}}$-matrix for $q \neq 1$. For $q = 1$ the eigenvalues degenerate and it is not possible to obtain all the projectors from $\hat{\mathcal{R}}$. But we could compute the projectors first and then put $q = 1$.

The projector P_1 projects on a one-dimensional subspace, we can define a metric from this projection:

$$\widetilde{X} \circ X = \widetilde{X}^0 X^0 - q\widetilde{X}^+ X^- - \frac{1}{q}\widetilde{X}^- X^+ = \widetilde{X}^A X^B \eta_{BA} \tag{2.430}$$

$$\eta_{00} = 1\,, \quad \eta_{+-} = -\frac{1}{q}\,, \quad \eta_{-+} = -q$$

With the metric η_{AB} we can lower vector indices:

$$X_B = X^A \eta_{AB}, \tag{2.431}$$

We raise them with η^{AB}

$$X^B = \eta^{BC} X_C \tag{2.432}$$

This defines η^{BC} by the equation

$$\eta^{BA}\eta_{BC} = \delta^A_C, \quad \eta^{AB}\eta_{CB} = \delta^A_C \tag{2.433}$$

The projector P_1 in terms of the metric is

$$(P_1)^{AB}_{CD} = \frac{q^2}{1+q^2+q^4}\eta^{AB}\eta_{DC} \tag{2.434}$$

In a similar way we can use the Clebsch-Gordon coefficients that combine two vectors to a vector for the definition of a q-deformed ε-tensor:

$$Z^A = \widetilde{X}^C X^B \varepsilon_{BC}{}^A \tag{2.435}$$

We find for the non-vanishing components:

$$\begin{aligned} &\varepsilon_{+-}{}^0 = q, \quad \varepsilon_{-+}{}^0 = -q, \quad \varepsilon_{00}{}^0 = 1 - q^2 \\ &\varepsilon_{+0}{}^+ = 1, \quad \varepsilon_{0+}{}^+ = -q^2 \\ &\varepsilon_{-0}{}^- = -q^2, \quad \varepsilon_{0-}{}^- = 1 \end{aligned} \tag{2.436}$$

The indices of the ε tensor can be raised and lowered with the metric:

$$\varepsilon_{ABC} = \varepsilon_{AB}{}^D \eta_{DC} \tag{2.437}$$

The projector P_3 in terms of the ε tensor is:

$$P_3{}^{AB}_{CD} = \frac{1}{1+q^4} \varepsilon^{FAB} \varepsilon_{FDC} \tag{2.438}$$

The projector P_5 can be obtained from the relation

$$P_1 + P_3 + P_5 = 1 \tag{2.439}$$

The $\hat{\mathcal{R}}$-matrix is a sum of projectors as well.

$$\hat{\mathcal{R}} = P_5 - \frac{1}{q^4} P_3 + \frac{1}{q^6} P_1 \tag{2.440}$$

We can ask for the most general linear combination of the projectors that solves the Yang-Baxter equation and obtain $\hat{\mathcal{R}}$ and $\hat{\mathcal{R}}^{-1}$.

A natural way to define the 3-dimensional Euclidean quantum space is:

$$\begin{aligned} \varepsilon_{FDC} X^C X^D &= 0 \\ X^0 X^+ &= q^2 X^+ X^0 \\ X^- X^0 &= q^2 X^0 X^- \\ X^- X^+ &= X^+ X^- + (q - \frac{1}{q}) X^0 X^0 \end{aligned} \tag{2.441}$$

Derivatives should be defined in the third part of this lecture:

$$\begin{aligned} \partial_B X^A &= \delta^A_B + q^4 \hat{\mathcal{R}}^{AC}_{BD} X^D \partial_C \\ \varepsilon^{FBA} \partial_A \partial_B &= 0 \end{aligned} \tag{2.442}$$

To summarize we have constructed an algebra freely generated by elements of the quantum space x and derivative ∂, this algebra is divided by the ideal generated by the relations (2.441) and (2.442). We have constructed it in such a way that the Poincaré-Birkhoff-Witt property holds and that the algebra allows the action of $SO_q(3)$.

For the definition of a conjugation that is consistent with (2.442) we first extend the algebra by enlarging it by conjugate derivatives. We use the notation:

$$\overline{\partial^A} = \overline{\partial}_A \tag{2.443}$$

and obtain from (2.442) and (2.402)

$$\overline{\partial}_C X_D = -\frac{1}{q^6}\eta_{CD} + \hat{\mathcal{R}}^{BA}_{DC} X_A \partial_B \tag{2.444}$$
$$\varepsilon_{FAB}\overline{\partial}^B\overline{\partial}^A = 0$$

The ∂, $\overline{\partial}$ relations are not yet specified. We can apply $\partial\overline{\partial}$ and $\overline{\partial}\partial$ to X as we did in (2.389) and find the consistent relation:

$$\varepsilon_{FAB}(\partial^B\overline{\partial}^A + \overline{\partial}^B\partial^A) = 0 \tag{2.445}$$

The X, ∂, $\overline{\partial}$ algebra divided by (2.441), (2.442) and (2.444) still satisfies Poincaré-Birkhoff-Witt, is $SO_q(3)$-covariant and consistent with (2.402).

Miraculously it turns out that $\overline{\partial}$ can again be related non-linearly to ∂ as it was the case in (2.393). This can be achieved by the scaling operator Λ:

$$\Lambda = q^6\{1 + (q^4 - 1)X \circ \partial + q^2(q^2 - 1)^2(X \circ X)(\partial \circ \partial)\} \tag{2.446}$$
$$\Lambda X^A = q^4 X^A \Lambda \tag{2.447}$$
$$\Lambda \partial^A = q^{-4}\partial^A \Lambda \tag{2.448}$$

If we now define an operator

$$\overline{\partial}^A = -\Lambda^{-1}\left(\partial^A + q^2(q^2 - 1)X^A(\partial \circ \partial)\right) \tag{2.449}$$

we find that this operator satisfies the relations (2.444) and (2.445). Thus we divide our algebra once more by the ideal generated by (2.449). This is exactly the same strategy we used in chapter 5. From (2.449) follows

$$\overline{\Lambda} = \Lambda^{-1} \tag{2.450}$$

As in (2.395) we define the momenta

$$P^A = -\frac{i}{2}(\partial^A - \overline{\partial}^A) \tag{2.451}$$

Now we are in a position to derive the q-deformed Heisenberg algebra for the three-dimensional quantum space defined by (2.402) and (2.441):

$$\varepsilon_{FAB}P^A P^B = 0 \tag{2.452}$$
$$P^A X^B - (\hat{\mathcal{R}}^{-1})^{AB}_{CD} X^C P^D = -\frac{i}{2}q^3\Lambda^{-\frac{1}{2}}\left\{(1 + \frac{1}{q^6})\eta^{AB}W - (1 - \frac{1}{q^4})\varepsilon^{ABF}L_F\right\}$$

The operators Λ, W and L^A at the right hand side of eqn (2.452) are defined as differential operators, eqn. (2.390) should serve as an example. The separation of the various terms has been done by transformation properties (singlet, triplet) and by factoring the differential operators into a product of a unitary and a hermitean operator. This is how W, L_A and Λ have been defined.

$$\overline{\Lambda} = \Lambda^{-1}, \quad \overline{W} = W \quad \text{and} \quad \overline{L^A} = L_A \tag{2.453}$$

It turns out that the differential operators form an algebra by themselves

$$\begin{aligned} \varepsilon_{BA}{}^{C} L^A L^B &= -\frac{1}{q^2} W L^C \\ L^A W &= W L^A, \\ \Lambda L^A &= L^A \Lambda, \quad \Lambda W = W \Lambda \end{aligned} \tag{2.454}$$

and in addition

$$W^2 - 1 = q^4 (q^2 - 1)^2 L \circ L \tag{2.455}$$

We can now take the relations (2.454) and (2.455) as the defining relations for the Λ, W, L_A algebra and consider X^A and P^A as models under this algebra.

In this way we arrive at the algebra that generalizes the Heisenberg algebra to a $SO_q(3)$ structure.

We summarize the algebra:
q-deformed phase space:

$$\begin{aligned} X^C X^D \varepsilon_{DC}{}^{B} &= 0, \quad \overline{X^A} = X_A \\ P^C P^D \varepsilon_{DC}{}^{B} &= 0, \quad \overline{P^A} = P_A \end{aligned} \tag{2.456}$$

$SO_q(3)$:

$$\begin{aligned} L^C L^B \varepsilon_{BC}{}^{A} &= -\frac{1}{q^2} W L^A, \quad \overline{L^A} = L_A, \quad L^A W = W L^A, \quad \overline{W} = W \\ W^2 - 1 &= q^4 (q^2 - 1)^2 L \circ L \end{aligned} \tag{2.457}$$

scaling operator:

$$\begin{aligned} \Lambda^{\frac{1}{2}} L^A &= L^A \Lambda^{\frac{1}{2}} \quad \overline{\Lambda^{\frac{1}{2}}} = \Lambda^{-\frac{1}{2}} \\ \Lambda^{\frac{1}{2}} W &= W \Lambda^{\frac{1}{2}} \end{aligned} \tag{2.458}$$

comodule relations for $SO_q(3)$:

$$\begin{aligned} L^A X^B &= -\frac{1}{q^4} \varepsilon^{ABC} X_C W - \frac{1}{q^2} \varepsilon_{KC}{}^{A} \varepsilon^{KBD} L_D \\ W X^A &= (q^2 - 1 + \frac{1}{q^2}) X^A W + (q^2 - 1)^2 \varepsilon^{ABC} X_C L_B \end{aligned} \tag{2.459}$$

The coordinates can be replaced by the momenta to obtain the module relations for P_A.

Scaling properties:

$$\Lambda^{\frac{1}{2}} X^A = q^2 X^A \Lambda^{\frac{1}{2}} \qquad (2.460)$$
$$\Lambda^{\frac{1}{2}} P^A = q^{-2} P^A \Lambda^{\frac{1}{2}}$$

generalized (mimicked) Heisenberg relation:

$$P^A X^B - (\hat{\mathcal{R}}^{-1})^{AB}_{CD} X^C P^D = -\frac{i}{2} q^3 \Lambda^{-\frac{1}{2}} \left\{ (1 + \frac{1}{q^6}) \eta^{AB} W - (1 - \frac{1}{q^4}) \varepsilon^{ABF} L_F \right\} \qquad (2.461)$$

Eqns. (2.456) to (2.461) are the defining relations for the algebra we will be concerned with. The third lecture of the Schladming Winter School should analyze this algebra in the same way as we analyzed the one-dimensional Heisenberg algebra in the first lecture. The representation theory of the new algebra has been thoroughly investigated, it leads to very similar phenomena as we saw in part I for the representations. The q-deformed Minkowski structure has been analyzed in the same way.

References

BA ss E. Sweedler, *Hopf Algebras*, W. A. Benjamin, New York, 1969.

BB Elicki, *Hopf Algebras*, Cambridge Tracts in Mathematics, Cambridge University Press, 1977.

BC Klimyk, K. Schmüdgen, *Quantum Groups and Their Representations*, Springer, Berlin, 1997.

BD Kassel, *Quantum Groups*, Springer, Berlin, 1995.

BE Majid, *Foudations of Quantum Group Theory*, Cambridge University Press, 1995.

BF Madore, *An Introduction to Noncommutative Differential Geometry and its Physical Applications*, Cambridge University Press, 1995.

BG Landi, *An Introduction to Noncommutative Spaces and their Geometries*, Springer Lecture Notes, Springer, Berlin, 1997.

BH Lusztig, *Introduction to Quantum Groups*, Birkhaeuser, Boston, 1993.

BI D. Faddeev, N. Y. Reshetikhin, L. A. Takhtajan *Quantization of Lie Groups and Lie Algebra*, Algebra Analysis 1, **178**, 1989; translation: Leningrad Math. 3.1., 193, 1990.

BJ G. Drinfeld, *Quantum Groups*, Proc. ICM Berkeley, 798, 1987, .

BK Jimbo, *A q-Difference Analogue of $U(q)$ and the Yang-Baxter Equation*, Lett. Math. Phys.**10**, 1985.

BL Lorek, A. W. Schmidke, J. Wess, *$SU_q(2)$ Covariant $\hat{R}$-Matrices for Reducible Representations*, Lett. Math. Phys. **31**: 279, 1994.

BM Snyder, *Quantized Space-Time*, Phys. Rev. **71**: 38, 1947.

BN B.Schmidke, J. Wess, B. Zumino, *A q-deformed Lorentz Algebra*, Z. Phys. **C 52**: 471, 1991.

BO Ogievetsky, W. B. Schmidke, J. Wess, B. Zumino, *Six Generator q-deformed Lorentz Algebra*, Lett. Math. Phys.**23**: 233, 1991.

BP Ogievetsky, W. B. Schmidke, J. Wess, B. Zumino, *q-deformed Poincaré Algebra*, Comm. Math. Phys.**150**: 495, 1992.

BQ Lukierski, H. Ruegg, A. Nowicki, V. N. Tolstoy, q-Deformation of Poincaré Algebra, Phys. Lett. **B 264**: 331, 1992.

BR L. Woronowicz, *Twisted $SU(2)$ Group. An Example of Non-commutative Differential Calculus*, Publ. RIMS-Kyoto **23**: 117, 1987.

BS Wess, B. Zumino, *Covariant Differential Calculus on the Quantum Hyperplane*, Nucl. Phys. B (Proc. Suppl.) **18 B**: 302, 1990.

BT Dimakis, J. Madore, *Differential Calculi and Linear Connection*, J. Math.Phys. **37**: 4647, 1996.

BU Carow-Watamura, M. Schlieker, S. Watamura, W. Weich, *Bicovariant Differential Calculus on Quantum Groups $SU_q(N)$ and $SO_q(N)$*, Comm. Math. Phys. **142**: 605, 1991.

BV Carow-Watamura, M. Schlieker, S. Watamura, *$SO_q(N)$ Covariant Differential Calculus on Quantum Space and Quantum Deformation of Schrödinger Equation*, Z. Phys. C **49**: 439, 1991.

BW Fiore, J. Madore, *Leibniz Rules and Reality Conditions*, Preprint 98-13, math/9806071, Napoli, 1998.

BX Fiore, *Harmonic Oscillator on the Quantum Euclidean Space $\mathbf{R}_q{}^{N-1}$*, Proceedings of the Clausthal Symposion on Nonlinear, Dissipative, Irreversible Quantum Systems, 1994.

BY Ogievetsky, B. Zumino, *Reality in the Differential Calculus on q-Euclidean Spaces*, Lett. Math. Phys.**25**: 121, 1992.

BZ A. Azcàrraga, P. P. Kulish, F. Rodenas, *On the Physical Contents of q-deformed Minkowski Space*, Phys. Lett.**B 351**: 123, 1995.

CA Fiore, J. Madore, *The Geometry of Quantum Euclidean Spaces*, Preprint LMU-TPW 97-23, math/9904027, 1997.

CB Pillin, *On the Deformability of Heisenberg Algebras*, Comm. Math. Phys. **180**: 23, 1996.

CD Zippold, *Hilbert space Representations of an Algebra of Observables for q-deformed Relativistic Quantum Mechanics*, Z. Phys. **C 67**: 681, 1995.

CE Ocampo, *$SO_q(4)$ Quantum Mechanics*, Z. Phys. **C 70**: 525, 1996.

CF Lorek, W. Weich, J. Wess, *Non-commutative Euclidean and Minkowski Structures*, Z.Phys. **C 76**: 375, 1997.

CG L. Cerchiai, J. Wess, *q-deformed Minkowski Space based on a q-Lorentz Algebra*, E. Phys. J. **C 5**: 553, 1998.

CH L. Curtright, C. K. Zachos, *Deforming maps for quantum algebras*, Phys. Lett. **B 243**: 237, 1990.

CI Pillin, W. B. Schmidke, J. Wess *q-Deformed Relativistic One-Particle States*, Nucl. Phys. B **403**: 223, 1993.

CJ Rohregger, J. Wess, *q-deformed Lorentz Algebra in Minkowski Phase Space*, E. Phys. J. **C 7**: 177, 1999.

CK Kiesler, J. Madore, T. Masson, *On Finite Differential Calculus*, Contemp. Math. **203**: 135, 1997.

Quantum Gravity with Matter Fields in Two Dimensions

Elmar Bittner[1,2], Wolfhard Janke[2], Harald Markum[1], Jürgen Riedler[1]

[1] Institut für Kernphysik, Technische Universität Wien, A-1040 Wien, Austria
[2] Institut für Theoretische Physik, Universität Leipzig, D-04109 Leipzig, Germany

Spin systems coupled to two-dimensional manifolds are studied as a simple example for matter fields coupled to gravity. We consider two versions of quantum Regge calculus. The Standard Regge Calculus [1] where the quadratic link lengths q of the simplicial manifold vary continuously and the Discrete Regge Model [2] where they are restricted to two possible values. These manifolds are coupled with Z_2 spins [3].

We simulated the partition function $Z = \sum_s \int \mathcal{D}\mu(q) \exp(-I(q) - KE(q,s))$, where $I(q) = \lambda \sum_i A_i$ is the gravitational action with the cosmological constant λ, and $E(q,s) = \frac{1}{2} \sum_{\langle ij \rangle} A_{ij} \frac{(s_i - s_j)^2}{q_{ij}}$ is the energy of the Ising spins s_i, $s_i = \pm 1$, which are located at the vertices i of the lattice. A_i and A_{ij} are barycentric areas associated with the vertices i and the edges $\langle ij \rangle$, respectively. We use the same path-integral measure $\mathcal{D}\mu(q)$ as in the pure gravity simulations [4], which is chosen to render the Discrete Regge Model particularly simple.

The lattice topology is given by the triangulated tori of size $N_0 = L^2$ with $L = 8, 16, 24, 32, 64, 96$, and 128. The simulations are performed at $\lambda = 1.0$ and at spin coupling $K = 1.040$. After an initial equilibration time we took for each lattice size about 50 000 measurements. For each run we recorded the time series of the energy density $e = E/N_0$ and the magnetization density $m = \sum_i A_i s_i / N_0$.

From the time series we computed the Binder parameter U_L, the specific heat C, and the susceptibility χ. By applying reweighting techniques we determined the maxima of C and χ, and extracted the critical exponents. The Standard Regge Calculus and the Discrete Regge Model produce the same critical exponents of the Ising transition which agree with the Onsager exponents for regular static lattices. The KPZ exponents [5] are definitely excluded. The gravitational action does not affect the critical exponents of the Ising phase transition.

References

[1] T. Regge, Nuovo Cimento 19 (1961) 558
[2] W. Beirl, P. Homolka, B. Krishnan, H. Markum, and J. Riedler, Nucl. Phys. B (Proc. Suppl.) 42 (1995) 710
[3] C. Holm and W. Janke, Phys. Lett. B335 (1994) 143
[4] E. Bittner, A. Hauke, C. Holm, W. Janke, H. Markum, and J. Riedler, Nucl. Phys. B (Proc. Suppl.) 63A-C (1998) 769
[5] V.G. Knizhnik, A.M. Polyakov, and A.B. Zamolodchikov, Mod. Phys. Lett. A3 (1988) 819

A Quantum Minkowski Space-Time

Bianca Letizia Cerchiai

Ludwig-Maximilians-Universität München, Theresienstr. 37, 80333 München and Max-Planck-Institut für Physik, Föhringer Ring 6, 80805 München

Abstract. The representation theory of a q-deformed Minkowski algebra, which is as quantum space a co-module of the q-Poincaré algebra, is studied. The spectra of the coordinates are discrete: space-time acquires a lattice-like structure. The eigenvalues grow exponentially and there are accumulation points on the light-cone.

The Poincaré symmetry of space-time is deformed to a quantum group symmetry [1] with a deformation parameter $q > 1$. In this way space-time acquires a lattice-like structure, which provides an ultraviolet cut-off and hence conceivably a regularized field theory.

The deformed Minkowski algebra is defined in [2]. It is generated by four coordinates and four conjugated momenta, the q-Lorentz generators, and an extra generator acting as a 'dilatator'. The spatial coordinates are not commutative, but time still commutes with them. The quantum space cannot be separated from the symmetry acting on it, because a realization of the q-Lorentz algebra appears in the q-Heisenberg relations between coordinates and momenta. According to [3,4] the eigenvalues of the invariant length s^2 and time t are discrete. They are determined by the integer numbers $M \in \mathbb{Z}$ and n:

$$\text{Timelike: } s^2 = t_0^2 q^{2M}, \quad t = t_0 q^M (q^{n+1} + q^{-n-1})(q + q^{-1})^{-1}, \; n \in \mathbb{N}$$
$$\text{Spacelike: } s^2 = -l_0^2 q^{2M}, \; t = l_0 q^M (q^n - q^{-n})(q + q^{-1})^{-1}, \quad n \in \mathbb{Z}$$

Here t_0 and l_0 are continuous real parameters labelling inequivalent representations. The operators t, s^2, the square of the angular momentum and its third component form a complete set of commuting observables.

The momentum operators are not self-adjoint, but a self-adjoint extension can be constructed by taking a direct sum of representations. Then the light-cone is only included as hyper-surface of accumulation points.

References

1. O. Ogievetskii, W. B. Schmidke, J. Wess, B. Zumino: *q-Deformed Poincaré Algebra*, Commun. Math. Phys.150, 492 (1992)
2. A. Lorek, W. Weich, J. Wess: *Non-commutative Euclidean and Minkowski Structures*, Z.Phys. **C** 76, 375 (1997)
3. B. L. Cerchiai, J. Wess: *q-Deformed Minkowski Space based on a q-Lorentz Algebra*, Eur. Phys. J. **C** 5 (3), 553 (1998)
4. B. L. Cerchiai: *Hilbert Space Representations of a q-deformed Minkowski algebra*, Ph.D. Thesis, Ludwig-Maximilians-Universität München, (1998)

Supersymmetry and Nonperturbative Aspects in Quantum Cosmology

Evgueni E. Donets and Mirian M. Tsulaia

Joint Institute for Nuclear research, Dubna, 141980, Russia

The important question of the modern superstring and M – theories is the problem of the spontaneous breakdown of the supersymmetry. We consider the dynamical (nonperturbative) breaking of supersymmetry, caused by gravitational and Yang-Mills (YM) instantons in quantum cosmology.

We use $N = 2$ SUSY sigma-model technique for the supersymmetrization and quantization of $SU(2)$ Einstein-Yang-Mills system in homogeneous axially-symmetric $Bianchi - I, II, VIII, IX$, $Kantowski - Sachs$ (KS) and closed $Friedman - Robertson - Walker$ (FRW) cosmological models, which really can pretend to describe the very early universe. It was found the desired supersymmetrization is possible and the explicit expressions for the corresponding superpotentials W are obtained (E.E.Donets, M.N.Tentyukov, M.M.Tsulaia (1999)). Superpotentials W in all these cases are found to be the direct sums of pure gravitational W_{GR} and Yang-Mills W_{YM} parts with W_{YM} parts being equal to the YM Chern-Simons functionals.

Wave function of "universe" in null fermion sector is expressed in terms of the superpotential as $\Psi_0 = const * e^{-W_{GR}-W_{YM}}$. This wave functions are normalizable for pure gravitational $(W_{YM} = 0)$ $Bianchi \; - I, II, IX_{(1)}$, KS and FRW models that means the supersymmetry is unbroken by the gravitational instantons. For $Bianchi \; - VIII, IX_{(2)}$ models the wave function is nonnormalizable that indicates the breakdown of SUSY in a such pure gravitational system. Further inclusion of the YM field always leads with necessity to the spontaneous breaking of SUSY by YM instantons, since W_{YM}, being the Chern-Simons term is not an even function hence leading to the divergency of the norm of the wave function.

The $N = 4$ multidimensional SUSY QM can shed the new light onto the problem of SUSY breaking in quantum cosmology since it allows various possibilities for partial breaking of SUSY (E.E.Donets, A.Pashnev, J.Rosales and M.M.Tsulaia (1999)). This work is in progress now.

Acknowledgment E.E.D. is grateful to Organizing Committee of the XXXVIII Schladming Winter School for partial support during the meeting.

References

E.E.Donets, M.N.Tentyukov, M.M.Tsulaia (1999): *Phys. Rev.* **D59**, 023515 (1-9).

E.E.Donets, A.Pashnev, J.Rosales and M.M.Tsulaia (1999) *Phys. Rev.* **D**, to be published; hep-th/9907224.

Noncommutative Supergeometry of Graded Matrix Algebras

H. Grosse[1] and G. Reiter[1,2]

[1] Universität Wien, Institut für Theoretische Physik, Boltzmanngasse 5, A-1090 Wien, Austria, grosse@doppler.thp.univie.ac.at
[2] Technische Universität Graz, Institut für Theoretische Physik, Petersgasse 16, A-8010 Graz, Austria, reiter@itp.tu-graz.ac.at

Generalizing the ungraded case one can build up $\mathbb{Z}_2$-graded differential calculi over arbitrary $\mathbb{Z}_2$-graded $\mathbb{C}$-algebras based on their respective Lie superalgebra of graded derivations. Especially for the $\mathbb{Z}_2$-graded $\mathbb{C}$-algebra $\mathbb{M}(n|m)$ of $(n+m) \times (n+m)$-matrices with block-matrix grading $(n, m \in \mathbb{N}_0, n \neq m)$ the resulting differential algebra $(\Omega^g(\mathbb{M}(n|m)), \mathrm{d})$ coincides - as far as we are interested only in its linear structure - with the cochain complex of the Lie superalgebra $\mathrm{sl}(n|m)$ with coefficients in $\mathbb{M}(n|m)$. Here we want to point out two remarkable facts about this differential algebra (for proofs see Grosse and Reiter 1999).

Proposition 1 *The canonical cochain map from the (intrinsically) $\mathbb{Z}_2$-graded differential envelope to $\Omega^g(\mathbb{M}(n|m))$ is onto and its restriction to the respective first-order differential calculi is an isomorphism. Moreover there exists an even graded 1-form $\Theta \in \Omega^{g,1}(\mathbb{M}(n|m))$ such that*

$$\mathrm{d}M = [\Theta, M]_g \equiv \Theta \wedge M - M \wedge \Theta \qquad M \in \mathbb{M}(n|m) \ . \tag{1}$$

The map $\beta : \mathbb{M}(n|m) \longrightarrow \mathbb{M}(\underline{n}), \underline{n} := \max(n, m)$, projecting out the greater diagonal-block, should be interpreted as noncommutative analogue to the body map: As in the case of graded manifolds (see Kostant 1977) it extends to a cochain map $\beta : \Omega^g(\mathbb{M}(n|m)) \longrightarrow \Omega(\mathbb{M}(\underline{n}))$ and

Proposition 2 $H(\beta) : H(\mathbb{M}(n|m)) \longrightarrow H(\mathbb{M}(\underline{n}))$ *is an isomorphism.*

Acknowledgment. The research of G. Reiter was supported by the "Fonds zur Förderung der wissenschaftlichen Forschung (FWF)" through the research project P11783-PHY "Quantum Field Theory on Noncommutative Manifolds".

References

Grosse H., Reiter G. (1999): *Graded Differential Geometry of Graded Matrix Algebras.* Preprint UWThPh-6-1999.

Kostant B. (1977): *Graded manifolds, graded Lie theory, and prequantization.* Lecture Notes in Mathematics **570**.

Duals for Nonabelian Lattice Gauge Theories

Harald Grosse[1], Karl-Georg Schlesinger[2]

[1] Institute for Theoretical Physics University of Vienna,
Boltzmanngasse 5
A-1090 Vienna, Austria
[2] Department of Mathematics, University of Wuppertal
Gaussstr. 20
D-42097 Wuppertal, Germany
e-mail: schles@math.uni-wuppertal.de

The usual duality between an abelian group and the group of its characters gives rise to the well known notion of the dual of an abelian lattice gauge theory which is of considerable importance. The development of a concept of a dual for nonabelian lattice gauge theories has long been prevented by the fact that the Doplicher-Roberts theorem shows that the dual of a nonabelian group is no longer a group but a certain monoidal category. After very briefly reviewing the needed concepts from lattice gauge theory (in order to make the talk selfcontained), we present a categorical construction of such duals, making use of a functorial reformulation of the notion of a lattice gauge theory (due to J. Baez). We show that the commutative tetrahedron of 2-category theory immediately leads to a gauge invariant action for the dual theories. As an example, we discuss the case of the gauge group SU(2) where we find that classical connections are labeled by spin networks, i.e. the theory as an "already quantized" form.

Absolute Conservation Law for Black Holes

Daniel Grumiller

Institut für Theoretische Physik, Technische Universität Wien, Wiedner Hauptstr. 8–10, A-1040 Wien, Austria

The spherically reduction of the 4d Einstein-Hilbert action reduces to a 2d dilaton theory[1], which can be shown to be equivalent to a first order action (with nonvanishing torsion)[2] being a special case of a Poisson-σ model[3].

A certain linear combination of the equations of motion with respect to the Cartan variables leads to a conservation law[4]

$$dC^{(g)} + W = 0 \tag{1}$$

with $C^{(g)}$ being a 0-form and $W = W\,(matter)$ a 1-form.

In (generalized) Schwarzschild gauge $(ds)^2 = \alpha\,(t,r)\,(dt)^2 - a\,(t,r)\,(dr)^2 - r^2\,(d\Omega)^2$ the conservation law (1) can be brought into the simple form[5] $\frac{\partial m}{\partial t} = A\,(t,r)$ and $\frac{\partial m}{\partial r} = B\,(t,r)$ with the "mass-aspect function[6]" $m\,(t,r)$ defined by $a^{-2} = 1 - \frac{2m}{r}$ and $A\,(t,r)$ and $B\,(t,r)$ being functions of the scalar fields first derivatives[5].

Using the fact that W is closed (and in our case also exact) one obtains:

$$m\,(t,r;t_0,r_0) = \int_{t_0}^{t} dt' A\,(t',r) + \int_{r_0}^{r} dr' B\,(t_0,r') + m_0 \tag{2}$$

which contains new information, namely the integration constant m_0 which labels a certain solution much like the total energy in an ordinary mechanical system.

Note that our result for $m\,(t,\infty;t_0,0)$ differs from the "total ADM mass[6]" not only by m_0, but also by the t-integral of eq. (2).

Generalizations to spherically reduced Einstein-gravity in $d > 4$ dimensions and to the Einstein de-Sitter case are straightforward[5].

---

1 P. Thomi, B. Isaak and P. Hajicek, Phys. Rev. D **30** (1984) 1168; P. Hajicek, ibid. D **30** (1984) 1178; S.R. Lau, Class. Quant. Grav. **13** (1996) 1541.

2 M.O. Katanaev, W. Kummer and H. Liebl, Phys. Rev. D **53** (1996) 5609; W. Kummer, H. Liebl and D. V. Vassilevich, Nucl. Phys. B **493** (1997) 491.

3 N. Ikeda and K-I. Izawa, Progr. Theor. Phys. **89** (1993) 1.; P. Schaller and T. Strobl, Mod. Phys. Lett. A **9** (1994) 3129.; T. Strobl, Phys. Rev. D **50** (1994) 7346.; T. Klösch and T. Strobl, Class. Quant. Grav. **13** (1996) 965.; W. Kummer and D. Schwarz, Phys. Rev. D **45** (1992) 3628.

4 W. Kummer and P. Widerin, Phys. Rev. D **52** (1995) 6965.; W. Kummer and G. Tieber, *Universal conservation law and modified Noether symmetry in 2d models of gravity with matter*, TUW-98-16, hep-th/9807122 (to be published in Phys. Rev. D).

5 D. Grumiller, W. Kummer, *Absolute conservation law for black holes*, gr-qc/9902074

6 C. Gundlach, Adv. Theor. Math. Phys. **2** (1998) 1; M. W. Choptuik, *The (unstable) threshold of black hole formation*, gr-qc/9803075.

Double Numbers and Two Dimensional Anomaly Free Field Models

V.Hrytsau and Yu.A.Kurochkin

Institute of Physics National Academy of Sciences of Belarus,
Fr. Scoryna ave. 70, Minsk 220072, Belarus

In ref.[1] it has been made the certain assertions concerning possibility of nonanomalous quantization of the string inspired model [2] of two dimensional gravity. In some variables this model can be described by the first class constraints as in the case of ordinary relativistic bosonic string in two dimensional space-time.The ring of double numbers is natural tool for woking with the two dimensional Minkowski space.We propose the model of the two-dimensional quantum gravity wich is invariant with respect to the two-dimensional superluminal Lorentz transformation.The Virasoro algebra of the quantum gravity model defined over double number ring is found to be without conformal anomaly.

Double (hyperbolic complex) number (see [3] and ref.therein) is a combination of the type $w = x + \varepsilon y$ with real x,y and $\varepsilon^2 = 1$ such that $x\varepsilon = \varepsilon x$, $y\varepsilon = \varepsilon y$. Double numbers form a commutative ring. Moreover $\mathbf{D} = \mathbf{R} \oplus \mathbf{R}$. In double numbers algebra conjugated element is defined by $\bar{w} = x - \varepsilon y$. The real and imaginary parts of the analytical (in double variables sence) function f obey to the two dimensional wave equation: $\partial\overline{\partial} f = 0$. The multiplication on double numbers of minus unit squared modulus defines a nonproper (superluminal) Lorentz transformation $w^{'} = \pm\varepsilon e^{\varepsilon\chi} w$. The model of two dimensional quantum gravity considered in ref.[1] can be described by the first class constraints as the relativistic bosonic string in 2D Minkowski space: $\mathcal{E} = -\mathcal{E}_0 + \mathcal{E}_1 \approx 0$, $\mathcal{E}_0 = \frac{1}{2}((\pi_0)^2 + (r^{0'})^2)$, $\mathcal{E}_1 = \frac{1}{2}((\pi_1)^2 + (r^{1'})^2)$, $\mathcal{P} = r^{a'}\pi_a \approx 0$ where r^a, π_a, $(a = 0, 1)$ are canonically conjugated fields. The Hamiltonian is bilinear combination of oscillators: $\mathcal{H} = \int |k| a_k^{a+} a_k^a dk$. The usual procedure of normal ordering for operators (a_k^{a+}, a_k^a) in quantities $\mathcal{E}, \mathcal{P}$ leads to the anomaly and means that all creation operators are placed at the left hand side of annihilation operators. The discrete superluminal transformation interchange t and x and exchange positive and negative frequences. The unitary operator M in states space of model corresponds to this transformation and acts on oscillators as follows: $M a_k^0 M = a_k^{1+}$, $M a_k^1 M = a_k^{0+}$. This one is automorphism of commutational relations. For compatibility with condition on vacuum one has to impose the new normal ordering prescription:

$$a_k^{0+}|0>= 0 \quad , \quad a_k^1|0>= 0 \tag{1}$$

Then normal ordering for $\mathcal{H}$ does not lead to c-numbers anomaly.

References

1. Jackiw, R.: E-print archive, gr-qc/9612052.
2. Callan, C., Giddings, S., Harvey, J., Strominger, A.: Phys.Rev. **D 45** (1992) 1005.
3. Hucks, J.: J.Math.Phys. **34** (1993) 12.

A Global Path Integral for Yang–Mills Theory

Helmuth Hüffel and Gerald Kelnhofer

Inst. f. Theor. Physik, Univ. Wien, Boltzmanngasse 5, A-1090 Wien, Austria

Abstract. We present a global path integral density for the Yang-Mills theory by summing over all partitions of the space of gauge potentials.

Upon suitable restrictions the set $\mathcal{A}$ of Yang-Mills fields A is a principal $\mathcal{G}$ bundle $\mathcal{A} \xrightarrow{\pi} \mathcal{M}$ over the paracompact space $\mathcal{M} = \mathcal{A}/\mathcal{G}$ of all inequivalent gauge potentials with projection π. We locally separate the Yang-Mills fields into gauge independent and gauge dependent degrees of freedom by choosing a fixed background connection $A_0^{(\alpha)} \in \mathcal{A}$ and considering a sufficient small neighbourhood $U(A_0^{(\alpha)})$ of $\pi(A_0^{(\alpha)})$ in $\mathcal{M}$. Then the subspace $\Gamma_\alpha = \{B \in \pi^{-1}(U(A_0^{(\alpha)}))/D^*_{A_0^{(\alpha)}}(B - A_0^{(\alpha)}) = 0\}$ defines a local section of $\mathcal{A} \to \mathcal{M}$. We first succeed in *locally* formulating a new generalized stochastic quantization procedure for Yang–Mills theory (Hüffel and Kelnhofer (1998a)) which leads to a modified Faddeev–Popov path integral density

$$\mu_\alpha \, e^{-S_\alpha^{\rm tot}}, \quad S_\alpha^{\rm tot} = S + S_{\mathcal{G}}. \tag{1}$$

Here $\mu_\alpha = \det \mathcal{F}_\alpha \sqrt{\det \Delta_{A_0^{(\alpha)}}}\ \nu$, where $\mathcal{F}_\alpha = D^*_{A_0^{(\alpha)}} D_B$ is the Faddeev–Popov operator; ν implies an invariant volume density on $\mathcal{G}$. $S_\alpha^{\rm tot}$ denotes a total Yang-Mills action defined by the original Yang-Mills action S without gauge symmetry breaking terms and by $S_{\mathcal{G}} \in C^\infty(\mathcal{G})$ which is an arbitrary functional on $\mathcal{G}$ such that $e^{-S_{\mathcal{G}}}$ is integrable with respect to ν. Note that locally all expectation values of gauge invariant observables coincide with the conventional Faddeev-Popov results. Finally we propose (Hüffel and Kelnhofer (1998b)) the definition of a *global* expectation value of a gauge invariant observable f by summing over all partitions of unity γ_α of the space of inequivalent gauge potentials such that

$$\langle f \rangle = \frac{\sum_\alpha \int\limits_{\Gamma_\alpha \times \mathcal{G}} \mu_\alpha \, e^{-S_\alpha^{\rm tot}} f \, \gamma_\alpha}{\sum_\alpha \int\limits_{\Gamma_\alpha \times \mathcal{G}} \mu_\alpha \, e^{-S_\alpha^{\rm tot}} \gamma_\alpha}. \tag{2}$$

We can prove that this expectation value is independent of the choice of the partition of unity and of the local gauge fixing conditions.

References

H. Hüffel and G. Kelnhofer (1998a): Ann. of Phys. **270**, 231 (1998).

H. Hüffel and G. Kelnhofer (1998b): Univ. of Vienna preprint UWThPh-1998-65; hep-th/9901165.

Anyonic Solutions to the Thirring Model

N. Ilieva and W. Thirring

Institut für Theoretische Physik
Universität Wien
Boltzmanngasse 5, A-1090 Wien
Erwin Schrödinger International Institute
for Mathematical Physics

Abstract. Solutions to the Thirring model are constructed in the framework of algebraic quantum field theory. It is shown that for all positive temperatures there are fermionic solutions only if the coupling constant is $\lambda = \sqrt{2(2n+1)\pi}$, $n \in \mathbf{N}$, otherwise solutions are anyons. Different anyons (which are uncountably many) live in orthogonal spaces, so the whole Hilbert space becomes non-separable and in each of its sectors a different Heisenberg's "Ungleichung" holds. This feature certainly cannot be seen by any power expansion in λ. Moreover, if the statistic parameter is tied to the coupling constant it is clear that such an expansion is doomed to failure and will never reveal the true structure of the theory.

On the basis of the model in question, it is not possible to decide whether fermions or bosons are more fundamental since dressed fermions can be constructed either from bare fermions or directly from the current algebra.

Twisting of Quantum Differentials

Shahn Majid and Robert Oeckl

Department of Applied Mathematics and Theoretical Physics,
University of Cambridge, Cambridge CB3 9EW, UK

We report on recent work (see Majid and Oeckl (1998)). Let H be a Hopf algebra over a field k. We recall that a unital 2-cocycle $\chi : H \otimes H \to k$ over H gives rise to a new Hopf algebra H_χ (the twist of H) with the same unit, counit and coproduct, but modified product. We show that a bicovariant bimodule V over H can be made a bicovariant bimodule over H_χ by equipping it with the same coactions but modified actions. The new (twisted) left action is

$$a \triangleright_\chi b = \chi(a_{(1)} \otimes v_{(1)})\, a_{(2)} \triangleright v_{(\underline{2})}\, \chi^{-1}(a_{(3)} \otimes v_{(3)}) \ ,$$

where the subscripts denote the coproduct or application of the left and the right coaction with the component remaining in V underlined. Similarly for the right action. We obtain in fact an isomorphism of bicovariant bimodule categories ${}^H_H\mathcal{M}^H_H \cong {}^{H_\chi}_{H_\chi}\mathcal{M}^{H_\chi}_{H_\chi}$. A bicovariant differential calculus Ω over H is a bicovariant bimodule with a bicomodule map $\mathrm{d} : H \to \Omega$ (the "differential") satisfying the Leibniz rule. The above construction gives rise to a 1-1 correspondence of differential calculi over H and H_χ where d is unchanged under the twist. This correspondence extends in fact to the whole exterior algebra. Many Hopf algebras are related by twisting, so this gives the corresponding twisting of their calculi. In particular, given a Lie group G, a cocycle $\chi_\hbar$ over its Hopf algebra of functions $H = C(G)$ defines a deformation quantization of G. Thus, our result allows us to correspondingly deformation quantize the entire exterior algebra.

We apply this to the Planck-scale Hopf algebra (see Majid (1988)), a toy model for quantization of a particle in a curved phase space in one dimension. This is a Hopf algebra over $\mathbb{C}$ generated by x and p with relations

$$[x,p] = \imath\hbar(1 - e^{-\frac{x}{\mathsf{G}}}), \quad \Delta x = x \otimes 1 + 1 \otimes x, \quad \Delta p = p \otimes e^{-\frac{x}{\mathsf{G}}} + 1 \otimes p \ .$$

where $\hbar, \mathsf{G}$ are Planck's constant and a radius of curvature. It is a cocycle deformation quantization in $\hbar$ of a Lie group. We obtain the differential calculus

$$a\mathrm{d}x = (\mathrm{d}x)a, \quad a\mathrm{d}p = (\mathrm{d}p)a + \frac{\imath\hbar}{\mathsf{G}}\mathrm{d}a \ .$$

This suggests a quantum geometric approach to quantum mechanics.

References

S. Majid, *Hopf Algebras for Physics at the Planck Scale*, Class. Quantum Gravity **5** (1988) 1587–1606

S. Majid, R. Oeckl, *Twisting of quantum differentials and the Planck scale Hopf algebra*, DAMTP-1998-118, math/9811054

Wigner Solid and Laughlin Liquid of Bose Condensed Charge-Vortex Composites

Guenther Meissner

A unified many-body approach [1] is shown to allow for studying the nature of two novel competing phases of Coulomb-interacting electrons in high magnetic fields B with a fractional filling factor $\nu < 1$, as e.g. 1/3, of the lowest Landau level (LLL). These condensed phases are the magnetic field-induced 2D Wigner crystal (WC) and the Laughlin liquid respectively, i.e., a lattice-periodic structure of the guiding centers of cyclotron orbits (radius r_L) of the electrons and an incompressible quantum liquid (IQL) of Bose condensed charge-vortex composites. The quasi-particles of the Laughlin liquid exhibiting the fractional quantum Hall effect [2] having been discovered 1982 carry magnetic flux quanta ch/e and magic fractions $e^* = \nu e$ of the elementary charge e [3]. The electron-electron interaction within the LLL plays an important role in these highly correlated phases due to the massive degeneracy associated with the 2D free-electron motion of cyclotron frequency ω_c and the possible cyclic irreducible quantum group representations. The collective exitations in the long wavelength limit of the IQL-phase being related to superconductivity and superfluidity thus can be shown to have a finite gap (magneto-rotons) contrary to those of the WC-phase being gapless due to broken magnetic translational invariance (magneto-phonons). In the single-particle spectral function the splitting of the LLL due to Coulomb interaction leads to a kind of doublepeak structure with a wide pseudogap at the Fermi level in both competing phases. An additional splitting into subbands in the WC-phase indicates the intricate structure of the Hofstadter butterfly. The triangular lattice of the classical Wigner crystal predicted [4] already 1976 and verified experimentally [5] on liquid helium films 1979 turns out to be the ground-state of this sort of a many-body system in the dilute electron ($\nu \to 0$) or high magnetic field ($B \to \infty$) limit with the Larmor radius $r_L \to 0$. In summary, a weakly first-order transition from the Laughlin liquid to the Wigner crystal is found for non-rational filling factors $\nu < \nu_c \simeq 0.202$ together with reentrant behavior around $\nu = 1/5$ from comparing their ground-state energies $E_L(\nu_c) = E_C(\nu_c)$ being in reasonable agreement with experimental findings.

References

[1] G. Meissner, Physica B 184 (1993) 66.
[2] D.C. Tsui, H.L. Stormer, and A.C. Gossard, Phys. Rev. Lett. 48 (1982) 1559.
[3] R.B. Laughlin, Phys. Rev. Lett. 50 (1983) 1395.
[4] G. Meissner, H. Namaizawa, and M. Voss, Phys. Rev. B 13 (1976) 1370.
[5] C.C. Grimes and G. Adams, Phys. Rev. Lett. 42 (1979) 795.

The Modular Closure of Braided Tensor Categories

Michael Müger

Dipartimento di Matematica, Università di Roma "Tor Vergata", Italy
email: mueger@axp.mat.uniroma2.it

With every monoid M one can associate its center $Z(M) = \{x \in M \mid xy = yx\ \forall y \in M\}$, which obviously is a commutative monoid. If M has inverses, i.e. is a group, then $Z(M)$ is a normal subgroup and one can consider the quotient group $M/Z(M)$. In nice cases, e.g. if M is a direct product of simple groups, $M/Z(M)$ turns out to have trivial center. Monoids being 0-categories, the work to be reported here can be considered as the analogous construction for 1-categories. We refer to Müger (1998) for a full account. Given a strict tensor category $\mathcal{C}$ with braiding ε, we define its *center* to be the full subcategory defined by

$$\mathrm{Obj}\, Z(\mathcal{C}) = \{\rho \in \mathrm{Obj}\, \mathcal{C} \mid \varepsilon(\rho,\sigma) \circ \varepsilon(\sigma,\rho) = \mathrm{id}_{\sigma\otimes\rho}\ \ \forall \sigma \in \mathrm{Obj}\, \mathcal{C}\},$$

which clearly is a symmetric tensor category. We would like to define a quotient "$\mathcal{C}/Z(\mathcal{C})$" and to prove that its center is trivial in the sense that every irreducible object in the center is isomorphic to the tensor unit ι. Contrary to the monoid case, this can not be done by identifying the objects in $Z(\mathcal{C})$ with the unit ι. Instead, we embed $\mathcal{C}$ into a braided tensor category $\mathcal{C} \times Z(\mathcal{C})$ which has more morphisms, such that the objects in $Z(\mathcal{C})$ become isomorphic to (multiples of) ι. As in the monoid case we need $\mathcal{C}$ to satisfy additional assumptions, viz. to be a complex $*$-category with duals. (If such a category is rational, i.e. has only finitely many isomorphism classes of irreducible objects, then triviality of the center is equivalent to invertibility of the S-matrix, thus to the category being modular, cf. Müger (1999). Therefore, our work has applications to topology.) The crucial ingredient for our construction is a theorem of Doplicher and Roberts (1989), which implies that the center $Z(\mathcal{C})$ is isomorphic to a category of representations of a unique compact group G. Among many other things we prove that $\mathcal{C}Z(\mathcal{C})$ has trivial center, whence the name *modular closure*. Our work has strong relations to algebraic and conformal quantum field theory (simple current extensions), see Müger (1999).

References

M. Müger (1998): The modular closure of braided tensor categories. (math.CT/9812040) To appear in Adv. Math.

M. Müger (1999): On Galois extensions of braided tensor categories, mapping class group representations and simple current extensions. In preparation

S. Doplicher and J. E. Roberts (1989): A new duality theory for compact groups. Invent. Math. **98**, 157-218

Clifford Algebra as a Useful Language for Geometry and Physics

Matej Pavšič[1]

Jožef Stefan Institute, Jamova 39, 1000 Ljubljana, Slovenia

Clifford numbers can be used to represent vectors, multivectors and, in general, polivectors. They form a very useful tool for geometry [Hestenes et al. (1984)]. The well known equations of physics can be cast into elegant compact forms by using the geometric calculus based on Clifford Algebra.

These compact forms suggest a generalization that every physical quantity is a polivector [Pezzaglia (1997)]. For instance, the momentum polivector is

$$P = \mu + p^\mu e_\mu + S^{\mu\nu} e_\mu e_\nu + \pi^\mu e_5 e_\mu + m e_5 \tag{1}$$

and the velocity polivector is $\dot{X} = \dot{\sigma} + \dot{x}^\mu e_\mu + \dot{\alpha}^{\mu\nu} e_\mu e_\nu + \dot{\xi}^\mu e_5 e_\mu + \dot{s} e_5$ Here e_μ, $\mu = 0, 1, 2, 3$ are basic vectors, satisfying $e_\mu \cdot e_\nu \equiv \frac{1}{2}(e_\mu e_\nu + e_\nu e_\mu) = \eta_{\mu\nu}$, and $e_5 \equiv e_0 e_1 e_2 e_3$, $e_5^2 = -1$, is the pseudoscalar unit. The free particle action is given by $I = \int \mathrm{d}\tau (P\dot{X} - \lambda P^2)$ where λ it the Lagrange multiplier giving the constraint $P^2 = 0$. From (1) we have $P^2 = p^2 - m^2 - \pi^2 + \mu^2 + 2\mu(p^\mu e_\mu + m e_5) +$ etc. This constraint is satisfied if the coefficients of the scalar, vector, bivector, etc., parts vanish.

In the quantized theory we assume that a state is represented by a polivector wave function satisfying $\hat{P}^2 \Phi = 0$ where $\hat{P}$ is now operator. A particular class of solutions satisfies $\hat{P}\Phi = 0$. When Φ is and eigenfunction of the operators $\hat{\mu}$, $\hat{\pi}^\mu$, $\hat{S}^{\mu\nu}$ with eigenvalues $\mu = 0$, $\pi^\mu = 0$, $S^{\mu\nu} = 0$ we have

$$\hat{P}\Phi = (\hat{p}^\mu e_\mu + m\, e_5)\Phi = 0 \quad \text{or} \quad (\hat{p}^\mu \gamma_\mu - m)\Phi = 0 \tag{2}$$

where $\gamma_\mu \equiv e_5 e_\mu$, $\gamma_5 \equiv \gamma_0\gamma_1\gamma_2\gamma_3 = e_0 e_1 e_2 e_3 \equiv e_5$. This is just the Dirac equation. Among polivectors Φ satisfying (2) there are spinors. That spinors are just special kind of polivectors (Clifford aggregates), namely the elements of left or right ideals of the Clifford Algebra, is an old observation. Now, scalars, vectors, etc., can be reshuffled by the elements of the Clifford Algebra. This means that scalars, vectors, etc., can be transformed into spinors, and vice versa. Within Clifford Algebra we have thus the transformations which change bosons into fermions. It would be very interesting to investigate whether such a kind of "supersymmetry" is related to the well known supersymmetry.

References

Hestenes, D., Sobczyk, G. (1984): *Clifford Algebra to Geometric Calculus* (Reidel, Dordrecht, Boston, Lancesteer)

Pezzaglia, W. M. (1997): Physical Applications of a Generalized Clifford Calculus. Preprint gr-qc/9710027

Fields on Noncommutative Manifolds

M. Pavšič[1], M. Chaichian[1], A. Demichev[2], H. Grosse[3], P. Prešnajder[4]

High Energy Physics Division, Department of Physics, University of Helsinki and Helsinki Institute of Physics, P.O.Box 9, FIN-00014 Helsinki, Finland
Nuclear Physics Institute, Moscow State University, 119899, Moscow, Russia
Institut for Theoretical Physics, University of Vienna, Boltzmanngasse 5, A-1090 Vienna, Austria
Department of Theoretical Physics, Comenius University, Mlynská dolina, SK-84215 Bratislava, Slovakia

The essence of the noncommutative geometry consists in reformulating the geometry in terms of commutative algebras and modules of smooth functions, and then generalizing them to their noncommutative analogs, [1]. Here we shall describe models on a Fuzzy cylinder $\mathbf{R} \times S^1 = \{\tau$ on real line interpreted as the time, $x_\pm = \rho e^{\pm i\varphi}$ identified with the circle $S^1\}$, for details see [2], [3].

In the noncommutative case we replace the commuting variables $\tau, x_\pm$ by the operators $\hat{\tau} = -i\lambda\partial_\varphi, \hat{x}\pm = \rho e^{\pm i\varphi}$. The field is now an operator of the form $\hat{\Phi} = \sum_{k\in \mathbf{Z}} c_k(\hat{\tau})e^{ik\varphi}$. The field action is defined as the trace of the corresponding (operator) Lagrangian density $S[\hat{\Phi}] = \lambda \mathrm{Tr}[-\hat{\Phi}^*\hat{\partial}_\mu^2\Phi - m^2\Phi^*\Phi - \ldots]$, where the dots corresponds to the interaction term. The corresponding equations of motion for the spectral coefficients $c_k(n\lambda)$ of the operators $c_k(\hat{\tau})$

$$-\lambda^{-2}[c_k(n\lambda+\lambda) - 2c_k(n\lambda) + c_k(n\lambda - \lambda) = k^2 c_k(n\lambda) + \ldots$$

describe the *discrete* time evolution. We have shown that the model is UV-regular, contrary to the models on a noncommutative planes, [3].

References

[1] A. Connes, Geometrie Noncommutative (Inter Editions, Paris 1990).
[2] H. Grosse and P. Prešnajder, Acta Phys. Slovaca *49* (1999) 185.
[3] M. Chaichian, A. Demichev and P. Prešnajder, *Quantum field theory on noncommutative space-times and the persistence of ultraviolet divergences*, hep-th/981280.

Geometry of 2-Fold Degenerated 2-Level System

M.V. Pletyukhov and E.A.Tolkachev

Institute of Physics National Academy of Sciences of Belarus,
Fr. Scoryna ave.70, Minsk 220072, Belarus

It is well-known [1] that during the cyclic adiabatic evolution multi-level (quantum) system acquires the geometric phase factor, or Berry's phase. B.Simon has shown [2] that it is precisely the holonomy in a Hermitian line bundle since the adiabatic theorem naturally defines a connection in such a bundle. In the case of degenerated systems this factor is non-Abelian [3].

We study the adiabatic evolution of 2-level system with 2-fold degeneracy of each level. Its initial Hamiltonian can be chosen as

$$H(0) = \gamma_5 = diag(1, -1, 1, -1) \tag{1}$$

After time t it is defined due to the evolution operator $U(t)$ [4]

$$H(t) = U(t)\, H(0)\, U^{\dagger}(t) = P_{+}(t)\, H_0(x(t)) + P_{-}(t)\, H_0(y(t)), \tag{2}$$

$$\begin{aligned} U &= P_{+}U_0(x) + P_{-}U_0(y), U_0(x) = x_a\gamma_a\gamma_5\left(1+x^2\right)^{-1/2}, \\ H_0(x) &= \left(\left(1-x^2\right)\gamma_5 + 2x_a\gamma_a\right)\left(1+x^2\right)^{-1}, \quad a = 1,2,3,4, \\ P_{\pm} &= 1/2\left(1 \pm \gamma_5 \Sigma_{ab} x_a y_b \left(x^2y^2 - (x_a y_a)^2\right)^{-1/2}\right), \Sigma_{ab} = \frac{1}{2}\left[\gamma_a, \gamma_b\right]. \end{aligned} \tag{3}$$

The geometric phase factors for upper and lower degenerated levels are [3]

$$V_{\pm}[C] = \mathcal{P}\exp\oint A_{\pm}. \tag{4}$$

Let us introduce the Maurer-Cartan form $\Theta = U^{\dagger}dU = \begin{pmatrix} A_{+} & -L \\ L^{\dagger} & A_{-} \end{pmatrix}$ which satisfies the Maurer-Cartan equation $d\Theta + \Theta \wedge \Theta = 0$.

Then, $F_{+} \equiv dA_{+} + A_{+} \wedge A_{+} = L \wedge L^{\dagger}$ and $F_{-} \equiv dA_{-} + A_{-} \wedge A_{-} = L^{\dagger} \wedge L$. This helps to calculate the components of connection and curvature forms. The explicit expressions for $A_{\pm}$ and $F_{\pm}$ are quoted in [4].

References

1. Berry, M.V.: Proc. R. Soc. **A 392** (1984) 45.
2. Simon, B.: Phys. Rev. Lett. **51** (1983) 2167.
3. Wilchek, F., Zee, A.: Phys. Rev. Lett. **52** (1984) 2111.
4. Pletyukhov, M., Tolkachev, E.: J. Phys. **A 32** (1999) 1073.

On q-Deformations and Dunkl-Deformations of Harmonic Oscillators

Andreas Ruffing

Technische Universität München, Zentrum Mathematik, Arcisstrasse 21, D-80333 München, Germany

It is a well established fact that there are deeper connections between the theory of orthogonal polynomials on the one hand and properties of Schrödinger operators on the other hand. Those operators are assumed to act in conventionally used Hilbert spaces like for example $\mathcal{L}^2(R^n)$. A prominent example for these connections are the classical continuous Hermite polynomials which correspond to Schrödinger operators with a quadratic potential. In the one dimensional case, the support of these polynomials is the real line. When dealing with a discretization of this support, one meets the next ingredient which enriches the investigation of orthogonal polynomials: It is the aspect of deformation. The idea of deforming polynomials plays a crucial role in the context of special functions. In the case of a $q-$deformation one sees that the deformation itself can be associated with discretizing the support for orthogonal polynomials. This allows to switch from polynomials defined in the continuum to polynomials defined on a geometric progression. Behind these observations seems to lie a more general concept: Deformations can be related to discretizations or quantizations and vice versa. This concept has also turned out to be an important guideline in structures of non-commutative geometry, thus also in areas quite far away from the theory of orthogonal polynomials.
Our purpose is to address to generalized Hermite polynomials on the lattice-like support $\{+q^n, -q^n | n \in Z\}$. These polynomials will turn out to be one-parameter generalizations of the $q-$discrete Hermite polynomials of type II. Our basic starting point is the introduction of difference operators which transform these polynomials into each other. These operators contain two parameters, the deformation parameter q which is related to the discrete support $\{+q^n, -q^n | n \in Z\}$ and a parameter γ which couples one-dimensional Dunkl like operators M_q, M_q^+ to the difference operators under consideration. The parameter γ can be regarded as a further deformation parameter.
Having derived the $q-$generalized Hermite polynomials in the course of the second chapter, we classify them. It turns out that they are related to the discrete monic $q-$Laguerre polynomials $L_n^{(\alpha)}(x;q)$.
We investigate the related difference operators in more detail and establish a link to the moment problem for the $q-$generalized Hermite polynomials.

References

Lorek, A., Ruffing, A., Wess, W. (1997): *A q-Deformation Of The Harmonic Oscillator* (Zeitschrift für Physik C 74), 369-377
Berg, C., Ruffing, A., (1999): *Preprint to appear*
Ruffing, A., (1999): *Quantensymmetrische Quantentheorie und Gittermodelle für Oszillatorwechselwirkungen* (Shaker, Aaachen, ISBN 3-8265-4605-9)

Quantum Field Theory in Non-Globally Hyperbolic Space-Times

Helmut Rumpf

Institut für Theoretische Physik, Universität Wien, A-1090 Wien, Austria

Although it has been known for a long time that causality violation is not a priori ruled out by General Relativity, its investigation became a concerted effort only about a decade ago. This may have been prompted by the realization that if a "wormhole" could be created it would be possible to build a "time machine" (a space-time with compact achronal region). It has turned out to be surprisingly difficult to prove the physical impossibility of such space-times (as most relativists would want to have it). The most convincing argument in this direction rests on the semiclassical instability of the Cauchy horizon (i.e. a divergent $\langle T_{\mu\nu}\rangle$ at the boundary of the achronal region) and underlies the "chronology protection conjecture" of Hawking (1992). But even if this is taken for granted, microscopic closed time-like curves may be allowed in quantum gravity. One is then led to consider as a tractable model problem quantum field theory on a background space-time with closed time-like curves.

In the "local" algebraic approach to quantum field theory, the vital requirement of microcausality is no longer tenable if space-time is not globally hyperbolic. Recently it was proven by Kay et al. (1997) that a substitute for microcausality, called F-locality, is also violated if space-time possesses a compactly generated Cauchy horizon. This might suggest that it is not even possible to **define** quantum field theory in such a space-time. However, as we have pointed out, a more general construction is possible whenever the space-time manifold M is a foliation, $M \cong T \times \Sigma$, where $T = \mathbf{R}$ or S^1 is timelike, whereas the hypersurfaces Σ need not be globally spacelike (this excludes topology change, but allows an achronal region). The construction is based on the "symplectic" quantization of a linear scalar field (note that in a non-globally hyperbolic space-time symplectic quantization is no longer equivalent to local algebraic quantization) and the "principle of self-consistency" which says that even locally only those solutions to a field equation can occur that are globally self-consistent. Since this principle will in general exclude solutions of spatially compact support, a "renormalization" of the symplectic form is necessary. This is made unique by the requirement of "effective microcausality", which is independent of, and weaker than, F-locality.

References

Hawking, S.W. (1992): Phys. Rev. **D46**, 603

Kay, B.S., Radzikowski, M.J., Wald, R.M. (1997): Commun. Math. Phys. **183**, 533

Steps Beyond the Standard Model in Noncommutative Geometry

Rich Schelp

Center for Particle Physics, University of Texas at Austin, Austin, TX, 78712, USA
E-mail: schelp@physics.utexas.edu

The finite part of the standard model (SM) spectral triple [1] is based on the algebra $\mathcal{A}_F = \mathbb{C} \oplus \mathbb{H} \oplus M_3(\mathbb{C})$. The gauge group is the group of unitary elements of this algebra (subject to an ad hoc unimodularity condtion as well).

An interesting observation of O'Raifeartaigh [2] regarding the SM is that the gauge group, usually written as $SU(3) \times SU(2) \times U(1)$, is better written as $S(U(2) \times U(3))$. This motivates the consideration of a model in which the finite part of the algebra is $\mathcal{A}_F = M_2(\mathbb{C}) \oplus M_3(\mathbb{C})$. One finds an obstruction to this model, though, in that in noncommutative geometry (NCG) the representation of the gauge group must be a restriction of the representation of the algebra. The only possible representation of the $M_2(\mathbb{C})$ piece of the algebra is the fundamental, which does not give a representation of the electroweak part of the SM.

Another model is motivated by a series of papers [3] by Frampton and others who introduced a unified model of particle physics based on the gauge group SU(15). This model is appealing because (1) it predicts light leptoquarks which could be detected at accelerators of the near future and (2) in it each generation of fermions comprises a basis for a fundamental representation of the gauge group, which fits well within the NCG framework. The drawback is that Frampton's model implements quite an elaborate symmetry breaking scheme involving extremely complicated Higgs representations. In NCG the Higgs sector is calculated, not put in by hand, and such representations cannot be accomodated.

The axioms of NCG are sufficiently rigid so as to rule out many extensions of the SM [4]. The addition of only right-handed neutrinos of each generation cannot be accomodated in the NCG version of the SM either [5]. It remains to investigate further extensions that yield the same low-energy physics.

References

[1] A. Connes, J. Math. Phys. 36 (1995) 6194; C. P. Martin, J. M. Gracia-Bondía, and J. C. Várilly, Phys. Rep. 294 (1998) 363.
[2] L. O'Raifeartaigh, Group structure of gauge theories (Cambridge U., Cambridge, 1986).
[3] P. H. Frampton and B. H. Lee, Phys. Rev. Lett. 64 (1990) 619; P. H. Frampton and T. W. Kephart, Phys. Rev. D 42 (1990) 3892; P. H. Frampton, Mod. Phys. Lett. A 7 (1992) 559.
[4] F. Lizzi, G. Mangano, G. Miele, and G. Sparano, Mod. Phys. Lett. A 11 (1996) 2561.
[5] J. M. Gracia-Bondía, Phys. Lett. B 351 (1995) 510.

Vacuum Polarization Effects in the Background of Nontrivial Topology

Yurii A. Sitenko

Bogolyubov Institute for Theoretical Physics,
National Academy of Sciences,
14-b Metrologichna str., 252143, Kyiv, Ukraine

Topological phenomena are of great interest and importance because of their universal nature connected with general properties of space-time, on the one hand, and their numerous practical aspects, on the other hand. Since the discovery of Bohm and Aharonov 40 years ago, it has become clear that topology has to do with the fundamental principles of quantum theory. At present much attention is paid to the study of nonperturbative effects in quantum systems, arising as a consequence of interaction of quantized fields with a topologically nontrivial classical field background. The dependence of the vacuum polarization effects on the geometry and topology of the base space was discovered by the present author 9 years ago. In particular, it was shown that there exists a field-theoretical analogue of the Bohm-Aharonov effect: singular configurations of the external magnetic field strength induce vacuum charge on noncompact topologically nontrivial surfaces even in cases when the magnetic flux through such surfaces vanishes. This is due to the fact that in some noncompact, essentially curved or topologically nontrivial, spaces the asymptotics of the axial-vector current becomes nontrivial and contributes to the induced vacuum charge. As a result, the latter depends on global geometric characteristics of space, as well as on global characteristics of external field strength (total anomaly). A similar result was later obtained for the induced vacuum angular momentum. Thus, singular configurations of the magnetic field strength are shown to induce vacuum charge and angular momentum on noncompact topologically nontrivial surfaces also in cases when the magnetic flux through such surfaces vanishes. Recently, a comprehensive and self-consistent study of nonperturbative vacuum polarization effects in the background of nontrivial topology has been completed. The method of self-adjoint extensions has been employed to determine the most general condition on the boundary between the spatial regions which are accessible and inaccessible for the quantized matter fields. In view of the wide use of such concepts as strings and p-branes in modern physics, of special interest is the background of a singular magnetic vortex (string). All vacuum polarization effects in this background are determined. The nonvanishing vacuum characteristics are shown to be charge, current, energy, spin and angular momentum. We find the dependence of local and global vacuum characteristics on the magnetic vortex flux and the self-adjoint extension parameter. These results yield a completely new realization of the Bohm-Aharonov effect in quantum field theory.

F-Theory and Toric Geometry

Harald Skarke

Humboldt Universität, Institut für Physik (Quantum Field Theory),
Invalidenstr. 110, D-10115 Berlin, Germany

Toric methods found their way into string theory in the context of discussions of mirror symmetry in terms of reflexive polyhedra, and in the analysis of the phase structures of string compactifications. More recently, it turned out that toric geometry is also a valuable tool for the discussion of geometric properties of manifolds that become important in the context of string and F-theory dualities.

F-theory was originally defined as a particular compactification of type IIB string theory on a $\mathbb{P}^1$, leaving 8 uncompactified dimensions. Some of the type IIB fields can be reinterpreted geometrically as the complex structure of a two torus T^2 in such a way that the $\mathbb{P}^1$ and the T^2 combine into an elliptically fibered K3 surface, thereby giving the theory formally a 12 dimensional structure. This theory is believed to be non-perturbatively dual to a T^2 compactification of the heterotic string. Upon further compactification, several other dualities (among them dualities with type IIA string theory) emerge. These can be used, for example, by exploiting well known mechanisms for the occurrence of enhanced gauge symmetries at singularities of the compactification spaces.

Toric geometry is an extremely useful tool for describing the spaces used for compactifications. Toric varieties are generalizations of weighted projective spaces. There exists a pictorial way of encoding the data in terms of simple diagrams in a lattice N whose real dimension is the same as the complex dimension of the variety. From these diagrams one can easily read off not only the weights involved in the construction as a generalized weighted projective space, but also information on singularities and fibration structures.

With the help of the dual lattice M, one can also describe functions and line bundles and thereby the data necessary for specifying hypersurfaces. Under conditions that are again very easy to state in terms of the lattice data, such hypersurfaces will be Calabi–Yau. The analysis of singularities is more involved, but leads to a beautiful result: Under certain conditions, the Dynkin diagrams of enhanced gauge groups occurring in type IIA or F-theory compactifications appear as parts of the toric diagrams. In this way it was found that already in 6 dimensions gauge groups as large as $E_8 \times (E_8 \times F_4 \times G_2^2 \times A_1^2)^{16}$, for example, can occur as non-perturbative effects in string compactifications.

For a more detailed exposition and many references, we refer to [1].

References

1 H. Skarke, "String Dualities and Toric Geometry: An Introduction", Chaos, Solitons & Fractals, **10** (1999) 543 (invited contribution), hep-th/9806059.

Actions for Duality–Symmetric Fields

Dmitri Sorokin[1 2]

[1] Humboldt-Universität zu Berlin, Institut für Physik, Invalidenstr. 110, D-10115 Berlin, Germany
[2] Alexander von Humboldt fellow. On leave from Kharkov Institute of Physics and Technology, Kharkov, 310108, Ukraine.

The problem of constructing models described by duality–invariant actions has a rather long history. It goes back to time when Poincaré and later on Dirac noticed electric–magnetic duality symmetry of the free Maxwell equations, and, Dirac (1931) assumed the existence of magnetically charged particles (monopoles and dyons) admitting the duality symmetry to be also held for the Maxwell equations in the presence of charged sources. To describe monopoles and dyons on an equal footing with electrically charged particles one should have a duality–symmetric form of the Maxwell action. In 1971 Zwanziger constructed such an action. An alternative duality–symmetric Maxwell action was proposed by Deser and Teitelboim in 1976. The two actions, which proved to be dual to each other by Maznytsia et. al. (1998), are not manifestly Lorentz–invariant. This feature turned out to be a general one. Duality and space–time symmetries hardly coexist in one and the same action.

Later on this problem arose in multidimensional supergravity theories in space–time of a dimension $D = 4p+2$ where self–dual tensor fields (chiral bosons) are present. One of the ways of solving this problem is to sacrifice Lorentz covariance in favour of duality symmetry. A non–covariant action for $D = 2$ chiral bosons was constructed by Floreanini and Jackiw (1987), and Henneaux and Teitelboim (1988) proposed non–covariant actions for self–dual fields in higher dimensional $D = 4p + 2$ space–time. In a context of modern aspects of duality Tseytlin (1990) considered a duality–symmetric action for a string. Finally, Schwarz and Sen (1994) constructed non–covariant duality–symmetric actions for dual tensor fields in any space–time dimension.

There have also been developed covariant approaches to the construction of duality–symmetric actions. These use auxiliary fields. The first covariant Lagrangian formulation of chiral bosons was proposed by Siegel (1984). Another covariant approach is based on the use of an infinite number of auxiliary fields [McClain, Wu and Yu (1990)].

The third formulation was proposed by Pasti et. al. (1995). In its minimal version only one scalar auxiliary field is used to ensure space–time covariance. This approach turned out to be the most appropriate for the construction of the worldvolume action for the M-theory five–brane [Pasti et. al. (1997), Aganagic, et. al. (1997)], duality–symmetric D=11 supergravity [Bandos et. al. (1998)] and D=10 IIB supergravity [Dall'Agata et. al (1997)].

References

Dirac, P. A. M. (1931): Proc. R. Soc. **A133**, 60; (1948): Phys. Rev. **74**, 817.
Zwanziger, D. (1971): Phys. Rev. **D3**, 880.
Maznytsia, A., Preitschopf, C. R. and Sorokin, D. (1998): Nucl. Phys. **B539**, 438.
Deser, S. and Teitelboim, C.(1976): Phys. Rev. **D13**, 1592.
Floreanini, R. and Jackiw, R. (1987): Phys. Rev. Lett. **59**, 1873 .
Henneaux, M. and Teitelboim, C. (1988): Phys. Lett. **B206**, 650.
Tseytlin, A. (1990): Phys. Lett. **B242**, 163 (1990).
Schwarz, J. H. and Sen, A. (1994): Nucl. Phys. **B411**, 35.
Siegel, W. Nucl. Phys. **B238**, 307 (1984).
McClain, B., Wu, Y. S. and Yu, F. (1990): Nucl. Phys. **B343**, 689.
Pasti, P., Sorokin D. and Tonin, M. (1995): Phys. Lett. **B352**, 59; Phys. Rev. **D52**, R4277; (1997): Phys. Rev. **D55**, 6292.
Pasti, P., Sorokin D. and Tonin, M. (1997): Phys. Lett. **398B**, 41 (1997); Bandos, I. et. al. (1997): Phys. Rev. Lett. **78**, 4332.
Aganagic, et. al. (1997): Nucl. Phys. **B496**, 191.
Bandos, I. Berkovits N. and Sorokin, D. (1998): Nucl. Phys. **B522**, 214.
Dall'Agata, G., Lechner, K. and Sorokin, D. (1997): Class. Quant. Grav. **14**, L195; Dall'Agata, G., Lechner, K. and Tonin, M. (1998): JHEP 9807, 017 .

Unitary Representations of the Quantum Anti–de Sitter Group at Roots of Unity and Elementary Particles

Harold Steinacker

Institut für theoretische Physik
Ludwig-Maximilians Universität München
Theresienstraße 37
D-80333 München, Germany

Abstract. There exist unitary representations of the quantum Anti–de Sitter group at roots of unity which correspond to elementary particles, with an intrinsic high-energy cutoff. They can be described by polynomial functions on quantum AdS space. A length scale is identified where the space is expected to become noncommutative.

It is known that there exist finite–dimensional unitary representations of the quantum AdS groups at roots of unity in any dimension [1,2], where the properties of the classical one–particle fields are consistently combined with a high–energy cutoff. Here we describe how they can be realized in terms of the algebra [3]

$$(P^-)^{ij}_{kl} t_i t_j = 0, \quad t^2 = t_i t_j g^{ij} = R^2 \tag{1}$$

of functions on the quantum AdS space; we concentrate on the 2–dimensional case. Let $q = e^{i\pi/M}$ with even M, and consider the irreducible polynomials $\mathcal{F}(k)$ of degree k in t_1, t_0, t_{-1}, which are irreducible representations of the quantized universal enveloping algebra $U_q(so(3))$ with spin k. It turns out [1] that if $k < M/2$, then $\mathcal{F}(k)$ is a unitary representation of the compact form of $U_q(so(3))$ with dimension $2k+1$. However if $M/2 \leq k < M$, then $\mathcal{F}(k)$ decomposes into 2 unitary representations of the noncompact form $U_q(so(2,1))$, the "quantum Anti–de Sitter group". The energy spectrum of $\mathcal{F}(k)$ is $M - k \leq E \leq k$, plus the corresponding anti–particle states. The classical case is recovered correctly as $M \to \infty$, if $M - k$ is kept constant. Upon closer inspection, the commutation relations (1) show an intrinsic length scale $L_{NC}/R \approx \sqrt{M^{-1}}$, which is expected to be the relevant scale where the "fuzzyness" of the space becomes important. A detailed description of the higher–dimensional case is given in [2].

References

[1] H. Steinacker, "Finite–dimensional Unitary Representations of quantum Anti–de Sitter Groups at Roots of Unity", *Comm. Math. Phys.* **192**, 687 (1998)
[2] H. Steinacker, "Quantum Anti–de Sitter space and sphere at roots of unity", hep-th/9910037
[3] L.D.Faddeev, N.Yu.Reshetikhin, L.A.Takhtajan. "Quantization of Lie Groups and Lie Algebras" *Algebra Anal.* **1** 178 (1989)

The Limits of D-Brane Action

Maxim Zabzine

Institute of Theoretical Physics, University of Stockholm, Box 6730,
S-11385 Stockholm, Sweden

Abstract. This talk has been based on the preliminary results from [1]. We consider the D-brane action which is Born-Infeld action with Wess-Zumino terms. The main goal of present study is to review the different limits of the D-brane action. Even exotic limits are considered such as the strong coupling and high curvature limits.

1 Basic Results

The long wavelength limit of D-brane dynamics is described by the Born-Infeld action

$$S = T_p \left[\int d^{p+1}\xi \, e^{-\Phi'} \sqrt{-\det(g_{ij} + B_{ij} + 2\pi\alpha' F_{ij})} + i g_s \int C \wedge e^{2\pi\alpha' F} \right] , \quad (1)$$

where $g_{ij} \equiv \partial_i X^\mu \partial_j X^\nu G_{\mu\nu}(X)$ and $B_{ij} \equiv \partial_i X^\mu \partial_j X^\nu B_{\mu\nu}$ are the pull-backs of the background metric $G_{\mu\nu}$ and background Kalb-Ramond field. $F_{ij} = \partial_{[i} A_{j]}$ is a world-volume field strength for $U(1)$ potential A_j living on the $(p+1)$ dimensional world-volume. C is a sum of Ramond-Ramond (RR) gauge fields. The p-brane tension T_p is related to the fundamental string tension $T \equiv (2\pi\alpha')^{-1}$ and the string coupling constant $g_s \equiv e^{\Phi_\infty}$ by

$$T_p \equiv \frac{1}{(2\pi)^p g_s (\sqrt{\alpha'})^{p+1}} , \quad (2)$$

with Φ_∞ being the dilaton expectation value and $\Phi' \equiv \Phi - \Phi_\infty$. One can consider the different limits with respect to g_s and α'. One should keep in mind that the real parameter is not α' but the ration of α' and some dimensional moduli parameter.

1.1 A strong coupling limit

Using the method employed in [2], we find the desired limit to be,

$$S = \frac{1}{2} \int d^{p+1}\xi \left[e^{-\Phi'} V^i W^j (g_{ij} + B_{ij} + 2\pi\alpha' F_{ij}) \right] + i g_s T_p \int C \wedge e^{2\pi\alpha' F} \quad (3)$$

where $V^i(\xi)$ and $W^j(\xi)$ are world volume vector densities. This may be considered as the zeroth order term in an expansion of the action (1) in powers of g_s^{-1}. One can study it in its own right.

Without RR and NS gauge fields. The analysis of action (3) shows that in fixed gauge the system can be reduced to string equation of motions for X^μ and the Virasoro conditions. The brane time direction can be identified with string time direction. The spatial string direction is given by electric field $\partial_\sigma = E^a \partial_a$ which is subject of equations $div\mathbf{E} = 0$, $\partial_0 \mathbf{E} = 0$. Using Gauss theorem one can argue that strings are closed.

With RR gauge fields. It is quite interesting to see what happens if we add the RR fields $C^{(n)}$ to the above picture. $C^{(p+1)}$ modifies only string equation of motion. However $C^{(p-1)}$ can be regarded as a source for electric field $\mathbf{E}$ and thus strings can end on the hypersurfaces wich carry $C^{(p-1)}$. Following this logic we can get (at least partially) the well-known picture of branes inside branes.

1.2 High curvature limit

One can look at the limit $\alpha' \to \infty$ which means that characteristic length becomes much smaller than the Planck length. Such limit is well-defined when p is odd. Using the expansion of the determinant of a $2l \times 2l$ skew matrix as the square of the Pfaffian we get following expansion of the action (1)

$$\int \left\{ (e^{-\Phi} + iC^{(0)}) F \wedge ... \wedge F + \alpha'^{-1/2} (e^{-\Phi} B + iC^{(2)}) F \wedge ... \wedge F + O(\alpha'^{-3/2}) \right\} . \tag{4}$$

where we do not care about numerical factors. Thus we see that in this expansion the leading and subleading terms correspond to some topological field theory living on world-volume and they are $SL(2, Z)$ invariant as it has to be for Type IIB theory.

2 Discussion

We have discussed the different limits of D-brane action. There are non-trivial relation among them. Thus for example the gradient expansion (which we use for derivation of the QFT approximation of D-brane) does not work for strong coupling limit and it leads to contradictions.

Apart from string theory this results are interesting as general results for the Born-Infeld action coupled to the different differential forms in the arbitrary dimensions.

References

1. Lindström U., Zabzine, M., Zheltukhin, A.: *preprint in preparation*
2. Lindström, U., von Unge, R. (1997): Phys.Lett. **B 403**: 233-238, 1997
   Gustafsson, H., Lindström, U. (1998): Phys.Lett. **B 440**: 43-49, 1998

Lecture Notes in Physics

For information about Vols. 1–504
please contact your bookseller or Springer-Verlag

Vol. 505: D. Benest, C. Froeschlé (Eds.), Impacts on Earth. XVII, 223 pages. 1998.

Vol. 506: D. Breitschwerdt, M. J. Freyberg, J. Trümper (Eds.), The Local Bubble and Beyond. Proceedings, 1997. XXVIII, 603 pages. 1998.

Vol. 507: J. C. Vial, K. Bocchialini, P. Boumier (Eds.), Space Solar Physics. Proceedings, 1997. XIII, 296 pages. 1998.

Vol. 508: H. Meyer-Ortmanns, A. Klümper (Eds.), Field Theoretical Tools for Polymer and Particle Physics. XVI, 258 pages. 1998.

Vol. 509: J. Wess, V. P. Akulov (Eds.), Supersymmetry and Quantum Field Theory. Proceedings, 1997. XV, 405 pages. 1998.

Vol. 510: J. Navarro, A. Polls (Eds.), Microscopic Quantum Many-Body Theories and Their Applications. Proceedings, 1997. XIII, 379 pages. 1998.

Vol. 511: S. Benkadda, G. M. Zaslavsky (Eds.), Chaos, Kinetics and Nonlinear Dynamics in Fluids and Plasmas. Proceedings, 1997. VIII, 438 pages. 1998.

Vol. 512: H. Gausterer, C. Lang (Eds.), Computing Particle Properties. Proceedings, 1997. VII, 335 pages. 1998.

Vol. 513: A. Bernstein, D. Drechsel, T. Walcher (Eds.), Chiral Dynamics: Theory and Experiment. Proceedings, 1997. IX, 394 pages. 1998.

Vol. 514: F. W. Hehl, C. Kiefer, R. J. K. Metzler, Black Holes: Theory and Observation. Proceedings, 1997. XV, 519 pages. 1998.

Vol. 515: C.-H. Bruneau (Ed.), Sixteenth International Conference on Numerical Methods in Fluid Dynamics. Proceedings. XV, 568 pages. 1998.

Vol. 516: J. Cleymans, H. B. Geyer, F. G. Scholtz (Eds.), Hadrons in Dense Matter and Hadrosynthesis. Proceedings, 1998. XII, 253 pages. 1999.

Vol. 517: Ph. Blanchard, A. Jadczyk (Eds.), Quantum Future. Proceedings, 1997. X, 244 pages. 1999.

Vol. 518: P. G. L. Leach, S. E. Bouquet, J.-L. Rouet, E. Fijalkow (Eds.), Dynamical Systems, Plasmas and Gravitation. Proceedings, 1997. XII, 397 pages. 1999.

Vol. 519: R. Kutner, A. Pękalski, K. Sznajd-Weron (Eds.), Anomalous Diffusion. From Basics to Applications. Proceedings, 1998. XVIII, 378 pages. 1999.

Vol. 520: J. A. van Paradijs, J. A. M. Bleeker (Eds.), X-Ray Spectroscopy in Astrophysics. EADN School X. Proceedings, 1997. XV, 530 pages. 1999.

Vol. 521: L. Mathelitsch, W. Plessas (Eds.), Broken Symmetries. Proceedings, 1998. VII, 299 pages. 1999.

Vol. 522: J. W. Clark, T. Lindenau, M. L. Ristig (Eds.), Scientific Applications of Neural Nets. Proceedings, 1998. XIII, 288 pages. 1999.

Vol. 523: B. Wolf, O. Stahl, A. W. Fullerton (Eds.), Variable and Non-spherical Stellar Winds in Luminous Hot Stars. Proceedings, 1998. XX, 424 pages. 1999.

Vol. 524: J. Wess, E. A. Ivanov (Eds.), Supersymmetries and Quantum Symmetries. Proceedings, 1997. XX, 442 pages. 1999.

Vol. 525: A. Ceresole, C. Kounnas, D. Lüst, S. Theisen (Eds.), Quantum Aspects of Gauge Theories, Supersymmetry and Unification. Proceedings, 1998. X, 511 pages. 1999.

Vol. 526: H.-P. Breuer, F. Petruccione (Eds.), Open Systems and Measurement in Relativistic Quantum Theory. Proceedings, 1998. VIII, 240 pages. 1999.

Vol. 527: D. Reguera, J. M. G. Vilar, J. M. Rubí (Eds.), Statistical Mechanics of Biocomplexity. Proceedings, 1998. XI, 318 pages. 1999.

Vol. 528: I. Peschel, X. Wang, M. Kaulke, K. Hallberg (Eds.), Density-Matrix Renormalization. Proceedings, 1998. XVI, 355 pages. 1999.

Vol. 529: S. Biringen, H. Örs, A. Tezel, J.H. Ferziger (Eds.), Industrial and Environmental Applications of Direct and Large-Eddy Simulation. Proceedings, 1998. XVI, 301 pages. 1999.

Vol. 530: H.-J. Röser, K. Meisenheimer (Eds.), The Radio Galaxy Messier 87. Proceedings, 1997. XIII, 342 pages. 1999.

Vol. 531: H. Benisty, J.-M. Gérard, R. Houdré, J. Rarity, C. Weisbuch (Eds.), Confined Photon Systems. Proceedings, 1998. X, 496 pages. 1999.

Vol. 532: S. C. Müller, J. Parisi, W. Zimmermann (Eds.), Transport and Structure. Their Competitive Roles in Biophysics and Chemistry. XII, 400 pages. 1999.

Vol. 533: K. Hutter, Y. Wang, H. Beer (Eds.), Advances in Cold-Region Thermal Engineering and Sciences. Proceedings, 1999. XIV, 608 pages. 1999.

Vol. 534: F. Moreno, F. González (Eds.), Light Scattering from Microstructures. Proceedings, 1998. XII, 300 pages. 2000

Vol. 536: T. Passot, P.-L. Sulem (Eds.), Nonlinear MHD Waves and Turbulence. Proceedings, 1998. X, 385 pages. 1999.

Vol. 537: S. Cotsakis, G. W. Gibbons (Eds.), Mathematical and Quantum Aspects of Relativity and Cosmology. Proceedings, 1998. XII, 251 pages. 1999.

Vol. 538: Ph. Blanchard, D. Giulini, E. Joos, C. Kiefer, I.-O. Stamatescu (Eds.), Decoherence: Theoretical, Experimental, and Conceptual Problems. Proceedings, 1998. XII, 345 pages. 2000.

Vol. 539: A. Borowiec, W. Cegła, B. Jancewicz, W. Karwowski (Eds.), Theoretical Physics. Fin de Siècle. Proceedings, 1998. XX, 319 pages. 2000.

Vol. 540: B. G. Schmidt (Ed.), Einstein's Field Equations and Their Physical Implications. Selected Essays. 1999. XIII, 429 pages. 2000

Vol. 541: J. Kowalski-Glikman (Ed.), Towards Quantum Gravity. Proceedings, 1999. XII, 376 pages. 2000.

Vol. 543: H. Gausterer, H. Grosse, L. Pittner (Eds.), Geometry and Quantum Physics. Proceedings, 1999. VIII, 408 pages. 2000.

Monographs

For information about Vols. 1–19 please contact your bookseller or Springer-Verlag

Vol. m 20: E. Abdalla, M. C. B. Abdalla, D. Dalmazi, A. Zadra, 2D-Gravity in Non-Critical Strings. IX, 319 pages. 1994.

Vol. m 21: G. P. Berman, E. N. Bulgakov, D. D. Holm, Crossover-Time in Quantum Boson and Spin Systems. XI, 268 pages. 1994.

Vol. m 22: M.-O. Hongler, Chaotic and Stochastic Behaviour in Automatic Production Lines. V, 85 pages. 1994.

Vol. m 23: V. S. Viswanath, G. Müller, The Recursion Method. X, 259 pages. 1994.

Vol. m 24: A. Ern, V. Giovangigli, Multicomponent Transport Algorithms. XIV, 427 pages. 1994.

Vol. m 25: A. V. Bogdanov, G. V. Dubrovskiy, M. P. Krutikov, D. V. Kulginov, V. M. Strelchenya, Interaction of Gases with Surfaces. XIV, 132 pages. 1995.

Vol. m 26: M. Dineykhan, G. V. Efimov, G. Ganbold, S. N. Nedelko, Oscillator Representation in Quantum Physics. IX, 279 pages. 1995.

Vol. m 27: J. T. Ottesen, Infinite Dimensional Groups and Algebras in Quantum Physics. IX, 218 pages. 1995.

Vol. m 28: O. Piguet, S. P. Sorella, Algebraic Renormalization. IX, 134 pages. 1995.

Vol. m 29: C. Bendjaballah, Introduction to Photon Communication. VII, 193 pages. 1995.

Vol. m 30: A. J. Greer, W. J. Kossler, Low Magnetic Fields in Anisotropic Superconductors. VII, 161 pages. 1995.

Vol. m 31 (Corr. Second Printing): P. Busch, M. Grabowski, P.J. Lahti, Operational Quantum Physics. XII, 230 pages. 1997.

Vol. m 32: L. de Broglie, Diverses questions de mécanique et de thermodynamique classiques et relativistes. XII, 198 pages. 1995.

Vol. m 33: R. Alkofer, H. Reinhardt, Chiral Quark Dynamics. VIII, 115 pages. 1995.

Vol. m 34: R. Jost, Das Märchen vom Elfenbeinernen Turm. VIII, 286 pages. 1995.

Vol. m 35: E. Elizalde, Ten Physical Applications of Spectral Zeta Functions. XIV, 224 pages. 1995.

Vol. m 36: G. Dunne, Self-Dual Chern-Simons Theories. X, 217 pages. 1995.

Vol. m 37: S. Childress, A.D. Gilbert, Stretch, Twist, Fold: The Fast Dynamo. XI, 406 pages. 1995.

Vol. m 38: J. González, M. A. Martín-Delgado, G. Sierra, A. H. Vozmediano, Quantum Electron Liquids and High-Tc Superconductivity. X, 299 pages. 1995.

Vol. m 39: L. Pittner, Algebraic Foundations of Non-Com-mutative Differential Geometry and Quantum Groups. XII, 469 pages. 1996.

Vol. m 40: H.-J. Borchers, Translation Group and Particle Representations in Quantum Field Theory. VII, 131 pages. 1996.

Vol. m 41: B. K. Chakrabarti, A. Dutta, P. Sen, Quantum Ising Phases and Transitions in Transverse Ising Models. X, 204 pages. 1996.

Vol. m 42: P. Bouwknegt, J. McCarthy, K. Pilch, The W3 Algebra. Modules, Semi-infinite Cohomology and BV Algebras. XI, 204 pages. 1996.

Vol. m 43: M. Schottenloher, A Mathematical Introduction to Conformal Field Theory. VIII, 142 pages. 1997.

Vol. m 44: A. Bach, Indistinguishable Classical Particles. VIII, 157 pages. 1997.

Vol. m 45: M. Ferrari, V. T. Granik, A. Imam, J. C. Nadeau (Eds.), Advances in Doublet Mechanics. XVI, 214 pages. 1997.

Vol. m 46: M. Camenzind, Les noyaux actifs de galaxies. XVIII, 218 pages. 1997.

Vol. m 47: L. M. Zubov, Nonlinear Theory of Dislocations and Disclinations in Elastic Body. VI, 205 pages. 1997.

Vol. m 48: P. Kopietz, Bosonization of Interacting Fermions in Arbitrary Dimensions. XII, 259 pages. 1997.

Vol. m 49: M. Zak, J. B. Zbilut, R. E. Meyers, From Instability to Intelligence. Complexity and Predictability in Nonlinear Dynamics. XIV, 552 pages. 1997.

Vol. m 50: J. Ambjørn, M. Carfora, A. Marzuoli, The Geometry of Dynamical Triangulations. VI, 197 pages. 1997.

Vol. m 51: G. Landi, An Introduction to Noncommutative Spaces and Their Geometries. XI, 200 pages. 1997.

Vol. m 52: M. Hénon, Generating Families in the Restricted Three-Body Problem. XI, 278 pages. 1997.

Vol. m 53: M. Gad-el-Hak, A. Pollard, J.-P. Bonnet (Eds.), Flow Control. Fundamentals and Practices. XII, 527 pages. 1998.

Vol. m 54: Y. Suzuki, K. Varga, Stochastic Variational Approach to Quantum-Mechanical Few-Body Problems. XIV, 324 pages. 1998.

Vol. m 55: F. Busse, S. C. Müller, Evolution of Spontaneous Structures in Dissipative Continuous Systems. X, 559 pages. 1998.

Vol. m 56: R. Haussmann, Self-consistent Quantum Field Theory and Bosonization for Strongly Correlated Electron Systems. VIII, 173 pages. 1999.

Vol. m 57: G. Cicogna, G. Gaeta, Symmetry and Perturbation Theory in Nonlinear Dynamics. XI, 208 pages. 1999.

Vol. m 58: J. Daillant, A. Gibaud (Eds.), X-Ray and Neutron Reflectivity: Principles and Applications. XVIII, 331 pages. 1999.

Vol. m 59: M. Kriele, Spacetime. Foundations of General Relativity and Differential Geometry. XV, 432 pages. 1999.

Vol. m 60: J. T. Londergan, J. P. Carini, D. P. Murdock, Binding and Scattering in Two-Dimensional Systems. Applications to Quantum Wires, Waveguides and Photonic Crystals. X, 222 pages. 1999.

Vol. m 61: V. Perlick, Ray Optics, Fermat's Principle, and Applications to General Relativity. X, 220 pages. 2000.

Vol. m 63: R. J. Szabo, Ray Optics, Equivariant Cohomology and Localization of Path Integrals. XII, 315 pages. 2000.

FSC
www.fsc.org
MIX
Papier aus verantwortungsvollen Quellen
Paper from responsible sources
FSC® C105338

If you have any concerns about our products, you can contact us on
ProductSafety@springernature.com

In case Publisher is established outside the EU, the EU authorized representative is:
Springer Nature Customer Service Center GmbH
Europaplatz 3, 69115 Heidelberg, Germany

Printed by Libri Plureos GmbH
in Hamburg, Germany